the COMPETITIVE AR15

builders guide

ZEDIKER
PUBLISHING

The COMPETITIVE AR15
builders guide
by Glen D. Zediker

Design, layout, illustration by Glen D. Zediker
Photography by Lloyd Allen, Glen Zediker, Matthew Zediker

Produced by Zediker Publishing, Oxford, Mississippi

First printing, May 2010

ISBN 978-1-4507-1607-9

PRINTED IN THE UNITED STATES OF AMERICA

To Matthew and Charlie...

As always, this book, along with everything else I do (including endless laundry...) is dedicated to my sons. They are starting their careers as involved kids and they have both chosen shooting as one of their games. They have other choices, and so do other kids. Help them choose shooting. Support local youth shooting sports programs so they'll tell the other kids about it at school. Make it cool.

It is!

"It doesn't matter if you win or lose, until you get a boatload of medals..."
Good job Boys!

Special thanks to Matthew for helping edit this book.

RULES (NEVERS, ALWAYS, AND NO MAYBES)

When you do up an AR15, there are two areas never to be lax about. Safety and barrel.

NEVER fire an AR15 unless you ALWAYS do what's next:

Check safety (selector) function. Be brutal with it. The hammer should never fall no matter how hard the trigger is pulled.

Check disconnector function in the trigger system. Following a shot, a cocked hammer should stay cocked no matter how slowly you release the trigger. The hammer can never, ever "follow."

Confirm correct headspace. Get the correct gage set for the chambering.

NEVER keep loaded (live) **cartridges in the vicinity of the work area.** There are times in a build process where we need to get our hands on cartridges. Checking magazine function, for instance. Make inert or dummy rounds and use these for that purpose. Size a cartridge case, seat a bullet over no propellant and no primer. Clearly mark these rounds (best is to drill out the primer pocket).

I had a friend once who always looked both ways before entering a one-way street. I asked him why he did that and he said because it's the one time you don't look that you'll get nailed... Just about everything I can think of regarding firearms safety is pretty much summed up in that belief.

0.1 CONTENT

WHAT, WHERE

7   *Introduction*

11   **Tools** (shop necessities)

33   **Lower Receiver** (assembly)
- 34   **Preparation**
- 38   **Step-by-Step**
- 55   **CAR Stock**

61   **Basic Trigger**
- 62   **Preparation**
- 66   **Step-by-Step**

75   **Upper Receiver** (assembly)
- 78   **Step-by-Step**
- 83   **Charging Handle** (seal modification)

85   **Barrel Installation**
- 85   **Barrel One** (configurations, preparations)
- 99   **Barrel Two** (installation)
- 102   **Step-by-Step**
- 105   **Gas Tube**
- 109   **Headspace**

111   **Bolt & Carrier**
- 112   **Components**
- 118   **Step-by-Step**

Content Note — *Within these segments are also specific installations that can apply to virtually any project. For instance, building the competition Service Rifle provided the best opportunity to also detail assembling an A2-style rear sight. These are listed below.*

161   **A2 Rear Sight**

171   **A2 Front Sight**

177   **Jewell Trigger**

227   **Geissele Trigger**

•••   **PROJECT BUILD SEGMENT CONTENTS ON NEXT PAGE**

0.2 PROJECT BUILDS

DIFFERENT TAKES ON THE SAME THEME

131 NRA/CMP Service Rifle
  131 Parts Selection
  140 Step-by-Step

155 Service Sights
  161 A2 Rear
  169 A2 Front

177 Jewell Trigger

185 Carbine Project
  186 Parts Selection
  196 Step-by-Step

209 NRA Match Rifle
  210 Parts Selection
  217 Step-by-Step

227 Geissele Trigger

237 Field Gun (varmint-style rifle)
  239 Parts Selection
  247 Step-by-Step

251 Magazines
  252 Parts Selection
  257 Step-by-Step

263 Troubleshooting

269 Sources & Acknowledgments

272 Spring & Pin Guide

No. There's not an index... I know everyone wants one but due in part to, for instance, the outrageous number of times I said "roll pin..." it just wasn't feasible. Instead, each project segment has its own content details right where it matters (right at the start of the segment) in addition to what's on these content pages, so flip to the front of the book instead of the back of the book, go to the segment, and you'll be very close to what you're looking for.

0.3 INTRODUCTION
WHYS + WHEREFORES

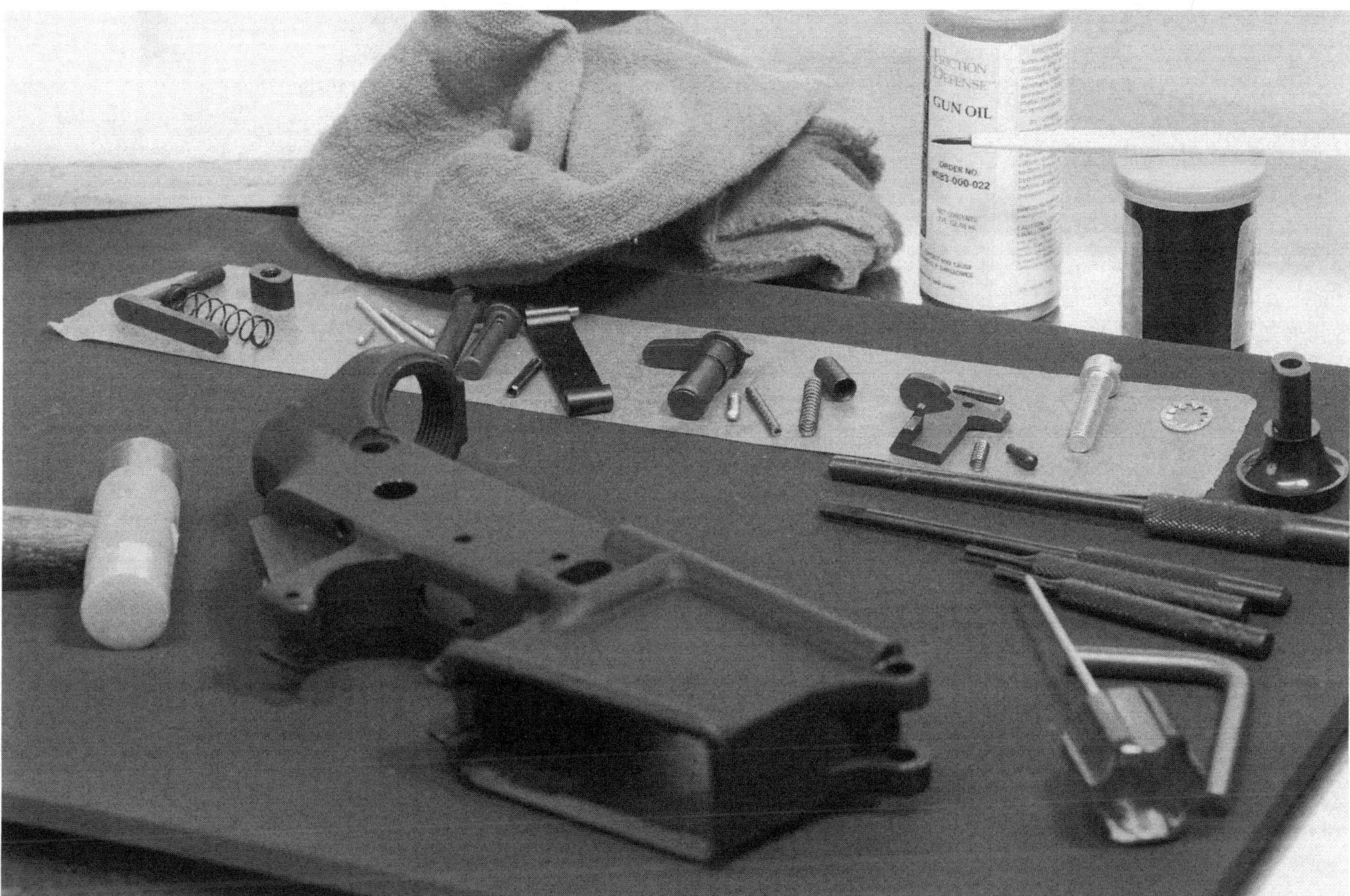

This book is for all who want to learn more about AR15 rifle systems operation, and, of course, who want to create their own manifestations of those assembled systems.

Putting together a full-functioning and target-worthy AR15 isn't what I'd call "hard," but it's not always all easy. The overall difficulty level in assembling a fully and correctly operational AR15 lies somewhere between changing a light bulb and fixing a lawnmower. As with many projects around the house or under the hood, when parts are what they should be and the proper tools are in hand, and wisdoms and precautions are followed, it's at the least easier, if not actually easy. It's when any number of mismatches, or oversights, or what can only be explained away as curses occur that it can become more difficult.

Yes, I used files building a couple of these rifles. Despite what many, maybe me included to at least some point, have led us to believe, AR15s don't always just thread and pin together like we want. Even more, some have suggested that AR15 builders don't really have to be gunsmiths, and, while largely true, there will be times when services of a trained 'smith or machinist will be needed, either eventually, or, depending on luck, immediately. With all the "standardization" (more about that throughout…) it seems that tolerances would be such that a little give or take between parts should go scarcely noticed. Sometimes there's going to be two parts demonstrating the maximum gives or a polarized give and take relationship, and, well, sometimes that requires files.

PREPARATION

Preparation first. Always. Nails make poor pin punches! There is a right tool for virtually any job. I don't know that all tools shown will be necessary for all projects, and know some will not be, but I can show a few (a many actually) that work and work well. Some specialty tools are indispensable, but the caveat is always the value of the investment. How much use will the tool get? Each separate project rifle segment will have its own tools and tricks list, and options will be discussed if they exist.

We'll talk a lot more, and all throughout, about tools in this book.

That's because I can tell you that it's tools, honestly, that make a world of difference between good and bad experiences (sometimes that means results) in threading and pinning parts in place. I've been building up AR15s for a few years now in all manner of configurations, but I am strictly not representing myself as a trained and titled gunsmith. I have a fair mechanical aptitude but, as said, it's not my trade. I wrench on racing motorcycles and do triggers as well as anyone. The first segment in this book will set out to show tools (hard and soft) that range from requisite to really cool. Again, others will be shown throughout as circumstance reveals their utility.

GAINING EXPERIENCE

Through experience comes knowledge and confidence. Once you know how to install a float tube, pin a trigger in place, screw on a pistol grip, and fit up a receiver extension tube, choosing each of those things changes your rifle into what you envision it to be. Of course, you'll also have a firm handle on assembling upper and lower receiver essentials, and addressing the gas system and action parts to ensure correct and complimentary function. Choices in barrels and chambers hopefully also will be fully comprehended and considered and you'll know what do there. Add-ons and accessories and unique-form enhancements, as you will see, first and always need to be workable within the whole of the rest you have chosen. Not all things work together and sometimes Plans B, C, and even D replace some or all of the parts group purchased or considered to complete original Plan A. Assembly and finishing tricks are treats, and through employing some of these you'll learn about make not only for a better rifle but also for a prouder owner.

The only real differences between a zoot-capri tactical carbine and a flush across-the-course NRA Match Rifle are in the configuration of what otherwise are the same components, because all the same components are in place and all matter. I'm talking about there being no difference, really, in how any top-shelf AR15 goes together beyond component selection. A carbine built with a 16-inch Krieger 1-7.7 twist barrel and one built with a 26-inch Krieger 1-7.7 twist barrel both shoot the same size groups, or should. I know they should, because mine did, and do.

SOURCES & RESOURCES

I made my best effort to make this book as complete as I could, without just being redundant in some of the project rifle builds. Please read through the whole book even if there's just one or two projects or jobs here you want to accomplish. The reason for that is because some specific operations or assemblies are detailed within the project segments. Major steps are given their own segments so those should be easy to find.

My bench manual has always been my old Colt's o-fficial M-16 A1 Armorers Manual, the only "slick" publication of its type possibly ever to exist. It's outstanding and although the specifics have changed, and some in a big way (like the rear sight), it's still the best resource for putting together one of these rifles I've yet seen. Noth-

ing new (except the rear sight) has to have additional help, and that help is as simple as a diagram and what's in this book. This publication is significantly superior to any of the "gun show" type mil-look books. I have no idea where a fellow might lay hands on one, but it's worth looking for. There are a few commercially produced books on the topic of assembling an AR15, but, as per usual, I thought I could do a better job and, this is the biggest difference, mine here talks about competition- and precision-use rifles, and, in keeping, purpose-built AR15s in differently purposeful configurations.

The rest of what I know and share has come from reading I've done on my own, and, for the extreme majority, experience in having done it now over and over and over. I've talked with all the builders who I've worked with on my own guns about all the ins and outs, and there are numerous tricks and tips revealed here in this book.

The goal, point, and purpose of this book — and I'll say this again and again — is not to see duty as a trade school textbook. It's for folks like you and me who are relegated to tackling projects on workbenches, not in machine shops. We're not going to talk about how to turn, chamber, and prepare barrel blanks or make machining cuts in receivers or create fixtures and jigs. I am, however, going to spend a lot of time talking about some of these topics with respect to arming you with some insights into what to request and address in a barrel and other aftermarket parts and where to get all the tools needed to make this as easy, efficient, and effective as possible.

WORKSPACE & TOOLS

Folks, this isn't really "real" what you see in this book. I set up a nice and new area to give a glean to these photos, and, ultimately, to make all more clear and pleasing to view. That's not at all dishonest, especially since I'm truthing up a storm right here and now. I usually build rifles on a wooden bench that's as cluttered as the underside of my truck seat, and you don't really want to see that. I don't want to see that.

Likewise for the tools. I got the best, or the most of the best, and tried, when possible, to use (actually reuse) examples of what I characterize as "standard" wrenching and pin-driving devices. I also wanted to show all the tools. That means you may see different ones in use in different projects. Just don't think you need all of them. I did a list of tools used and options possible when I did the hands-on segments. Some ops are so easy with specialty tools, and are so difficult without, and there are a few places I show or at least discuss how to get the same end result with a lesser tool set.

I can tell you that, overall, completing a project once is one thing and doing the same operation over and over again is entirely another. That's where the value of specialty tools comes into question, and that's where it's answered. There are some ops that I honestly detest, or mess up (related), without purpose-built tools. They are, therefore, worth the money spent if used only once. There are some jobs that demand ownership of a peculiar tool. A pivot pin installation tool or rear sight spring tool come instantly to mind. Beyond that first use, however, purpose-built tools are more and more valuable to anyone who either wants to ply this as a trade or has plans for reuse. I used to hate taking handguards off Service Rifles, for instance, until I got a handguard removal tool. I can't say that I now do it for fun when bored, but can say that it's become easy.

This really isn't much, if any, different from any other facet of shooting. I think most shooters who like to handload, and I can certainly draw inference from that little segment of shootership, appreciate tools and see them in the same way. I do. As with handloading operations, the more any certain bench function is done, the easier it is to justify the value of a tool that reduces stress on the self. That may sound odd, but it's really what it comes down to: whether physical or mental, time, ease, and comfort are bliss. The reason I said justify is because I have invested way on more in the tools to assemble it than I ever have in any one rifle. That's where the do-it-yourself plan gets tested. Relate to household projects. If you feel need to purchase a nail gun set-up to finish a deck, you might just want to hire it done. I can't advise there. I do like tools.

Tools used tend not to stay shiny and corrosion-free for long, or at least not used here in my neck of the woods, and some of my tools have had hard use while others were unpacked for their first spin in these photos. I'm just pre-excusing myself from criticism of photographing nappy looking appliances. Yes. I do take care of my tools, but I've got the Curse of the Orange Touch to live with. Some folks can handle things nilly-willy and never worry, but if I so much as brush against a piece of carbon steel it turns bright orange overnight.

DIVING IN

I learned all this stuff over years and years and years from other people, and then tuned much of it through my own experience. I don't think anyone really learned anything without somehow messing up. That's everything, not just building rifles. Mistakes lead, or should, to knowledge. I've had the advantage, or think it's an advantage, of having input on many operations from many different professional builders. This is knowledge that they tend not to share with each other but do with me. It's not that they wouldn't offer what they know to a peer, just that it's not nearly as likely to come up as part of a conversation. There will be a few things you'll see here that are, as a result of that, not going to agree with other people's books or takes on this topic. That's expected. After having read between the lines, and then having had to write right over them to fill those gaps in the best I could, I also have seen that there is more than one way to do most anything, and also that, seeing respected and successful builders tackle the same thing in what seems like contradictory ways, doing something to address what let's call a problem otherwise is handled nicely enough by doing anything suggested. Few, for

instance, seem to agree on the proper use and formula of adhesive applications, but their rifles all shoot well and nothing comes loose from them.

Enough on that.

As said, I've read most everything I could get my hands on over many years with respect to the content of this book, taken as a whole, and can honestly say that I'm proud of what's between these covers.

This is the first book I know that addressed the means and methods of home-done, but professional-quality, AR15 construction, solely. That's my focus, and my field, and all that follows is what I have to say about it.

If you want to learn way on more from a gunsmith's perspective, and this is now from someone who owns large and loud metal-working equipment pieces and has an "open/closed" sign on the shop window, and a cash register, check into Derrick Martins's book, *The Accurate AR15*. Derrick will show you how to use lathes, drill presses, milling machines, and the like, to work with barrel blanks, change parts geometries, and even do the upper receiver and bolt carrier mods shown on one of these projects.

ABOUT THE BOOK HISSELF

As is plain by now, this one has larger-format pages than my previous works and also a different binding. I had some industry (publishing industry) folks tell me not to do a spiral binding on a high-quality book. Well. I just don't listen. The reason I did it this way was so it could open up on a workbench or rest nearby and stay flat. This is an instruction manual with step-by-step photos, illustrations, huge and clear photos, and the whole reason behind all that is to help someone do something. In this case, assemble an AR15. It really helps to be able to see it as you're trying to follow it along.

Years and years (and years) ago I wanted to learn how to do golf club repair. I bought an outstanding manual that was done in this format. Like this book, it was high quality with crystal-clear photos and the instructions right on the same page with each photo. It also has wood stain, epoxy, smudgy fingerprints, and many, many coffee cup rings on virtually every page. I want my book to turn out looking the same for you and, again, the only way I could attain that was to make sure you can perch this on your bench and see it hands-free.

The larger pages simply let me run the photos larger, and increase the number of sequential illustrations on each page. That was easy. Big pictures, small pictures? Four pics per page, six pics per page? That decision required virtually no thought.

Thanks for reading this far, and now we'll get started doing what you really bought this book to accomplish: building your own ultimate AR15.

Take care of your tools. Give them a good wipe down after use and try to keep them from falling on concrete. Try.

ME It's the same me, just older. And that's a good thing! I know more than I used to, and, believe me, it's all been learned the hard way. As always, I hope to spare you such misfortunes. That's never been more true than in this book!

Just in case this is the first of my books you've read, I best tell you who I am and what I do.

I've been a competitive shooter since age 7; I started up in a DCM program we all called "Gun Club." I started writing about shooting in 1988, and math says that was 22 years ago. My main event is NRA High Power Rifle. I have primarily been a Service Rifle shooter and earned my NRA High Master classification running a smoke pole. I've participated in other sports, shooting sports of course, and also played as a pro in a couple outside of shooting (golf and motocross). I still race motocross (in the 40+ class). That only matters to know that I'm no stranger to competition, or to taking on the task of self-improvement. All that has taught me a lot.

I am a single man and am raising my two boys, who are just now starting their game-playing careers. Both are champion air rifle and air pistol shooters, my eldest three times in his first three years of competition, and this year my little man won both his events on his first go at targets for record. I've learned a lot from all that too!

I grew up on a cattle ranch in Colorado and live in Mississippi for no real reason. It's all good, and it's getting better. Keep in touch. **www.zediker.com**

TOP TEN REASONS TO CALL A GUNSMITH

You need to call a gunsmith if...

...you don't have a workbench

...you are using a Dremel tool in place of an end mill

...any carpentry tools are on your workbench

...you ever consider glue for metal repair

...you ever then use glue for metal repair

...you consider and then choose a nail for a pin punch

...you've tried installing roll pins using pliers covered with tape

...you've never ordered from Brownell's before

...there is a dishwasher within sight of your work area

...you ever try to fix something without first taking it apart

...you ALWAYS say, ***"I don't care how it looks..."***

1.0 TOOLS

YOU KNEW IT WOULD START THIS WAY

[No one has to look farther than Brownell's for virtually all of the following tools and supplies, and I recommend that they don't. Brownell's sells no junk. Midway USA carries a nice line too. Some of these things can come from other sources also, and these may be places where aisles frequently close and everything beeps when it backs up, but don't drive yourself to exhaustion, or get your hopes too high, when you're shopping away from one of our industry sources. Locate a machinist's supply, and that's a good tip, because you may need some help from them, especially on fasteners.]

Hold it. *A sturdy vise and workbench are mandatory. My setup as shown in this book honestly isn't my reality. I figured no one wanted to see anything beat to pieces so I set up a very spic-and-span environment to collect these photos. Attach the workbench to a wall if possible. The best vise I've used is the Brownell's Multi-Visc. It's worth every penny, especially if you use it every day.*

<u>SEGMENT CONTENT</u>

<u>12</u>	**Indispensables** (bolt-downs)
<u>16</u>	**Pins & Punches**
<u>19</u>	**Adhesives**
<u>21</u>	**Lubricants**
<u>22</u>	**Specialty necessities**
<u>24</u>	**Tools, tools, and more tools...**

PREFACE

Okay. I got a lot of tools. Therefore, I have a lot of tools. Good tools, and the right tools, are so important not only to the ultimate job result but also to the ease in which it's completed. Of course they are. That doesn't matter whether we're talking about mounting shelves, rebuilding a transmission, cleaning a kitchen, or, yes, building an AR15. Many of the tools you'll see are the product of creative and I'd venture also frustrated minds. After needing four hands to perform a routine task long enough, or after having to say, "Well, it doesn't look too good, but it works…" that's when things like pivot pin pushers and carrier key stakers and magazine latch pushers are born.

Part of the reason you'll see so many specialty items in these pages is because, one, I'm a tool junkie. Freely admitted. Next, and more closely related to the content-utility of this book, is that you should know they exist. There are also some proprietary tools engineered to handle specific assemblies. Good examples are unique wrenches that mate with various free-float tube assemblies.

I'm not about to say that you need each and every one, but some are indispensable, or will become indispensable, and, as could be said when justifying the purchase of any specialized tool, the frequency of its use largely determines its value.

Given that, when there are ready (maybe) alternatives to the specialized appliances they'll be discussed. Once you experience how some work, though, you'll need a bigger toolbox.

INDISPENSABLES

The can't-do-without list is long, and length varies, of course, depending on how many things you find out you can't get along main vise. It's going to have demands put on it that may not seem feasible until you're in process. I've damaged them. I've also learned to get better ones. I got the "best" version at a home-improvement store and after a week of use, which decidedly was not full of 8-hour days, it was out of whack and not closing properly. It wasn't cheap, and it was a waste of money. Alls I did was give the handle a whack with a hammer… It wasn't that big a whack either.

My favorite is the Brownell's Multi-Vise. This is a truly well-engineered tool that's worth its cost. It can easily swivel and allows for vertical or horizontal jaw orientation. Again, you might not think it's worth its cost until a lower quality vise lets you down. Of course, Brownell's has a custom-fit set of pad inserts available for it too, and those are recommended.

Like the vise, the workbench should be bigger and stronger than you might first think. One of the reasons a workbench needs to be over-engineered is actually because of the vise. I'm not about to try to tell anyone how to set up a full-fledged gunsmithing operation,

without. Chances are that some wouldn't be missed, but, as just stated, after using a few you might think not.

The mainest things are a place to work, something to do the work on, and, then, the tools specific and absolutely necessary to AR15 construction.

BUILT-IN TOOLS

Also, heavy tools... A good vise is absolutely necessary. You'll need a heavy-duty model with at least a 4-inch jaw opening to hold things like receiver blocks and fixtures and barrels. Do not scrimp on the because that's not really what this book is about. I will say, however, that I've never been to a full-time builder's shop and seen a wobbly workbench. Carpeted work surfaces are nice, and if you don't go that far, go as far at least as to get a rug or mat to help avoid scratching up the expensive and visible stuff. My favorite workspace topper is from Brownell's but even a discount-store bathroom rug is better than nothing.

Same thing also goes for shelves, storage, and organization. The more you do and the more different things you get into should lead you to determine what you'll need to stay on track. I'm the last per-

There are blocks that fit up into and inside the upper, and that style is my preference. This style is externally formed like the upper receiver internally, and uses takedown pins so it is, in effect, like clamping the lower with the vise. Derrick Martin gets credit from me for this invention. Others are "clam-shell" parts that encase the upper and also fill the space where the bolt carrier normally occupies. There's really no difference in how well they work.

The big difference, however, and pay close attention, is that the clamshell-style blocks are a skin-fit to an A2-configuration upper, whether it's a flattop or carry-handle. That's all they'll work with. If you need to hold an upper that's dimensionally different, such as one of the DPMS "competition" style uppers, they won't work at all. Those require the use of the "plug-in" style support just mentioned.

Holding the barrel itself, normally using a pair of barrel vise jaw inserts, is a popular way to secure an upper, but I so do not recommend it. The note against holding the barrel and for securing the upper is that the indexing pin on the barrel extension that fits into the upper can lever against its slot when the barrel, not the upper receiver, is held in place. This can elongate the slot. There's too much stress on the upper. Barrel clamping block sets are numerous and it's hard to find any that actually work all that well. This is

son who should offer advice on staying organized because I have been known to lose the very things I bought to put parts in, but the less time spent hunting for anything the way on easier any project goes.

especially true if you're installing a barrel that's even a little different dimensionally than issue-spec standards (which is what the vise inserts are engineered around). Rosin can help stiffen the grip. It will take a good deal of pressure to secure a barrel from slipping when an inordinate amount of pressure has to be applied to a barrel nut to align a stubborn gas tube hole. Don't worry, by the way, you're not going to bend or crush a barrel. I don't like clamping the barrel because of what was just said. It also mars the fool out of them, no matter what I've tried.

CLAMPS AND WRENCHES

To do any barreling work, you'll need a means to hold the upper receiver securely, and without damage. The receiver should never be held directly with even padded vise jaws. Protect it using a go-between. Upper receiver blocks come in essentially two formats.

I do, though, think such appliances are wise investments. They're handy for installing, or removing, muzzle attachments and securing a barrel for a chamber polish or for thorough cleaning. Especially in an op like removing a flash suppressor, I strongly prefer clamping the barrel itself rather than using an upper-receiver clamp to secure the works. There's a lot of leverage working all the way out at the end of the barrel and clamping the barrel itself takes that advantage away from what might be levied against the upper receiver.

Holding the lower is painless and downright convenient with one of the blocks that substitutes for a magazine. There's little stress ever levered against a lower that in any way compares to that raged against an upper during a barrel installation. One of these blocks is all that's necessary. They are also daggone handy for cleaning, sight installations, and general maintenance work. I keep one clamped in a vise most all the time.

To install a barrel, any barrel using virtually any retention means (meaning free-float tube or not) you'll need a spe-

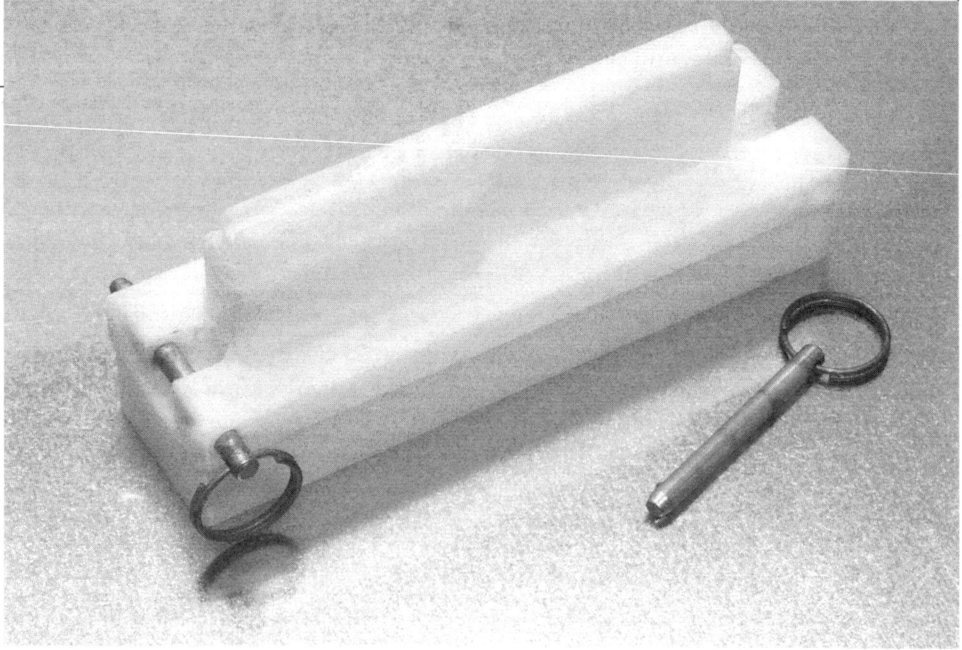

Upper Receiver Clamps.

Never, ever, vise up an upper without either of these devices. The "clamshell" type is the most popular and is a skin-fit to an A2 upper with or without its charging handle. My preference is the lower photo that fits exactly inside the upper receiver. The upper receiver cannot move using this block. The limitation of the clamshell is that it only fits an A2, not any of the "other" configuration uppers, such as the newer billet-made uppers. In other words, the clamshell works off the exterior and the insert works off the interior. The interiors don't change.

cialty barrel nut wrench. There are many available.

The "armorer's combo-tool" standard has three prongs that engage the scallops in a barrel nut. This will work with any barrel nut arrangement that's configured like a standard nut, including those on most free-float tubes. I prefer the "full-round" style that engages each and every scallop. This wrench, however, has to be able to fit over the muzzle and be brought back to the barrel nut, so can't be used if installing a barrel that already has the front sight housing affixed.

I do not recommend purchasing one of the multi-purpose tools that has its own handle and is a self-contained tool. (By the way, the reason I'm referring to these as "multi-tools" is that they have cutouts that fit, sort of, the receiver extension and flash suppressor nut. More later.) The variety of barrel nut wrench you need has an attachment for a 1/2-inch drive, into which goes a wrench handle, and now we'll talk about those.

A breaker bar and torque wrench are important tools, especially for those unfamiliar with barrel installations.

If you don't have a torque wrench then you're guessing, but after a few installations you'll be guessing on the safe side of error. The reason for the breaker bar is because rumor has it that a torque wrench should only be a one-direction tool: it's only used to tighten, never loosen. I couldn't really confirm that rumor, even after asking a few race mechanics, but let's assume it's true. The breaker bar is therefore only needed to loosen this nut.

These tools can come from any good mechanic's supply. I like torque wrenches with longer handles because sometimes the amount of force needed for a barrel installation gets on up there and the longer handle makes subtle shifts toward tight easier to control.

As mentioned, the combination tools have additional cutouts to fit the receiver extension tube and muzzle brake nut, but the fit is

Barrel clamps. These are not to be used for installing barrels, but do come in handy when it's necessary to work on the barrel itself. Rosin helps increase the grip if necessary.

Because of the torque forces at work, it's better to hold the upper receiver for barrel installation. This reduces stress on the barrel indexing pin. I use barrel clamps for ops like polishing chambers and feed ramps, installing (and also removing) muzzle attachments, and they're really good just to give the barrel a thorough cleaning.

Always clamp the "fat" part of the barrel (the area that won't show). These will mar.

Cushion. Vise jaw pads are a good purchase. These are made for the Multi-Vise from Brownell's and must be filed to a perfect fit. One set (red) is a soft composition, the other hard (blue). Very sano.

usually poor. It's far preferable to have handy a good quality adjustable wrench, or better still, open end wrenches that are correctly sized for these fasteners.

Building a bench. For what it's worth, I have had really outstanding results with creative use of the steel specialty brackets from Z-Max. These are at places like Lowe's and The Home Depot. Oh, and take a long, hard look at what you'll bolt to the bench and make sure there's adequate overhangs and clearance. Don't get a support beam right in the way of where you need to drill a hole!

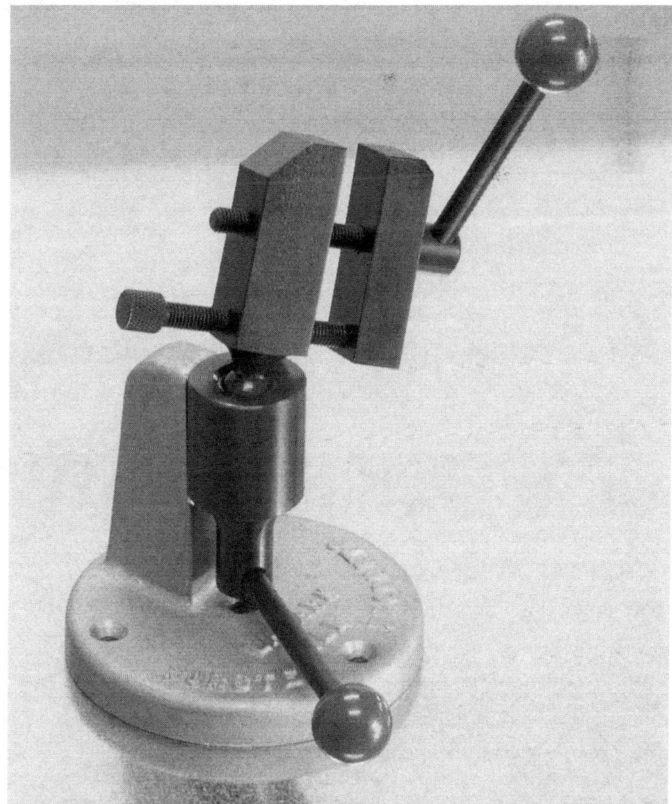

Small parts. A small vise like this Swiv-o-ling from Forster provides a good hold for doctoring tiny items. I use mine often to help position a dial indicator to check rear sight run quality. Panaflex is another.

TORQUE TECH

As you can see, I like big torque wrenches. They're just easier to work, and, mostly, make it easier to make little shifts in rotation to align the barrel nut. Also shown is a companion breaker bar. This tool is used only for loosening the barrel nut during assembly. Some maintain that torque wrenches should be strictly one-directional tools — they should only tighten — so I replace the barrel nut wrench onto it during

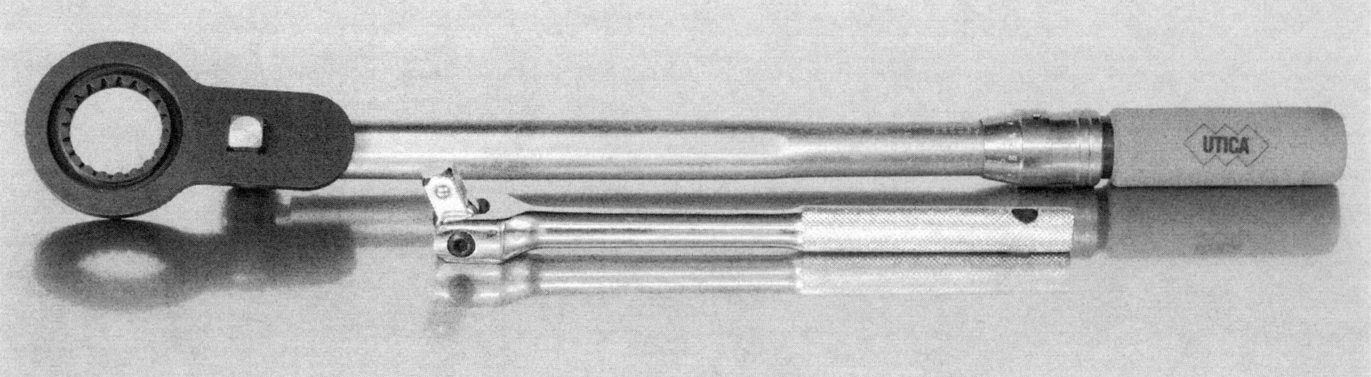

assembly as needed. They both have 1/2-inch drives.

Now, just as a matter of noteworthy fact, when the wrench end extends beyond the drive center on a torque wrench, it increases the leverage of the wrench and therefore affects the net amount of torque applied to the fastener. There's an equation that should be used to determine this increase, and it can be significant. Here's how it works...

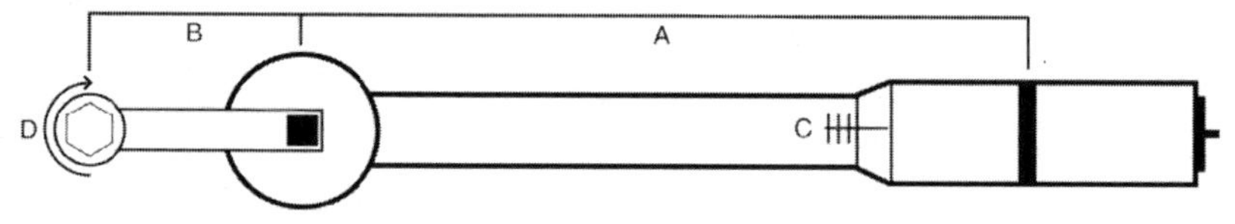

A = length from center of handle to center of drive

B = length from center of drive to center of extension

C = wrench setting

D = desired torque applied to fastener

Here's the formula

$$C = D\frac{A}{(A+B)}$$

Here's the answer
for this combination

$$39.96 = 35\frac{15}{(15+2.125)}$$

That's a fairly significant amount, call it 5 foot-pounds, and mostly it's clear that setting the wrench to 35 foot-pounds would mean it would click over short of what most agree on as the minimum torque setting for an AR15 barrel nut. So now you know...

PINS & PUNCHES

Assemblies ranging from the gas tube to magazine catch, bolt stop, bolt components, detents, sight parts, and more are secured using roll pins. A roll pin is a hollow pin with a split. It's oversized to the hole it fits into by about the gap width of the split. It squeezes down as it enters the hole and this tension keeps it in place. They are beveled on their ends but that's often not nearly enough to get one started gracefully, and that is the trick — gracefully or not, getting one started. Of course there is a specialty tool that helps, and that is a roll pin punch. Get some. For a basic build, you'll need #s 1, 2, 3, and 4.

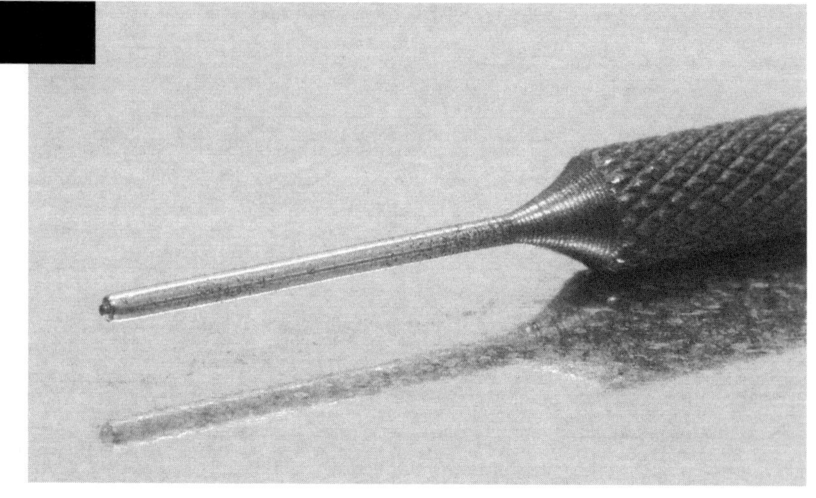

Hammers. I use a 3/4-inch brass-headed tap-hammer for punching punches. There are times when a bigger hammer is called for, such as removing, or installing, taper pins, but a slip-hit from a light brass hammer doesn't cause near the chagrin as unintentionally whacking something with a steel hammer. The steel hammer shown is plenty big enough for AR15 work. It's a 12-ounce ballpeen.

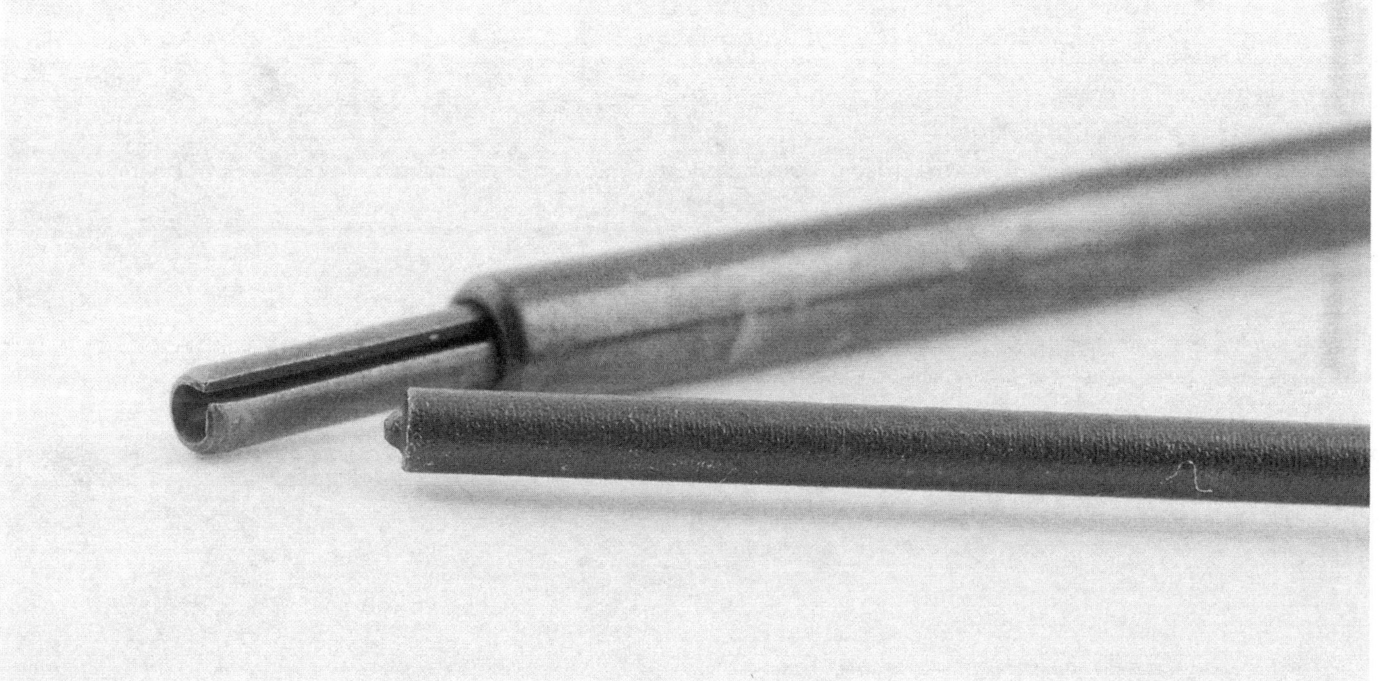

Pin Punches. You really need a set of these. Really need a set of these. They are roll pin punches, one to start and one to finish. The starter punch has as its sole function and favor getting the oversized pins started into the hole. It's sized to fit the outside of a pin. After it's in and fully on its way, then switch punches to the one with the little nib on its end and send it home. The nib fits into the hollow in a roll-pin and helps "grip" the pin so it can be seated to flush, plus without unnecessary marring. Seating roll pins is a difficult and scratch and ding producing job without these punch pairs. Tip: drive the pin as far as you can using the starter punch, without contacting the part itself with the punch end.

Consider your sources. I just have to comment on this because it's stupid. I've seen many places on the internet where folks are showing all about how to install pieces-parts on AR15s and they suggest a pair of pliers with its jaws coated over with tape to install roll pins. The idea is to press the pin in place. That will bend the pin and no matter how much tape is used mar the opposite point on the rifle part. Just don't do that. Pin punches aren't expensive. Rifle parts can be.

> ***Pre-Pining.*** *Get some emery or a hard Arkansas stone to run roll pin ends over before assembly. The ends are usually rough and irregular. It's easy and makes for easier work. Assembly is greatly assisted but it's also disassembly that improves when the ends are deburred. Just chuck the pin, lightly, in a drill and run on an angle for a few seconds.*

The ends of roll pins are often craggy or out of round, or both. These are not precision-made parts. Smooth and polish the ends of every roll pin you install. This doesn't take much effort or time and is a worthwhile step. The easiest thing is to lightly chuck one into a drill and spin it against some emery or a stone. I do both ends but only the entry is necessarily smoothed. Removal is easier too when both ends are polished. Steel pins going into aluminum holes make life way harder on the holes than on the pins. I can also tell you that a drop of oil helps and will never diminish the effectiveness of a roll pin. That also reduces any corrosive "sticktion" potentials. That makes them come back out easier.

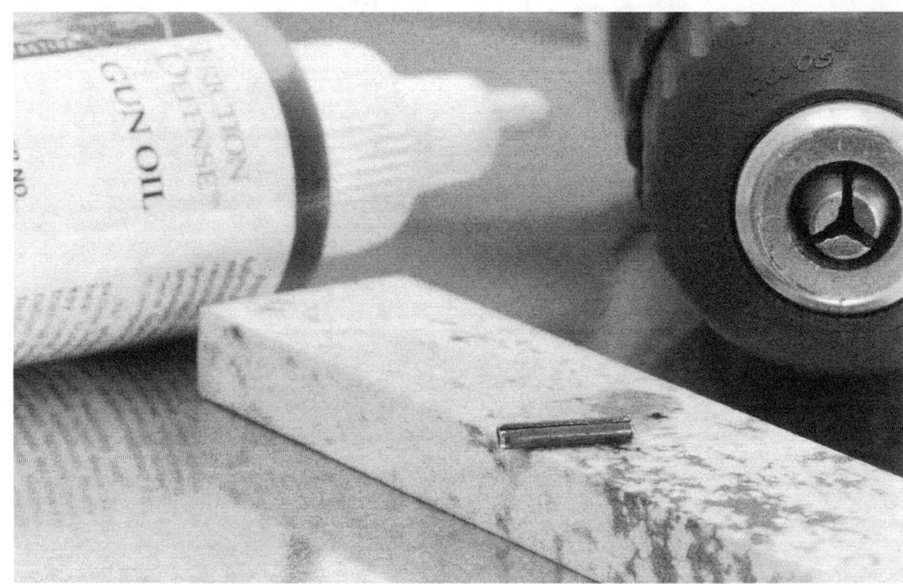

It doesn't take much effort to drive a roll pin (with a couple of exceptions), but true hits count. They can bend. Most roll pins are a little shorter than the full span of the hole so a sano job finishes with the pin ends at equal depths, and that should have each end a tad below

> ***Cushion.*** *A few pieces of wood come in handy, if not necessary, to back up parts being pinned. The 3/8-inch dowel fits the bolt face for ejector installation. A short chunk of 2x4 and maybe a few short lengths of 1/2x1-inch stock serves all.*

flush with the part surface. A roll pin should never protrude above the surface to ensure no snagging potential.

You'll notice that there are punches of varying lengths used in the work shown and the shorter ones are a little easier to operate but the longer ones are necessary for some installations simply because they protrude freely and clearly beyond rifle parts you don't want to accidentally mis-hit with a hammer, or have the larger diameter handle portion in contact with a rifle receiver.

I often chuck up a small punch and polish its outside. They're not all perfect, and sometimes these little imperfections are annoying if not damaging. This is especially true when using one as a capture

> **Use the right punch.** I'll list the needed punches for each pin in each project. There are specialty punches also and these will be shown and described when they're needed. All the punches shown in this book come from Brownell's. They are numbered as well as described in inches. Make sure the punch fits into the hole correctly, not just fits onto the pin, because it's the hole that ultimately determines the correct size for the punch. That's easy, just check it with the punch before you seat the pin.

punch such that it has to extend fully through the hole set. Same sort of procedure for polishing most anything: spin against emery cloth until the metal is smooth. Likewise, with use they often get a little deformed around their edges, and that's easy enough to true back up with a stone.

The pin punches I used for each job are listed. Assuming you have a full set you should have all you need to at least replicate any use of them shown here. A punch that's too small for the pin will tend to deform and also expand the pin end. One that's too large may do the same and also won't ultimately enter the pin's hole to seat the pin-end correctly.

Use a brass-headed tap-hammer for punching punches. It's plenty enough power and a slip won't cause undue maring.

I can't tell you much about running a punch that you won't learn on your own, but make sure the end is centered and the punch is in-line with the pin. A "follow-through" sort of strike is usually better than mimicking a woodpecker.

As mentioned, oil pins before installation. It sure makes them seat easier and will in no way affect their holding power.

Oh, and tape. Use masking tape all over the immediate work area. It's not a sign of weakness. Tape the fool out of everything around the installation and it's less likely to need touch up afterward.

HANDY HELPERS

These things may not be strictly necessary for assembly of an AR15, but come in handy nonetheless. They are more or less shop supplies rather than working apparatus.

ADHESIVES

[To hopefully head off some confusions, we've been using this stuff for eons now and it's had different brand names and different ways of being identified. First it was Loctite and had numbers and colors to identify it; then it was Permatex and just had numbers; now, at the time of this first publication, it's still Permatex and no longer has numbers, just colors. No matter how they're calling it, it's still going to be the same stuff.]

You do need to know your glues to do a good job assembling an AR15. Regardless of its intended use, everyone wants to know they did all they could to make their rifle shoot its best. Believe it or don't, but glues help.

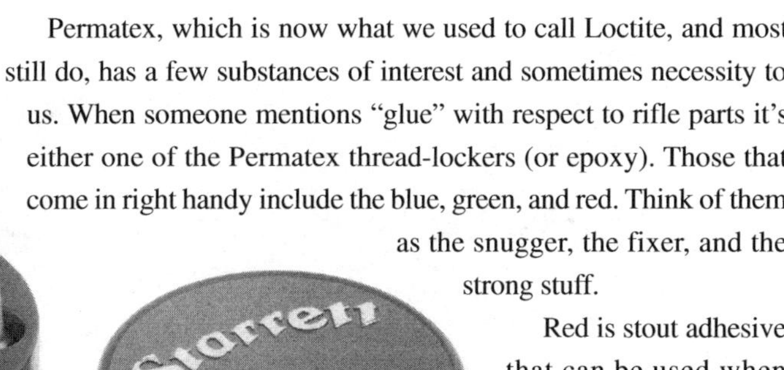

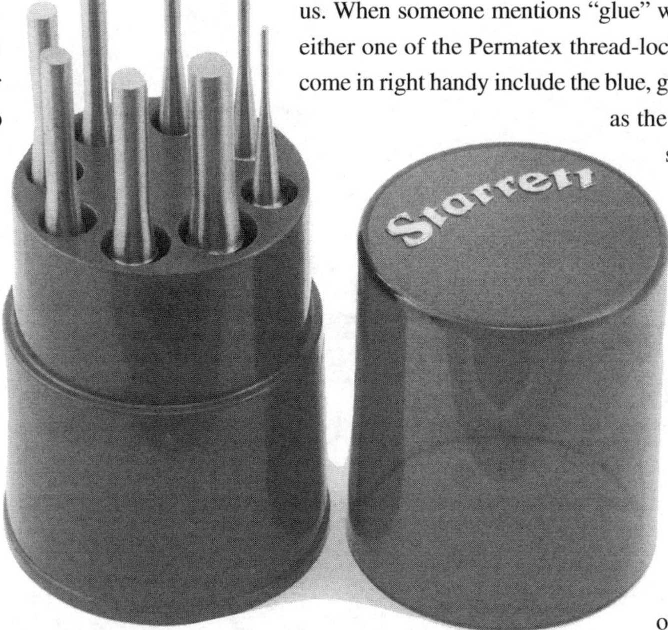

These punches are virtually perfect. It's a set from **Starrett**. I use them primarily as capture punches but sometimes are useful otherwise.

Permatex, which is now what we used to call Loctite, and most still do, has a few substances of interest and sometimes necessity to us. When someone mentions "glue" with respect to rifle parts it's either one of the Permatex thread-lockers (or epoxy). Those that come in right handy include the blue, green, and red. Think of them as the snugger, the fixer, and the strong stuff.

Red is stout adhesive that can be used when there's no need, and perhaps no possibility (keep in mind), to ever remove or separate the affixed parts. Comparing parts sizes with amount of area glued, sometimes the Red actually takes more force to surrender its hold than the part itself can take. In other words, if you put it on a small set screw, it's pretty much over with at that point; the screw has become a part of the object it's threaded into. Red is applied prior to assembly and gives a fairly long working time before it starts to set. A common use is to fix barrel extensions into upper receivers. Red takes up small gaps pretty well. Under the Devcon brand (same company) there's a maximum-strength version (#2277) that will take up a large gap, which they claim is up to 0.015 inches. High-temp Red is not necessary. The high-temp holds to 450° F and the standard variety is good to 300°. I don't know of anything on a rifle that could conceivably be glued that gets hotter than that.

Blue is a lighter-duty substance that's sort of like sticky-notes, in a way. It has the same effect as a locking nut. It also takes up gaps in threads and makes the threaded part resistant to unthreading, but can be loosened without heat or solvent (usually). It sort of gums it up. It's great for assurance, but not intended or advised in the role of an adhesive. It sets relatively more quickly so test fit before applying it.

The Green is wondrous stuff. Green is a glue, almost as strong as Red, but can be applied to parts after they've been assembled. It's penetrating or "wicking," like a light oil. It works well to fix parts together to save reassembly, like when alignment or orientation is difficult to attain. It can be used on front sight housings and gas blocks, for two good instances, or most any threaded-together or press-fit part combination. It takes heat up to 300° F.

Anything assembled with Red or Green needs heat to come back apart. It's sometimes enough to use a heat gun (not a hair dryer), but usually requires a torch. These both can take up to 450° to come loose. If flame has to get involved, wrap the part in aluminum foil and keep the torch moving to help avoid discoloration or general ruination of the finish. A heated wrench tip placed into a fastener head is a good trick to try on stubborn set screws.

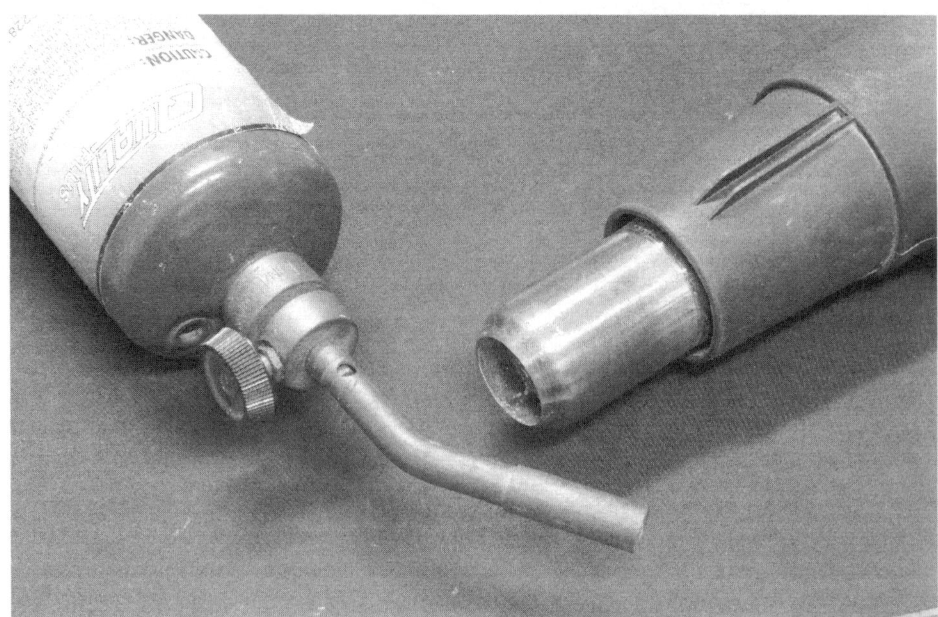

These thread-locking adhesives will not set in air. Right. Unlike epoxy or most other glues, Permatex thread lockers will remain in their liquid states in the presence of air. If it gets outside the parts you want to glue, just wipe it off.

Make sure to use enough of the compound to avoid gaps in threads, which trap air, to ensure the thread locker will set and cure as expected. That doesn't mean glob it on, but it does mean to make sure a large enough threaded area is covered by the liquid to do a 360-degree seal. Same goes for adding Permatex to a barrel extension exterior. Get full coverage circumferentially over a generous area of the extension and the glue will set and secure just fine.

Allow time for setting and curing. Overnight is usually enough, but overnight is also necessary. A little heat helps it set more quickly, and a heat lamp is the right tool.

Again, just make sure you give it plenty of time to cure.

Getting it unstuck. *Sometimes you'll need one of these to free the bonds of adhesives. Careful with the torch. Cover the parts with foil to help reduce discoloration. The heat gun shown is a heavy duty model intended primarily for paint stripping.*

Masking tape has more than one good use, and the most common I put it to is protecting ancillary surfaces from pin punches and hammers. A strip of tape laid down to back a bolt stop installation, for instance, usually excludes, and certainly diminishes, the potential for a receiver scratch. A strip laid out on the bench to hold small parts in place, and in order, is sometimes a huge help, especially if you need to leave the work for a while. Get the low-tack variety from any home-improvement store.

LUBRICANTS

If you have rifle maintenance equipment, you should have just about all you need already. Grease and oil appropriate for use on firearms should always be on the workbench. When assembling an AR15 there are small parts going into small recesses and right then and now is not only the best chance to lubricate them long-term, it may be the only chance. Grease also functions as a glue, of sorts, that secures ball detents and the like for assembly.

I admittedly (and literally will admit where appropriate) didn't "show" every single instance of the lube being applied to some of the parts installations demonstrated in this book, and that's only because it obscured the details. I will, and did, point out what should be greased or otherwise lubricated prior to assembly. In particular are detent assemblies, the "hidden" parts.

My preferred grease is one I've been using that contains boron nitride. Anything, including "Grease, Rifle" works well too. I use Plastilube for A2 rear sight assemblies because it's tenacious. The oil

Making it stick. You will absolutely need something to serve as a contact cleaner to get correct adhesion from any adhesive. Either brake parts or electrical cleaner works the same and comes from auto parts stores. This strips a part down to nothing but its surface, and always, always use it on both surfaces that meet glue.

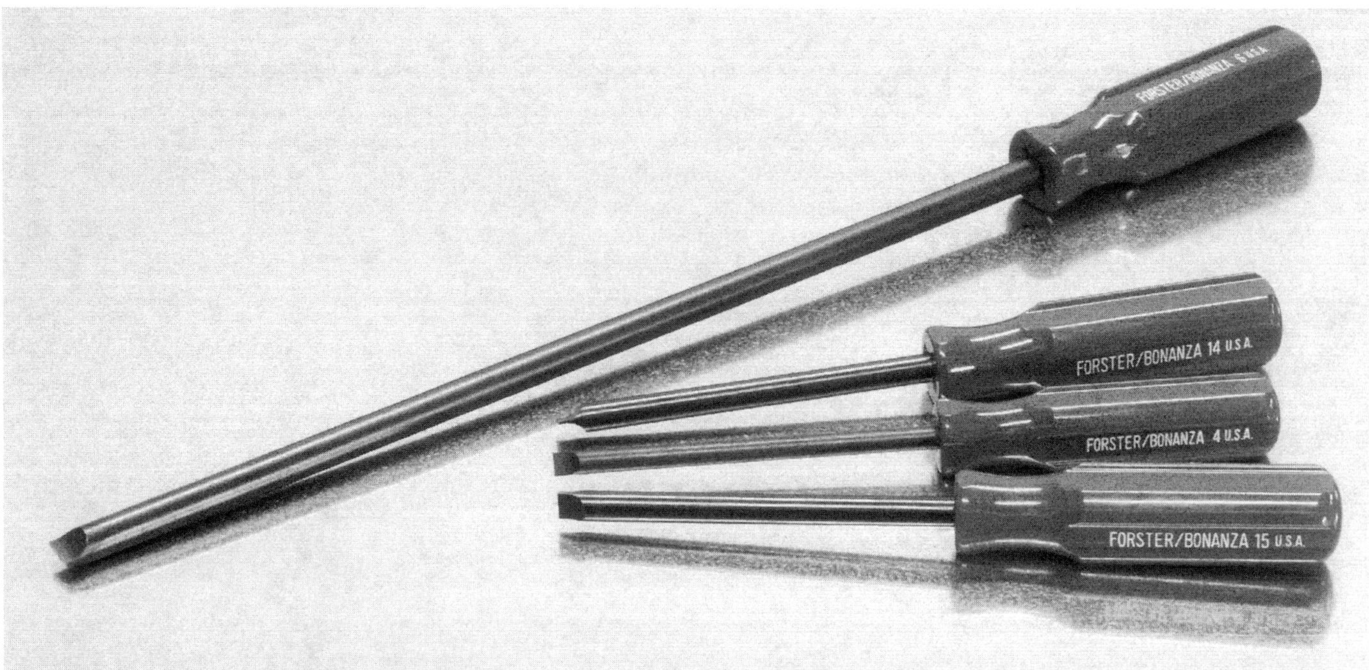

Screwdrivers must be the sort that feature hollow-ground blades. And they must be carefully chosen for slot width lest the slot take on a nappy appearance. This is especially true when installing a buttstock. Brownell's and Forster make great tools for us. The Forster set up top covers virtually any and all screws encountered. The honking driver in the top photo is what I use for grip screws. A pair of Brownell's Magna-Tip handles (long and short) is probably in every professional gunsmith's shop along with a tip set (shown on right).

I picked for use here was a bottle of Brownell's gun oil. It's good oil with a good viscosity, but honestly any gun oil works about as well for parts assembly.

Another lubricant I strongly recommend for barrel nuts is anti-seize. Loctite C5-A is a good choice. It's copper-based. The point and purpose to anti-seize is to prevent galling and parts "sticktion" such as can occur between the steel barrel nut and the aluminum receiver threads. It in no way diminishes the hold or grip of this assembly, but does in effect increase the ease of adding that little (to a lot) additional tightening that is sometimes necessary for proper assembly. It also makes it more easily possible to unscrew the assembly should that be wanted.

SPECIALTY NECESSITIES

Some of these things are just cool. I'm not at all impressed with electronic gadgets but do appreciate "why didn't someone think of that before" tools. If routine jobs, in particular, can be helped along they can be made less bothersome. Here's a collection of those I will never say can't be

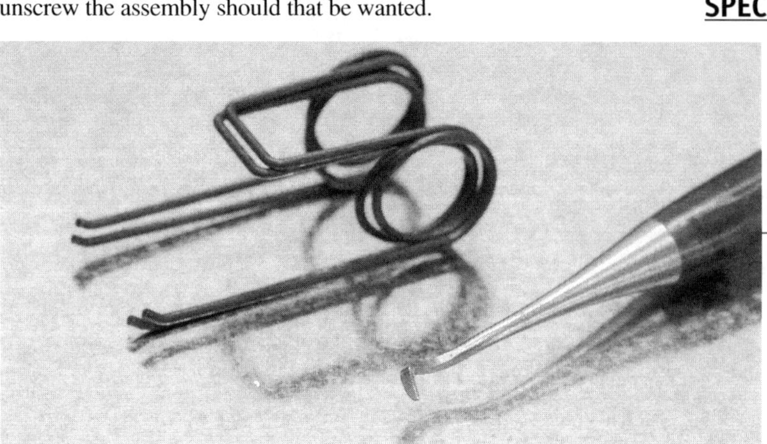

Little things become invaluable, or at least their value is commensurate with their cost and ease to obtain. A dental pick set is a good example. They're handy for shifting springs around like when you want to double up trigger return springs to get a higher first-stage pull for a Service Rifle.

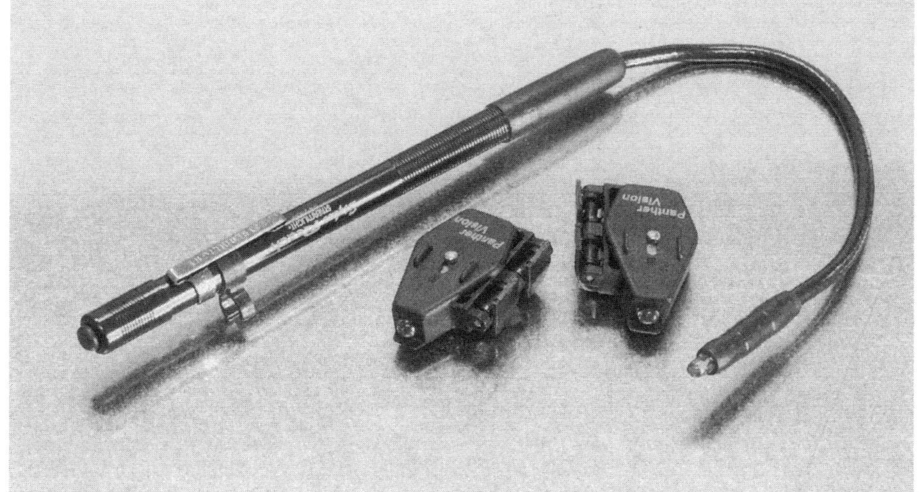

Seeing. *Good lighting is valuable, when it can be put where it needs to be, especially to direct light between the tools and the fingers and the parts. I found these handy means at my local auto-parts store. A plain old flashlight has helped find more loose springs and detents on my shop floor than anything else. Illuminating the area will show up things not noticed otherwise, including how dirty the floor is, but that takes another tool I've never liked using. However! Carefully sweeping up the suspicious area into a dustpan is my version of panning for gold.*

substituted for creative use of common devices or perseverance, but will say that they are well designed to do what they do.

Others will be shown when used to build the project rifles. Again, they're not all exactly mandatory but the more the op is performed, the more valuable they become. I can't imagine a pro not eventually having each and every one, or, maybe as likely, manufacturing things similar for himself.

A good corded drill is a good tool to have, and don't get a cheap one. I use this to spin things, of course, and usually it's something that's being polished, like a roll pin end. A drill with a rheostat-style trigger helps control its speed. Spinning pins and controlling cutting tools is even easier with a collet-end flexible shaft installed. Corded drills are way on better than anything cordless when actually used as shop tools.

There are always uses for tiny screwdrivers and, of course and as mentioned, gun cleaning equipment is part of the tool set as well.

SERIOUSLY MISCELLANEOUS THINGS

A larger magnifying-type light is also valuable. Brownell's stocks some very nice ones but it's not really necessary to spend all that much on one. Most bigger chain discount stores will have something adequate. Sometimes there's no substitute for a closer, brighter look at a part or assembly. I get a lot of benefit from such lights in reloading also.

A set of ball-end allen wrenches is a wise purchase. These let you work the screw without having the wrench directly or fully engaged, and also can run securely as shallow as a 20-degree angle, making them the only option sometimes for sneaking a wrench into an obscured fastener. They come in handy on accessory installations that are usually et up with set screws. These kind also go with me to the firing line to run stock adjustment pieces. They are easier to use than conventional tools. If you run allen-head screws in place of slotted pistol grip screws, which is wise, a t-handle allen wrench makes for lighter work.

Vision can be an issue and magnification usually clears it up. I can't see well up close anymore with my daily glasses on. I have a pair of reading glasses that I wear in the shop too.

Of course you'll need shop rags, paper towels, and other such products for clean up detail. A bottle of Simple Green found in anyplace's automotive section can take the place of most solvents to rid a surface of an oil-based mess, and it's good on and for hands too. Use it in the washer with shop rags, if you bother to clean them, to get out the grease.

Big plastic boxes sometimes come in handy to keep project guns parts together before the parts have gone together.

Other containers without a doubt can be invaluable, as long, of course, as what came out of them gets put back after use. Fishing tackle boxes, as well as those made for jewelry making, can house pin punches, parts, and so on.

Eating away at it. *This stuff removes adhesive residue better than anything I've yet tried, including residual locking compounds from screw threads. It's Permatex Gasket Remover. Strong stuff.*

Well, I ran out of things to say here in the main text but not nearly out of things to show! Over the next few pages are some photos and captions depicting and describing things that will get use in a shop dedicated to assembling AR15s.

Cutting Tools

These things scare me, but they're necessary sometimes. It's a Dremel with a flex shaft. They scare me because it's all too easy to go overboard with one, or has been for me. This one doesn't have this feature, but well worth the extra for it is an on-board rheostat to adjust its rotational speed. I operate mine off an external rheostat which is a decidedly duct-tape-class solution; it's an in-line plug-in dimmer switch. Otherwise, the revs are very high (35,000 rpm) and, therefore, material removal often happens at the *oopsie* level. Super high speed makes for clean cuts, though, and there's little torque in these tools so speed is necessary. One trick is choosing a bit-tool that moderates capacity to reduce plastic or aluminum to dust and then a very light touch under very firm control. I use this Dremel to polish select pieces-parts (with felt and paste only), trim plastic, and cut springs. For spring cutting get the reinforced cutting wheels. And wear safety glasses. I once had a cutting op pit the fool out of my routine specs. Sparks fly like little meteors cutting springs and screws.

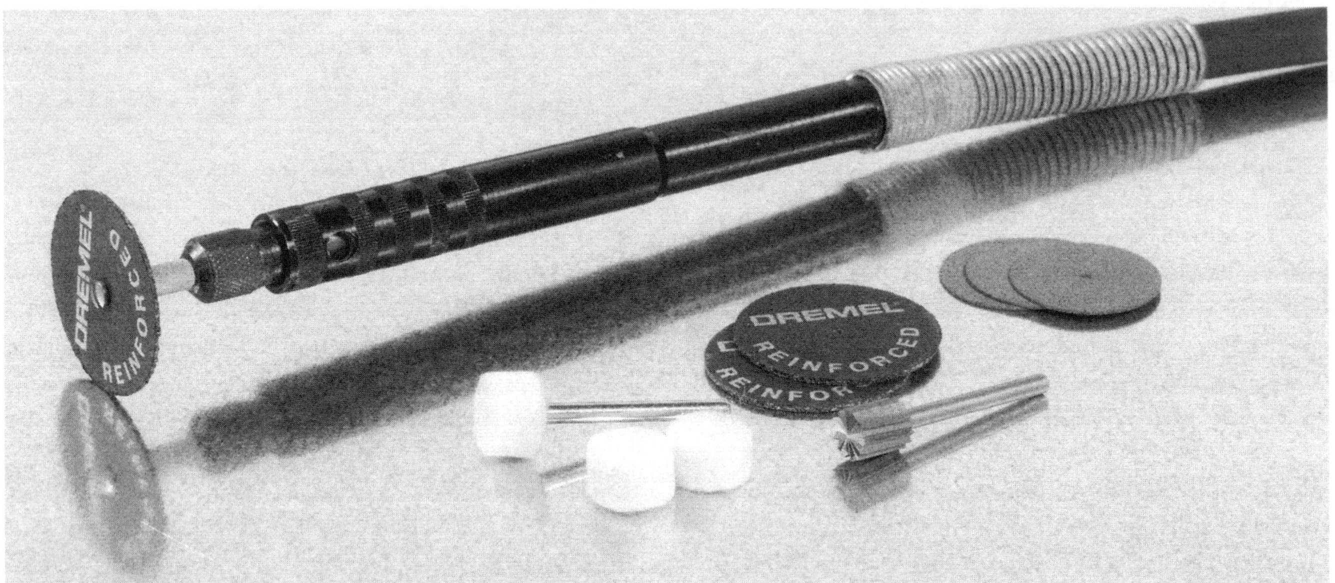

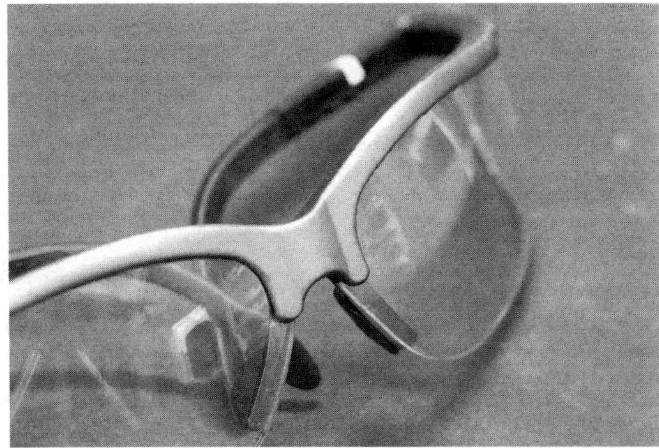

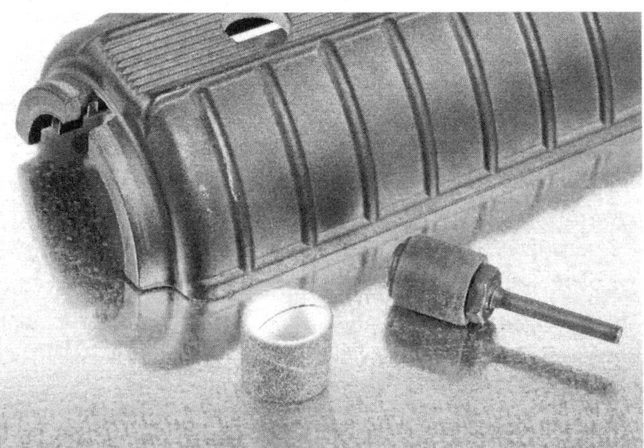

For plastic removal post-haste, rotary rasp-style cutters do well. I have good luck with sandpaper wheels too. As with mentions along with other abrasives, these won't get much use by many but work wonders modifying plastic handguards for use with CMP-legal free-float tubes.

When things really go wrong. There aren't a lot of "common" gunsmithing tools needed to assemble an AR15, but when you need them there's no substitute for having them. Get them from Brownell's. Yes. You will pay more but the tools they carry are worth the extra they cost. I've searched enough home improvement and auto parts stores to tell you that, for instance, it is unlikely to find a good quality and correct file at those places. Screwdrivers are another good example. So are stones, abrasives, taps, drill bits, and the list goes on. Brownell's carries things that are useful to a gunsmith, and when that is the role you are playing in your project then that is what you should have on the bench and in your hands. Pray to the deity of your choice that you live long enough not to use a file on an AR15.

Abrasives

Abrasives are a part of working with gun parts. Emery paper (always use wet) in varying grits is a staple. More aggressive cutting I say best comes in the form of aluminum oxide paper. A trick on stubborn fine aluminum threads is to very lightly run some 600-grit emery around them. Just sort of twirl the threaded area over a hand-held piece of emery paper, and then thoroughly "dust" it out. Some float tube pieces, as well as competition sight parts, have very fragile, very fine (related) threads and when these have even the least bit of roughness they can be a bear to get started and then to thread in fully, even after they're meshing properly.

For polishing and lapping work, aluminum oxide abrasive works better than silicon carbide because the later tends to imbed into metal. Aluminum oxide cleans up nice too. These are lapping compounds from Brownell's. Such will get infrequent use by most, but those who like to polish under power can do no better.

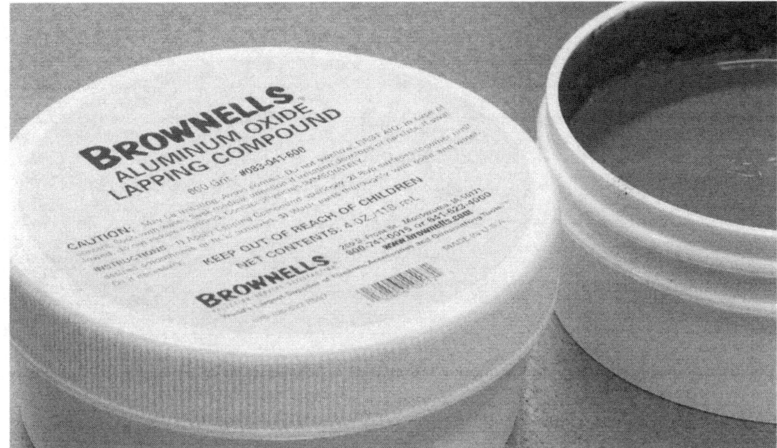

If you're going to get files, get good ones and I can tell you you won't find them at a brightly-lit discount store. A fine "second-cut" file, especially any at a larger size, comes from gunsmithing outlets. Accessory handles are patently necessary. Bigger files work better on smaller parts. It's easier to keep a wider, longer file level and stroking flat and true. Needle files will see use...

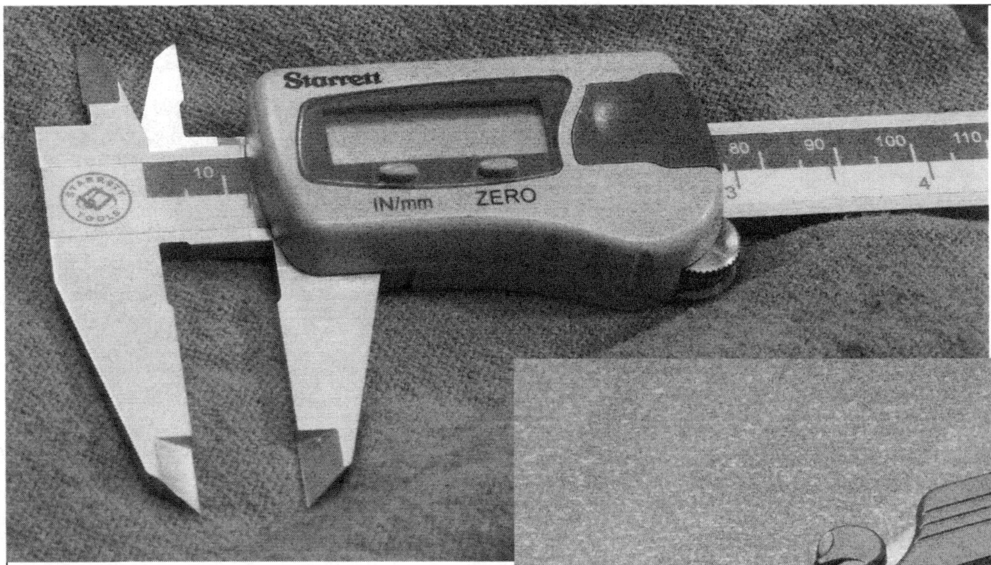

Measuring Tools

You may already own these if you're a handloader. If not, or maybe even if you are, a good quality micrometer and caliper pretty much constitute the collection necessary. If you don't have either of these, I strongly recom-

mend either brand. You really do get what you pay for.

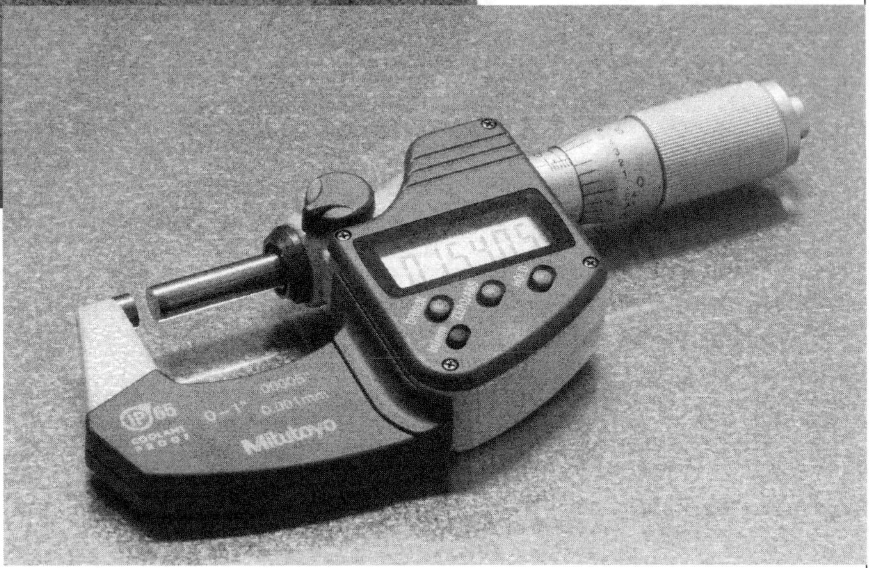

Take careful care of such tools. Store them in their cases and never with the measuring surfaces flush. Leave a little gap between the jaws.

The most use either is likely to get is when installing specialty parts, like accessory gas manifolds, or quality checks on trigger pins.

Paint and Pens

Got to have this stuff. Shiny scratches and dings really stand out on a black rifle… A flat-black paint touch up marker makes most any roll pin installation look factory. Aluminum Black tackles other such surface glitches, and a good cold steel blue such as Oxfo-Blue from Brownell's ought to be next to these products on their shelf.

Sharpies come in handy too. The metallic variety is great for sight indexing or referencing alignment marks on black parts.

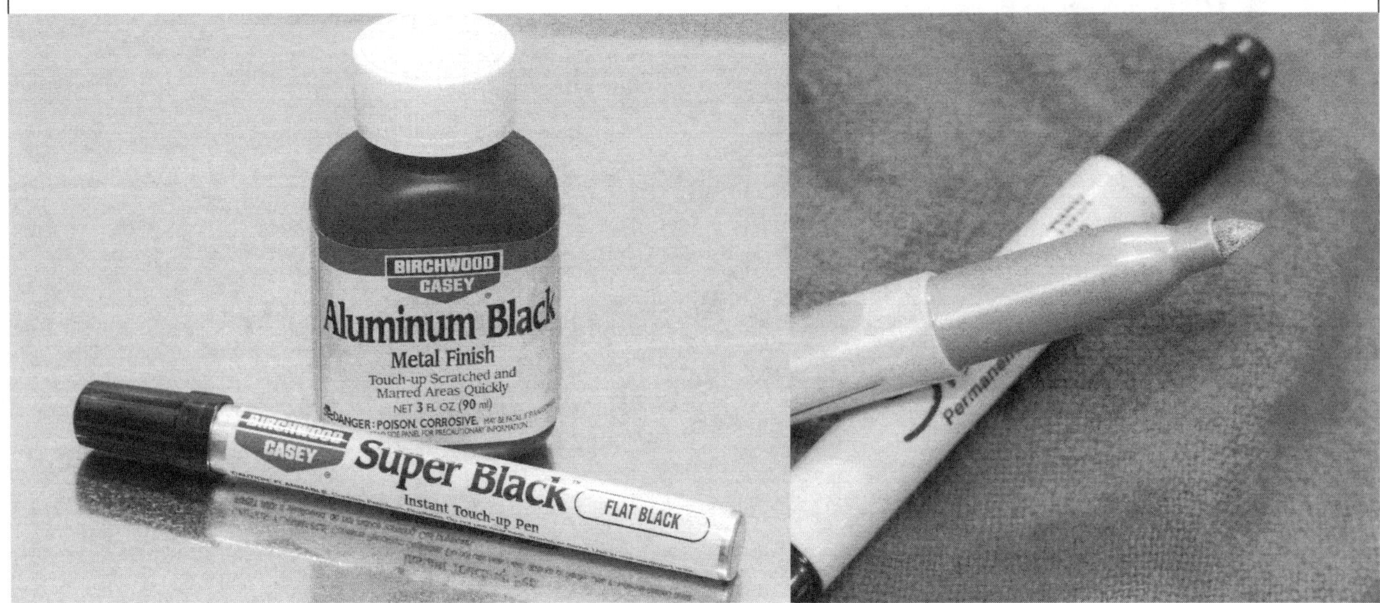

Cushions and Cover-Ups

A scrap of carpet atop the workbench all but eliminates scratches and blems but this shop mat from Brownell's is a decidedly nicer product.

I have a pair of Mechanix-brand gloves and use them frequently. They're no good for small parts manipulations like starting roll pins but help a little on the more manly tasks. They come in dang handy for installing triggers. The cushion makes it a sight easier to position a hammer to take its pin. These come in different specialty forms, of course they do, and my favorite is a "high dexterity" pair intended for use by electricians.

Pearls are optional.

"Beaver! Wally! Time for dinner!" A shop apron makes me feel like June Cleaver, but I have to admit they come in handy because they have tool pockets. There's little more frustrating than finally getting something aligned, reaching for a pin punch to keep the parts put, and then seeing it way out of reach. Just put them in your apron pockets. Men, just don't wear one of these when you grill. Just don't.

A way to hold the rifle...

I keep one of these in a vise in my shop. It's a Delrin block from Brownell's that fits in place of the magazine to hold a lower receiver, with upper attached if needed. It's too handy to even talk about, but of course I will. It makes it so much easier to install triggers, stock parts, and you-name-it, and never any worries about padding vise jaws and the like.

More shop essentials

Real shop rags work better than anything. That's probably why they are what they are. Get them at most any automotive parts store. They're cheap too. Paper towels, of course, can't really be substituted for much else so always have a roll handy. The "blue" kind sold for use by mechanics are especially suitable.

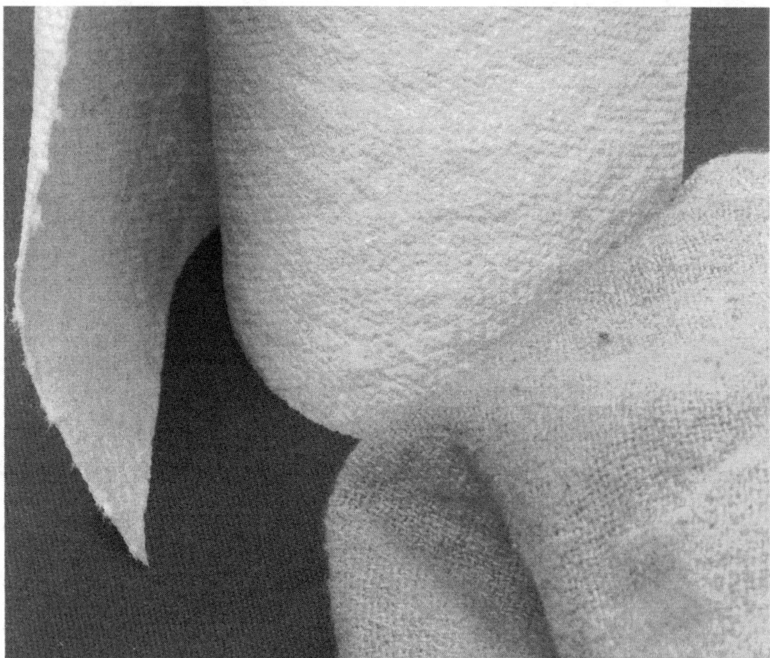

A good solvent-style oil might not see a lot of use on rifles but will otherwise around the shop, especially if corrosion is a problem. This product gets a lot of use by me. It's a penetrating oil called PB-laster and the comic-book style claims made on the can exterior are true. For those of you familiar with Kroil-brand penetrating oil, I can tell you this is nearly its equal at a significantly lower cost. Get it at any auto-parts supply.

Needing a bigger toolbox

A correctly-fitting trigger- and hammer-pin-sized rod is invaluable doing trigger tuning. Shown is brass assembly punch from Brownell's. With such a tool it's possible to do test fitting without installing the actual pin because the tool dimensions are the same as the pins themselves, and that's 0.154 inches. Such otherwise correctly-sized tools come in handy for other operations, like pivot pin detent assembly.

Frustrations, I venture, lead to development of cool little tools like the little takedown pin pusher shown in the center photo. It's dang handy on stubbornly tight pins. If needed, position it, give it a little thump, and the pin will pop free. Always keep front and back pins greased and push the upper down onto the lower when removing the back pin. If the back pin is stubborn, do the front pin first.

Here's something I think everyone ought to purchase for each rifle they put together. It's a DPMS pin kit. It has one of each for a complete A2. Get spare springs too. I know that parts kits also come complete but I still keep the DPMS baggies-full of spares around because I know all the springs and pins are correct. That's not always the case in some of the lower and upper parts kits I've gotten, and it's likewise, and unfortunately, not true with some of the rifles I've disassembled. I've seen detent springs in place of ejector and extractor springs, roll pins that weren't correct, and so on.

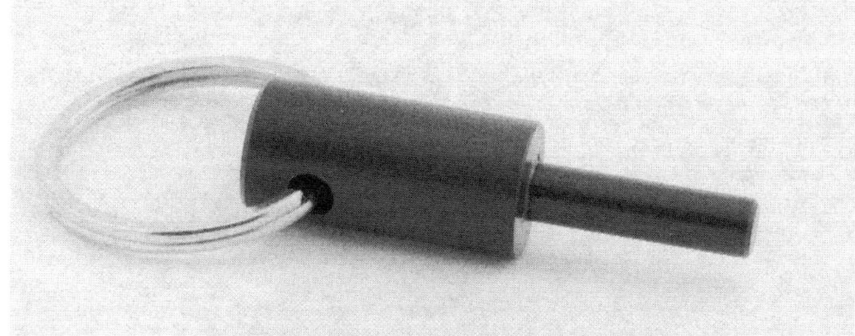

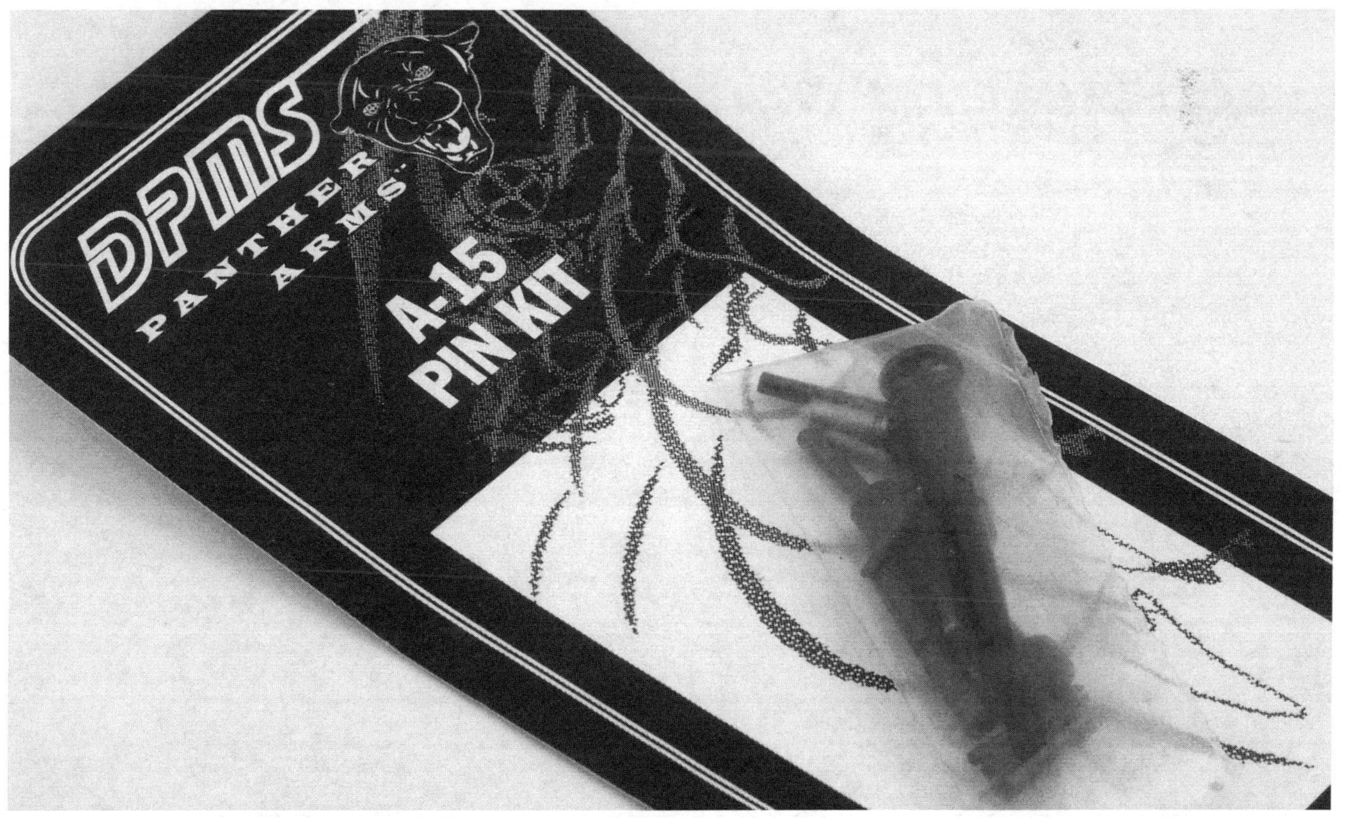

Some I don't want to do without

Really, that could be a lot of the tools in my shop, but a few in particular just aren't options for me. Any time a specialty tool reduces or removes the need for the extra hand, it's invaluable. Shown below and left is a good example. Its job is to depress the magazine catch button so the assembly can be easily threaded together.

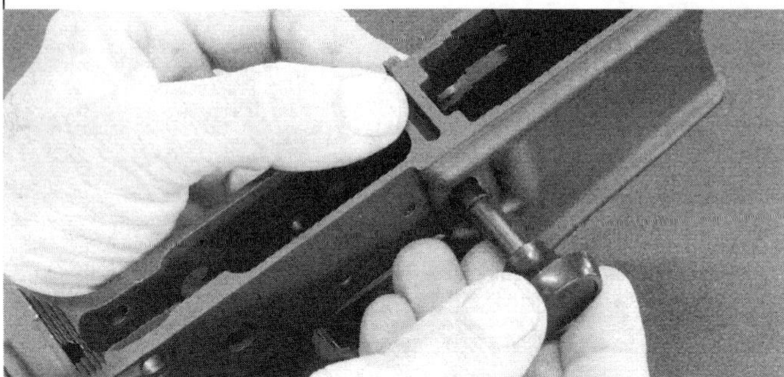

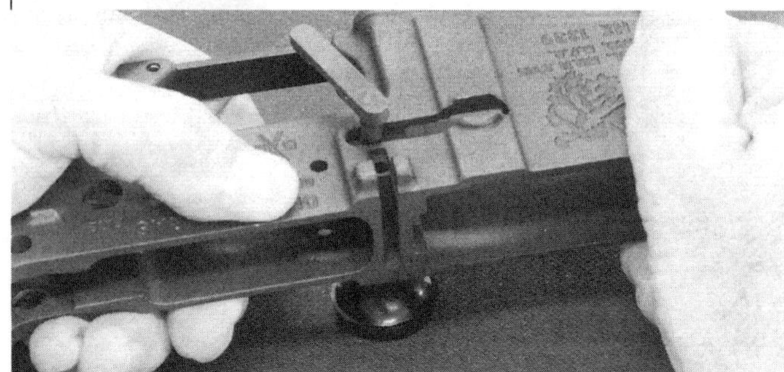

Shown below is a handguard removal/installation tool. Need I say more? This job is unpleasant, and this tool makes it pleasant.

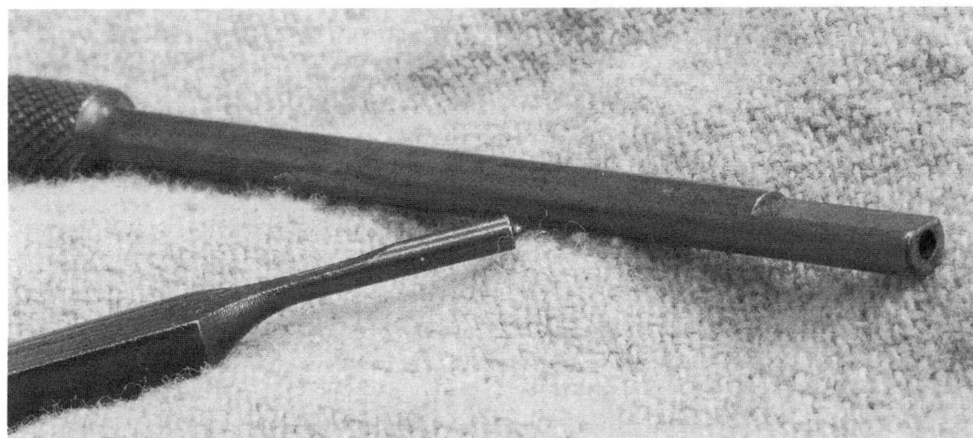

Shown at left floating around on this page is a tool that ranks up there with a hammer. It's a **pivot pin installation tool** and is virtually a required purchase. The hole in its end lets you install the detent and its spring and then capture it. Exceedingly difficult task without this gadget.

Brownell's makes **specialty punches** like this pair in the middle photo that can get "closer" to the receiver without as much risk of scratching its finish, and also allow a little more straight-line drive. They are no substitute for masking tape though.

The tool in the screwdriver handle is designed to compress the big spring in the rear sight so the pin that holds the sight base in can be installed. There are far fancier versions, and they are nice to work with, but this one or something very similar is mandatory to complete this operation.

Folks, there are always more and more tools. The more you do this and the easier you want it to be, the better value many become. Some tools that initially seemed superfluous to the point of indulgence have since become treasured in my shop. This is just the same as working on anything mechanical.

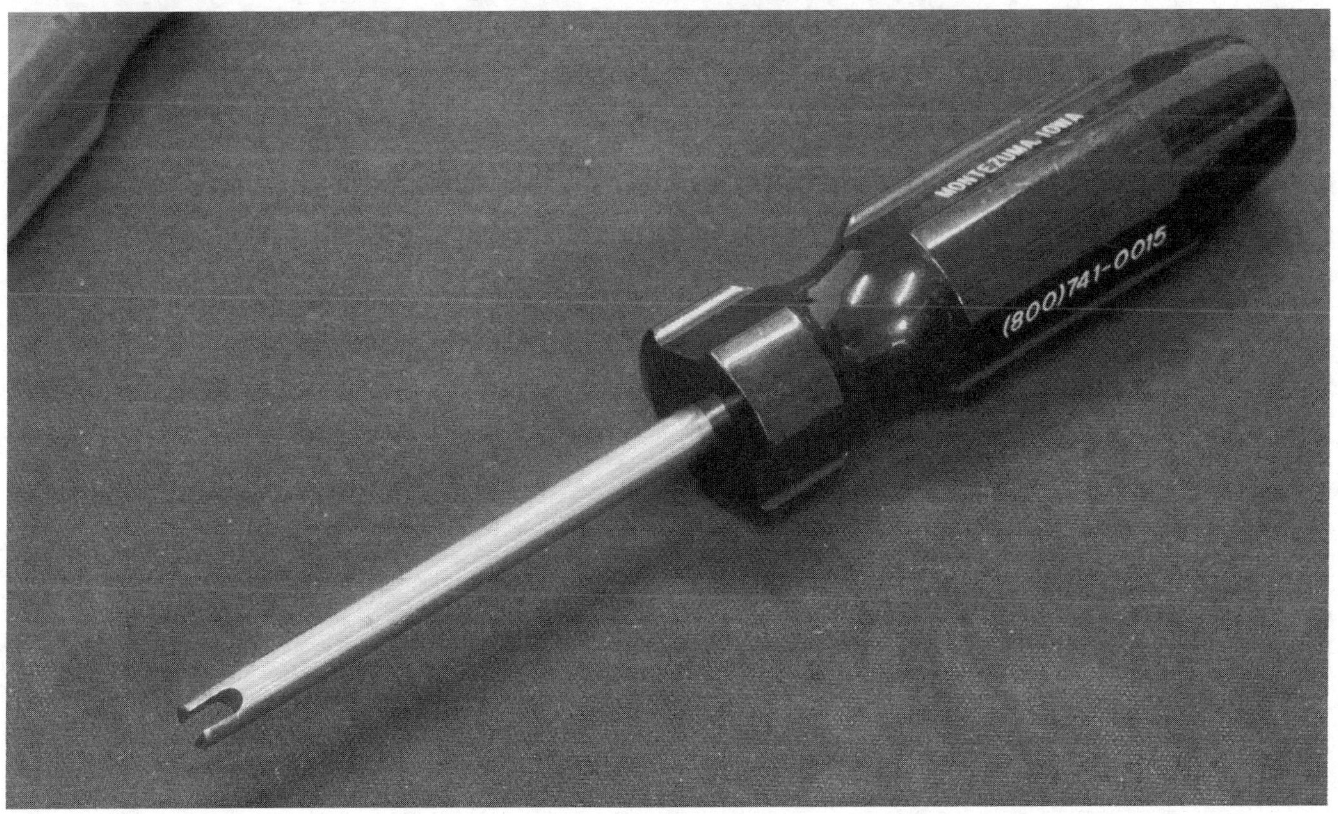

Getting ready to get ready to get going...

The first project detailed in this book is building a lower receiver assembly, and it can involve a few pieces that serve to sound the culmination of the "tools" segment. **Again, more (and more) specialty tools will be shown and demonstrated as these projects unfold.**

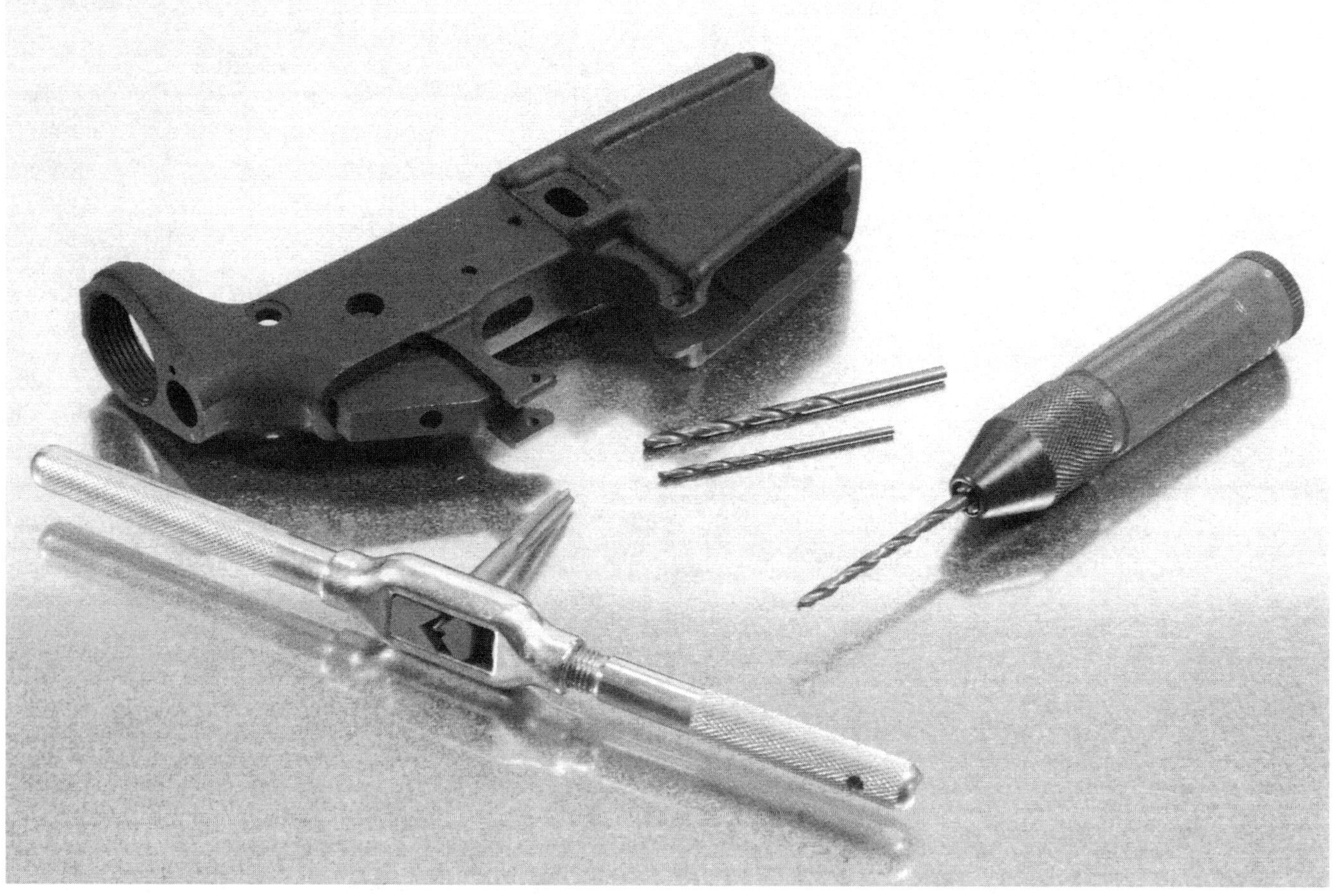

An assemblage of **machinist's drill bits** is a worthwhile purchase, as is a tap for the grip screws. Get the bits from a machinist's supply, or outlet like Brownell's. Machinist's bits are identified by letters and numbers and are exceedingly fractional in sizing steps.

One of the advantages everyone gets from doing their own work is taking time to address details. Some details, like ensuring component correctness to start, make assembly that much easier. Use these tools to chase out burrs and finish blems from pin holes prior to assembly. Don't run them under power. Just a spin with the fingers, if that much, is usually all that's necessary to clear a hole. The screwdriver-style chuck tool shown is a good way to do it. The last thing you want is to unnecessarily enlarge a hole.

Here's the list and what they're for:

#22 (safety detent hole)

#23 (0.154-inch trigger and hammer holes, also bolt stop plunger hole)

#40 (pivot pin and takedown pin detent holes)

1/4-28NF tap (grip screw threads)

Use the bits, along with canned air and a pipe cleaner, to dislodge debris from holes. There will be debris in the holes, in varying amounts of course.

One note: The #22 listed for use in the safety detent hole won't protrude clear through; just run it lightly until it stops, and, of course, check for free function using the detent itself.

2.0 LOWER RECEIVER

THE MOST WORK

[If you don't have a lower receiver chosen, choose a good one... That's easy to say and not so hard to do, not really. These come good from the major outlets, and it's honestly not going to matter which you go with for a forged 7075 part, as long as you've heard of them. If you haven't heard of them, well, get one you've heard of. Exceptions are the billet-made milled lowers from smaller shops like Titan, TKS, Sun Devil, and others. Those, by and large, are outstanding. It's wise to "match" brands of uppers and lowers to get a finish match.]

SEGMENT CONTENT

34	**Before you start...** (receiver preparation)
36	**Parts & Tools**
38	**Trigger Guard**
40	**Magazine Catch**
42	**Bolt Catch**
44	**Pivot Pin** (front pin)
46	**Safety & Pistol Grip**
48	**Extension Tube Installation**
48	**Buttplate & Buttstock** (assembly and installation)

Here's where we'll start. *I'm starting this book with this project because it made some sense to start from the ground up, and the lower receiver assembly is the foundation for any rifle. Doing yours yourself is not difficult (just need the tools) and you will, no doubt, ensure that what you have will work.*

PREFACE

There are a lot of parts here. First thing is making sure you have them all. Refer to page 36 to locate and identify them.

The majority of lower parts are pinned in place and left in place. Unless something breaks or otherwise quits working, it's unusual, for instance, to encounter a need to disassemble the magazine catch assembly.

I lay out the assemblies on strips of wide masking tape to keep all the pins, springs, and detents in place. This can really help and you'll know the value of this step if you've ever looked for a ball bearing on a shop floor. It's amazing how far a tiny inanimate object can travel, and so often not in the anticipated direction.

As suggested many times, I tape the fool out of the area where the parts will be installed. This cuts down the chance for unintentional and often otherwise unavoidable finish damage.

Before you start...

If you have them, run the drill bits shown on page 32 carefully through pin holes prior to assembly — by hand — to remove any debris or burrs that might interfere. There is also never anything wrong with breaking edges on most parts. Anything that's rough usually smoothes out on its own, in time, but metal-to-metal contact points should have smooth surfaces. I'm not talking about changing the dimension or "shape" of an edge, just breaking it. Use emery or a hard stone or a soft polishing attachment under power. Magazine release and bolt stop assemblies get the most attention from me. Don't hit the "working" edges though, those which are responsible for the engagement with other assemblies. Keep the bolt stop engagement surface, for good instance, sharp and square. It might not work if it's not. Smooth the sides so it functions freely.

While there's a little space here to devote to this topic, the more you do this and the more parts you have on hand as a result the more you'll see that unknown variances exist, and primarily there's coincidental success with coincidental parts. And what I mean by that is seeing a magazine catch assembly, for instance, install easy and work perfectly in one lower and not in another, and the one that worked for that one may or may not work as well in the other. It's often a matter of degrees, though, which is to say smoothness of function.

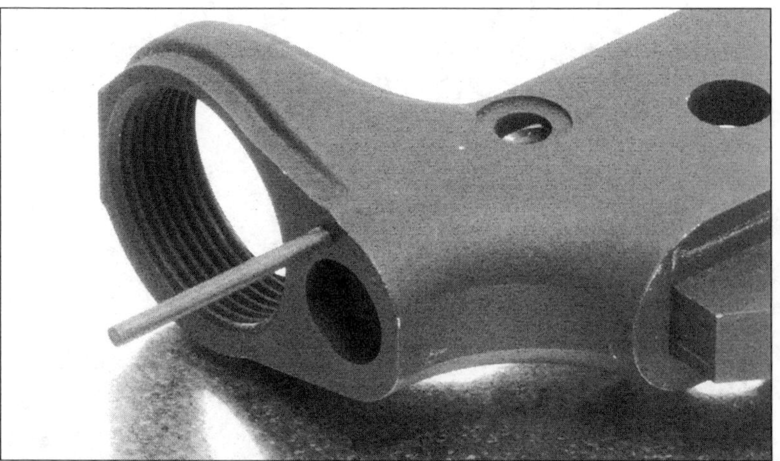

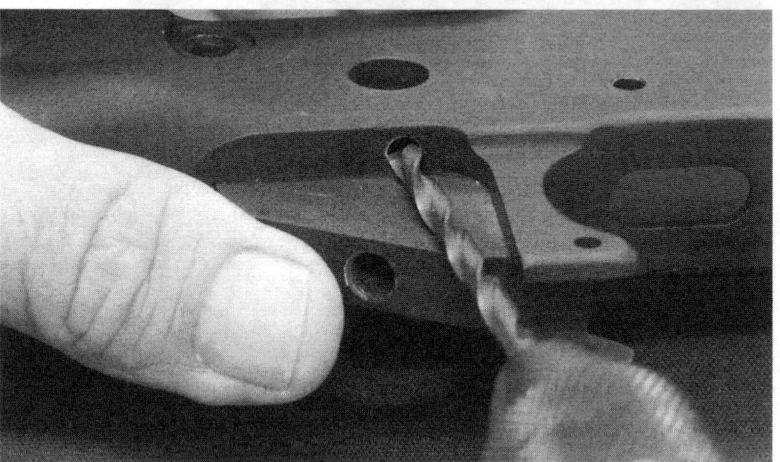

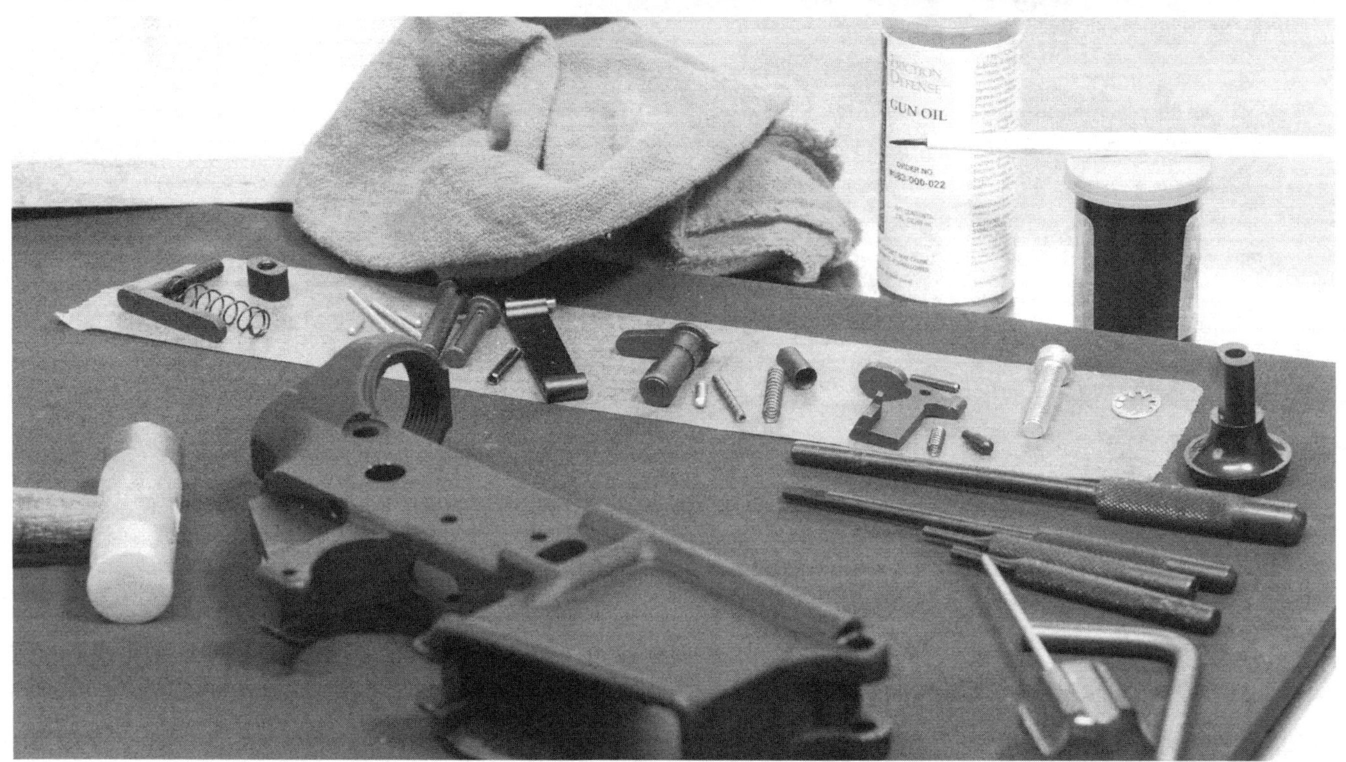

Grip screw tap

It's extra work to be sure, but after you have experienced its effect you'll see it's well worth the effort. This one is a 1/4-28NF from Brownell's.

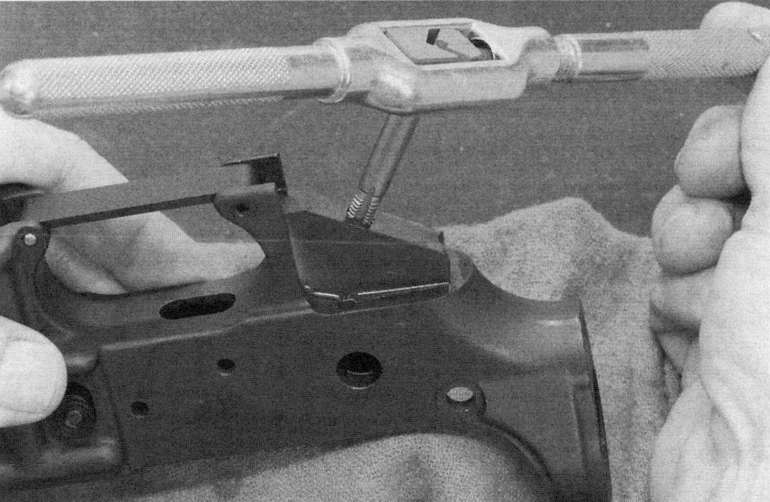

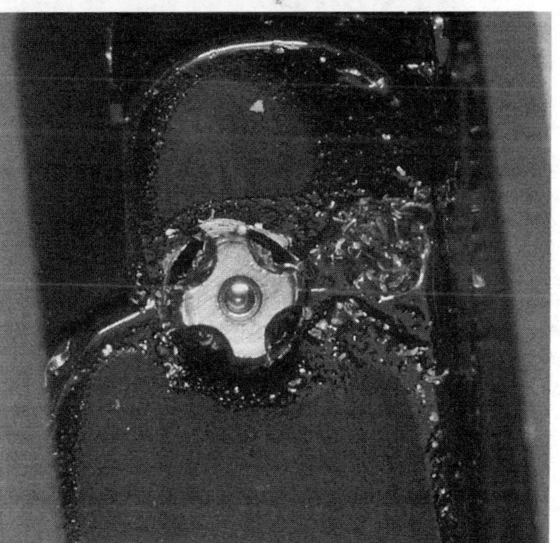

Alls we're going to do in this op is touch up the threads to what they should be. Just run it through, one or two turns at a time, backing it all the way out a few times to clear the debris. Don't get greedy with tapping, even on something that looks this simple. By the way, the only thing that works like cutting oil is cutting oil. Any tapping demands this product, not just plain oil. Nothing really substitutes because cutting oil is specially formulated for, well, cutting. Believe me and I know the hard way…

You will be astounded at how much easier it is to start and continue a grip screw after doing this. I've encountered some that were otherwise just horrid. Take a look at the amount of chips that resulted from this operation on this lower and it's no wonder why. Clean it up thoroughly.

PARTS & TOOLS

Needed for the essential build. Trigger and stock installation tools and parts are shown in their own segments that follow. Make sure you're looking at all these things before starting. If you're not, well then get what's missing.

PREPARATION

Tap hammer
Roll pin punches and roll pin starter punches
Pivot pin installation tool
1/16-inch plain punch
Wood block
1/4-inch dowel piece
Gun oil
Gun grease
Screwdriver or allen wrench (grip and buttplate)
1/2-inch spanner wrench (or equivalent)
Options (good ideas)
   Magazine catch assembly tool
   Flat black touchup paint

PRECAUTIONS

Use the masking tape on the receiver
Select the correctly sized punches

PARTS

Stripped lower receiver

Trigger guard
Trigger guard pivot roll pin

Magazine catch
Magazine catch button
Magazine catch spring

Bolt catch
Bolt catch roll pin
Bolt catch plunger
Bolt catch spring

Pivot pin
Pivot pin detent
Pivot pin spring

STOCK PARTS

(shown page 52)

Receiver extension tube

Buttplate
Trapdoor
Trapdoor hinge
Trapdoor hinge pin

Buffer retainer
Buffer retainer spring

Sling swivel (rear)
Sling swivel screw

A2 stock shell
A2 stock spacer
A2 upper buttplate screw

LET'S START BANGING...

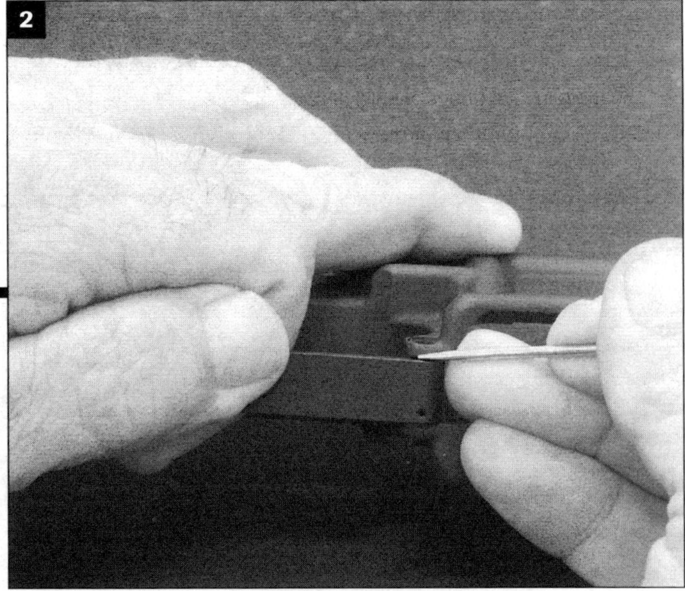

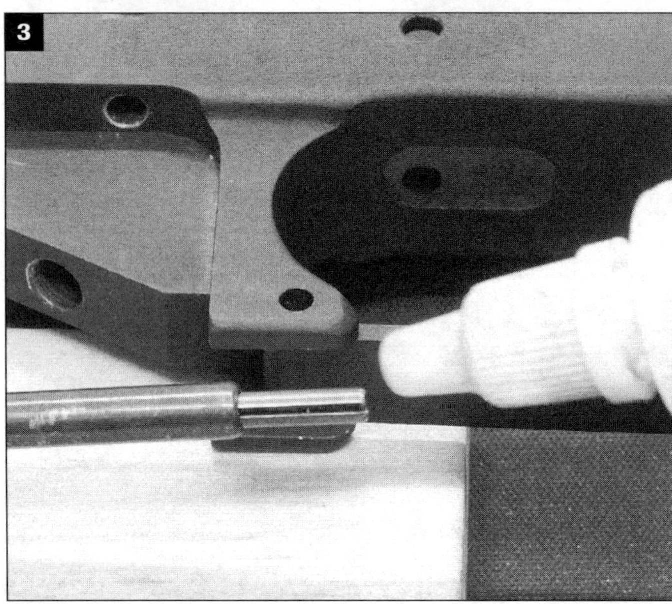

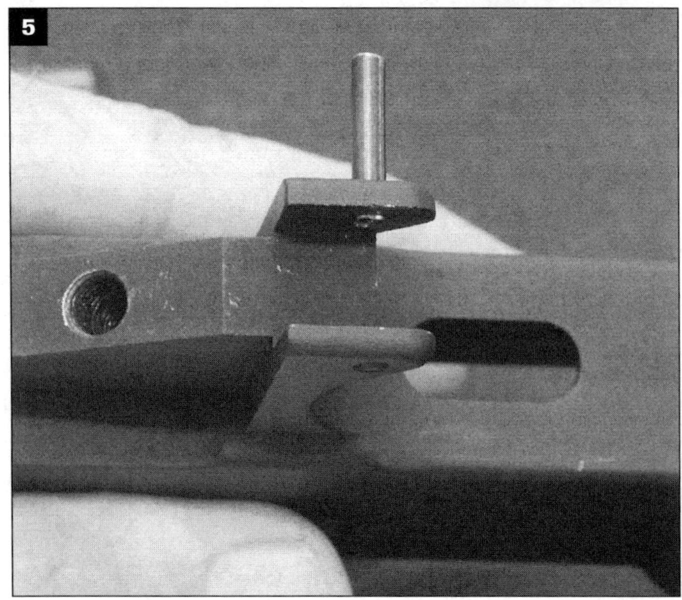

TRIGGER GUARD

The trigger guard installs only one way. **1.** Engage the latch-end into its tabs at the back of the mag well. **2.** A small screwdriver can depress the plunger to help get this piece in place.

3. Pivot the guard into alignment with the receiver pin holes. Back up this piece with a wood block. The little "ears" can and will easily break if you don't.

Oil this pin. **4.** Start the pin using a roll pin starter punch #4S (1/8-inch) and keep the holes in alignment. **5.** It helps (a lot) to tap the pin in just far enough to engage the hole in the trigger guard so it "clicks" into position before seating the pin through.

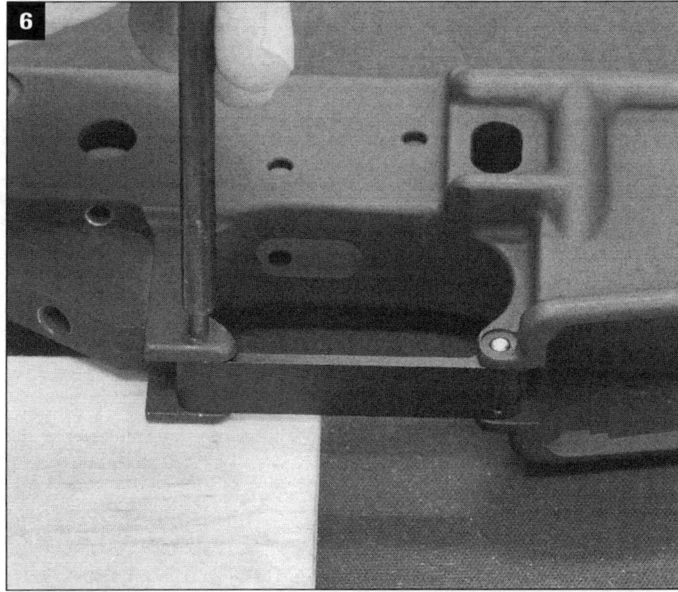

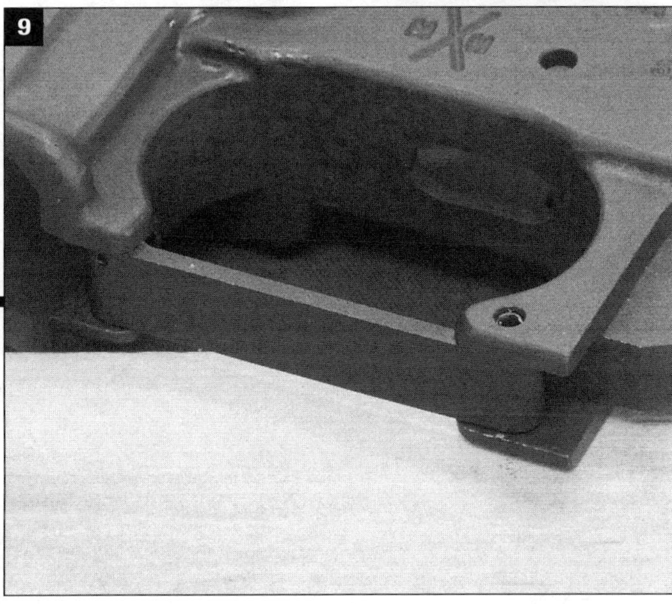

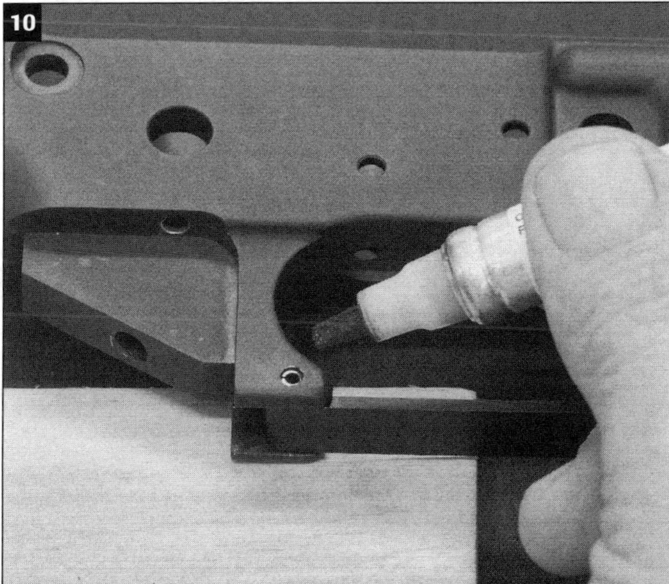

6. Drive the pin until the starter punch bottoms out. **7./8.** Switch to a roll pin punch and tippy-tap the pin through until it's just below the surface; **9.** check the other side and make it the same.

There is a good deal of force required to seat this pin. It's a big one. Just make sure that the guard piece itself is in place to take up the gap between the ears, and also that the ears are backed up by the wood block.

10. This pin will need a touch up when finished so dab a little flat black paint on it to hide the extreme force with which it was inserted and you're good to go (onto the next step).

> **When you drive roll pins,** tap them to just below flush and try to make them even in depth on both sides. It's easy to touch up the ends of the pins with a flat black touchup stick or marker.

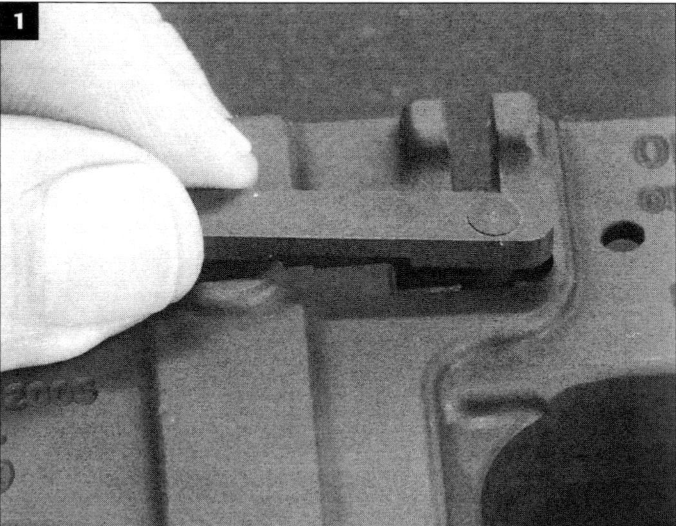

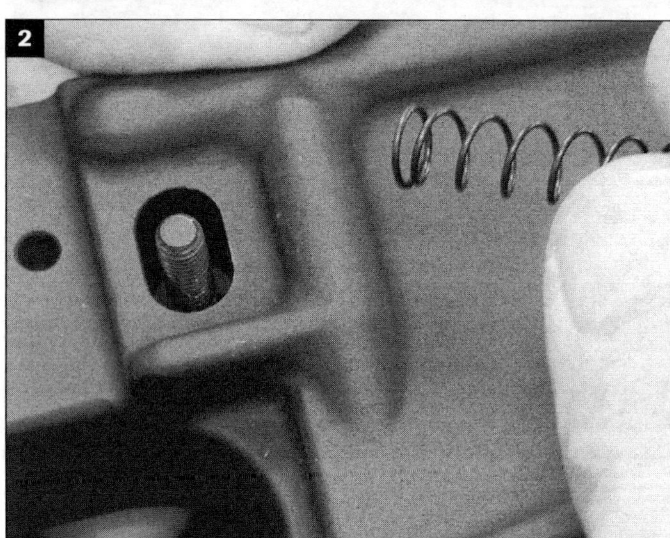

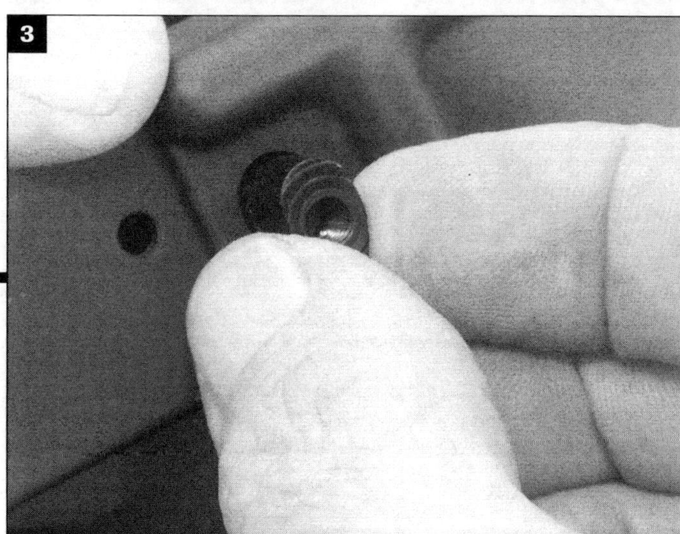

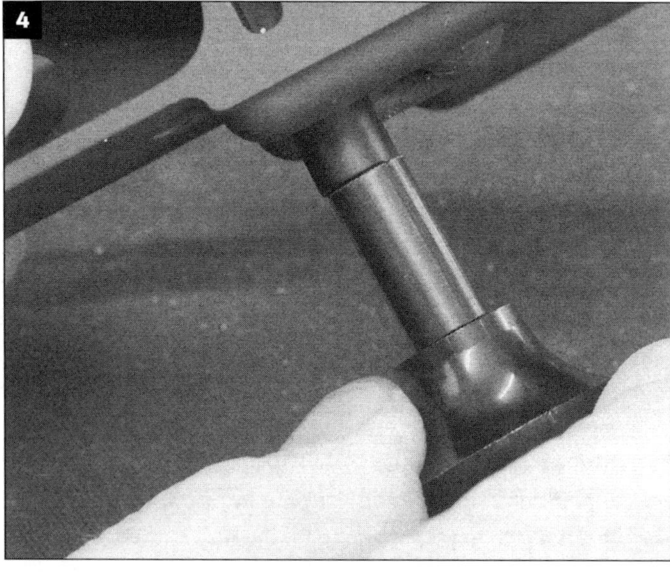

MAGAZINE CATCH

1. Insert the catch into its place in the left side of the receiver (note the threads on its end). **2.** Place the spring over the catch and into its recess on the opposite side of the lower. **3.** Take the mag catch button and position it against the end of the catch, turn it onto the threads enough to get it started.

4. Then push the button in so it engages the threads. Keep the button depressed (use a dowel or, way better, the specialty tool shown), and **5./6.** screw the catch in until the end of the catch shaft **7.** is just flush with the "inside" mag button surface (see detail photo **8.**). Using the specialty tool makes it easy to keep the button fully compressed by using the butt end of the tool like a pedestal.

This is one of those parts it never hurts to break edges on. I've had to do that before to get them to work smoothly in some receivers. Usually stoning the "nose" end where the shaft attaches makes it a smoother operator, as well as deburring the edges along its length.

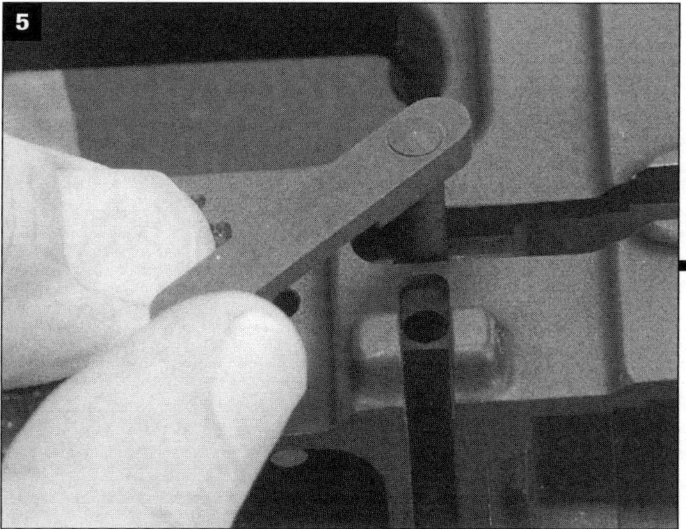

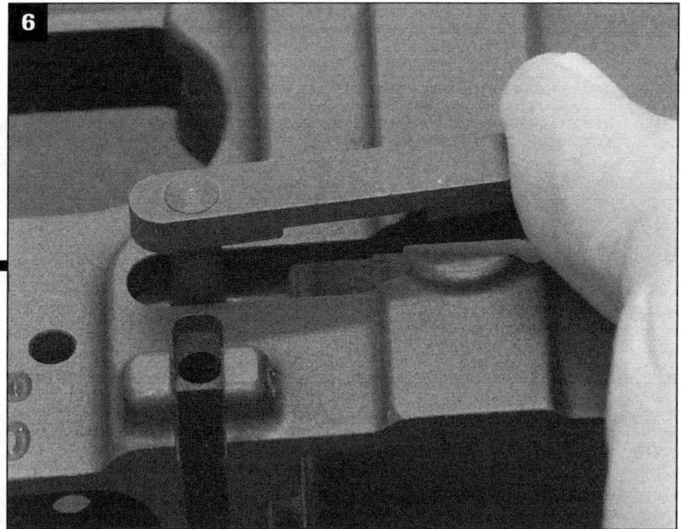

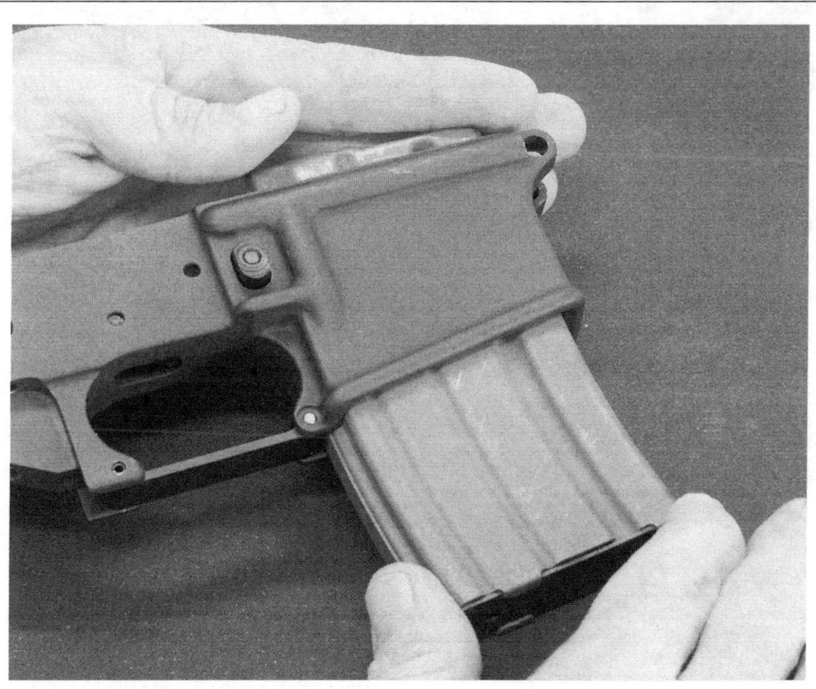

Test the installation by inserting a magazine. It should click into place smoothly and solidly. Try to pull it out. If it won't latch or stay latched, the catch may not be screwed in far enough, so depress the button again and take up another turn on the catch shaft. If the magazine latches, now it should also release, so give that a try. If it latches and then doesn't release, the catch shaft may be screwed too far in, so back it off a turn. What happens if it's "in-between"? The latch has to hold...

Now, it is possible to have a fully-functioning assembly that's still not what it should be. If the shaft on the catch is not threaded far enough into the button it's possible for the catch to come out so far when the button is depressed that the catch can rotate out of place and get hung on the outside of the receiver. Make sure that can't happen and if it can, take up another turn.

Try it with more than one magazine.

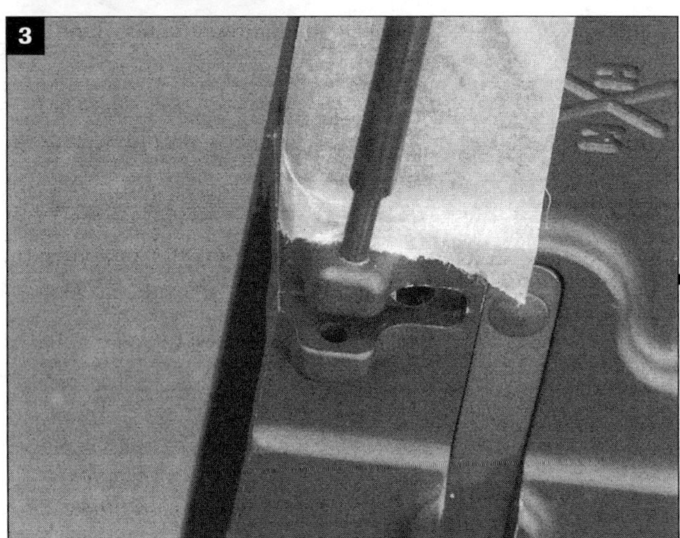

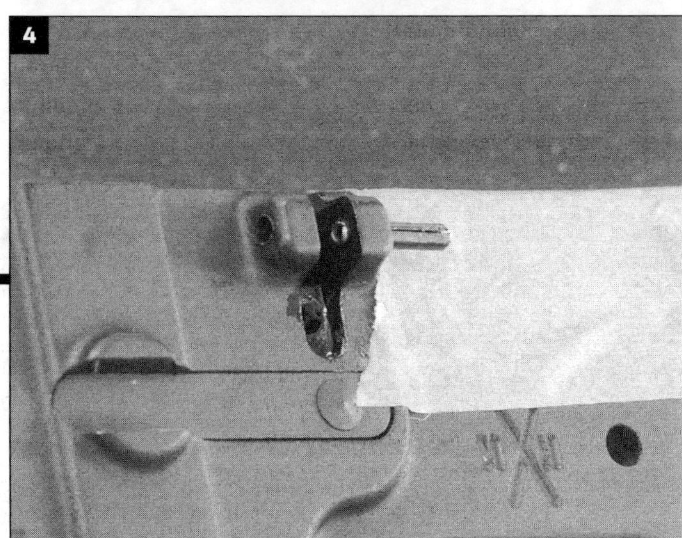

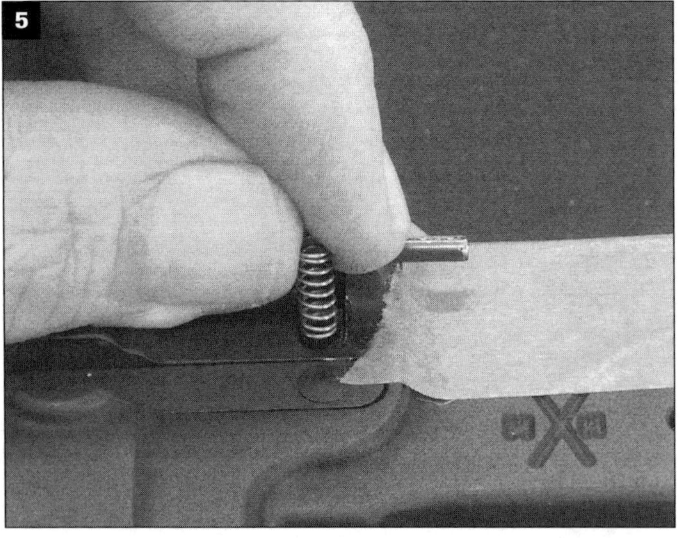

BOLT CATCH

1. Tape over the receiver surface. **2.** Get a roll pin starter punch to start the roll pin (oil the pin). **3./4.** Tap it carefully and lightly to seat it so it's just protruding enough to insert the piece. This makes alignment easier when it's time to send the pin through.

5./6. Insert the spring into its recess located in the left side of the lower and cap it with the plunger (grease this part). Make sure the larger end is facing up. Use a punch to check that the plunger can be depressed below the surface of the receiver and moves smoothly. If it doesn't check the hole for debris. The easiest way is with a #23 drill bit. This hole often has junk in it and I've seen some that weren't drilled deeply enough.

7./8. Orient the piece itself and push it down against the plunger. Hold the plunger down and tap the pin through, switching, of course, **9.** to a roll pin punch after the starter punch bottoms out. "Pre-installing" the roll pin so it engages the catch reduces the need for an extra hand for much of this operation. Make sure it's all lining up before starting the pin through the second recess. **10.** Seat the pin below the surface, equal depths on each side.

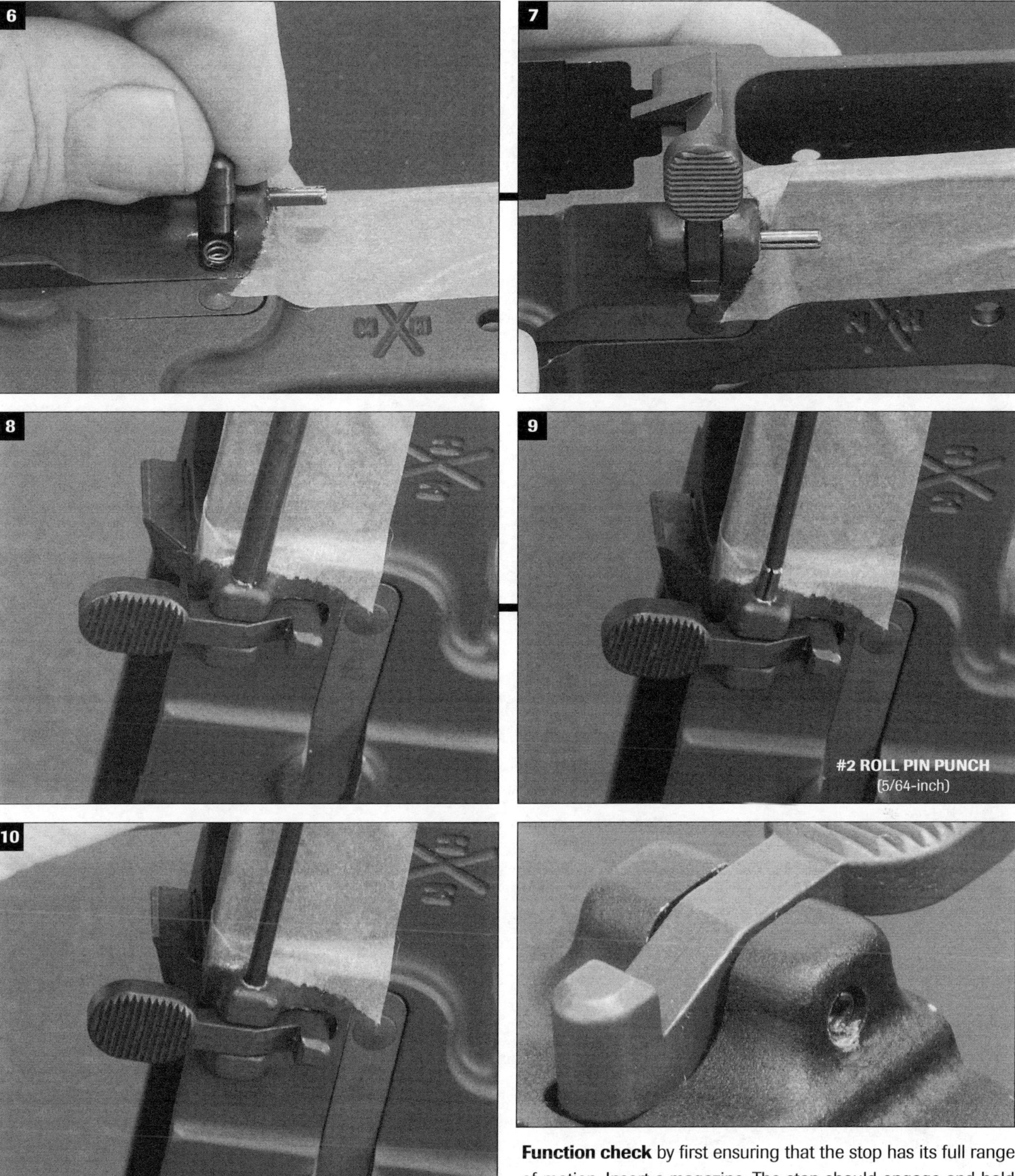

Function check by first ensuring that the stop has its full range of motion. Insert a magazine. The stop should engage and hold put even when you're tugging back on the magazine body.

Pin punches for this op need to be what they need to be, and that is long enough to clear the back of the receiver and not even a little bit bigger diameter than the pin. Brownell's makes a specialty #2 with a flat side to clear the receiver. Position the punch carefully and strike it deliberately to avoid dinging the area above the pin recess (like I did here a little more than I'd like...). Touch up paint!

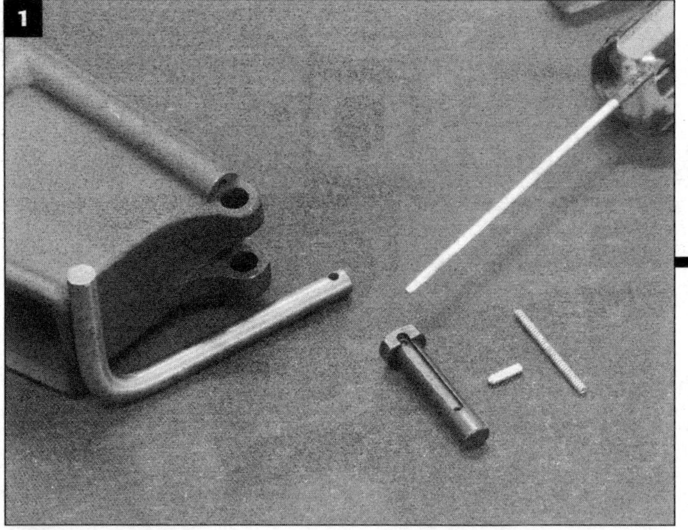

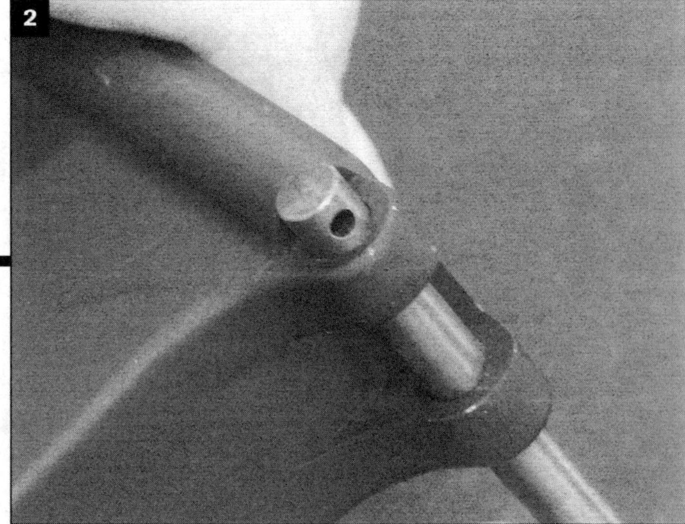

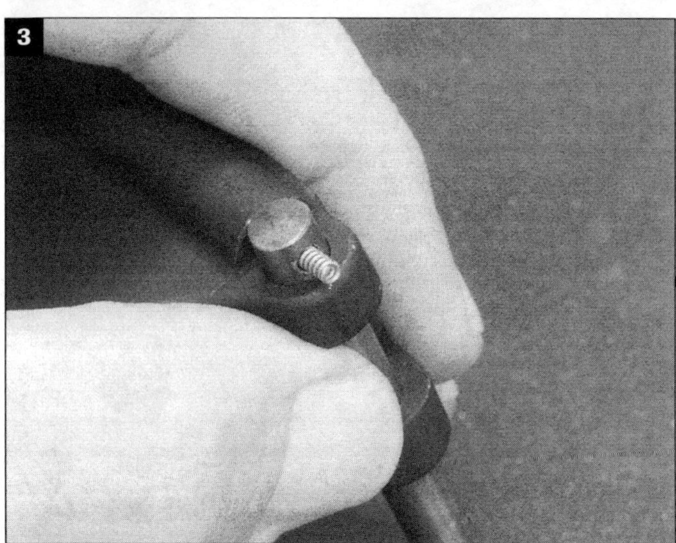

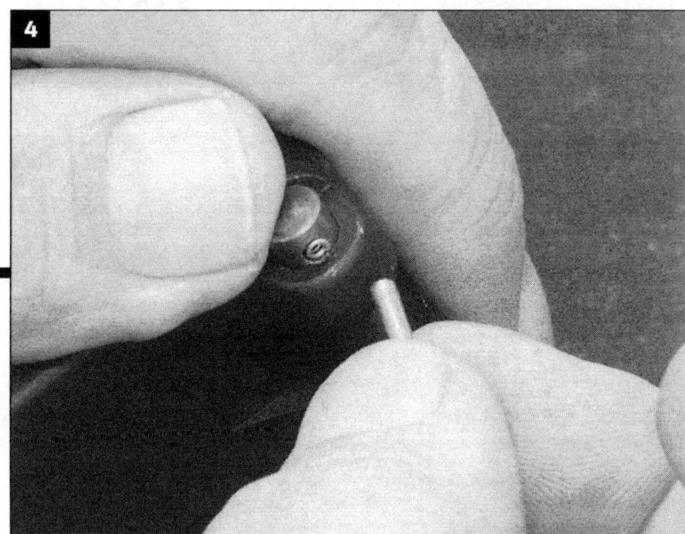

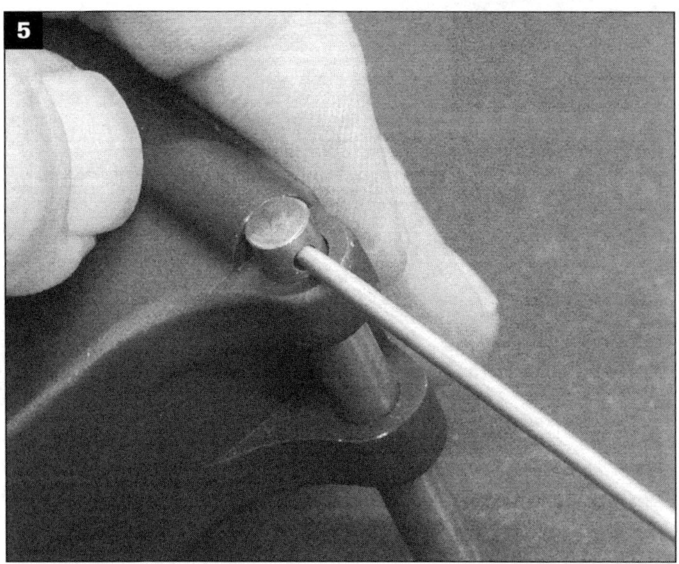

PIVOT PIN

The pin installs from right to left. Test fit the pin first. The pin should go in and out easily, and sometimes needs a little smoothing via emery polish to operate smoothly. Breaking the edges will not, in any way, negatively affect fit or performance (that's true for the receiver holes as well).

First, make sure you lube all the assembly parts, especially the detent recess because it's not likely to see light of day for the life of the rifle. If you have the drills shown earlier, make sure you run one into this hole to dislodge any debris that might (very likely) have been there.

1. Get your pivot pin tool (no substitute for this thing) and insert it through the holes from left to right. **2./3./4.** Turn the tool into place so its hole lines up with the lower receiver detent hole. Put the detent spring into the tool and cap it with the detent itself. **5.** Use a smaller-than-hole-size punch to depress the detent all the way into the receiver hole. I used a small assembly punch. Keeping the detent depressed, **6.** rotate the installation tool so it captures the detent (I use the punch to rotate the tool).

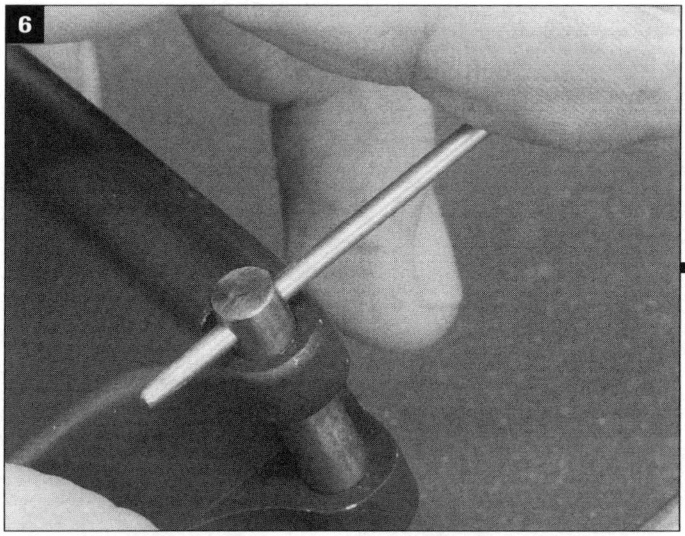

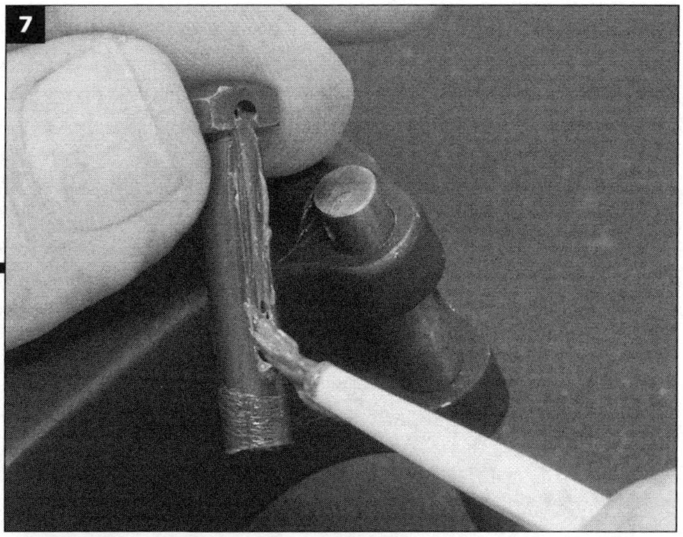

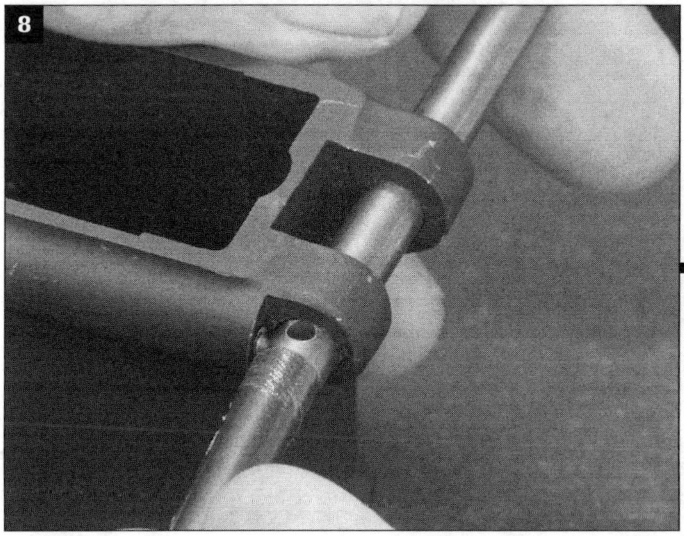

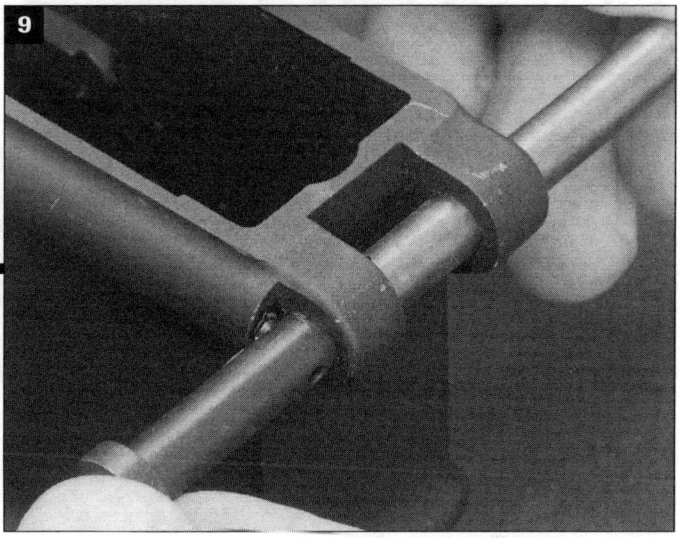

7./8. Grease and position the pivot pin itself against the end of the installation tool so its groove is facing the detent and carefully push the pivot pin in. Go slowly and keep firm pressure against the tool end. **9.** The detent should pop right into the groove on the pivot pin. It's sort of a "hand off." If this little hand off gets fumbled, chances are remote you will ever find both the spring and the detent that will have launched nilly-willy into Lost Parts Heaven. Spares are wise.

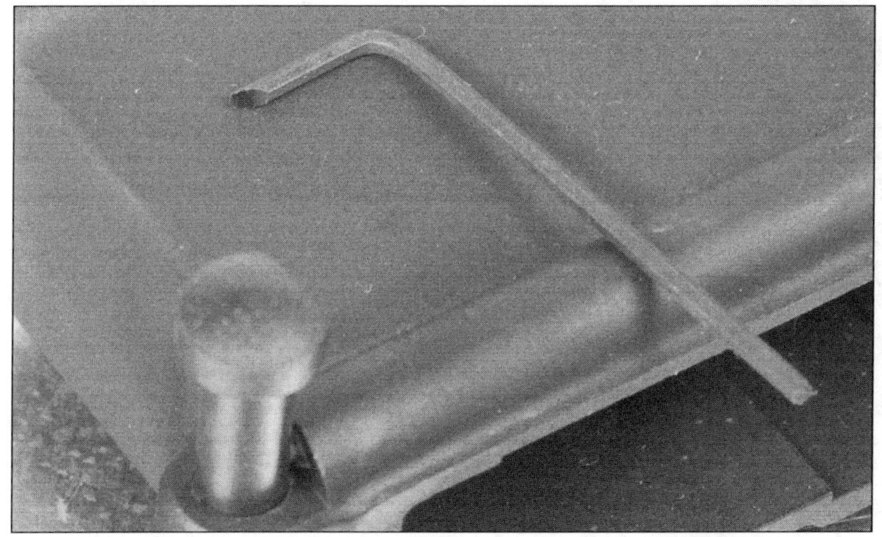

This detent captures the pin and should only be removed using, you guessed it, a **pivot pin detent removal tool** (shown at left). **Function test** by slipping the pin in and out. The pin should not rotate, and won't if the detent is into its slot.

I really don't like this job. It's not always as straightforward as it seems. It's very easy for the whole thing to come unglued just when you move the pin through in place of the tool, and zinging detents and springs ensue. Some place the lower in a large plastic bag to contain the works should that happen. Just a tip.

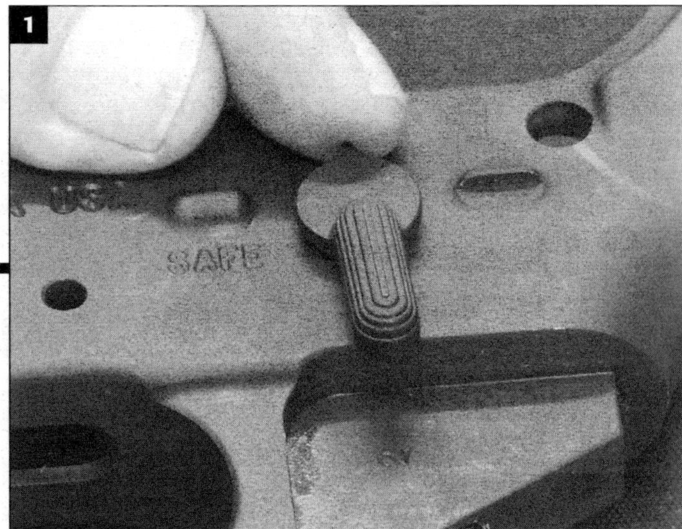

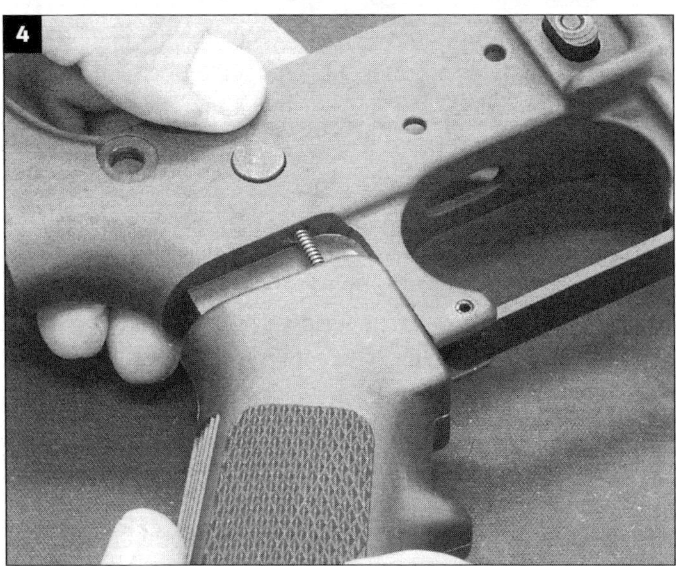

SAFETY & PISTOL GRIP

These jobs go together because the grip provides the hole for the safety detent spring. First, I kicked around whether to show this before or after trigger installation and decided to do it before. One reason was that it's more clear, I think, this way, and the other is that it usually doesn't matter if the safety is installed before or after. I say "usually" because some aftermarket triggers need to have the safety removed then replaced after the trigger assembly is in position inside the lower. This has to do with the way some triggers are constructed.

Test fit the grip first and then remove it. Check the threads by running the pistol grip screw into them and then back out. Sometimes finish on the threads makes this tight. That's the whole reason for the tapping operation shown earlier in this book, and that operation is very strongly recommended.

If the trigger is installed, cock the hammer and **1.** push the safety itself through its receiver holes from left to right. Either with or without the trigger in place, keep the indicator pointing up at "Fire."

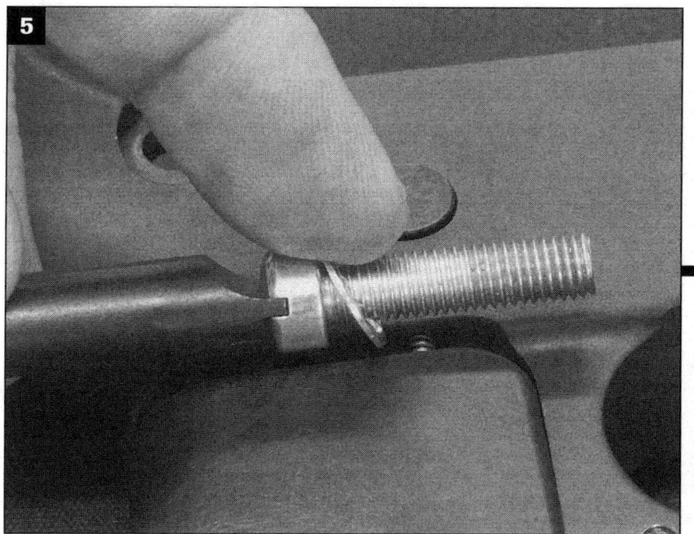

2. Invert the lower and insert the detent, pointed end first, into its hole in the receiver. Lube the hole first with grease. Check for free movement. **3.** Place the spring into the pistol grip. About a third of the spring will be sticking up. **4.** Set the pistol grip screw (don't forget the washers, at least the star washer; raised points go against the grip) into and through its hole in the grip. **5.** Note the nice fit between the screwdriver blade and screw head slot. That's important to help keep the screw on the driver. Man, do allen-head screws ever make this easier. **6.** Keeping everything upside down and retaining the grip screw in place, align the spring with the detent hole (it is really easy to kink this bad boy) and scooch the grip up. If you keep the front of the grip against the receiver channel it will help keep the spring from bending. The grip goes up at an angle and that's what makes it easy to kink the spring right between the receiver and grip. Thread in the grip screw and tighten it securely. Sometimes a new grip takes a little extra to seat fully because they're not nearly perfectly formed. They are plastic after all. I hope the photos don't confuse anyone because the detail of the screw head is shown just for detail. Again, push the grip screw through its hole in the grip before installing the grip onto the receiver. It can be tedious getting the grip screw into its hole if you wait until the grip has been pushed up into place on the receiver. Keep the screwdriver in place as you seat the grip, and once the grip is fully seated the screw threads and hole should be aligned. Should be. Careful here because it is very easy to cross these threads. All the more reason to do the tap.

Function checks come after the trigger is in. For now just make sure the safety lever rotates as it should.

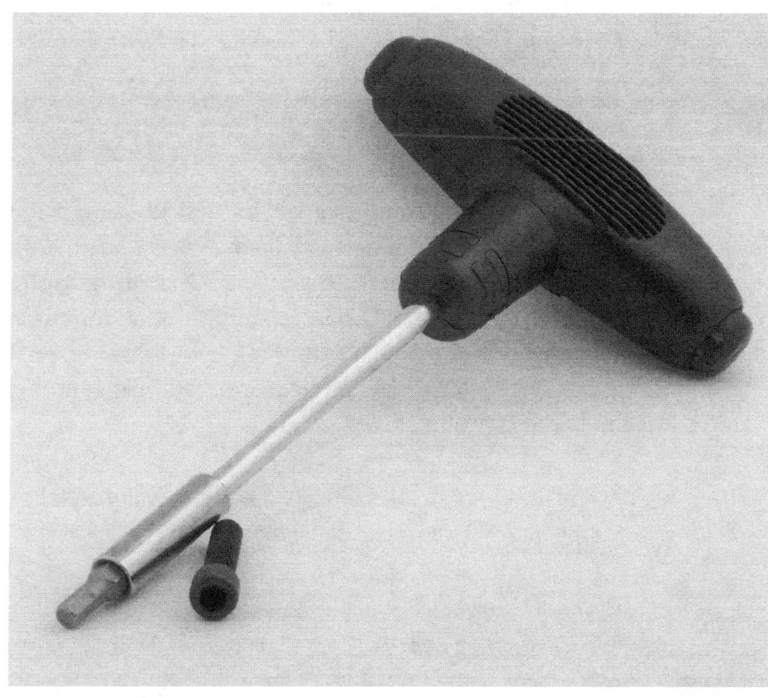

Grip screws *should be allen-head. Most are also overly long. Go to a hardware store and get one 1.125 long at 1/4-28 size and life is easier. Always use the washer!*

EXTENSION TUBE (STANDARD)

There are a few steps in this job. The first is getting the spring and buffer retainer body in place. Some, me usually, call this the "receiver extension tube." Assembling the buttplate itself is another. And then putting it all together, along with the takedown pin (rear pin), completes the op. I added this step prior to trigger installation because that is shown separately in its own segment.

Secure the lower. I use a Brownell's magazine well block in a vise. It's the most effective and convenient means I've found.

1. Thread the receiver extension tube into place to test it. Notice how it extends just a little over the retainer hole. I snug it down and then loosen it, and then **2.** run it back off about a turn.

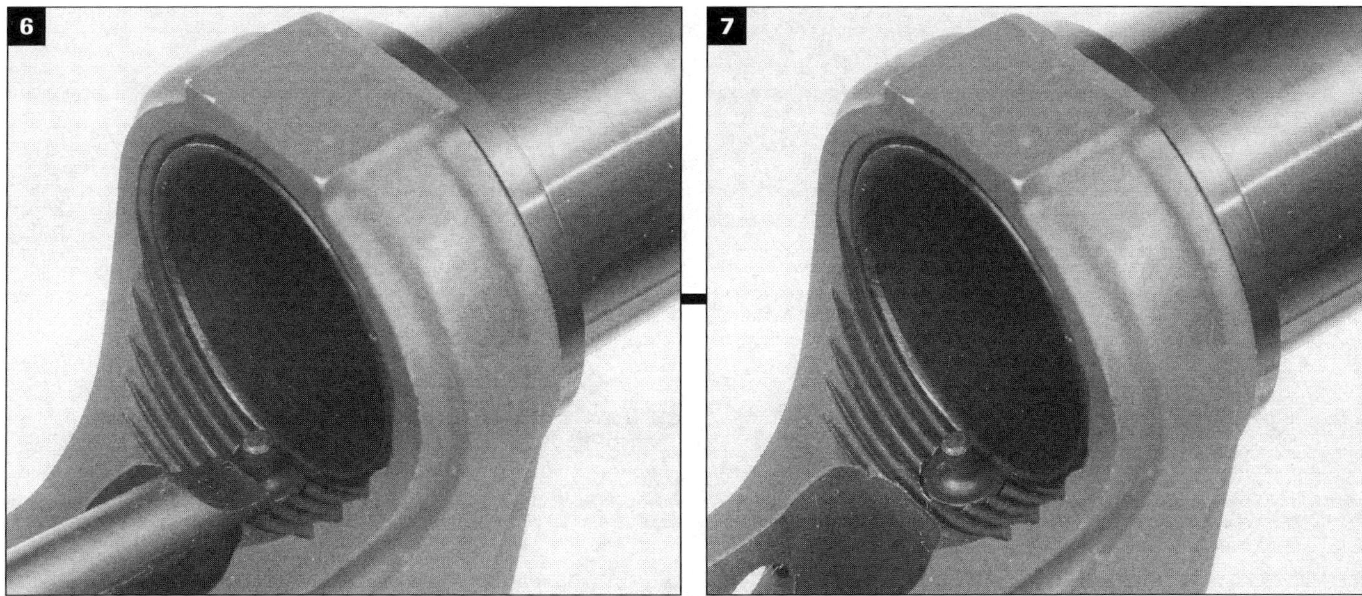

Here's how this works, and has to work. **3./4.** Put the spring and buffer retainer into their hole. **5.** When the tube is nearing the retainer, depress the retainer and continue threading the tube. **6./7.** When the tube is threaded in all the way to its shoulder so it engages the flange on the receiver, the end of the tube should capture the retainer but not be onto it so far that it interferes with free movement of the retainer into its recess. **As long as the retainer is free to move and is indeed retained, it's okay.** If the tube extends too far, it has to be shortened at its mouth. That's easy. If it's too short, that's a problem beyond our scope here. Try another tube. If that's too short then there's a problem with the lower hisself. Send either bad part back.

Run the tube back out and get ready to install it for good. That means glue. A light duty threadlocker, like Permatex Blue or even Purple (lighter duty) is the right choice. Apply contact cleaner to threads inside and outside to degrease. I usually use CRC Brakleen. Tighten the tube with a firm tug against your wrench handle, but don't apply too much pressure. It's not necessary and makes it way hard to remove if you want to replace with another stock assembly. Set the works extension-tube-down until the glue dries. You don't want adhesive getting into the retainer. My great honking green-handled adjustable wrench does a way better job than any of the cutouts on the combo tools. It fits better. Or you can use a 3/4-inch open-end wrench.

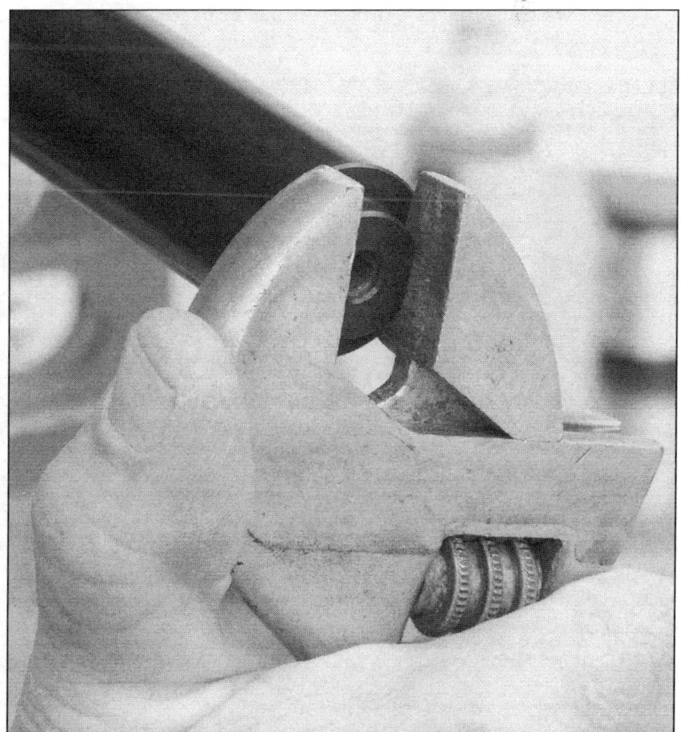

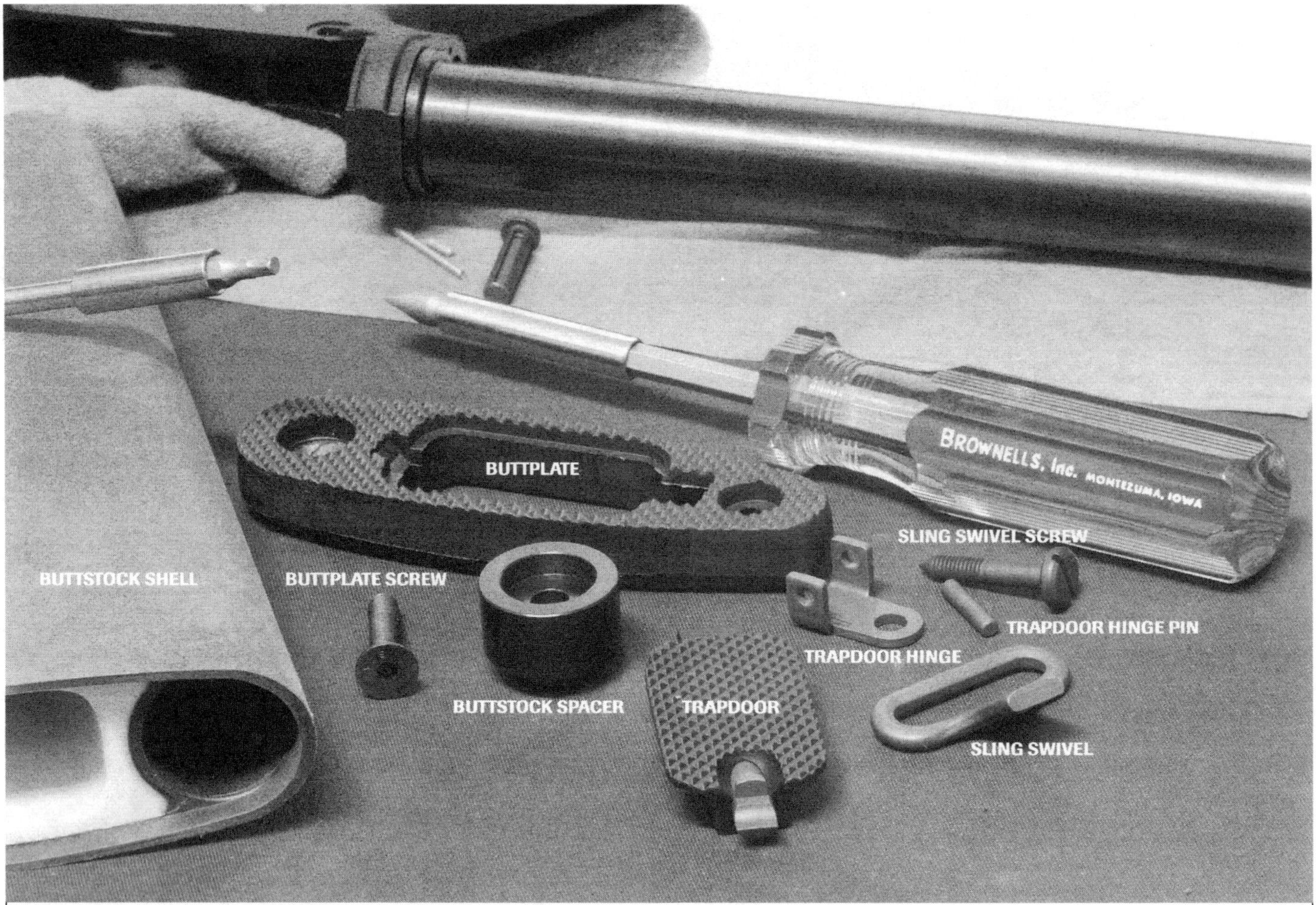

To finish the installation you'll need all the pieces shown above and a suitable screwdriver. The buttplate itself requires no tools to assemble.

Buttplate/buttstock screws are easy to identify. The sling swivel screw is pointed (left). The upper buttplate screw that holds the stock to the receiver extension comes in a couple of forms (right). Some have a hole drilled through, some don't. The hole is there to release pressure from buffer compression, and that always seemed like a good idea, although I have to say I sure can't tell any difference in firing one with a solid screw... The shorter screw shown is for an A1-style stock, which is 5/8-inches shorter than the newer A2, and that is the reason for the aluminum stock spacer shown. The receiver extension tube is the same length for each rifle, which is to say that the longer stock was a retrofit. Just don't install an A1 stock with an A2-length screw! Just cut the screw down if you don't have one made for the A1.

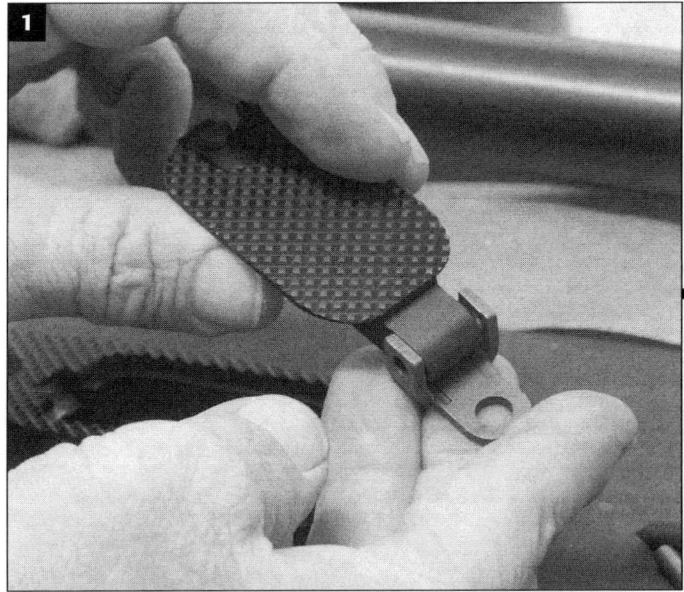

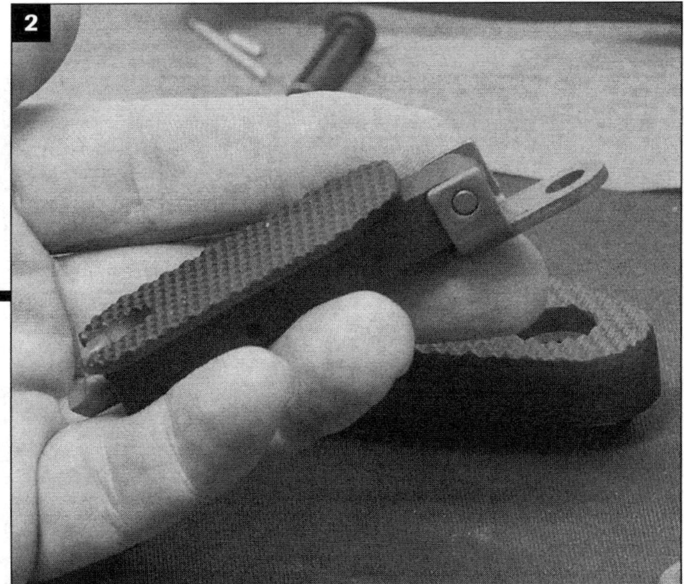

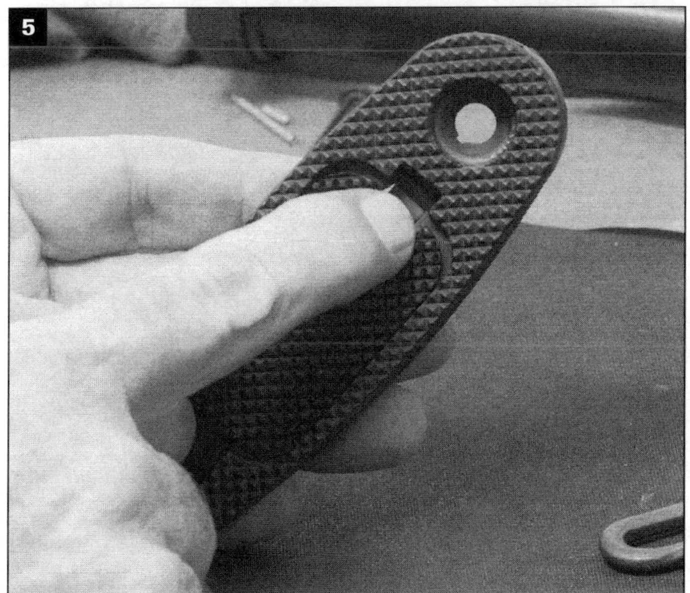

BUTTPLATE (STANDARD)

A2-style buttstocks are are frequently pre-assembled, and that's a good way to get them. Nothing tricky here, just more time. If the whole thing is in pieces, first parts to consolidate are those for the buttplate assembly.

1./2. Take the buttstock "door" and pin its hinge in place with, no shock, the hinge pin.

3./4. Take this assembly into the plate, fitting the hinge into place at the rear of the plate.

5. Secure the hinge in place and close the door, depressing the latch to the door snaps into its locked position. Done with that.

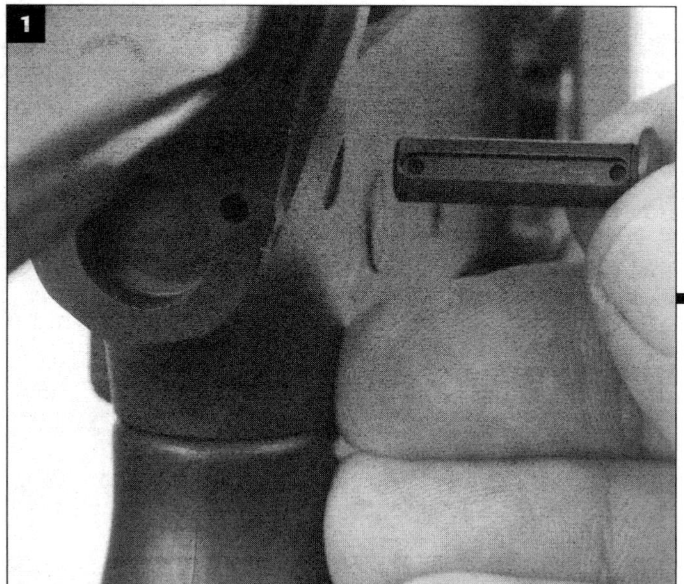

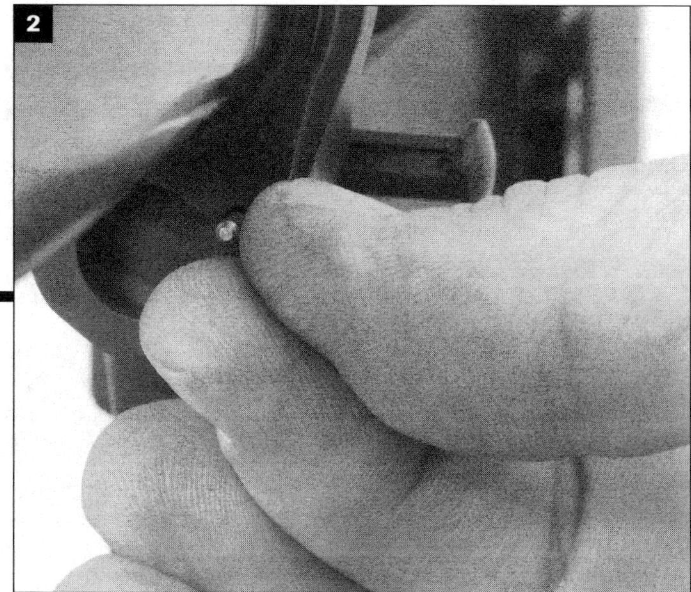

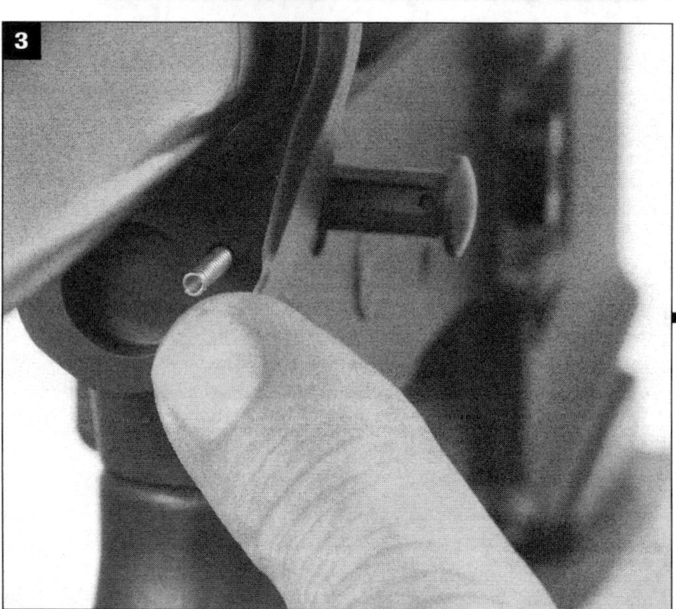

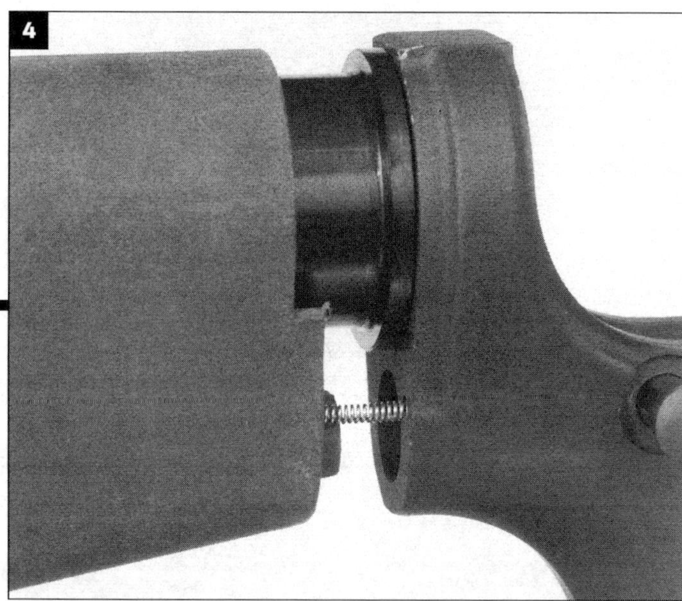

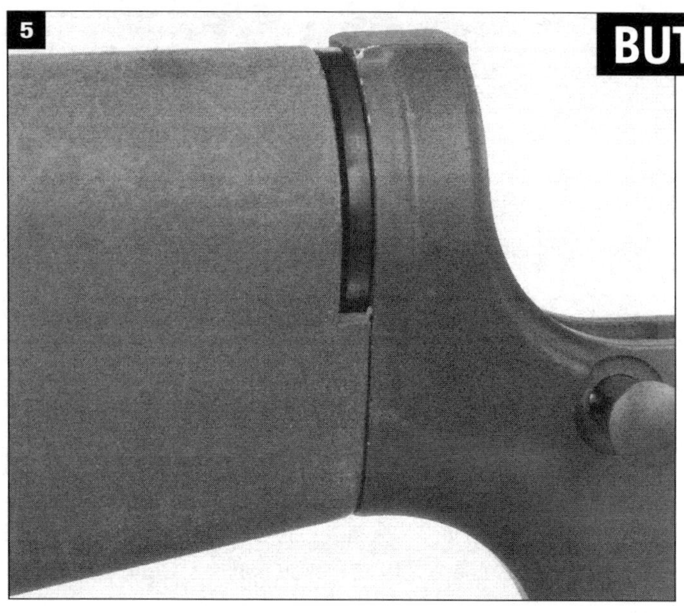

BUTTSTOCK SHELL/TAKEDOWN PIN

Takedown Pin (rear pin)

1. Insert the takedown pin into the receiver from the right side. Groove facing to the rear, push the pin in far enough so the groove extends beyond the hole where the detent and spring go. **2./3.** Insert the grease-lubed takedown pin detent into its receiver hole, followed by the takedown detent spring.

4./5. Slip the stock shell over the receiver extension tube and line it up with the recess in the receiver. Careful not to kink the spring! Push it all the way home and check that the detent has indeed engaged the groove in the takedown pin (the pin won't rotate). Keeping everything pushed together, next step is to affix the buttplate. (Tip: a short strip of duct tape will keep it together if you want to keep your hands free.)

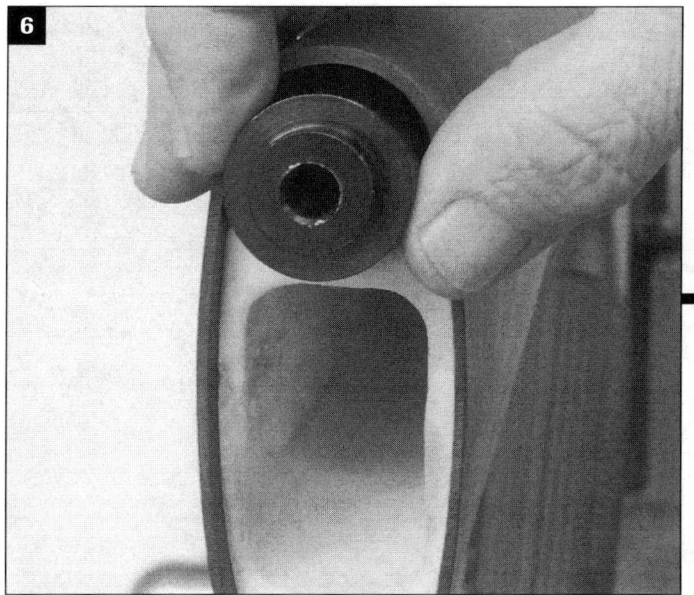

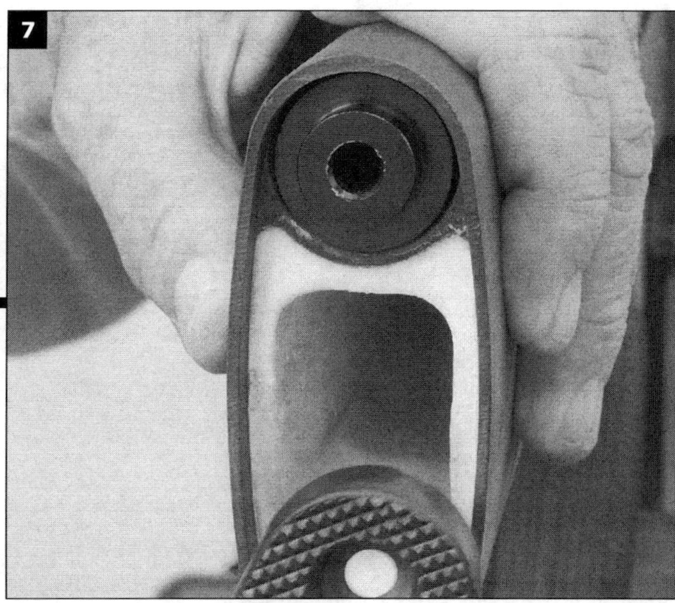

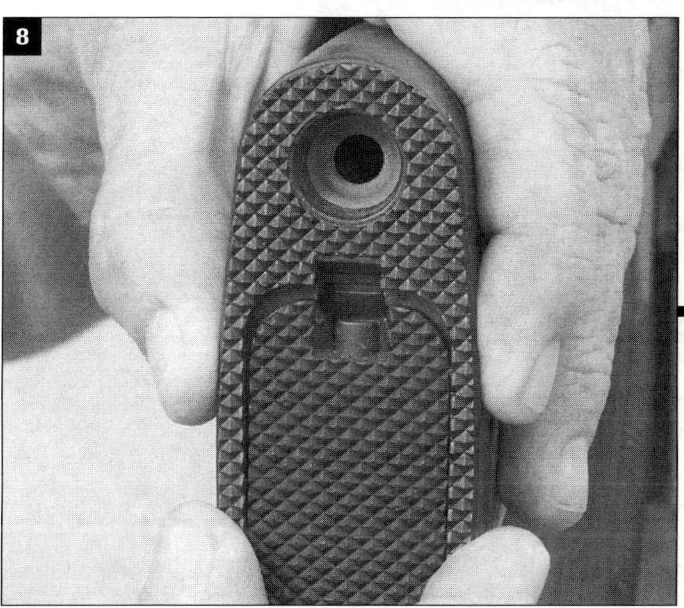

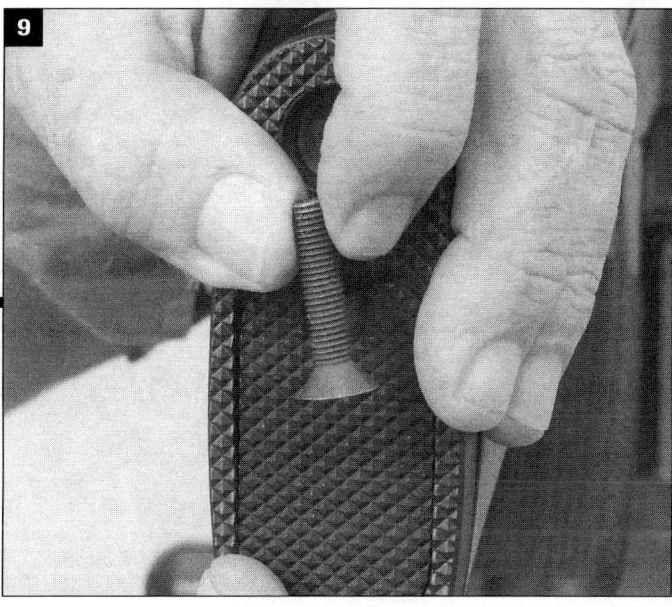

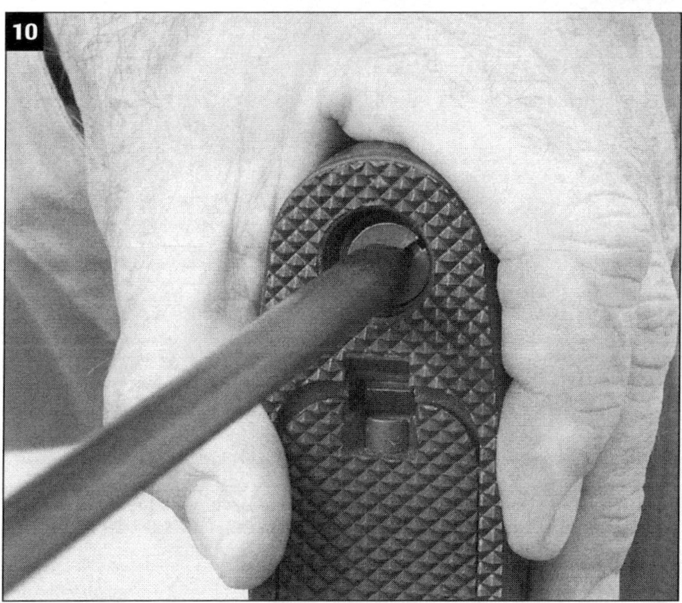

Buttplate

6./7. /8. Position the spacer over the end of the extension tube (only one way it can go), and position the buttplate on the stock shell. **9./10.** Thread the upper buttstock screw into the end of the tube and commence to twisting it on home.

The upper stock screw needs no help to stay put. Matter of fact, some have elastomer inserts or locking compound that can actually, and commonly, prevent fully tightening the screw. Get that mess off of there for an easier life, and better installation (Perma-tex Gasket Remover). **Make sure the stock is tight on the rifle.**

[This op can be done by positioning spacer and buttplate into the stock shell first, and that is easier, but I did it this way for the sake of step-by-step clarity.]

Swivel installation is shown on the next page.

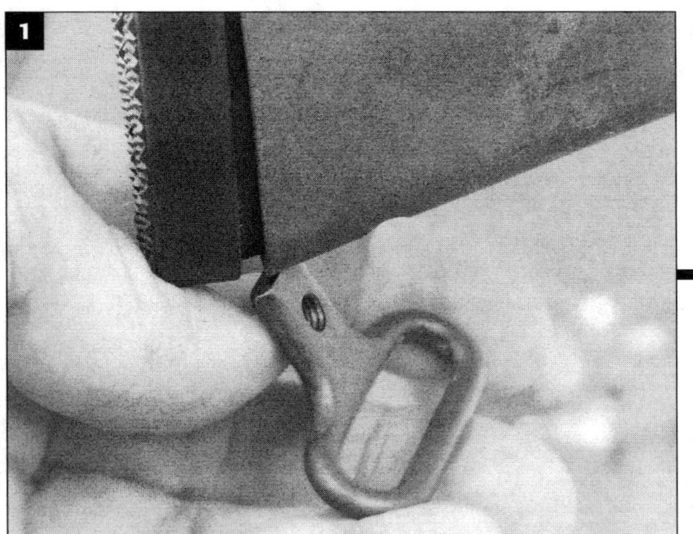

Insert the rear sling swivel (angled forward) into its slot in the bottom of the stock shell, catch it with the swivel screw (that's why this screw has a pointed end), and thread in the screw. Make sure it's firmly tightened.

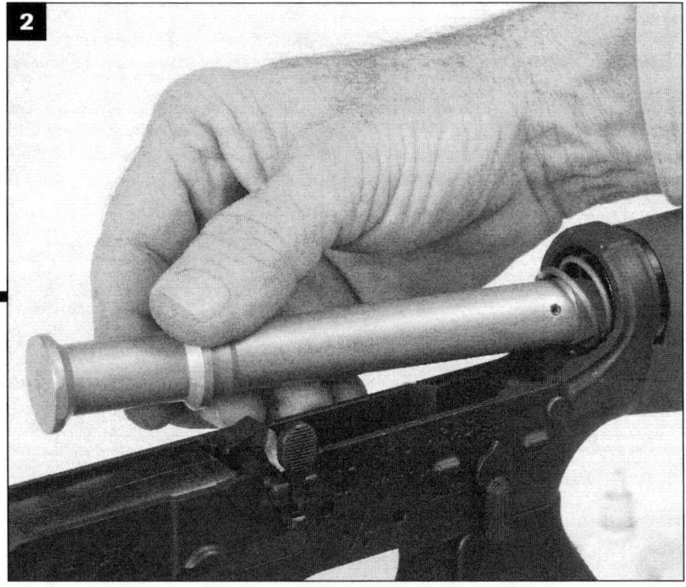

Buffer spring and buffer are the last parts, for now. If a trigger is installed, cock the hammer. **1.** Lubricate and insert the buffer spring, pushing it fully back into the tube so it's held there by the retainer detent. **2./3.** Insert the buffer and rotate it so its flange is held in place by the retainer detent. This can also be assembled with the buffer already in the spring. Doesn't really matter, but some think it's easiest to do spring first.

That's that. The lower is ready to go, after, of course, we select and install a trigger, and that's coming on page 61.

Other Stocks

Other buttstock styles install about the same way, and most are actually easier than the standard A2. There are examples of collapsible (CAR-style) shown next, and aftermarket tactical and competition stocks illustrated in the project segments to come.

3.0 CAR STOCK
STANDARD SHORT GUN FURNITURE

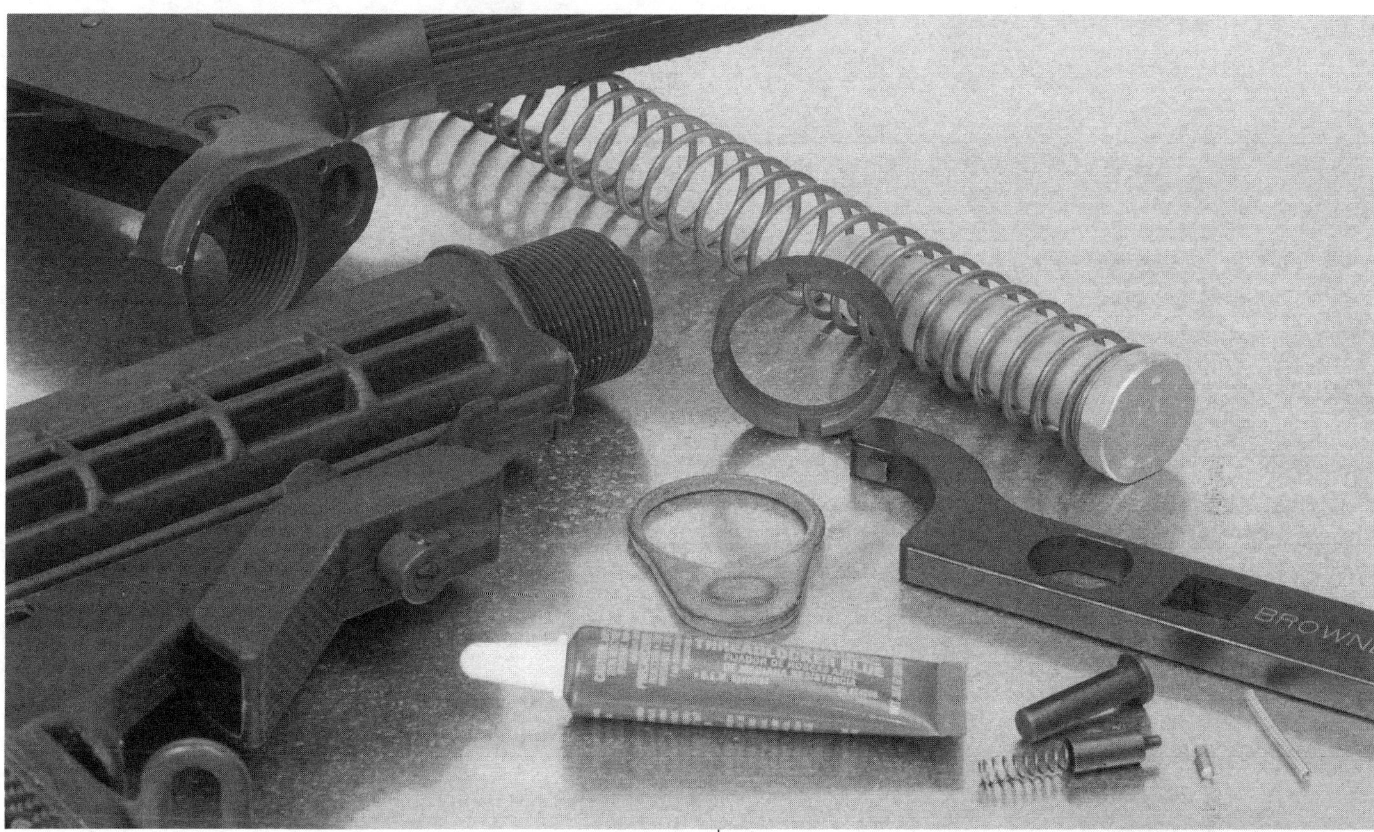

[We just got through assembling a conventional A2 buttstock, and here's another convention. The essential CAR collapsible buttstock is almost self-explanatory after doing the rifle version, but since so many things don't truly explain themselves, here's the step-step on it. If you can install this stock, most competition-style assemblies are a snap as well. They all go together about the same way. That's true for the Medesha, Eliseo, Magpul, and on down the list. This is actually easier than the A2 we just studied.]

About the only difference *to install a collapsible stock is that you'll need a spanner wrench to cinch up the collar. Of course, the buffer will be shorter, as will the buffer spring. There is a plate that covers the back end of the receiver. Most all I've seen ship complete, and the common parts you'll need at hand are takedown pin, detent and spring. You will encounter different representations of available CAR stocks. Some are better than others...*

SEGMENT CONTENT

56   **Stock Components Assembly**

57   **Installation** (plus buffer detent, takedown pin)

OVERVIEW

Okay. Nobody likes this stock, but a lot of short guns have it. Nobody likes it because it's uncomfortable, flimsy, and generally uncomfortable on the self and on the shoulder to manipulate.

There are far better alternatives on the market, and my favorite is shown in the carbine project build segment.

Honestly, the significance of this portion of the book is showing how to put together and install a CAR stock because most all other aftermarket stocks, including those intended for competitive use, go together in virtually the same way.

As said, most all (honestly all) I've seen are complete in what we could call sub-assembly. All the pieces unpack from the same box.

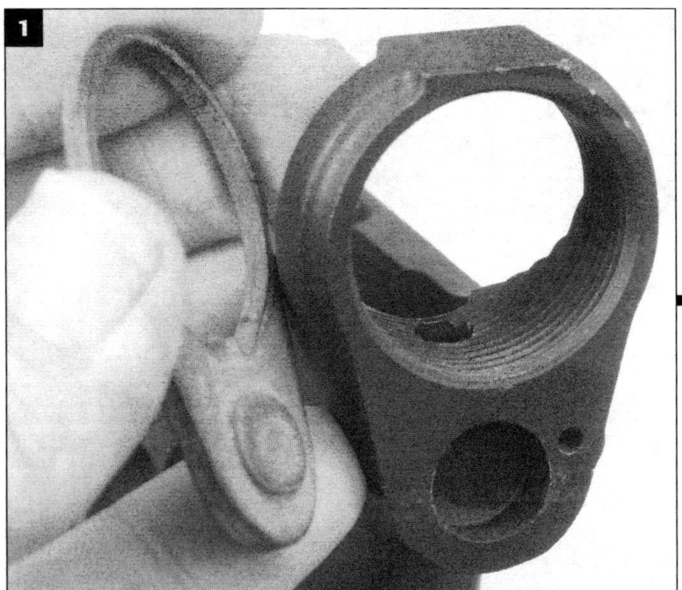

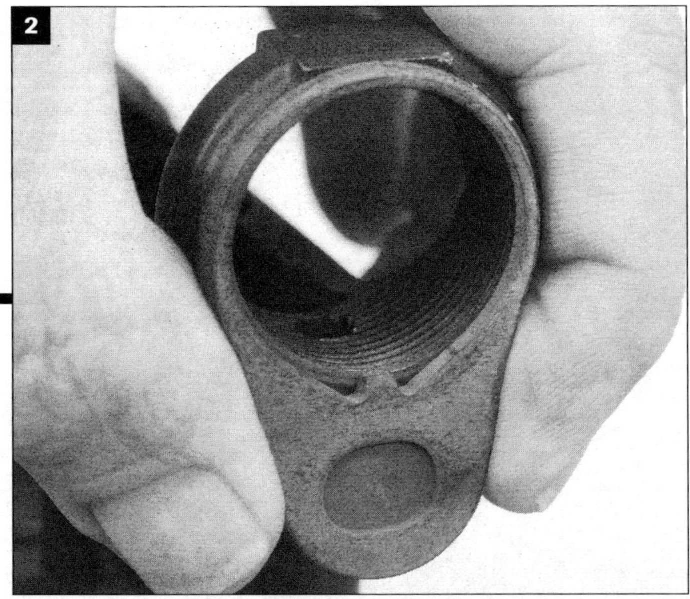

ASSEMBLY & INSTALLATION

1./2. The plate covers the open area at the back of the receiver and functions to secure the takedown pin spring and detent. Notice the orientation of the raised area.

3. Thread the locking collar fully onto the receiver extension tube, followed by the plate. The notches on the collar face back.

4./5. Extend the stock fully. Start the threads into the receiver. After degreasing all threaded areas, apply about a 1/2-inch wide dose of thread locker (Permatex "blue" in this example) to the threads on the tube. That's enough to hold it without leaving a mess. Some of the threads will be exposed after installation so there's no need to douse it. Thread the tube on until its leading edge is just shy of the buffer retainer hole in the receiver.

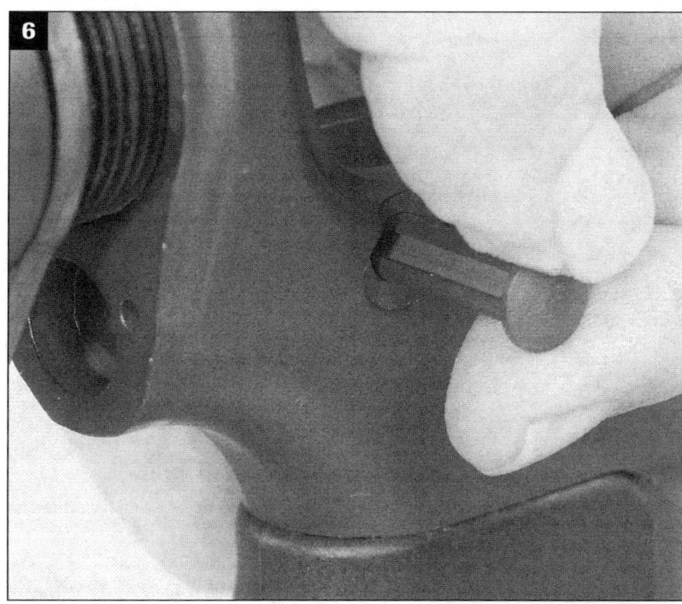

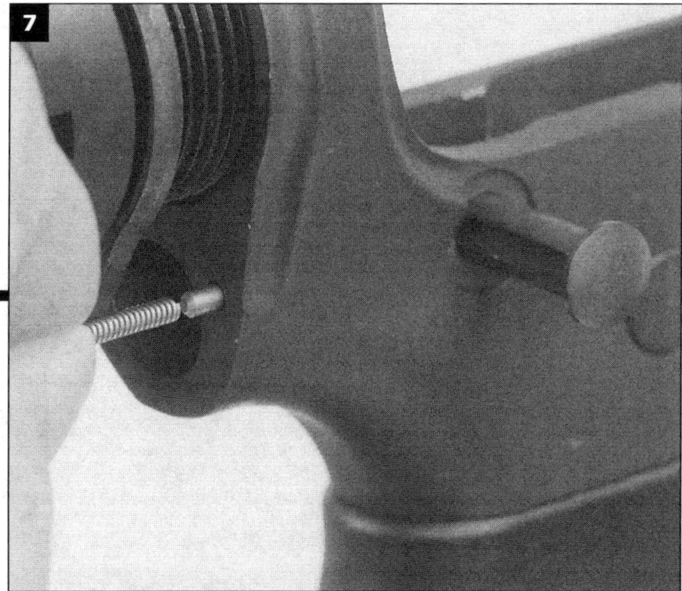

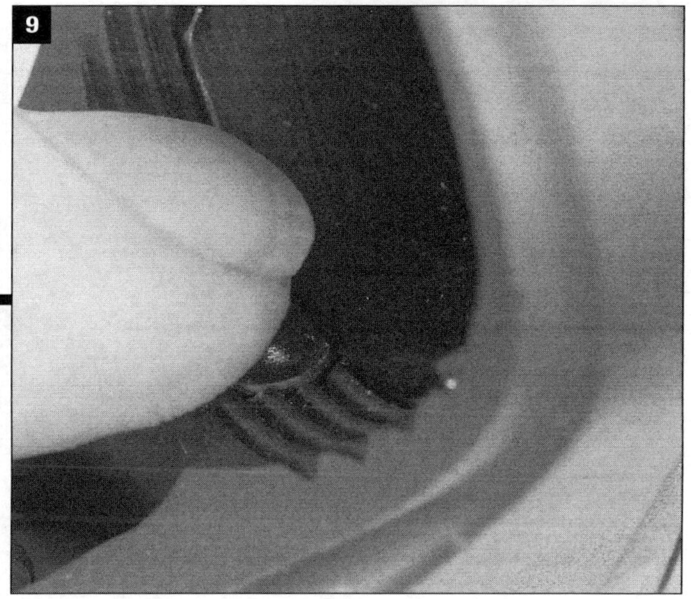

6./7. Insert the rear takedown pin, detent, and spring. Grease these parts prior to installation. *Do as I say, not as I show!*

8./9./10. Insert the buffer retainer and its spring and, just as with the conventional stock, the retainer can be depressed and the tube threaded in just a little more to secure the buffer retainer. Make sure the retainer is free to move downward but remains captured. The leading edge of the extension tube should be over the flange on the buffer retainer but not touching the retainer nib.

There's a notch on the underside of the extension tube that corresponds with a pin on the back plate. That has to be straight up and down or the plate won't fit into the recess at the back of the receiver. The plate can be pushed back against the locking collar to give enough room to install the takedown pin detent and spring after the buffer retainer is set, or do it before.

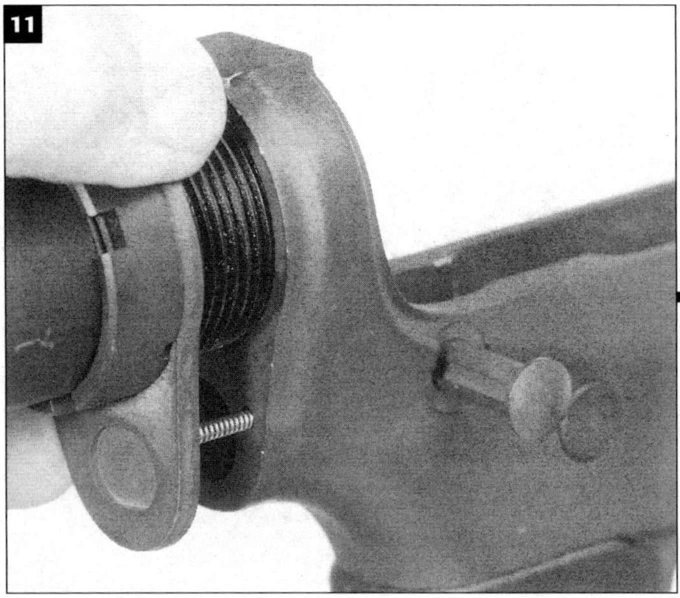

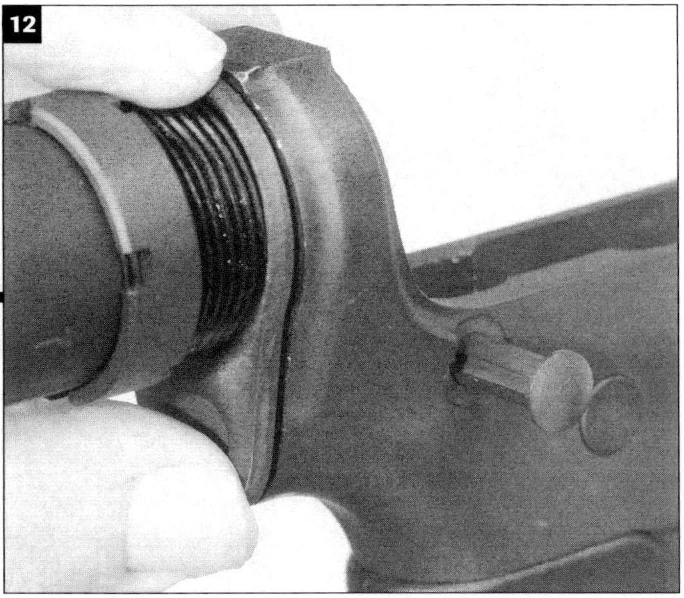

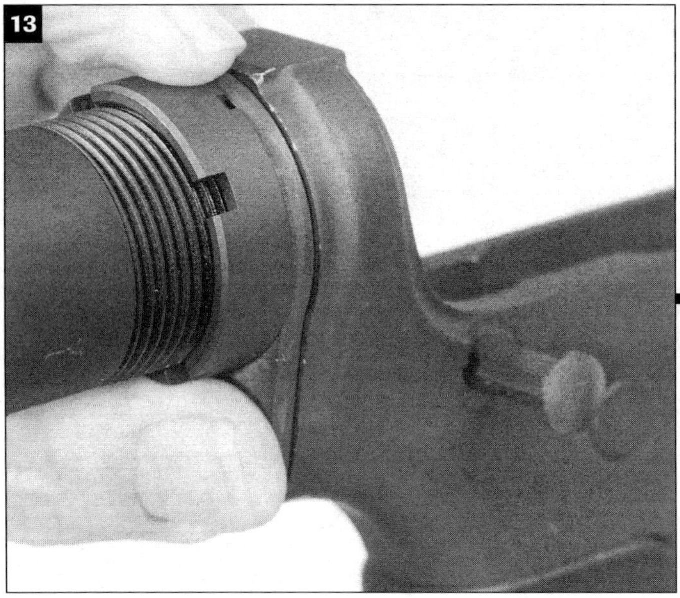

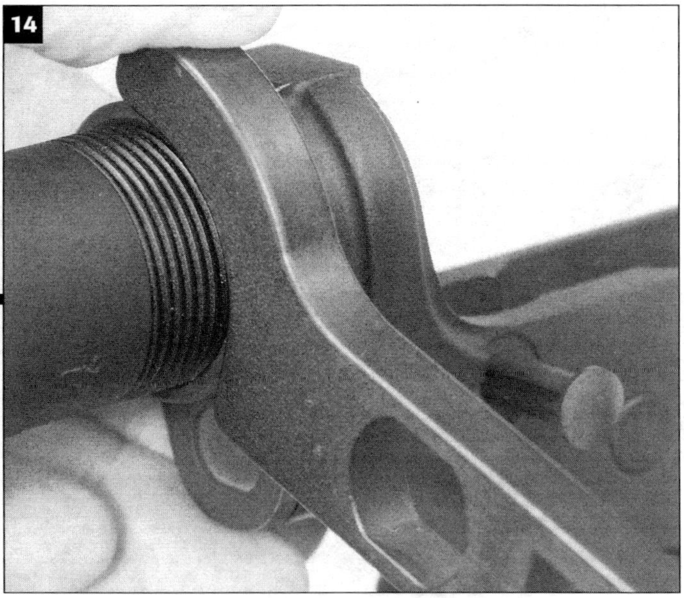

11/12. After making sure the takedown pin is captured, push the back plate fully forward so it is sitting flush against the receiver. Take care not to kink the spring. **13./14.** Thread in the locking collar and snug it down with a spanner wrench. If you don't have a spanner wrench, use a flat-ended punch (then touch it up...).

This **spanner wrench** is from Brownell's. The engagement stud is interchangeable from square to round to suit the stock assembly you have. Most take the square stud.

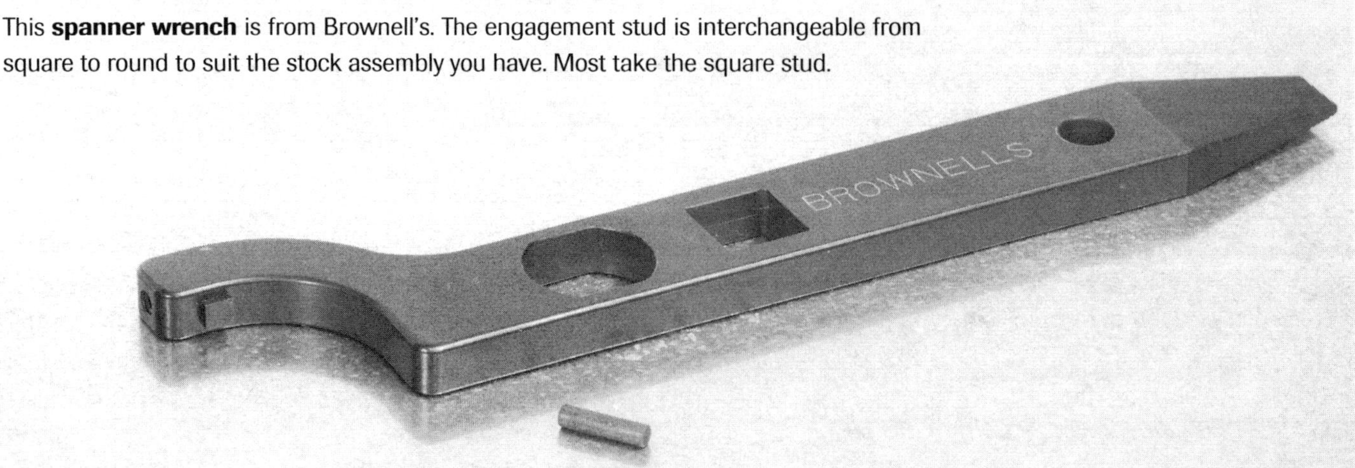

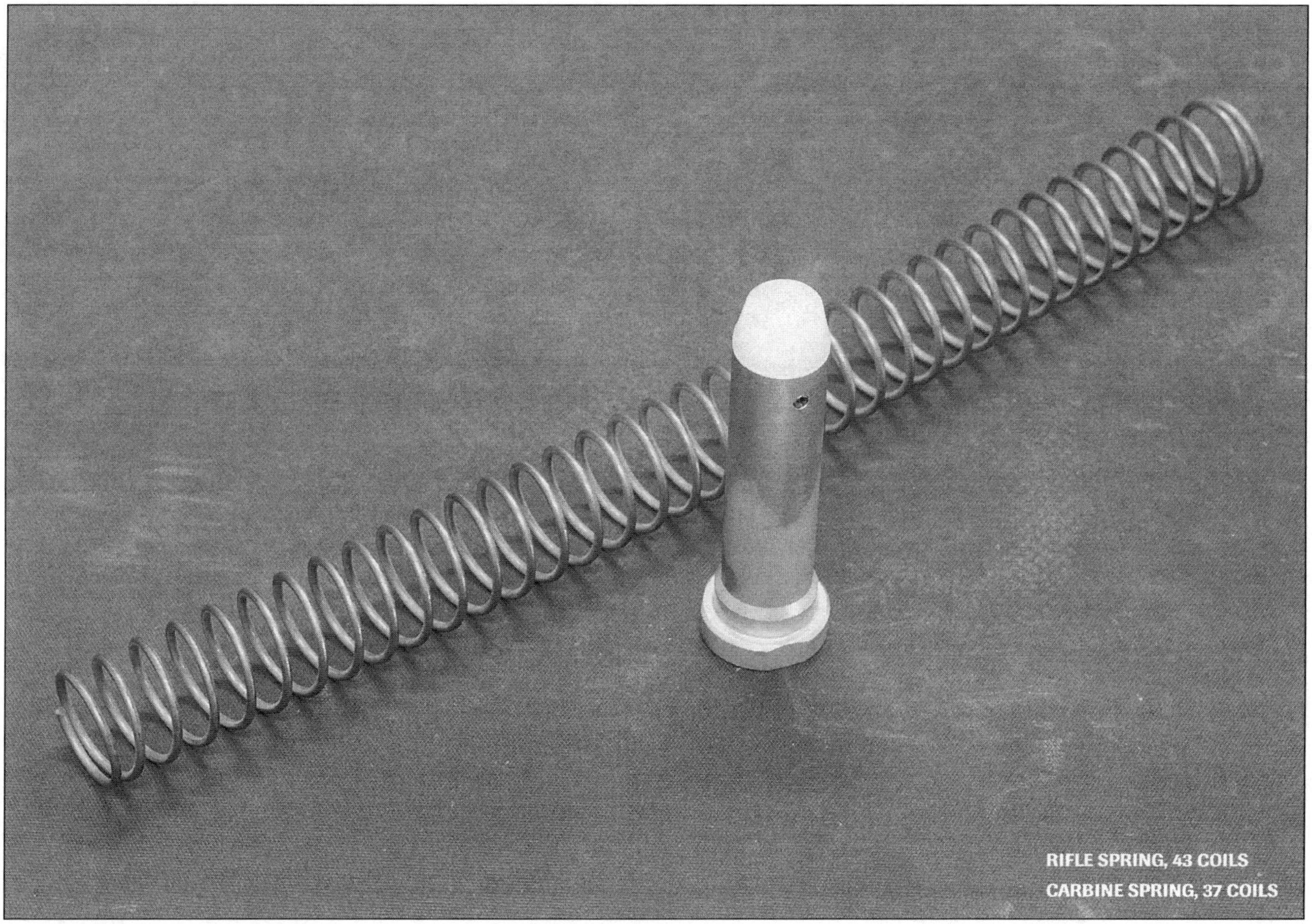

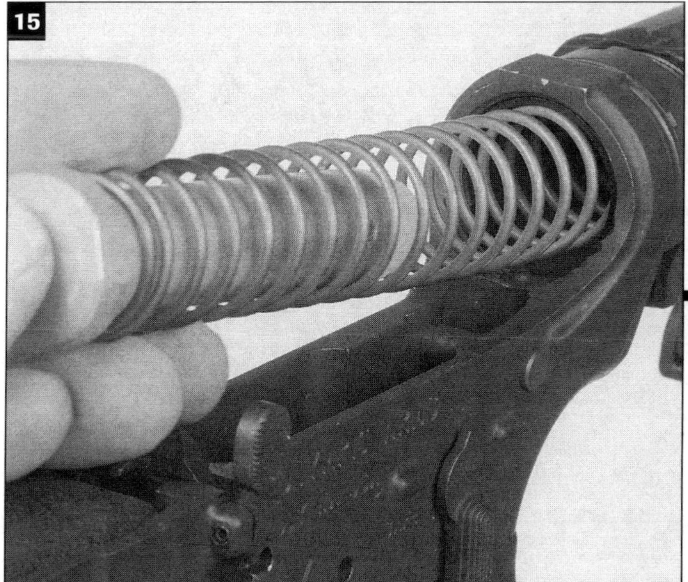

A carbine-length stock takes a shorter spring and shorter buffer. Make sure you have the right parts on hand.

16/17. Last step is simply to cock the hammer, insert spring and buffer, and latch the buffer flange in behind the buffer retainer.

As said, some people suggest inserting the buffer spring first and latching it behind the retainer, and then installing the buffer, but it can be done all at once. Cannot for the life of me see the difference, but it's a little easier to do spring first.

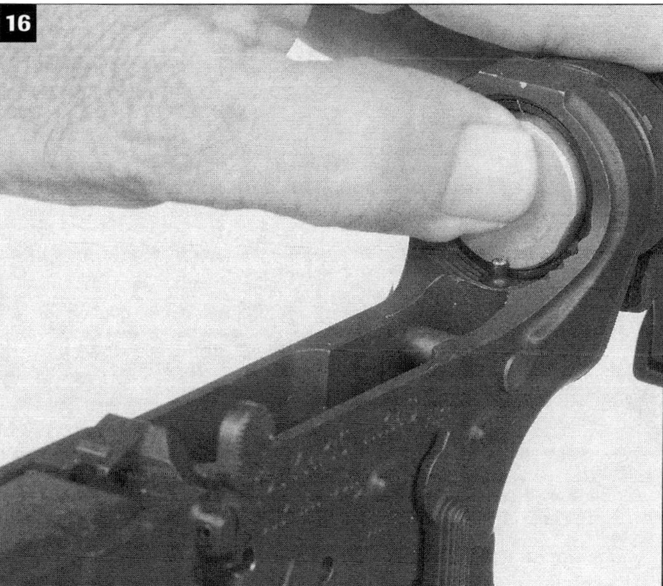

PARTS TO WHOLE (PLUS)

We just did a lot of work... Now that we have a lower receiver assembly, the fun can begin (well, just as soon as we get a trigger in it, and that's next). The fun, of course, is making the rest of the gun what you want it to be. That, in essence, is now what the rest of this book is about.

4.0 BASIC TRIGGER
ESSENTIAL INSTALLATION

[There are a lot of aftermarket triggers available for this rifle platform and chances are very good to outrageously good that you'll end up with one. Choice and installation examples are shown elsewhere in this book ranging from the very best of the match two-stage units to drop-in assemblies.]

If you can install all this, then other triggers will fall into place for you also. Don't expect a good pull from any stock system. Engagement surface polish and lighter springs are about the only ways to make a standard trigger work better.

The trigger shown above is mounted into a Brownell's fixture. Look closely at the spring orientations.

SEGMENT CONTENT

62   **Before you start...** (receiver preparation)

63   **Trigger Pins**

65   **Parts & Tools**

66   **Trigger Construction** (parts assembly)

70   **Trigger Installation**

71   **Hammer Installation**

73   **Function/Safety Checks**

74   **Disconnector Function Check**

TRIGGER

There are a few different trigger assemblies shown in this book. It's very unlikely that you'll be installing a stock trigger in your rifle, but most aftermarket triggers, although differently designed, still use the same essential parts in the same ways as the GI system. Springs, pins, and installation are therefore patently the same.

Study the photos carefully noting especially the orientations of the trigger return and hammer springs, and, later, their assembled or installed relationships.

Before you start...

Check the trigger and hammer pin holes against burrs using a #23 machinists drill bit. Test fit the pins. They should press into place through the receiver holes and the trigger and hammer should rotate freely.

Important! Lube all parts — pins, holes, engagement surfaces. I use grease with boron nitride along with an oil containing the same additive.

I normally use a Brownell's mag well block to support the lower for this installation. It's a lot easier to keep both hands free. The only problem with the block is that it doesn't work with another favorite, the Brownell's hammer drop block, which fits into the mag well area to cushion the impact of a falling hammer. Oh well. A small piece of wooden dowel or a scrap piece of leather (old sling keepers work) serves.

Pins

There are two. A trigger pin and a hammer pin. With rare exceptions that are usually encountered only in aftermarket sets, they are the same. Unless it's a Colt-brand, trigger and hammer pins are 0.154 inches diameter. Colt takes 0.173.

Shown below is a set of KNS-brand pins. They're the best. They are sized correctly and true. Same can be said for most pins that are supplied with aftermarket triggers, like Geissele, but a set of KNS will improve whatever a trigger pin set can improve used on any other installation.

They always install from right to left, ungrooved end first. (Some aftermarket trigger systems and pins are excepted from this rule.)

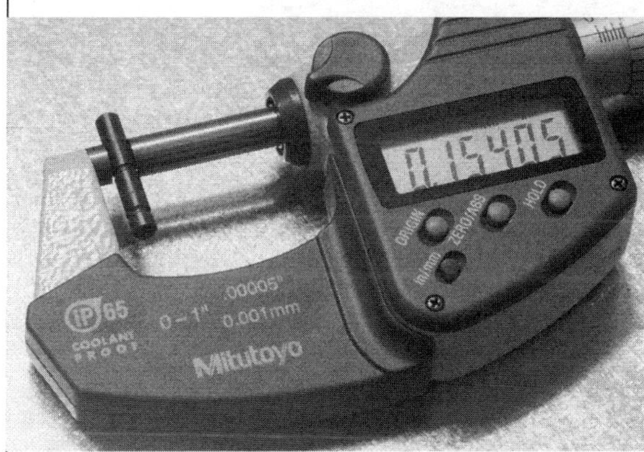

Unless you know better, and we'll talk about that in a bit, **oversized pins** can create problems. Can. Depends on whether one is really needed, and if it is it usually will be for the hammer. The first check, of course, is measuring the receiver holes. Extra-power hammer springs can create enlarged holes in the receiver and cause excessive movement. Pounding an oversized pin through a hole can do the same thing... Oversized pins also won't always work with all hammers. Sometimes the hole in the hammer is too small. It's possible to take down the center area of a pin by spinning it under drill power on emery. Shown is a DPMS standard (upper left) and oversized (lower left) pin.

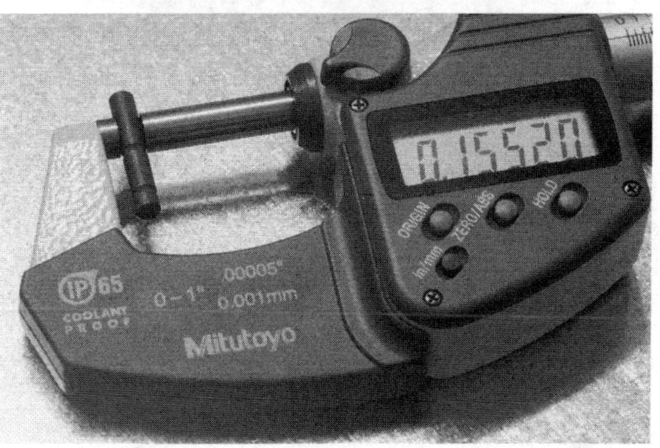

The photo on the right shows a specialty **oversized locking pin set from KNS** that comes with its own reamer. Be careful with this addition. It's irretrievable. The only time something like this might be warranted is if you know the pin holes in the receiver are larger than normal.

Brownell's Trigger Fixture

This little gadget lets you do trigger work on the outside of the gun before putting the trigger inside your gun. There could be change going from jigged to installed but I didn't see any such to not at all much in my experiences using this tool. It saves a lot of in and out on trigger pins, and that's a good thing.

It's really handy in doing adjustable triggers, like JP and equivalents. You can actually see the effect the adjustments are having on system operation.

It works simply: the pins run through from the right side of the receiver and the trigger installs then on the outside of the left receiver wall, springs and all under tension just like if it was sitting inside the receiver. Flexible washers and nylon wing nuts hold it to the receiver firmly with no marring.

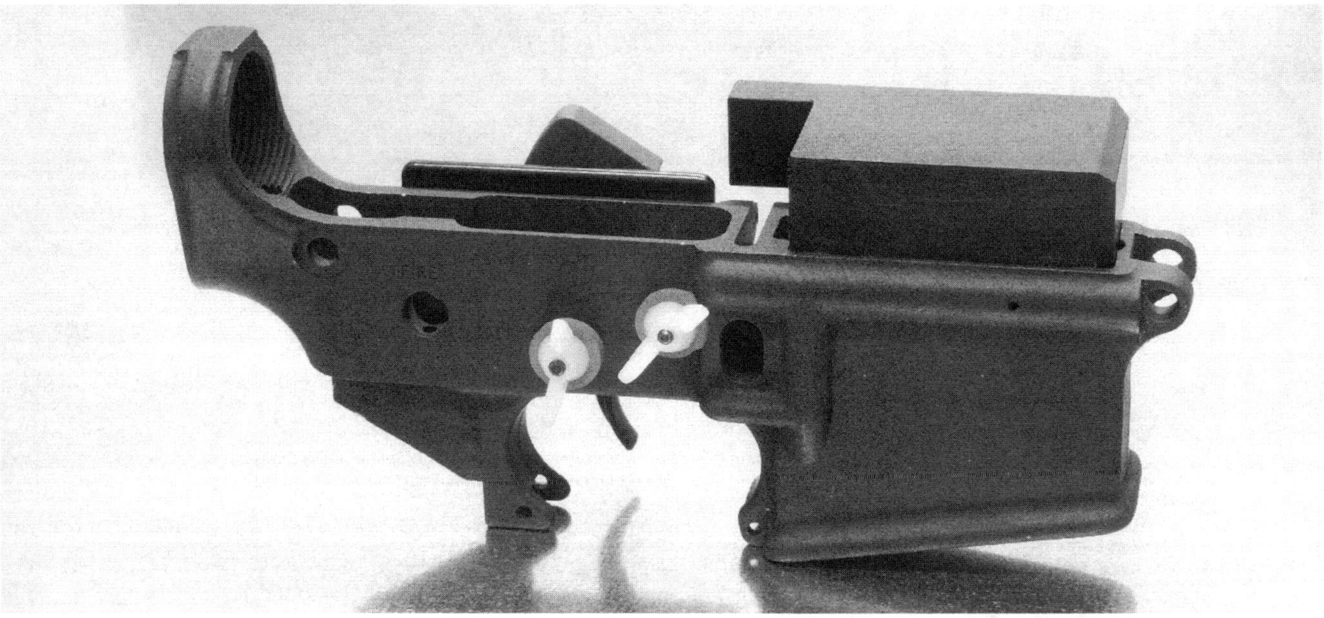

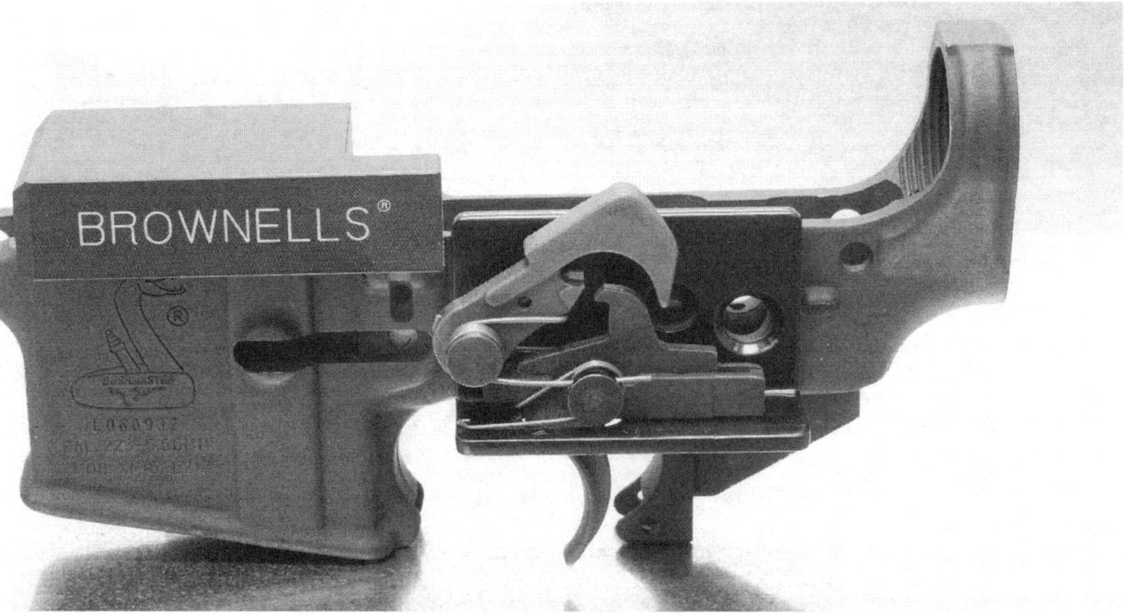

PARTS & TOOLS

This is, or should be, a hand-done installation so no special tools are really needed. As always, and of course, there are several helps available. One thing, however, is that you do not need a hammer... If a pin won't fit into the receiver hole using pressure alone, either pin or hole needs a close look and perhaps an edge broken, or better alignment from you. Do not tap in these pins!

PREPARATION

Gun oil

Gun grease

Options (good ideas)

    Assembly punch (0.154 size)

    Hammer block (to cushion the impact from a falling hammer)

    Mechanics gloves

    Trigger pull weight gage, or weight set

PARTS

Trigger

Trigger return spring

Disconnector

Disconnector spring

Hammer (with "J" spring installed)

Hammer spring

Hammer pin

Trigger pin

PRECAUTIONS

Test fit. Pins first through the parts to see that there is free movement. I first pin the hammer into the lower without the hammer spring to make sure the hammer body moves freely within the receiver.

Grease the pin holes. This helps especially overcome the resistance from the hammer pin fitting over the J spring.

Whatever you do, do not let the daggone hammer fall under power against the lower! It can easily crack the receiver.

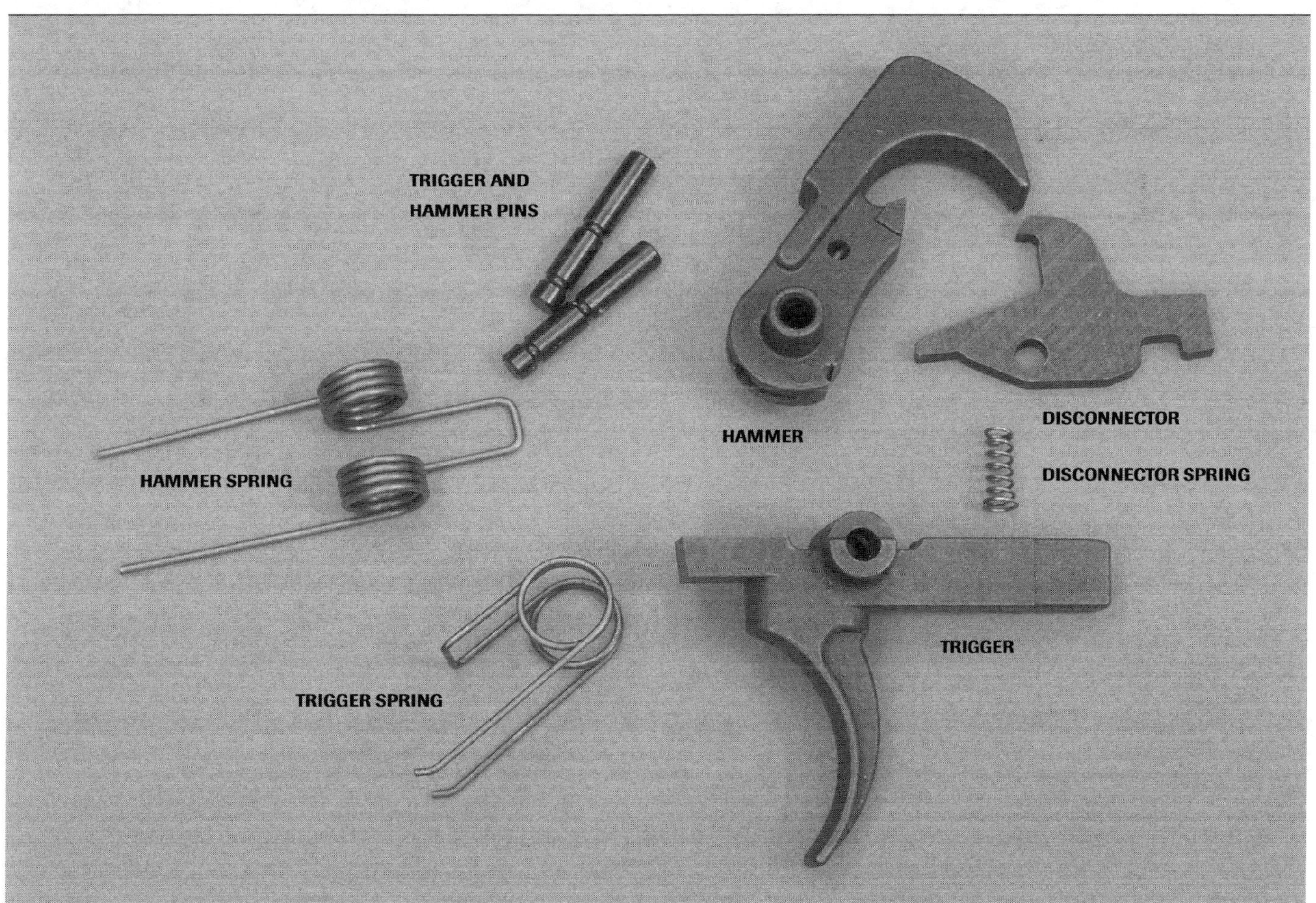

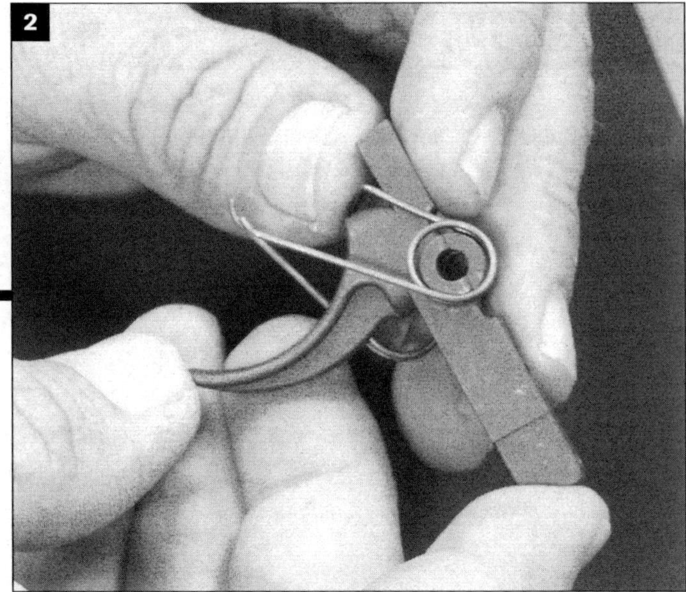

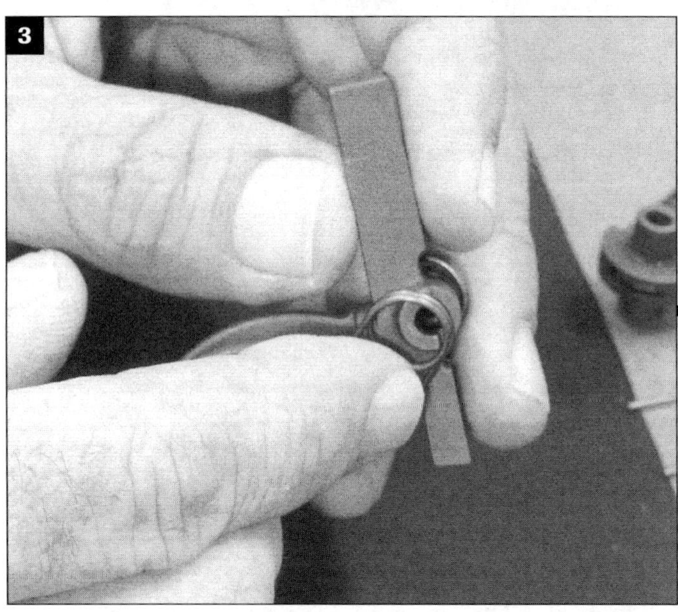

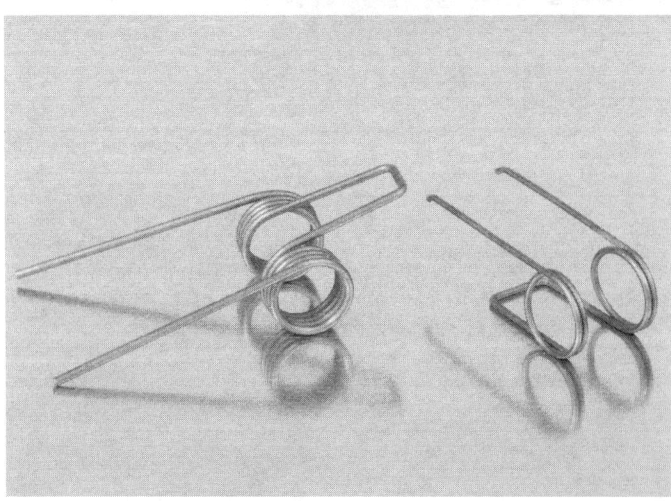

TRIGGER CONSTRUCTION

Put the pieces together. It hain't brain surgery but it sorely matters that the springs are positioned correctly.

1./2./3./4. The **trigger spring** loops just snap fit into place over the bosses on the trigger. Note the orientation. The little feet are pointing, well, like little feet would point, and the square portion of the wire is under the sear bar.

5./6./7./8. The **hammer spring** installs in the same way. Orient the piece correctly as shown and then snap one spring loop over one side and the other spring loop over the other side.

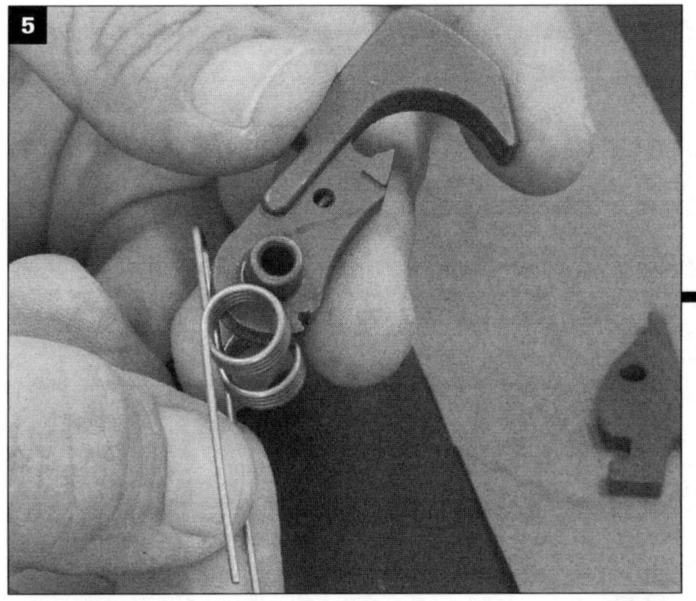

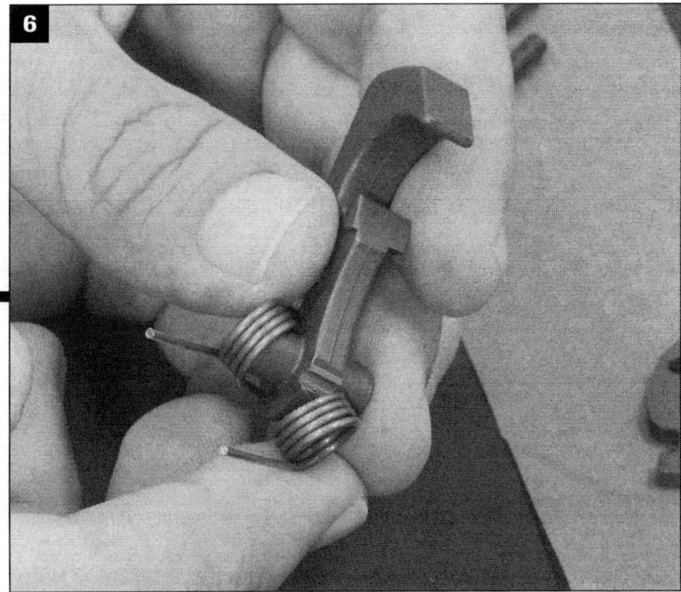

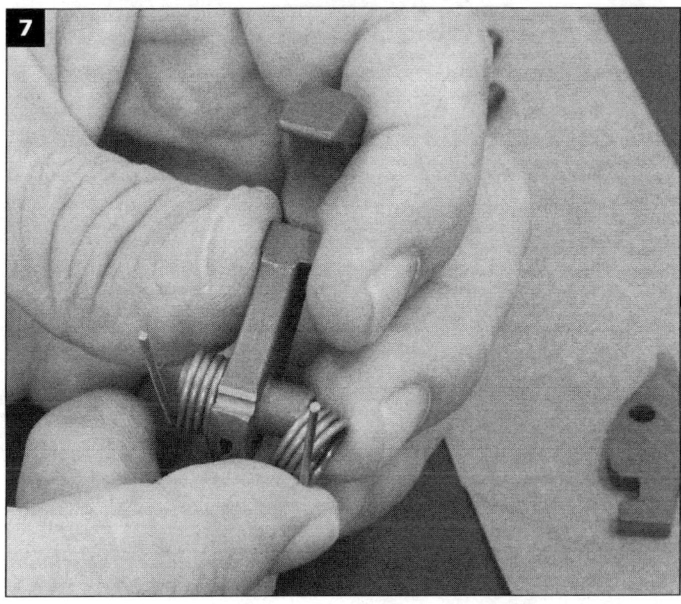

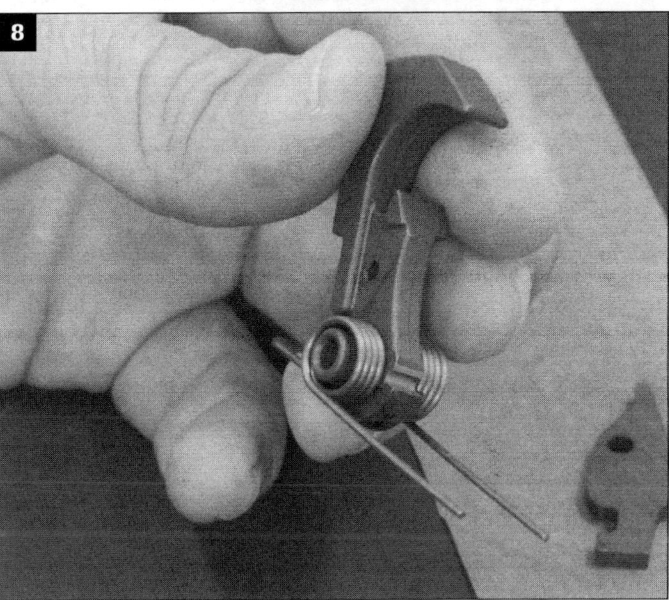

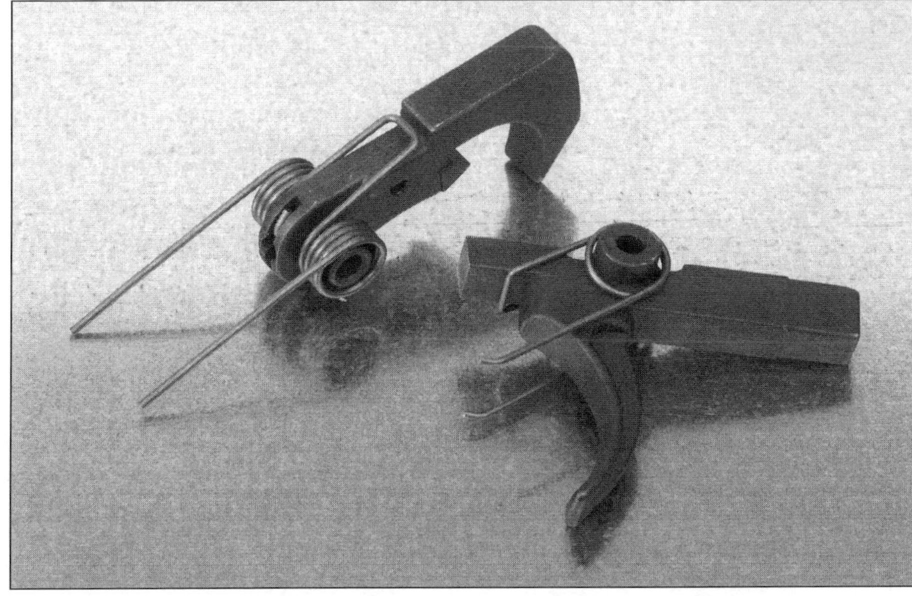

Get them looking like this and you're good to go, onto the next step...

One or two words on springs. I am a believer in **chrome silicon** hammer and trigger springs. This material shows far greater resistance to change than does the usual music wire that kit springs are made from. That means the trigger performs consistently over time. Chrome silicon also has compression and rebound characteristics that enhance performance. Essentially, a chrome silicon spring has effectively lighter compression and faster rebound compared to a music wire spring of the same rating. That's all good.

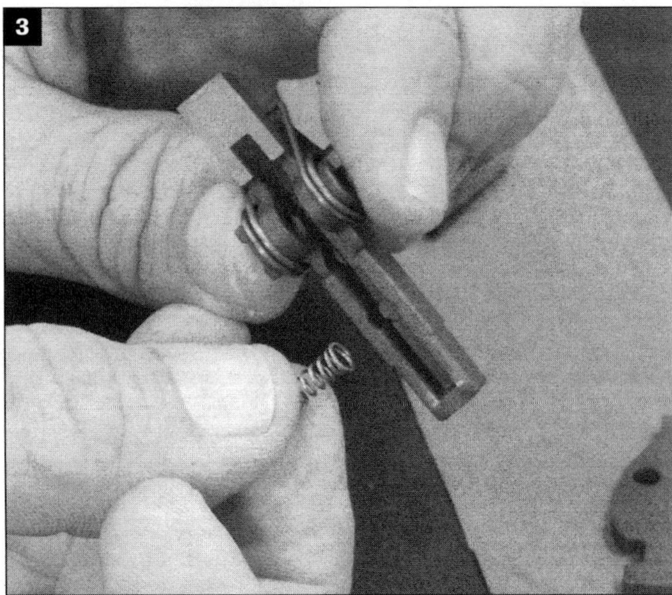

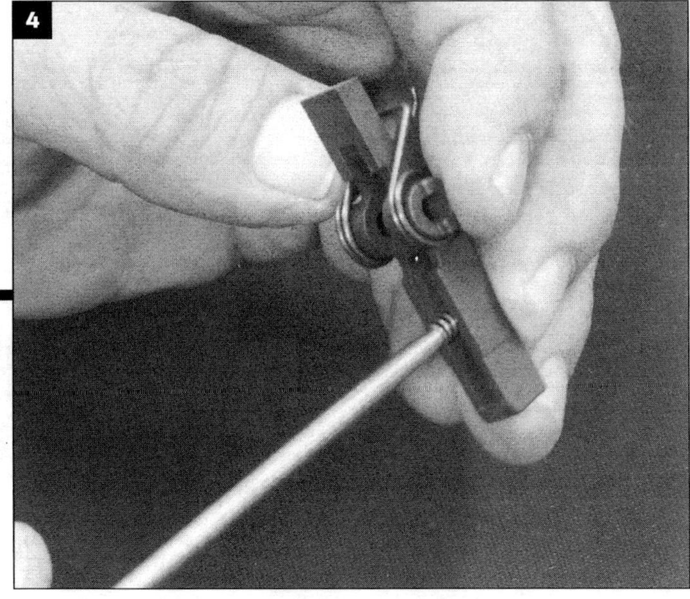

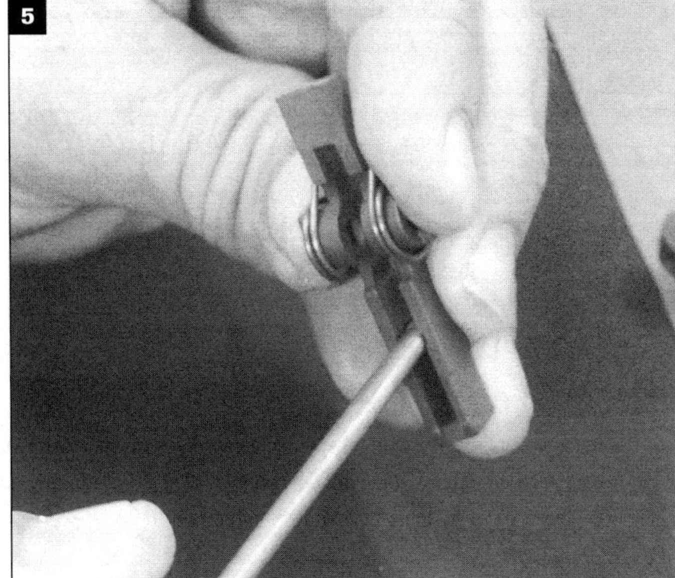

1. Don't confuse the bolt catch plunger spring with the disconnector spring. They are just about the same size, but the disconnector spring has a larger end. **2.** This larger end goes into the recess located in the channel on the trigger body. Some GI triggers have two recesses, and if so use the front hole.

3./4./5. Put the spring into the recess and then push it fully home using something like this assembly punch. Push it hard. You will feel it seat.

I break the edges on the disconnector and shine up its underside. I've seen many that were rough. Polish the underside of the disconnector hook but don't change its angle or dimension. Likewise for the underside of the "nose" that presses down behind the sear area on the trigger body.

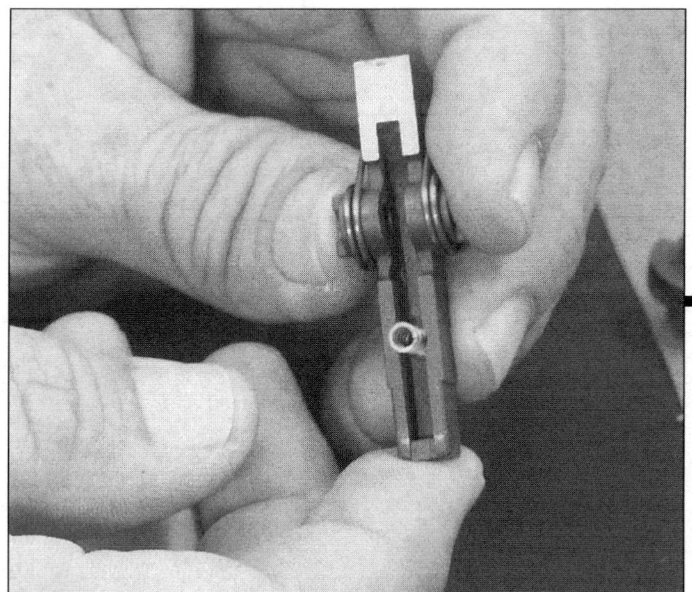

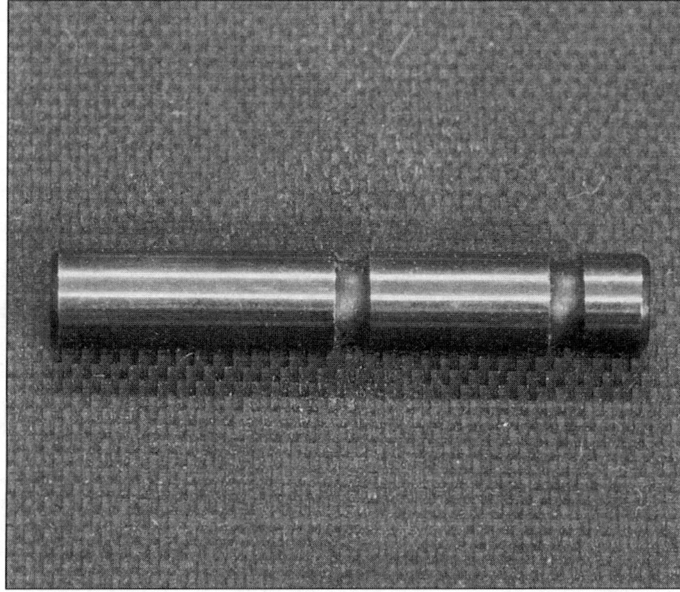

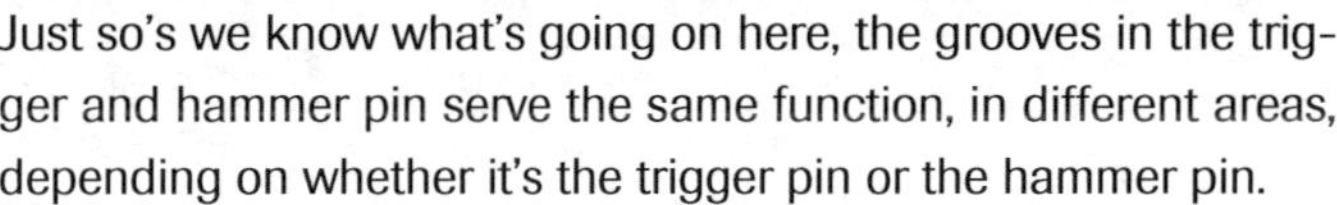

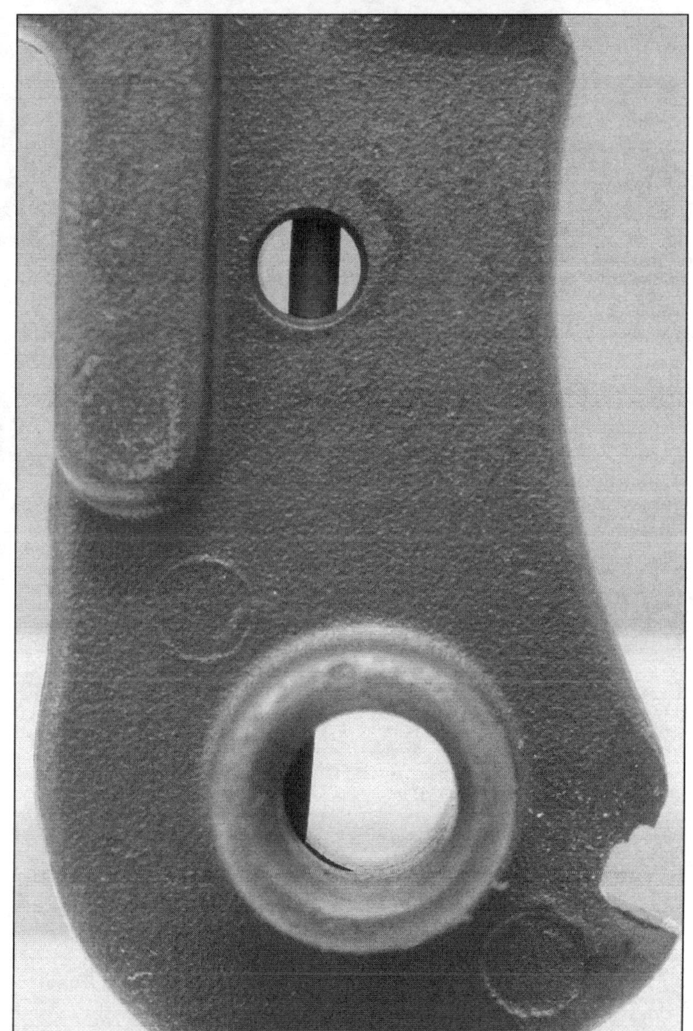

Just so's we know what's going on here, the grooves in the trigger and hammer pin serve the same function, in different areas, depending on whether it's the trigger pin or the hammer pin.

The above pin is shown in its correct orientation as installed (through the right side of the receiver and then into the left side of the receiver; ungrooved end first). That means the first groove of the trigger pin will be engaged by one of the legs on the hammer spring. That keeps it in the rifle. The second groove in the hammer pin engages the "J" spring in the hammer, and that keeps that pin in the rifle. The J spring is shown close up on right. The tension in the J spring is pretty stout and that's one reason for lubricating the pin prior to assembly. That part is almost always, and sure should be, already installed. Do make sure! I've seen one hammer sans J spring.

Now lets get that bad boy pinned all in place.

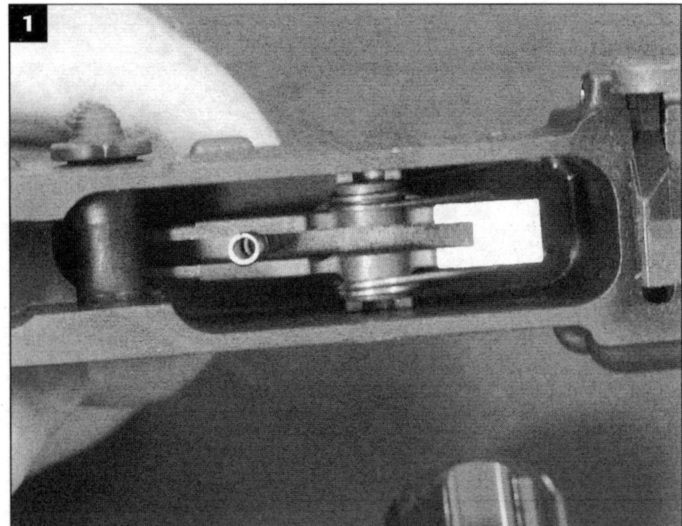

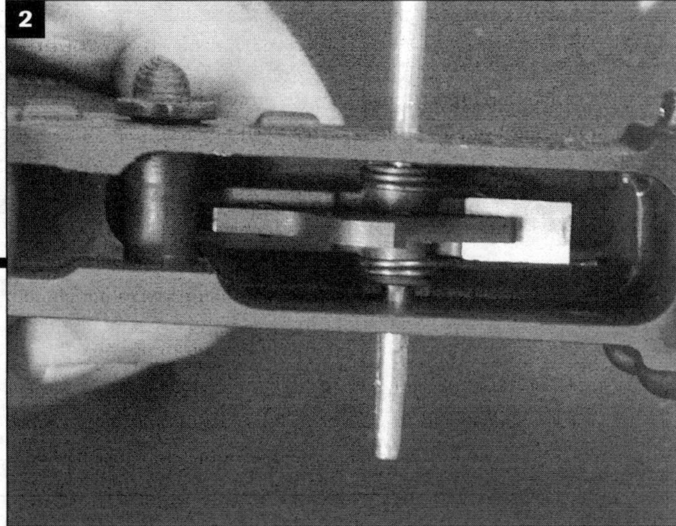

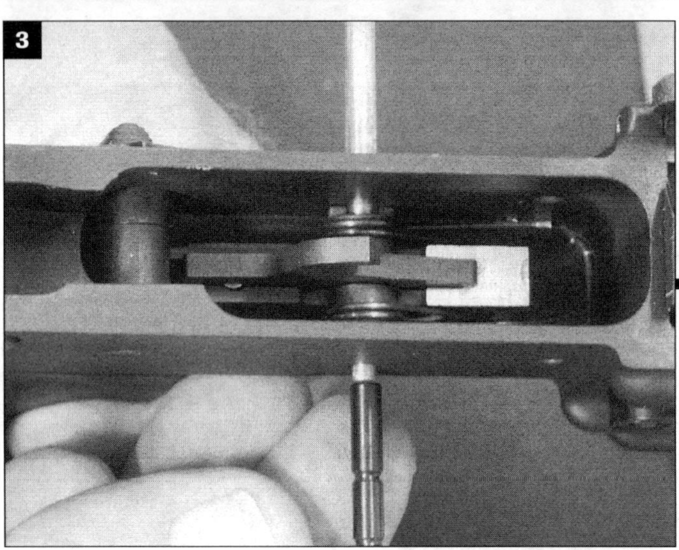

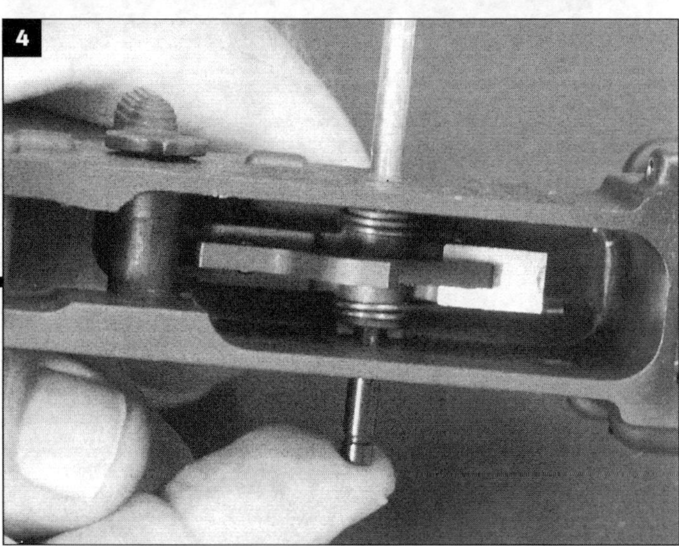

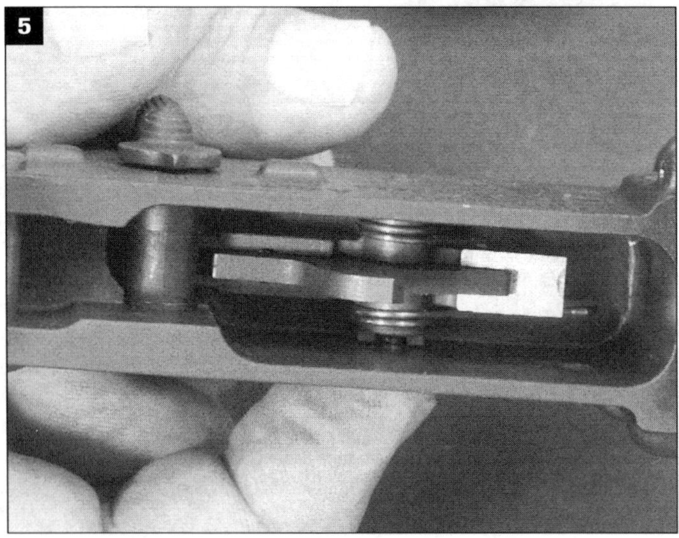

TRIGGER INSTALLATION

1. Put the trigger into the lower making sure the return spring stays in position. The disconnector spring is in its place. The tail of the trigger assembly goes under the safety bar, and the safety should be in the "FIRE" position.

2. Set the disconnector into its place and line up the holes. An assembly punch makes this much easier. Push the punch in from left to right. I run the punch into the trigger but not into the channel where the disconnector sits and then position the disconnector. The little notch should be right over the disconnector spring. Push the disconnector down so its hole lines with the trigger pin hole. It can take some wiggling.

3./4./5. Start the trigger pin from right to left, ungrooved end first. If you're not using a punch, use the trigger pin in the same way: push the pin into the trigger but not into the channel. Put the disconnector into place and push down on it to get alignment. Next is just continuing on until the pin enters the receiver hole on the left side of the receiver and comes to flush with its exterior surface. Done. Check that the trigger moves freely and the disconnector is free to rock to and fro. If it won't, make sure its spring is seated.

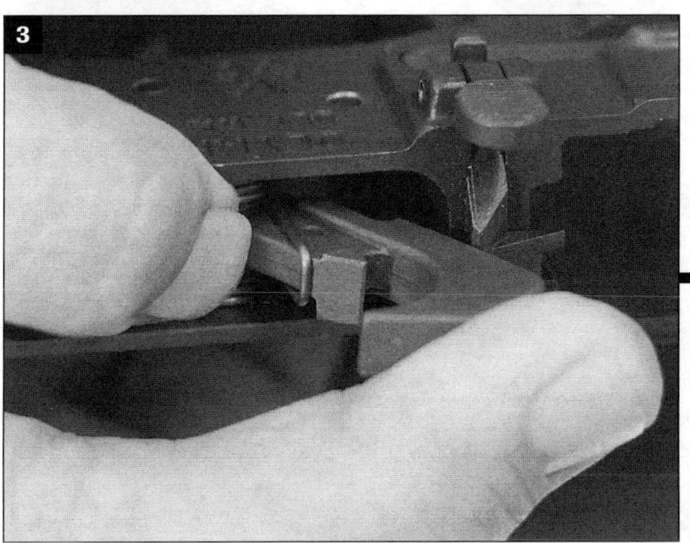

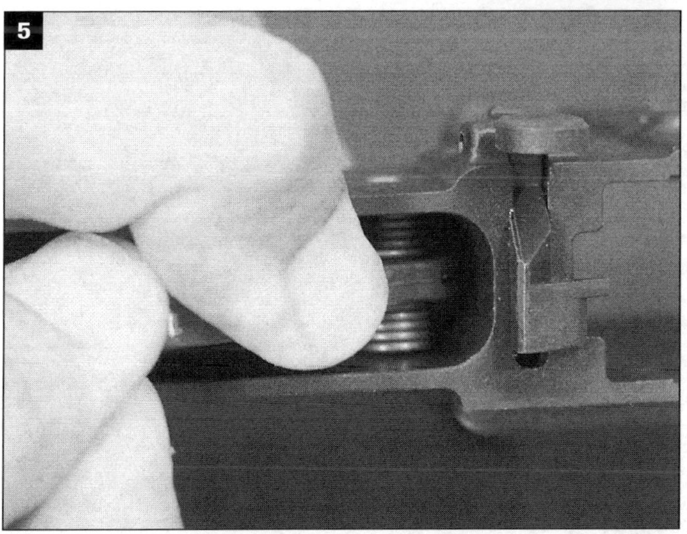

HAMMER INSTALLATION

1./2. Here's how the hammer goes. Pay attention to how this looks and keep it looking that way as you begin the installation. The legs of the spring should be pointing toward the rear of the receiver. Both legs of the hammer spring have to be **on top of** the trigger pin. Again, both legs of the hammer spring have to be on top of the trigger pin. One leg fits into the groove on the right side of the trigger pin to keep the trigger pin from slipping out. Check the photo on page 61 for a good close-up detail. Here's a time when a little help comes in handy. This is a strong spring, very strong, and we're having to do what I call delicate work working around its strength. I often use "mechanic's gloves" to take the stress off my fingers and always an assembly punch run through the left side of the receiver to position the hammer in place. Speaking of, here we go. **3./4./5.** Push down and somewhat forward on the hammer to compress the spring and lower its hole into alignment with the receiver holes. It will want to push back and up and the action that prevents that from winning this battle is a sort of pseudo-cocking force applied to the hammer with the fingers. The hammer is pushed down first and also then forward into position. Sometimes it's tedious. **6./7./8.** The punch helps! Pressure down on the hammer to get the pin fully through the left-side receiver hole.

[CONTINUED NEXT PAGE]

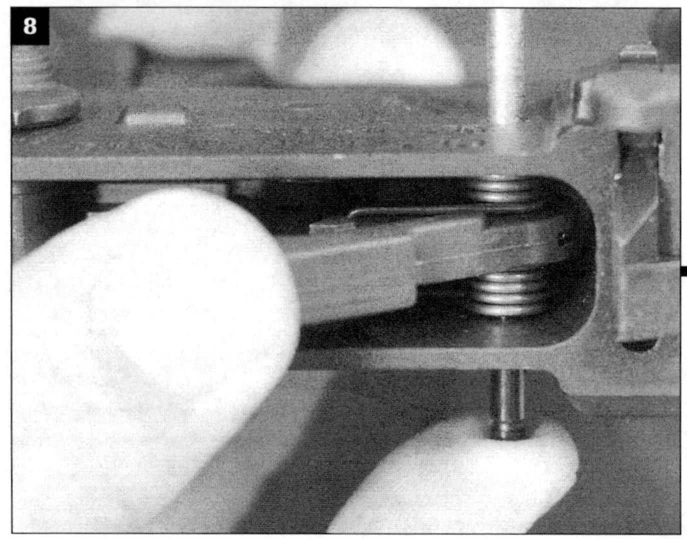

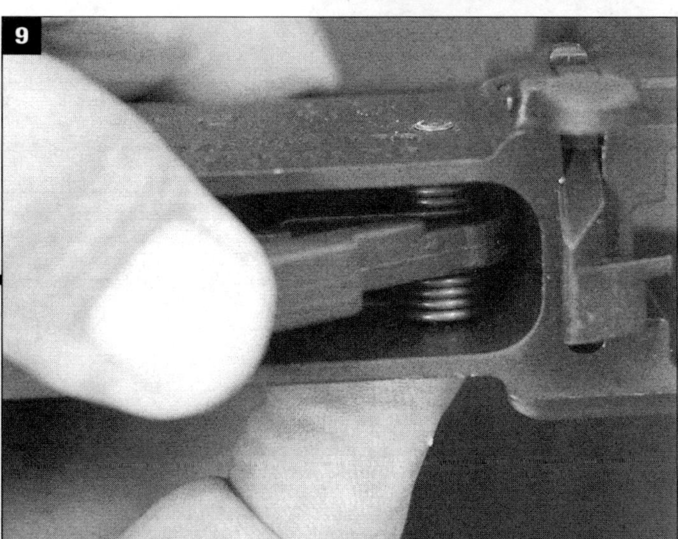

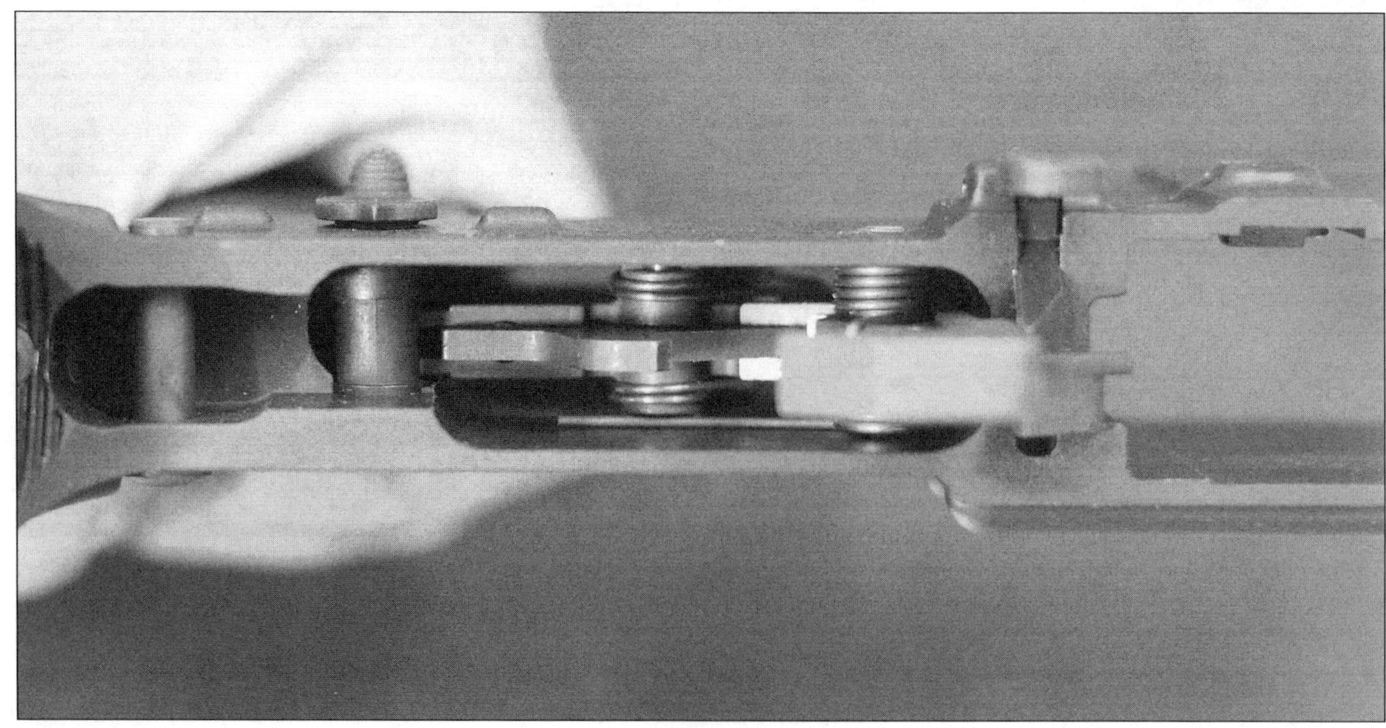

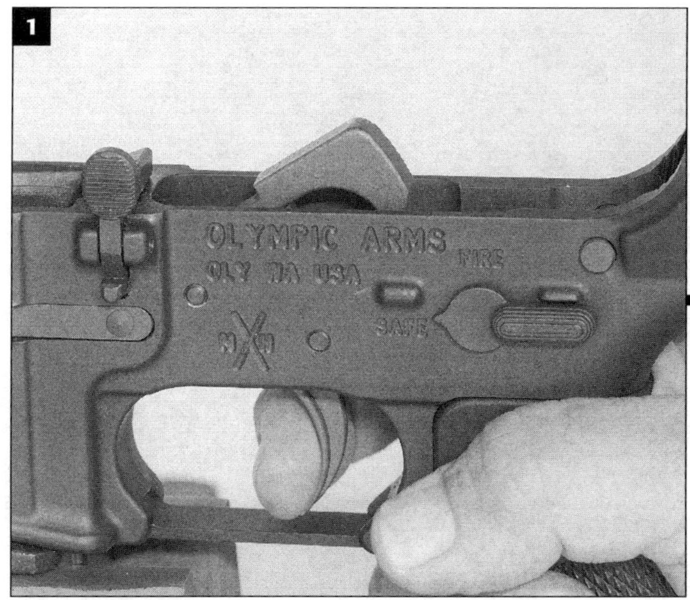

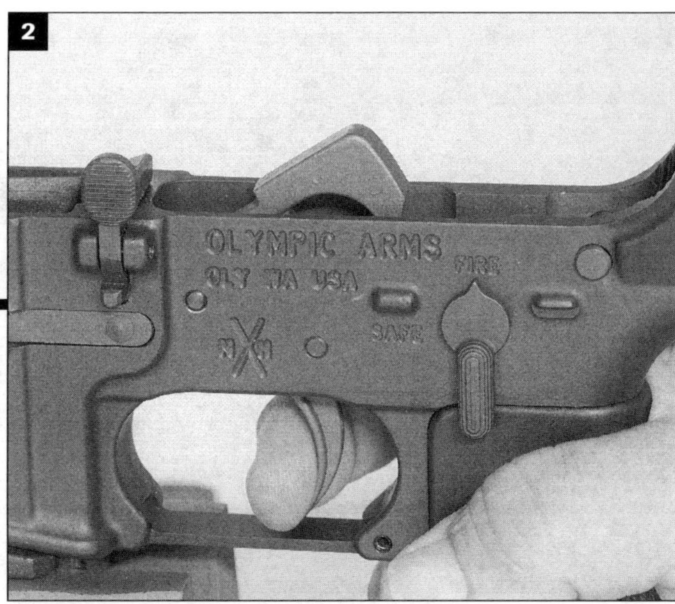

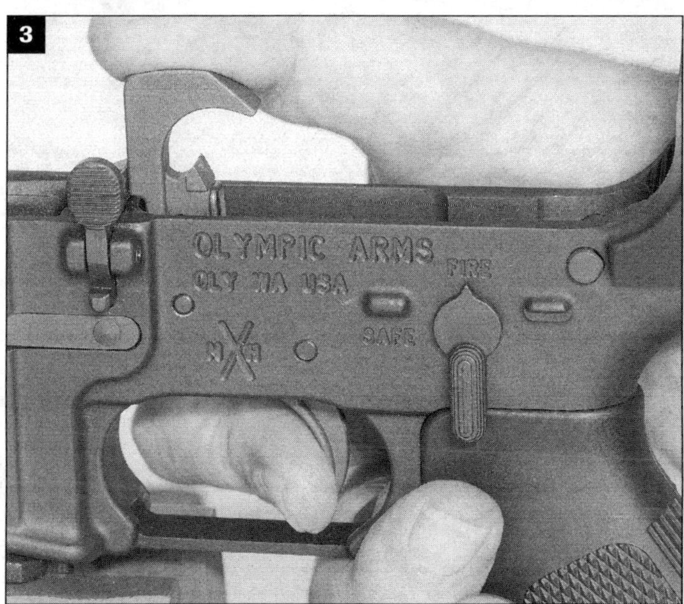

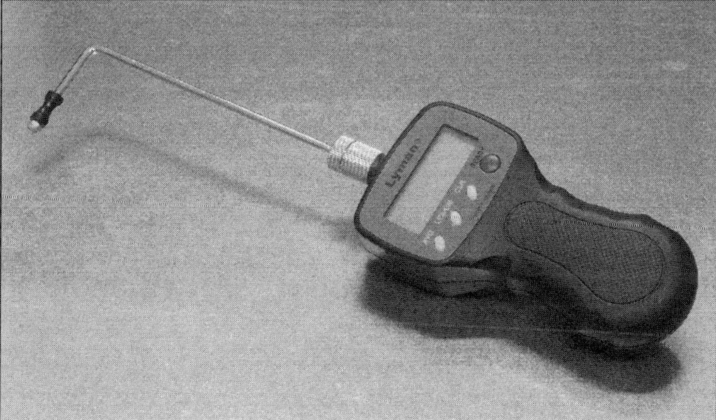

If you have a nice gage like this Lyman, don't break it checking a stock trigger! Sorry, but you're not going to get a good pull until you install something from the aftermarket. One thing good about stock triggers, though, is that they function...

FUNCTION/SAFETY CHECKS

Essential function checks have already been done on installation — making sure the hammer can swing freely, that the trigger moves freely, that the safety switch can rotate from stop to stop. Now that it's all together, it's a system and it all has to work together. Don't even **think** about firing the rifle without confirming that all is correct.

The safety switch should only move from position to position when the hammer is cocked. When the hammer is forward, it should not go to the "SAFE" position.

Cock the hammer. Just pull it back and push it down until it clicks into place. **1.** Rotate the switch to "SAFE" and pull the trigger. Pull it hard. Pull it harder. Hammer should not fall. **2./3.** Rotate the switch to "FIRE." Hammer should not fall. Now pull the trigger and the hammer should fall.

Very important! Do not let the hammer hit the bolt stop and receiver. Cushion its fall with your thumb or otherwise cushion the receiver with a dowel, piece of leather, or, even better still, a purpose-built block.

The lower receiver can and will crack if the hammer hits it!

Honestly, it's very unusual for a standard trigger to fail any of these tests. Aftermarket triggers, though, are a different story, and I've seen some that required mild to wild amounts of fiddling. Don't sacrifice anything for reliability.

[CONTINUED NEXT PAGE]

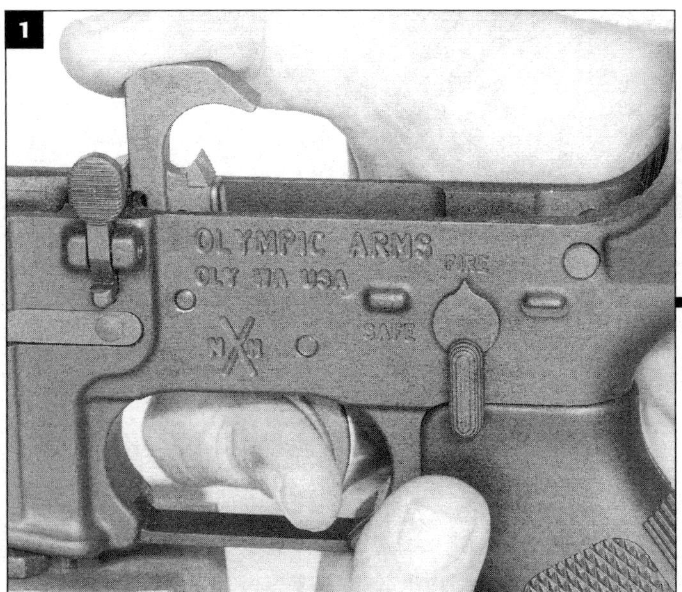

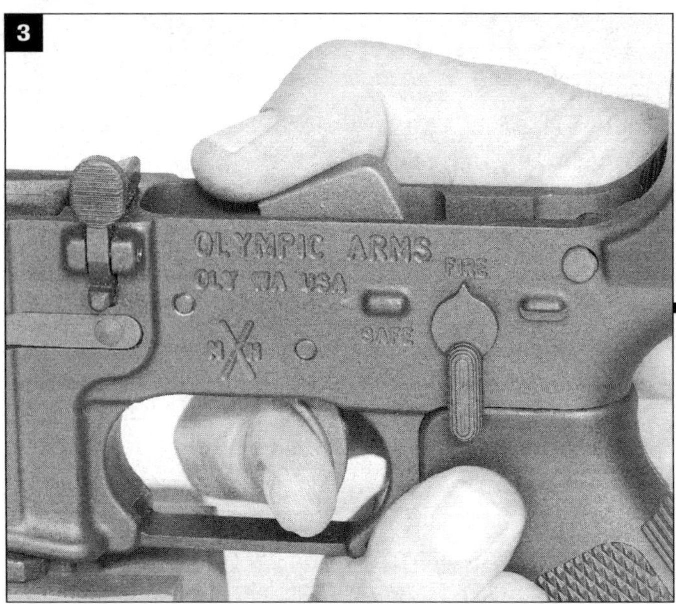
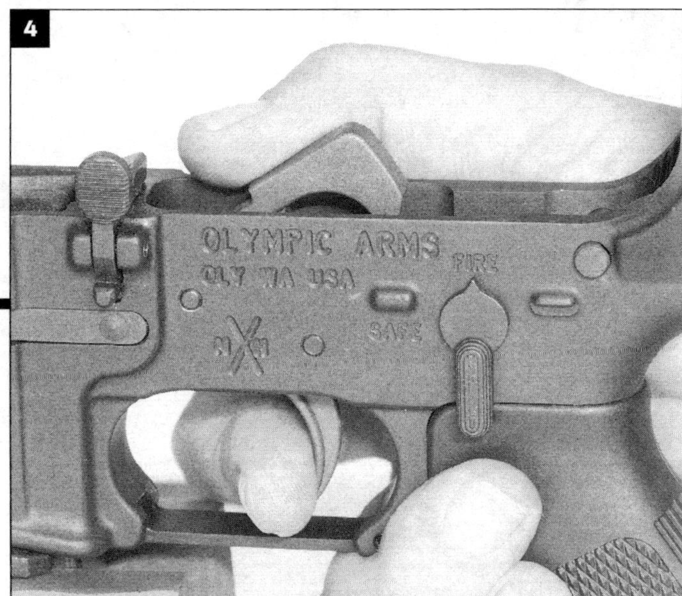

DISCONNECTOR

Do not fire the rifle unless the disconnector is functioning.
The **disconnector** prevents the hammer from falling until the
trigger is released forward enough to "reset" the sear into the
hammer sear notch.

Here's how it works: The hammer is cocked by rearward movement of the bolt carrier (either after a shot or by retracting and
releasing the bold carrier using the charging handle). As the carrier moves back forward, the hammer is caught and held in the
cocked position by the hook on the disconnector, which is tipped
forward and in position to catch the hammer. This stops the hammer from following the bolt carrier forward. If you hold the trigger
back after a shot has fired, the sear located on the forward part

of the trigger will not catch in the hammer sear notch when the
hammer cocks because the sear is depressed below the arc of
the hammer notch. When the trigger is released forward, that
allows the hammer to slip from under the disconnector hook and
be caught by the trigger sear engaging the hammer sear notch.

1./2. Cock the hammer and then release the hammer forward by
pulling the trigger. **Keeping the trigger held fully to the rear,**
cock the hammer again.

3./4. Now, **just as slowly as you can**, release the trigger forward. The hammer should "jump" up as it's handed off from the
disconnector to the trigger sear. I redo this check on an assembled rifle by letting the bolt carrier slam home to better duplicate
firing. Then, of course, come range tests. Fixing a faulty disconnector is covered in aftermarket trigger installations.

5.0 UPPER RECEIVER
BASIC BUILD

[All the parts for a complete upper assembly can be purchased by anyone, no FFL needed. Once you have a lower that suits, it too can be modified to be more suitable for the demands you may put on your new upper assembly. This short segment will detail the two essential steps required to assemble a traditionally configured upper, whether it has a carry handle attached or not.]

__The effort is variable.__ If you're building an A2 upper as shown, it's the rear sight that makes for the most work. If you're building a flat-top the amount of construction effort runs from very light to none at all, depending on the model you choose. Here we'll just cover the essentials. Rear sight gets its own segment.

SEGMENT CONTENT

75   **Upper Receiver Assembly** (component selection)

76   **Parts & Tools**

78   **Ejection Port Cover**

80   **Forward Assist**

82   **Charging Handle**

83   **Charging Handle Seal** (Permatex RTV trick)

UPPER RECEIVER ASSEMBLY

Most competition-style flat-top upper receivers don't need much component assembly, and some need none. The "A2" configuration (whether detachable or fixed carry handle), on the other hand, has two parts sets, not counting the rear sight, that have to be installed. Actually, they really don't for a target rifle with an "A3" upper (flat-top with forward assist and trapdoor ejection port cover). Of course, they have to be there if we're building a Service Rifle. Something like a DPMS "competition" upper requires no assembly work at all. No port cover, no forward assist.

The forward assist and ejection port cover are both a little tricky, but follow along and neither is hard to get through.

PARTS & TOOLS

Just read along and follow the photos. Don't lose the port cover clip, or your cool trying to keep its spring wound!

PREPARATION

Tap hammer

Roll pin punch and roll pin starter punch

Capture punch (just use another roll pin punch)

Gun oil

Gun grease

Options (good ideas)
  1/2-inch wide masking tape
  Flat black touchup paint

PRECAUTIONS

Use the masking tape on the receiver for the forward assist op, and drive the pin from the underside.

An upper receiver kit will have all these parts, plus the rear sight, but not the charging handle! Order that separately.

PARTS

Stripped upper receiver (A2-style)

Ejection port cover
Ejection port cover pin (rod)
Ejection port cover clip
Ejection port cover spring

Forward assist knob and pawl assembly
Forward assist spring
Forward assist roll pin

Charging handle body
Charging handle latch
Charging handle latch spring
Charging handle latch pin
  (This part is almost always sold as an assembly)

This upper from EGW doesn't have a carry handle attached but assembles in the same way as an A2. If you're building a competition or otherwise "precision" rifle, there's no need for the forward assist or ejection port cover. I like to install them just for appearances sake. From time to time EGW has a little round insert that covers the forward assist opening.

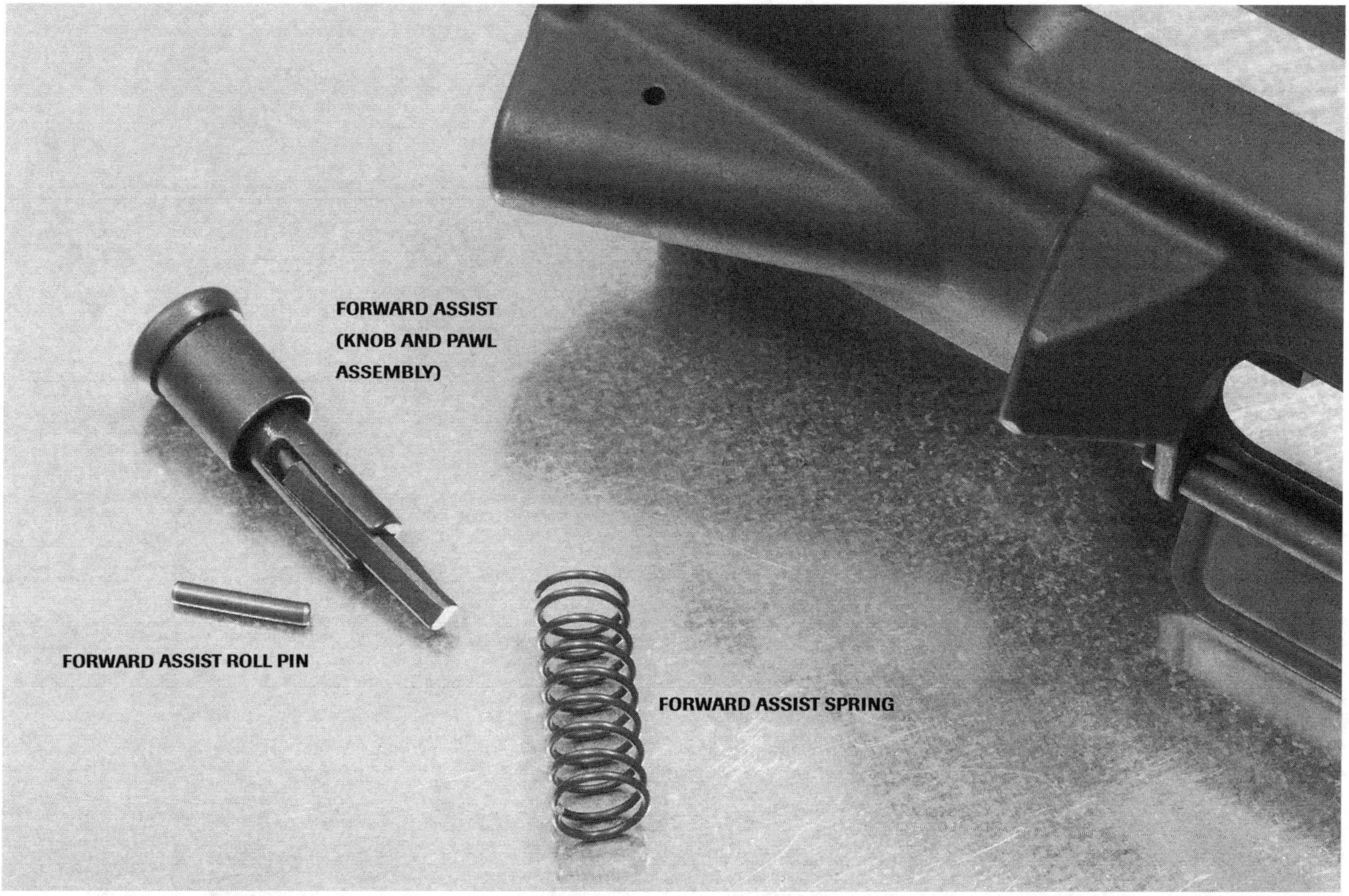

FORWARD ASSIST
(KNOB AND PAWL
ASSEMBLY)
FORWARD ASSIST ROLL PIN
FORWARD ASSIST SPRING

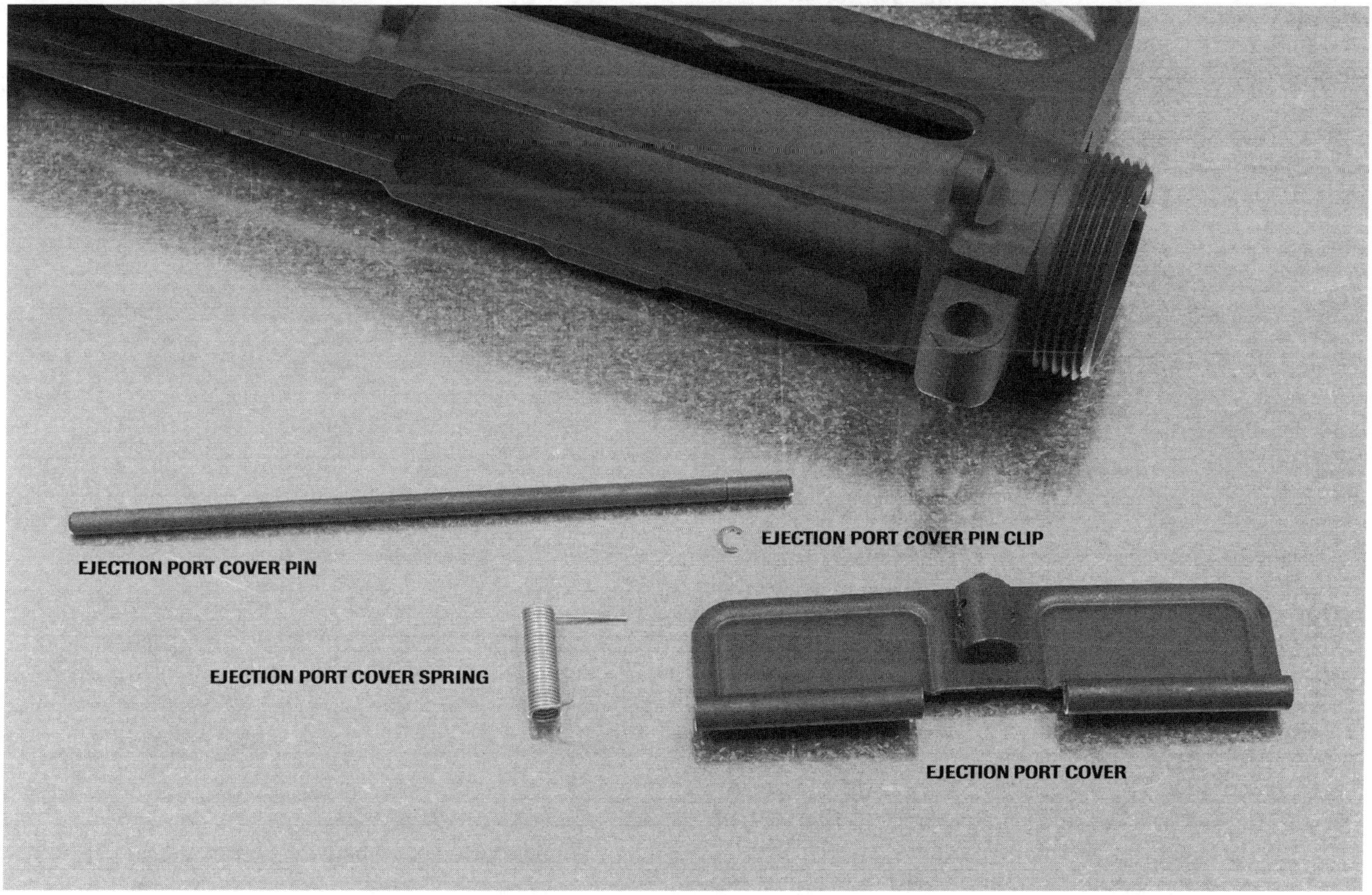

EJECTION PORT COVER PIN
EJECTION PORT COVER PIN CLIP
EJECTION PORT COVER SPRING
EJECTION PORT COVER

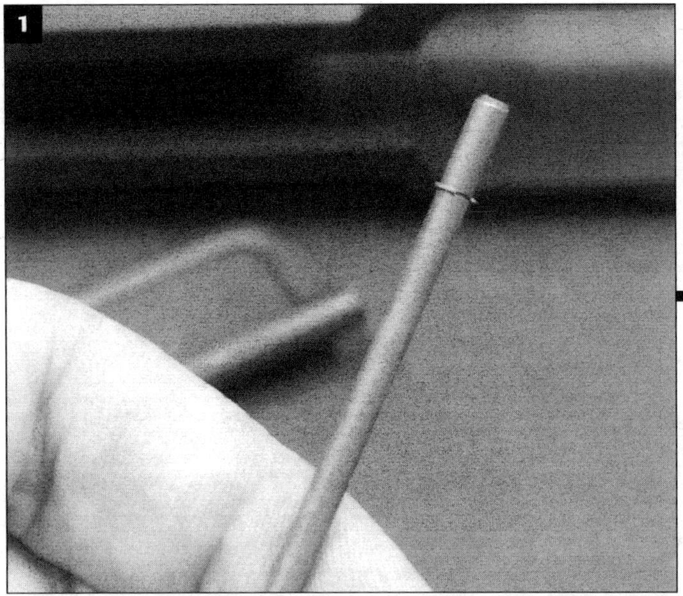

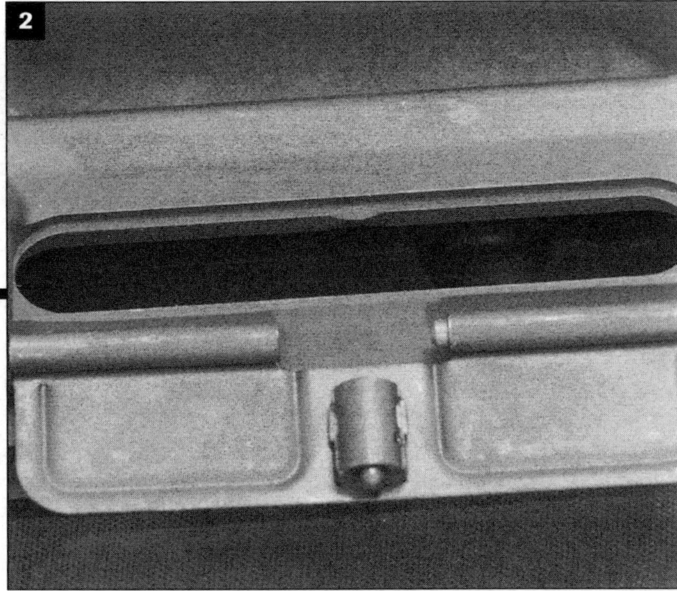

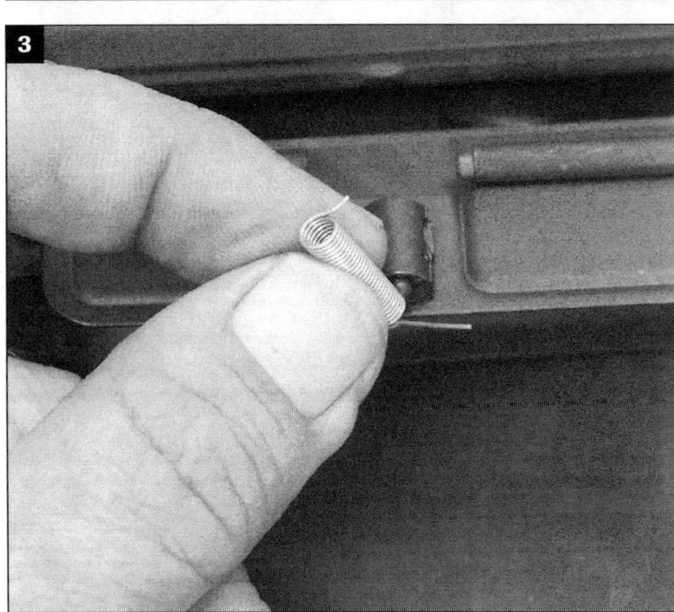

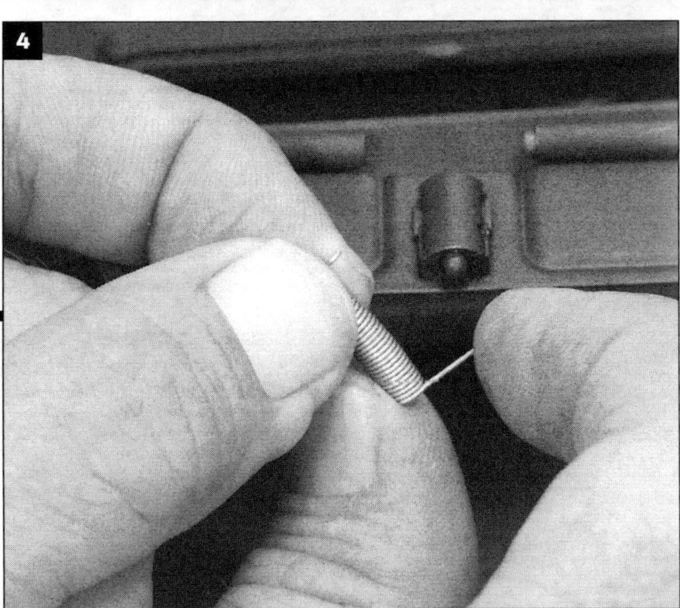

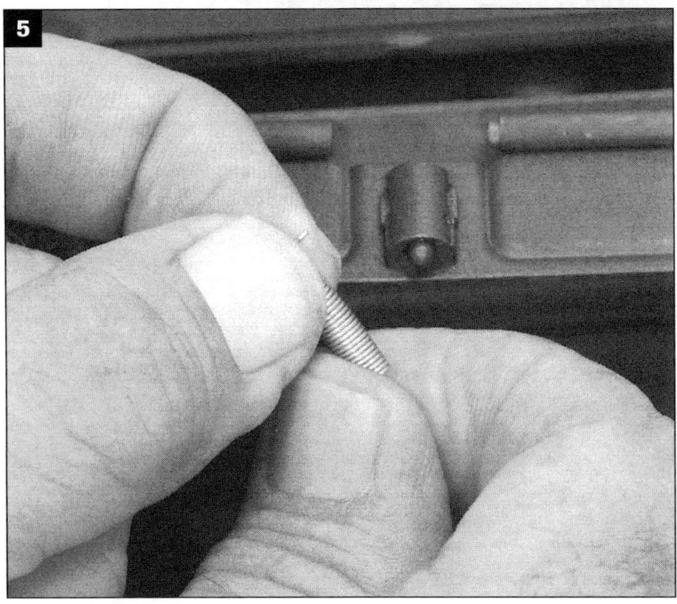

EJECTION PORT COVER

1. Clip the ejection port cover pin clip onto its groove on the end of the ejection port cover pin. This is a very small part. It doesn't take much to get it into place, and needle nose pliers may help. It's easiest to hold the clip bottom against a flat surface and push the pin down into its open end. Make sure it's "in."

2. Position the port cover against the receiver in its "open" position and align the holes in the cover where the pin will go with the openings in the bosses in the receiver. Insert the pin from front (right side) to back through the cover. Circlip is to the right.

Take a good look at this. 3./4./5. This spring is not easy. It has to be "wound" to work right and kept wound during pin installa-

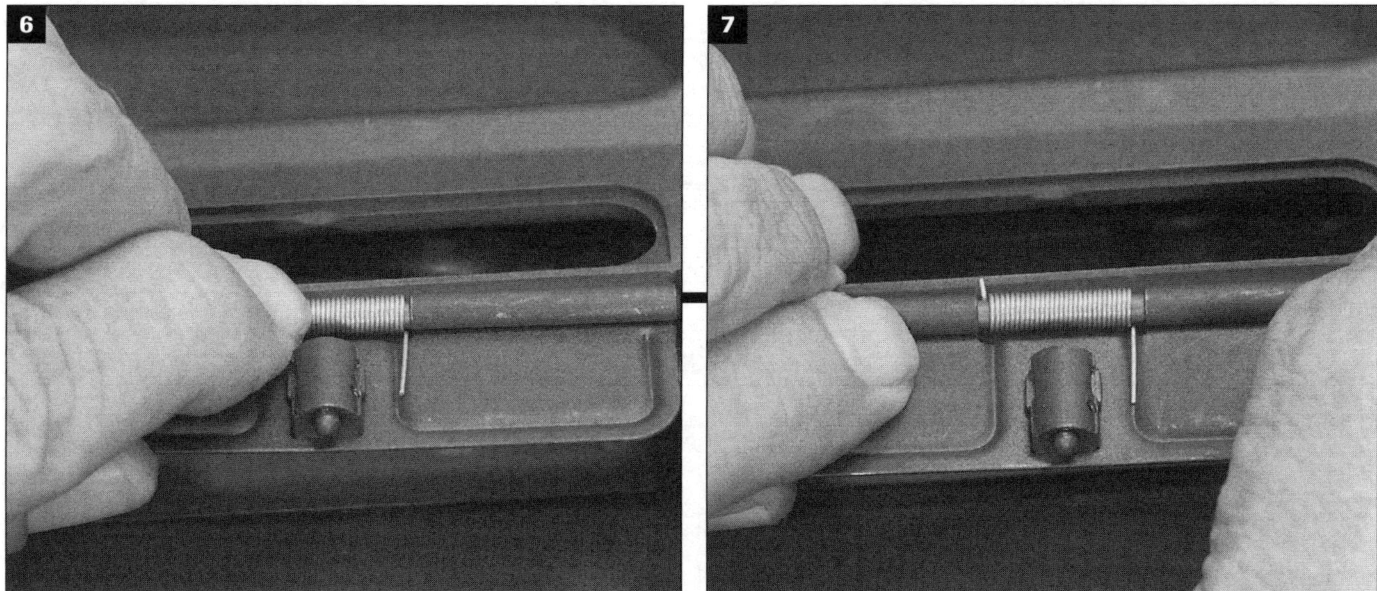

tion. The spring has to be tensioned by rotating its legs in opposite directions so the legs lie on the same plane. Hold the little leg put and rotate the big one 180-degrees. Notice how my finger moves in photos 4. and 5. That's the right direction. To keep it wound, hold the little leg against the receiver; the longer leg will rest securely against the port cover. It's hard to hold the little leg because it hurts!

6./7. So wind the spring, keep the tension on it, and push the pin through the spring and on through the left hand (rear) pin boss.

Test it by shutting it and then press against it from inside the receiver. It should snap down smartly.

Not hard, but the spring winding makes it tricky.

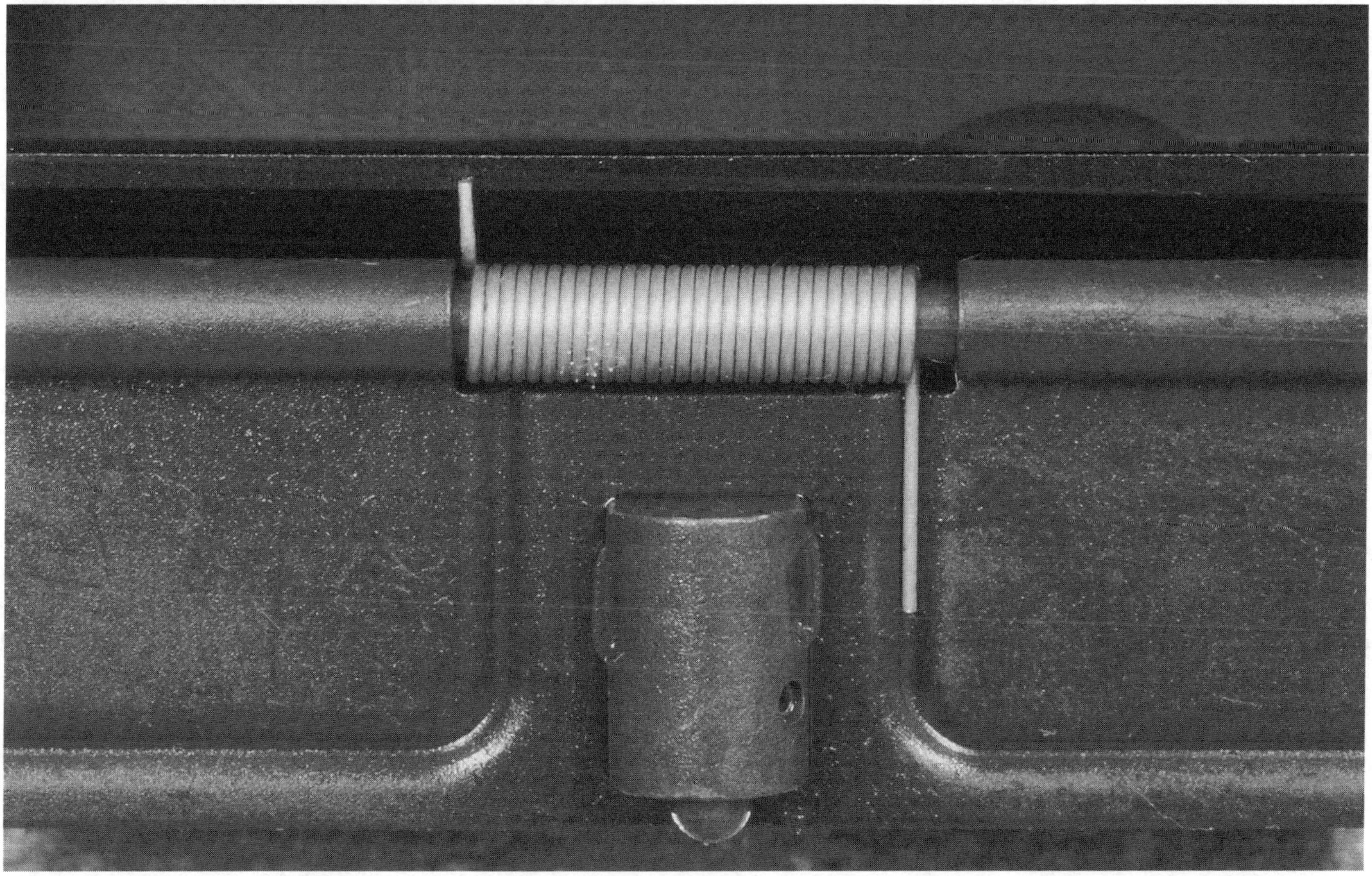

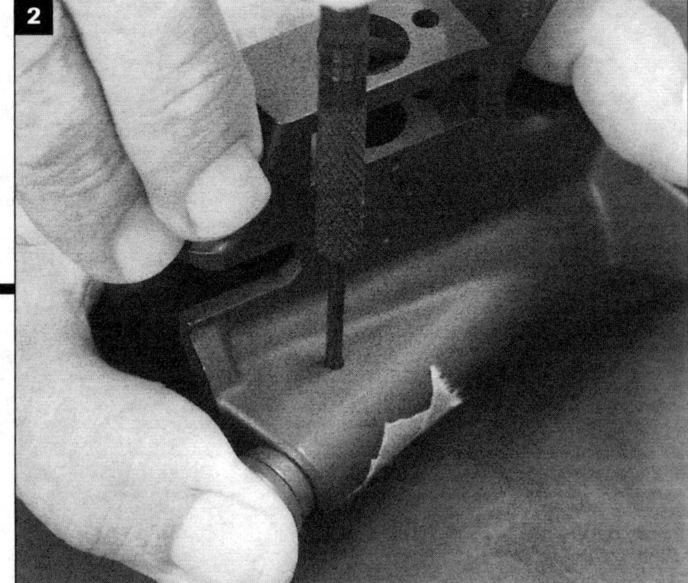

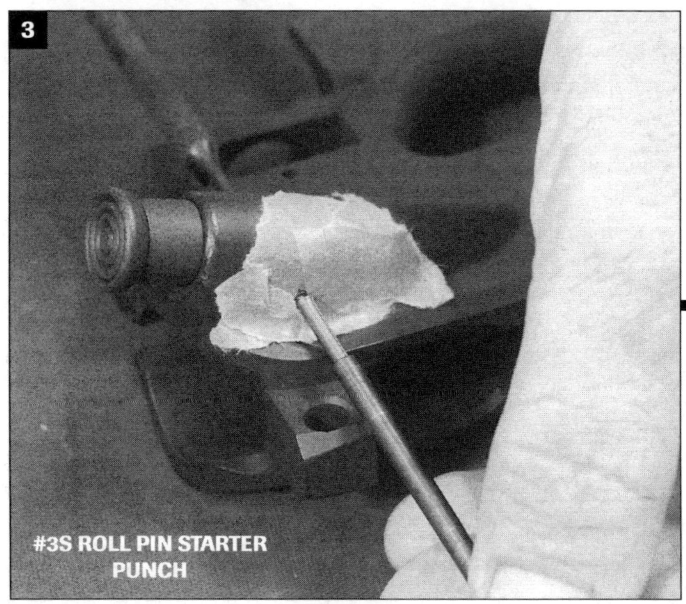

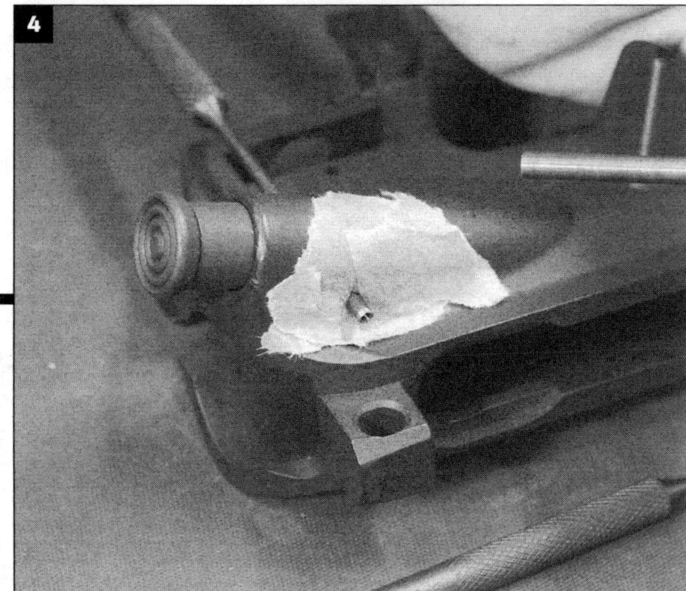

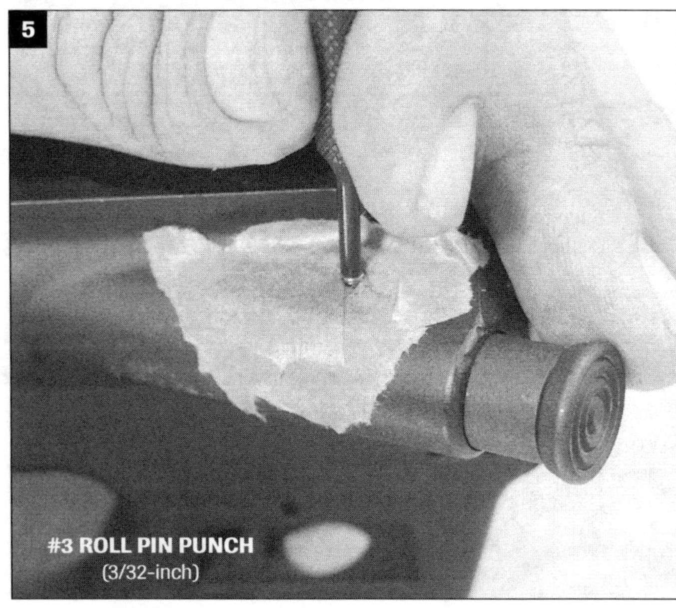

FORWARD ASSIST

Examine the forward assist knob and pawl for burrs and smooth any rough areas. Grease all parts prior to assembly.

1./2. Put the spring over the assist body. Study the orientation of the pawl. It should be facing in. This edge is what engages the notches in the bolt carrier to make this part operational. Notice also the clearance flat on the side of the assist. This is for the roll pin. Get a 3/32-inch diameter plain punch handy, or another #3 roll pin punch. You'll use this to capture the assembly in place so you can pin it there. Slip the spring over the pawl and Insert the assembly into the upper. Compress it inward enough to get the punch through the pin hole to capture the assist in place. Push this in from the top. Turn the receiver over. Notice the masking tape. Use it!

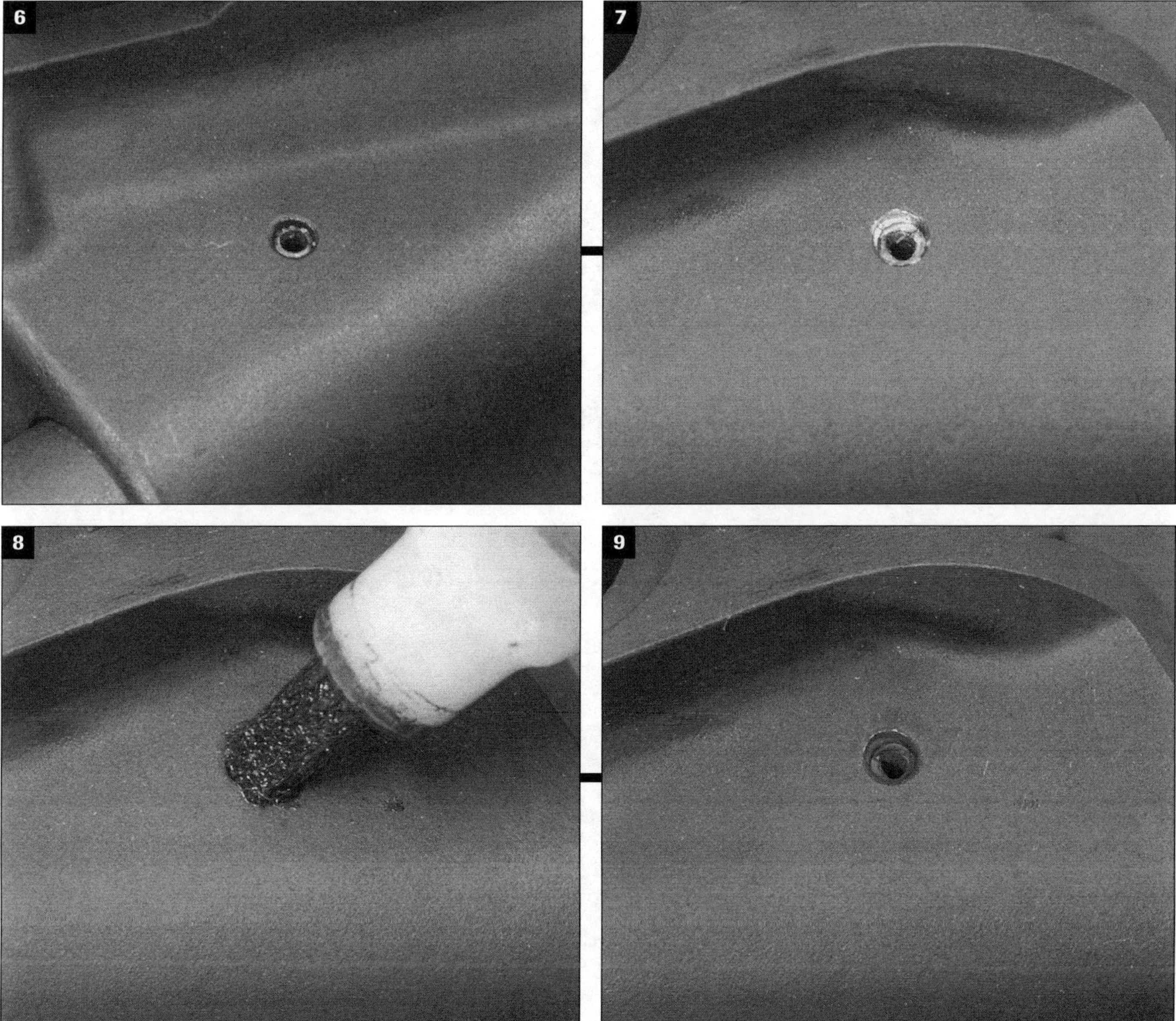

3./4. Get a roll pin starter punch and drive the roll pin in far enough to take over for the punch. Notice this is done from the underside of the forward assist housing.

5. Take it to flush with a roll pin punch. **6.** Check the top side to ensure that the pin is just at or below the receiver surface. You don't want it even a little bit sticking up from the surface.

7./8./9. Even with the tape, it's hard not to get a little dinging around the pin hole. That's why you drive the pin from the underside. Presto toucho!

Check by pushing it. Free movement, move on.

> **Style options.** Should you get a round- or teardrop-shaped forward assist? Doesn't matter. I think the teardrop variety is actually easier to work, if you ever need to work it. I know of no "correct" assembly to meet Service Rifle regulations.

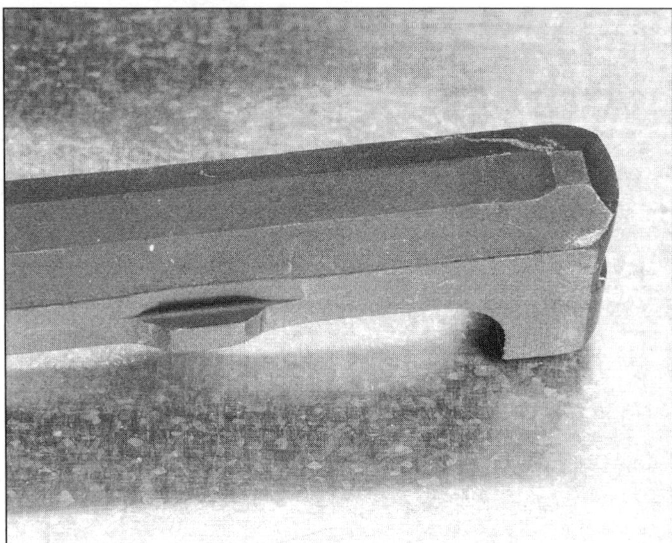

CHARGING HANDLE

I included this part here only because it made some sense where we are discussing upper receiver parts, and it is, I say, a part of the upper receiver. It's fairly self-explanatory, and I've yet to see one that wasn't already assembled so there's no work really save for just slipping it in place (eventually) once the upper receiver is ready to be put to use. There is, of course, zero reason to have it in the upper unless the bolt carrier is housed. The little ears on the charging handle fit up into their slot located at the rear sight post hole in the upper. Once they click up into the channel inside the upper, the charging handle is riding in its slot. Push it home until it latches.

COOL TRICK (JUST MESSY)

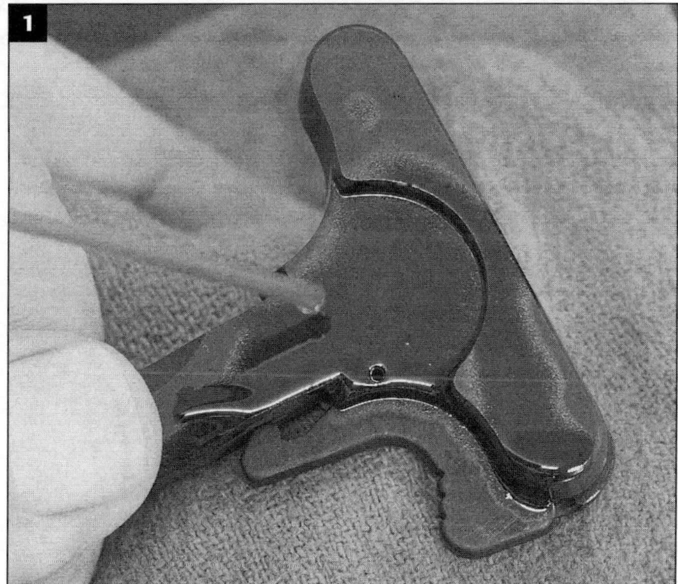

Only race-guns may need this, but it's an easy way to protect the self from any expelled gases following a primer mishap (like a piercing). Gas will get right past the gaps in the top side of the charging handle, and in your eye, so we'll fill that gap. Not pretty work.

Get **Permatex RTV** at any auto parts store. It's tough stuff that is intended for use as a gasket-maker. It sticks pretty well to any unprotected surface, even a very thin coat, so protect surfaces where you don't want it to stick using a release agent. Brownell's carries that, or you can use Johnson's Paste Wax just about as well. The idea is to get the RTV to stick only to the charging handle "horseshoe" so you can't be too thorough with release agent. If you ever want the RTV gone, Permatex Gasket Remover will get it gone.

1. Degrease the horseshoe area with contact cleaner.

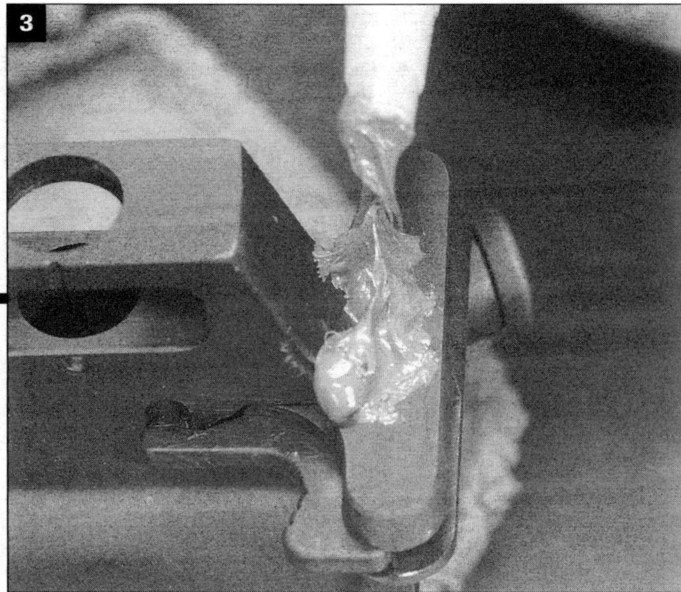

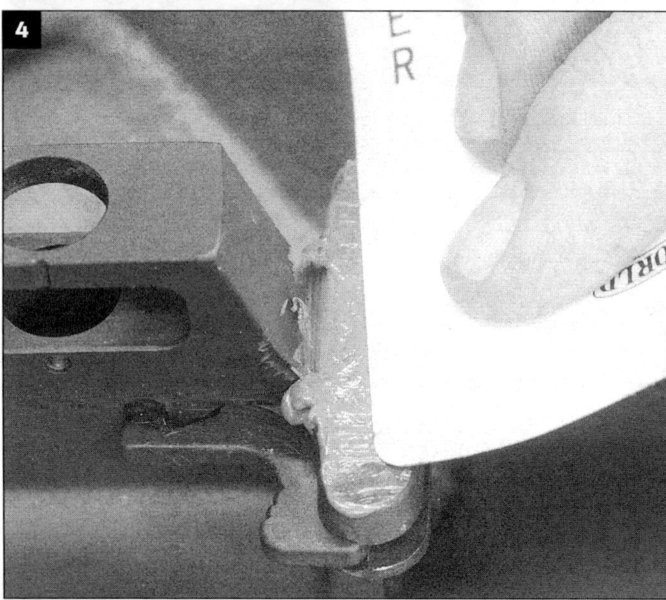

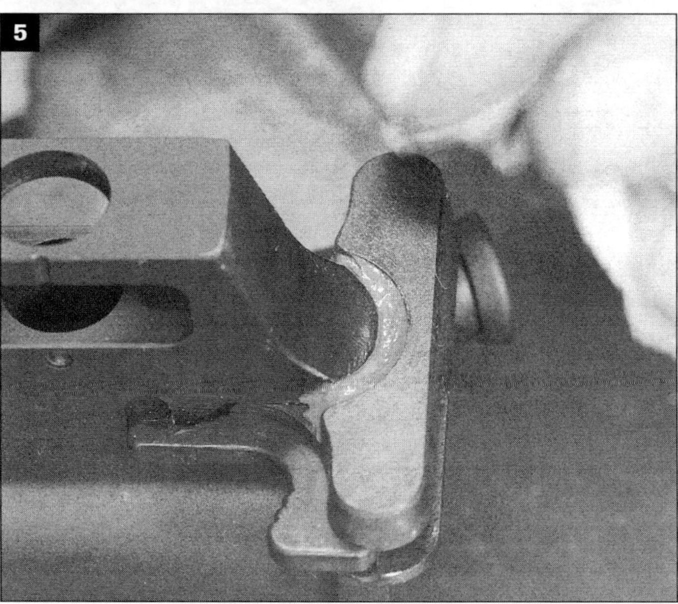

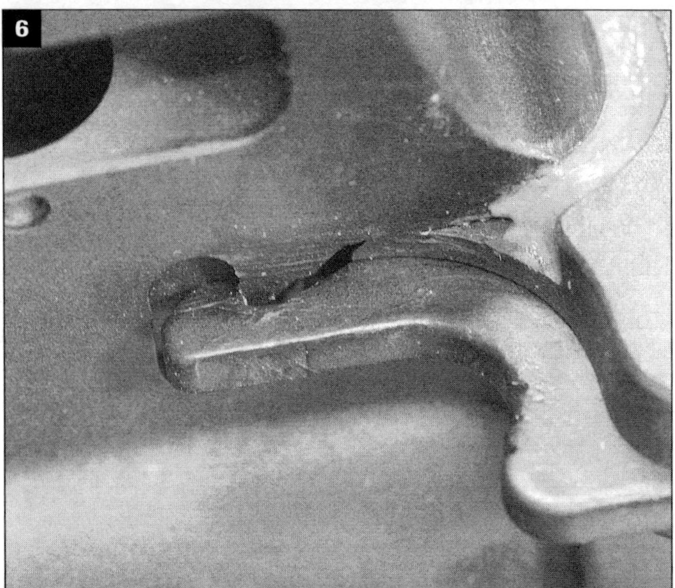

2. Apply release agent *everywhere* nearby... Be especially confident you have the charging handle track covered, both on the inside of the upper and on the handle itself. **3./4./5.** Gorp RTV in there, more than you need. Push it down so it gets fully to the bottom of this area, and then strike it off. Do that a couple of times, add if needed. A playing card works well. **Right away,** use paper towel to remove any excess. **6. Important**: Pull the handle back fully against its latch. This will increase tension and ensure a seal later. Run a little bead of RTV, about as much as went into the handle, to check its curing progress. That way there's no "oopsie" if you check too early...

6.0 BARREL ONE

CHOICES AND PREPARATION

Here is a DPMS carbine barrel I got from Brownell's. It's a good start and good part for building up a short gun. It came out of its wrapper with everything but the Delta assembly. To install two-piece handguards and keep it OEM, all we need to do is affix a Delta assembly (that retains the back of the handguards). Otherwise, we have to get all this off the barrel...

[If you learn how to do all this next, it actually gets easier when we get into anything custom. This segment, and the next, will outline installing a standard-form barrel, and related parts, onto an upper receiver.]

ADDING TO OR TAKING FROM?

First, it wasn't truly possible for me to anticipate what each of you wants out of this book, but the healthiest assumption I could make, given my readership in the past, is that most will be doing a custom-level installation rather than a conventional or standard-form assembly. That means, essentially, that you'll be installing a free-floating-style forend tube. If that's the case, it's taking parts off a standard barrel, rather than putting them on, that will be the first focus of the project. Most ready-to-go barrels are sold with the barrel nut installed and also the front sight housing pinned in place, along with the handguard retaining cap. This is how a barrel from Brownell's, DPMS, Rock River, or most other outlets, will unbox. To that you'll add a Delta assembly, gas tube, two-piece molded handguards and, often, a threaded muzzle attachment (assuming the threads are there).

So, if you're looking to build up a standard-form rifle or carbine, by all means get as complete a barrel assembly as you can. If you're not, then all you really need is the barrel.

SEGMENT CONTENT

86 **Barrel Basics**

88 **Barrel & Chamber Choices**

91 **Chamber Polish**

93 **Parts & Tools**

94 **Disassembly** (standard configuration)

95 **Reassembly** (standard configuration)

96 **Delta Assembly**

BARREL BASICS

This topic is killed to death, buried, dug up, killed again, and buried again in *The Competitive AR15: the ultimate technical guide.* There's a lot of discussion and opinion (a ton of opinion) in that book.

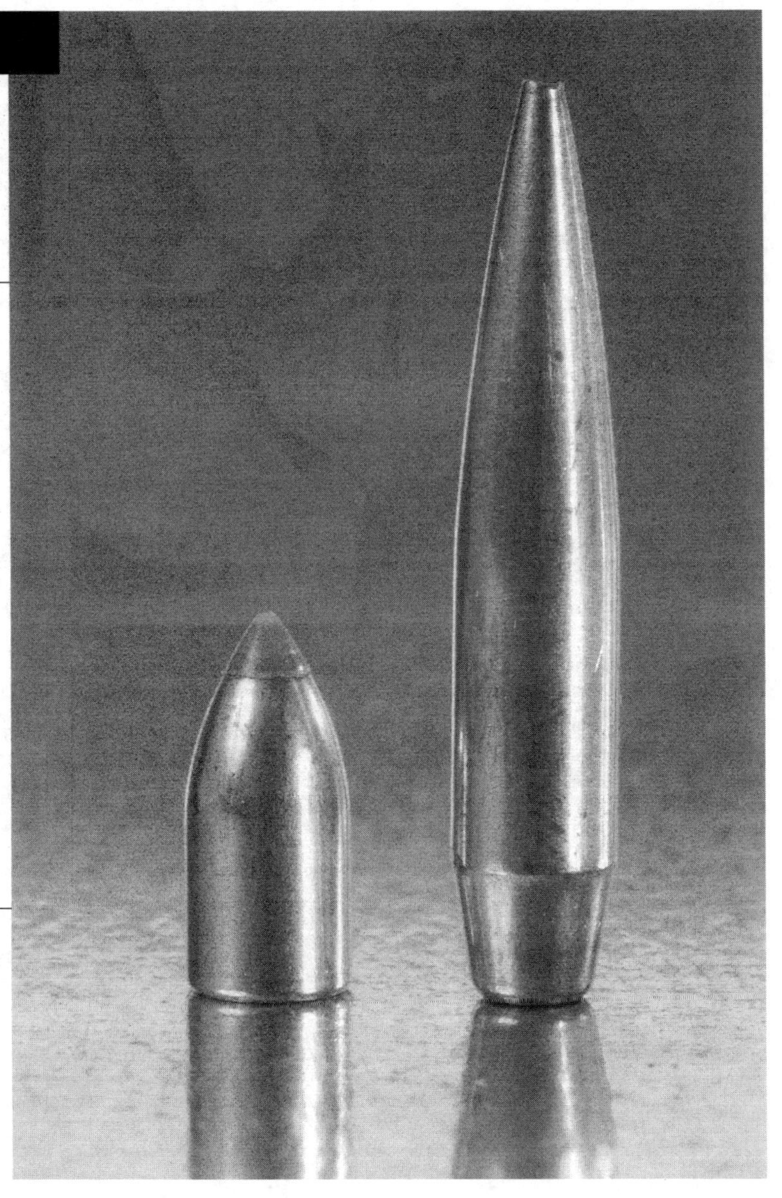

Twist rate

There is a huge spread of available bullets in .224 caliber. Perhaps more than in any other, comparing proportions as percentages. Factory loadings are less spread out, but the handloader has options that run from 35 grains up to 90s. Sure, no one twist rate is going to perform perfectly with anything close to that sort of spread, but 1-8 handles the most of them, including 80-grain bullets most use for across-the-course work. Most readily-available barrels are still 1-9 twist, and that's too slow for 80s, and it's too slow for 75s too. A 1-7 works well also and provides a little extra rotation to stabilize the VLD-form 80s, which are a little longer than, say, a Sierra or Hornady. If you want to experiment with 90-grain bullets, then a 1-6.5 twist is necessary for reliable stability. I wouldn't recommend choosing that one, though, unless you know 90s are in your future.

The essential advice given there is to start with a blank from one of a handful of sources known to produce reliably good barrels, get the right twist rate, and then have the finish work and chambering done by a competent gunsmith. That is the ticket to happiness. That is also expensive. That is also harder to do when shopping for a barrel that's ready to meld with your upper receiver.

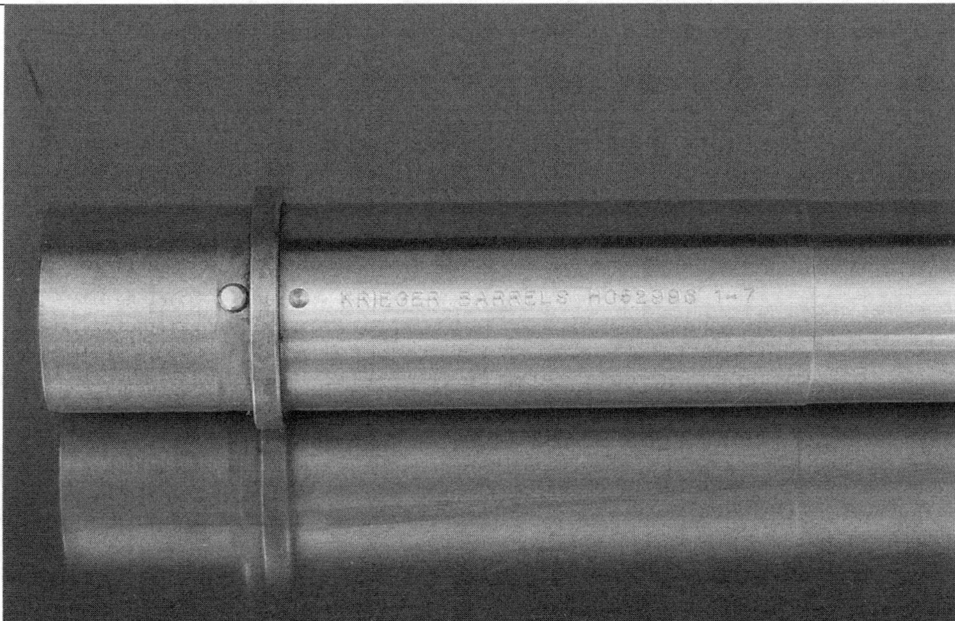

Getting what you really want.

Pac-Nor and Krieger both offer "finishing" services whereby you can supply the specs you want and they'll fix it right up for you. Or not. It can take a while to get delivery. Others may work with you on this. Some custom builders offer this service as well and keep "drop-in" barrels in stock. If anyone is working with you from a barrel blank, send in a stripped bolt and request that it be headspaced using that part. Then it will be truly ready to go using that bolt in your build. The barrel shown here is a 26-inch from Krieger that arrived to my spec ready for installation.

Since we're now at the mercy, to a point at least (and more later), of those who supply ready-to-go barrels, there's still as much to say but not all of it is going to matter. This topic, and the part chosen, is discussed in each project segment.

NATO. *Unless you can or want to specify something custom, try to get this chamber in a ready-to-go. It's just easier. NATO will digest any factory ammo and I think provides more flexibility in handloads. Usually, "5.56" stamping on a barrel means "NATO," as is the case with the DPMS carbine barrel shown on the left-hand page. That is opposed, by the way, to a ".223 Remington" stamping.*

The news is that you don't have to settle for less barrel. What you may have to settle for is not getting precisely what you want, if you want something different than those suggested in this segment.

Backing up a bit, there's no more important part if anyone wants to realize the full potential of a rifle build. I won't say that a goal gun six-and-a-half pound carbine is going to shoot like a house afire at distance, but, otherwise, anyone's competition, varmint, or "general purpose" AR15 variant should bring an uncontrollable smile after grouping it on target.

Here are some of the basics, and essential advice, in condensed form.

Just because it says "match" doesn't mean it's good. There are no objective standards other than those adhered to by the maker.

Twist rate should be, I think, at least 1-8. Read that as "one turn in eight inches" and know it means that a bullet will make a full rotation having traveled that distance through the bore. A tighter or faster twist (shorter distance for one rotation) means heavier, longer bullets will be stable. Many available are 1-9 and that's okay for anything up to and including a 69-grain Sierra MatchKing. That twist won't shoot a Sierra 77 MatchKing, or anyone else's bullet in that weight and length range. Although armchair experts have claimed for years that the 1-7 twist rate specified by our military many, many years ago was too fast, it's really not. It works fine with anything from 55 grains up to and including the Sierra 90. Many carbine shooters like the harder hit of the heavier bullets too, so same advice holds there as well. If all anyone is ever going to shoot is factory-boxed ammo or bullets under 70 grains weight, then 1-9 is fine. As a matter of fact, if it's a varmint hunter's rifle and all those he's ever going to run through it max weight at 55 grains, a 1-12 might be just perfect.

Overall, though, faster is wiser. Faster means you can shoot more nearly everything out there. What's the worst thing that can happen if the twist rate is too tight for a bullet? In an extreme combination, such as a 40-grain in a 1-6.5 twist, the bullet might sustain jacket

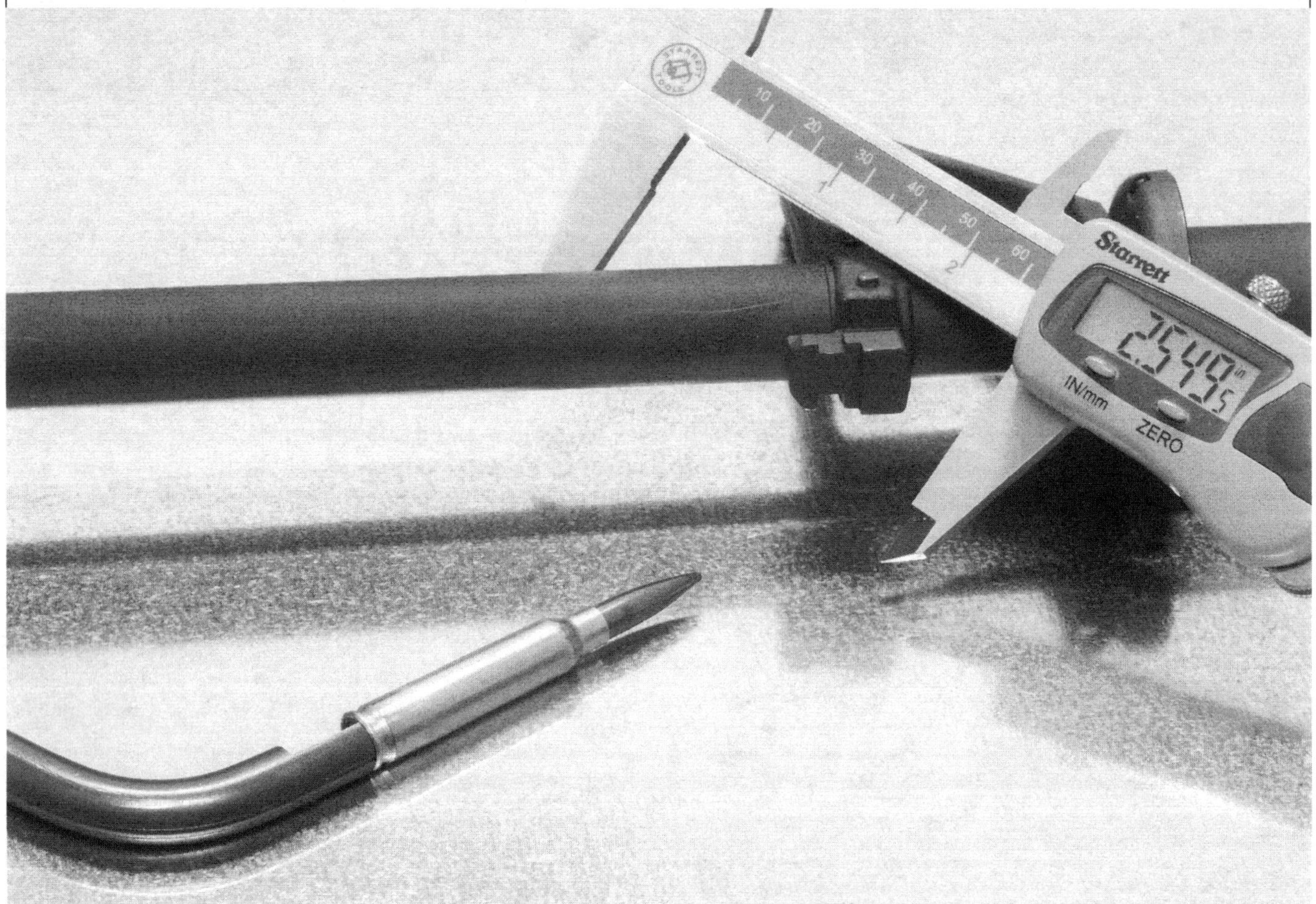

Figuring out what you have. *The only way to really know what you have is to check it. This tool does it easy. It will show you the overall length of a round that has the bullet touching the lands or rifling. Invaluable. It's a Hornady LNL gage. The reading displayed on the caliper is with a Sierra 80-grain MatchKing, and it indicates a NATO chamber throat. I've seen too many variances in SAAMI chambers to tell you what it should read with this bullet, or with any other. A "Wylde" competition chamber normally reads 2.475 inches with this combination.*

damage in the barrel and come apart or "blow up" in flight. Not a dangerous condition from the shooter-side, but it won't hit nowheres close to target center. The other way around, say a 90 in a 1-9 twist, the bullet will wobble like a circus bear on a bike and also not likely to get to where you want it to hit. The tell-tale sign of a rate that's too slow is oblong holes on the target, sometimes leaving the silhouette of the bullet because it hit sideways. Funny, though, I've had them hit, more than once from the same shot string, in the X-ring that way. Just a comment.

I do not recommend a SAAMI minimum .223 Remington chamber. Again, there's more about this in the build segments, but that is, essentially, the "tightest" chamber out there for .223 Remington. Tight, mostly, refers to the throat or free-bore, which is the space between the case mouth and first point into the bore where the lands start. More or less. Such a chamber will, yes, perform well with commercial or commercial-equivalent ".223 Remington" ammunition loaded to magazine-box overall length of 2.250 inches. This chamber is commonly and I say incorrectly referred to as a "match" chamber by many barrel suppliers and is only suited at that suggestion for

lighter-weight loadings for someone who, like the varmint hunter example, will fire only commercial-type loads. Such a chamber may be dangerous if NATO-spec ammunition (commonly encountered as military surplus) is fired in it. I strongly suggest getting a barrel with a "NATO" or "5.56x45mm" chamber. It will safely digest any and all factory-loaded ammo and allows all the flexibility in custom-done ammunition for the handloader. The SAAMI-minimum chamber is also tighter in other dimensions too, and that's another reason I say go with the NATO. It will handle anything.

Now, there are "match" chambers available that are cut with reamers actually engineered for competition ammunition use. Again, specifics are in the race-gun builds. Of these, the "Wylde" is my favorite.

[There is more about this topic — chamber specs — on page 133 in the NRA Service Rifle project segment.]

Chrome-lined barrels don't shoot all that well, usually. They last longer, or can, but more rounds through a bad barrel isn't the sort of

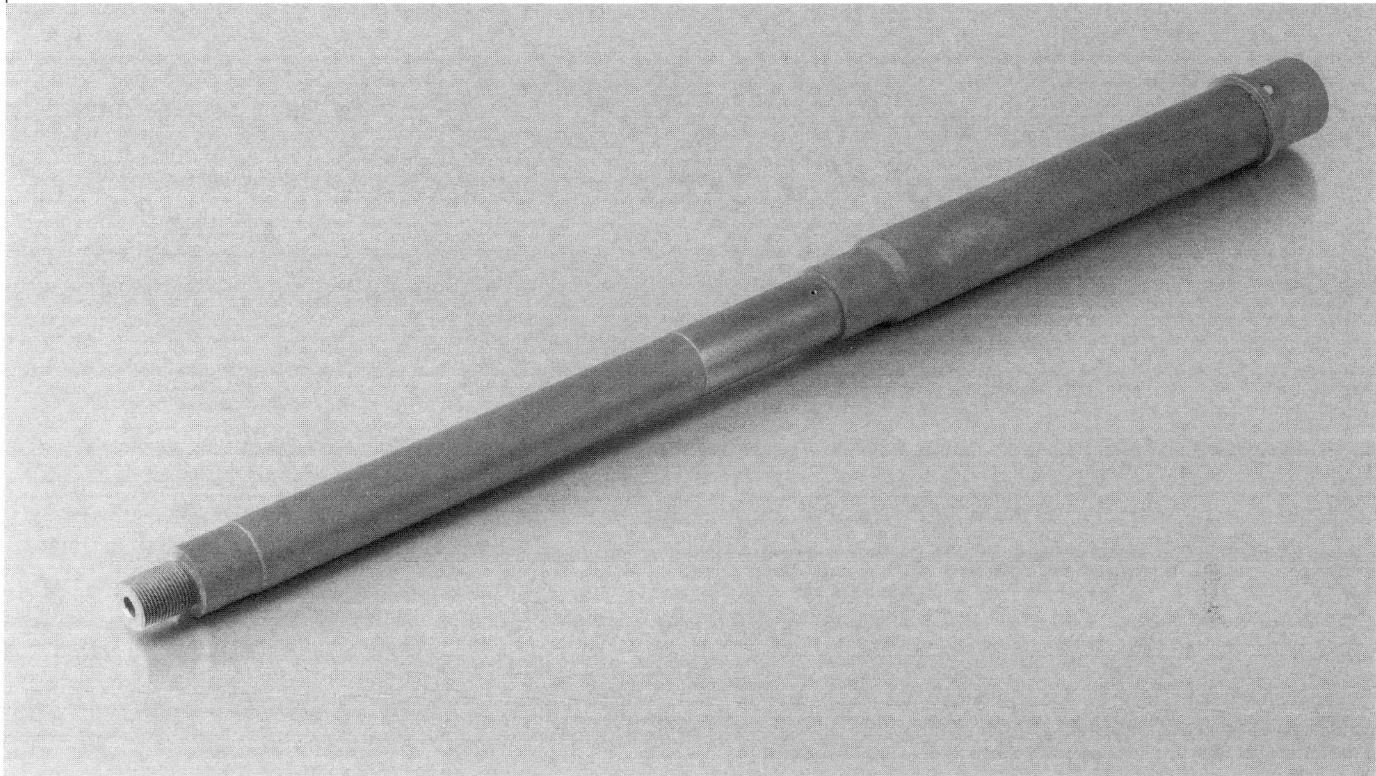

Getting what you really want, again.

Here's a custom carbine barrel I got from Krieger with a 1-7.7 twist. It wasn't cheap, but then none of the components for that project were.

Barrels are two pieces. *An AR15 barrel has this part, the barrel extension, threaded on its back. We won't get into that in this book, but make sure you tell a maker to fit an extension onto your barrel!*

*And a little **polish on the feed ramps** won't hurt a thing before installation. Just break the edges on the bridge and rub out the roughness in the two dished out areas that contact the bullet.*

value I shop for. They were engineered for use on full-autos. Plus, .223 Remington isn't hard on barrels, or at least no more so than .308 Winchester. They last a long time anyhow. Otherwise, chromemoly and stainless steel shoot equally well, if they are manufactured equally well. Stainless steel is usually recommended, and by me, because it does shoot its best for a longer time (about 15-percent longer at top accuracy) and also because of its corrosion resistance. Chromemoly often shoots longer at a "good" level than stainless. Stainless tends to just flat quit when it's hit its limit. Chromemoly is usually less money and will have a finish better suited to "dark" work, although there are stainless barrels with black finishes too.

Back to the point here, and that is finding a truly good barrel to incorporate into your rifle project. Can and should be done, and, again, the only thing is that you may not find precisely what you might specify to a custom builder. Your rifle, however, will shoot just as well on target. As said, there are custom barrel makers that will turn, chamber, and otherwise incorporate your specs into a barrel you can unbox ready for installation, for a price of course.

Enough on that. Let's get to the shop and get to work on this project. No matter which barrel you choose, or for which purpose, the basic installation is the same. All the details will be shown in other segments that involve match barrels and free-float tubes.

Before you start...

Barrel installation takes tools and supplies. Some things are options but the list of necessities can't really be worked around. You'll have to have some means to securely retain the upper receiver, against damage, and some means to apply adequate force to the barrel nut. Those two points are usually satisfied by a torque wrench and an armorer's barrel nut wrench, and something like this clamshell-style receiver clamp from Brownell's. And you'll need some way to know when the gas tube outlet is aligned. An old gas tube, 0.177-diameter drill rod, or a purchased specialty tool will all work. Also shown as part of the routine package I use is contact cleaner, glue, anti-seize, and a breaker-bar-style wrench handle. Oh, and you'll need access to a sturdy vise, 4-inch opening minimum.

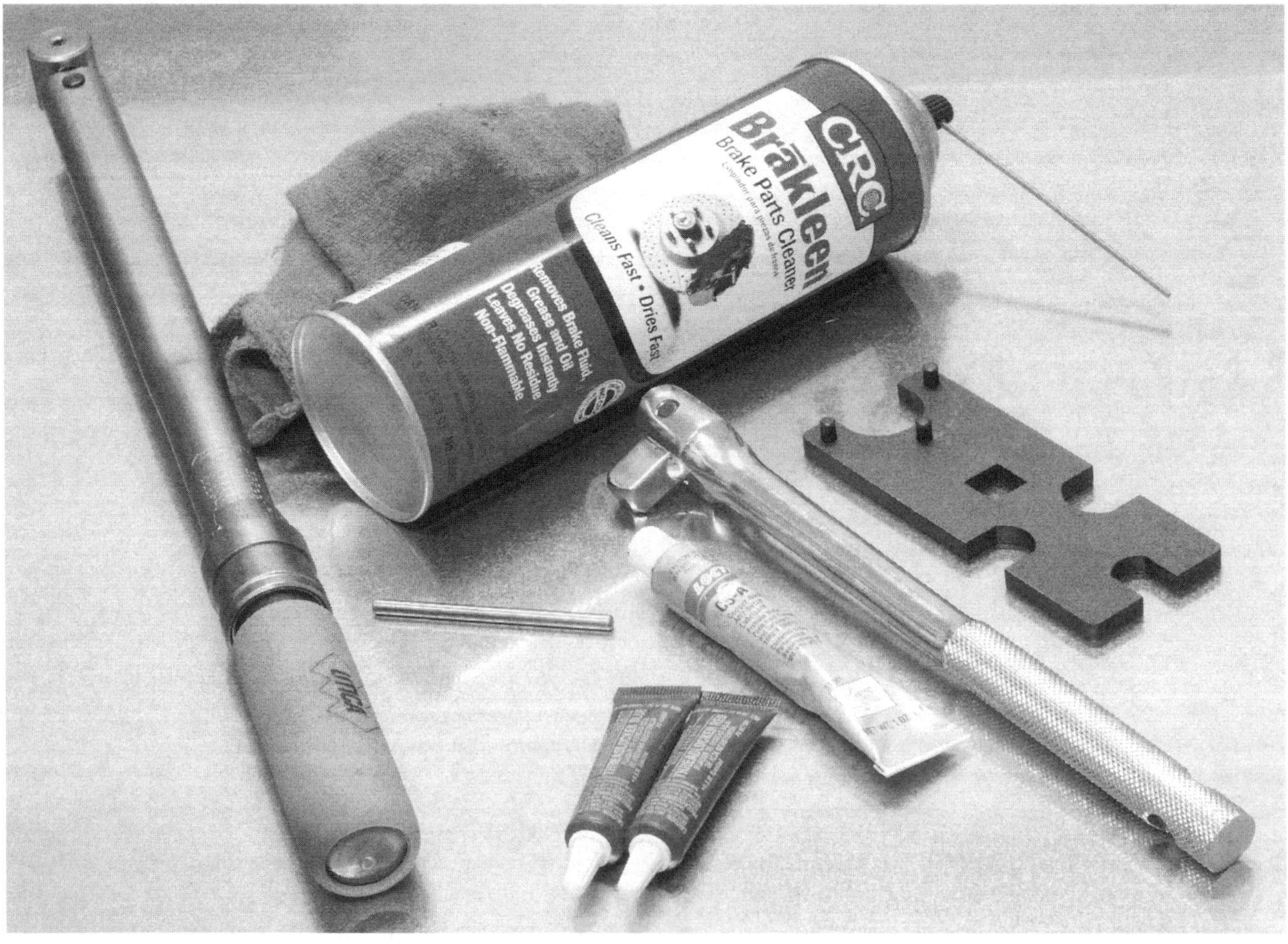

Chamber polish

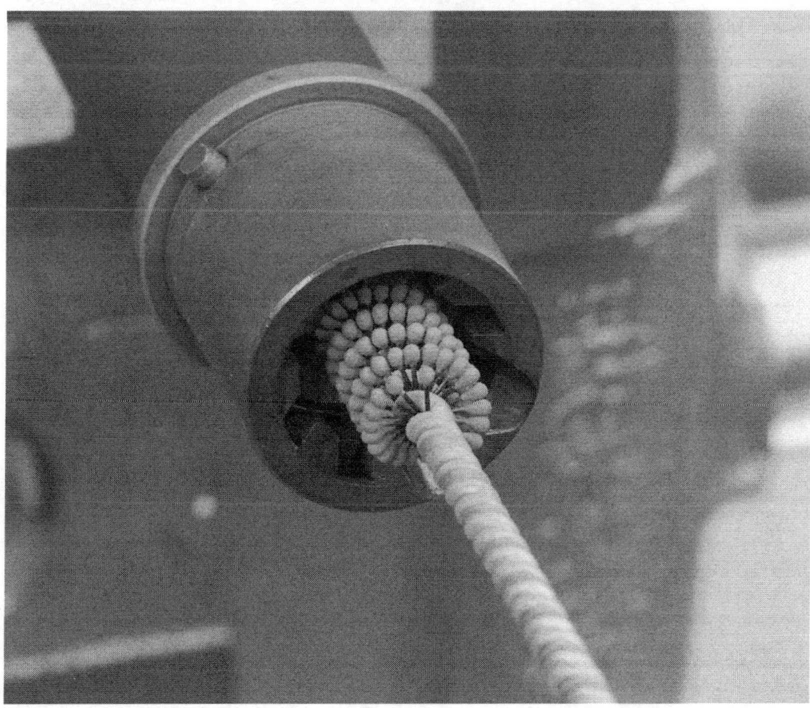

The best time to do this is when the barrel is out to itself. I think every AR15 should have its chamber polished. It's one of the surest means to guard against malfunctions. There are different ways to do it, and I've liked this one. It's a Flex-Hone for revolver cylinders. This is sized for a .357 Magnum, and it's within 0.001 inches of being just right for a .223 Remington chamber. Use the oil, chuck the rod in a drill, and run it no more than 1100 RPM. It doesn't take long. Don't push the tool up into the case shoulder area. The op can also be done by making up a split dowel and using emery cloth wraps or by lapping compound on a bore mop (also .357-sized). Use 320-grit. The idea isn't even to make it a mirror finish, although that's possible, but to smooth out the tool marks and get the roughness gone. I don't know that it matters a whit to feeding, but extraction force on the cartridge case is considerably reduced. It's a function solution for a lot of carbines. Makes it easier to clean the chamber too, and do clean it!

A few tips...

As said, I really do not suggest clamping the barrel to install the barrel. Use a receiver-style clamp set into a sturdy vise. The reason is that tightening the barrel nut against a secured barrel can and will increase the pressure on the alignment pin and also, of course, its corresponding slot in the receiver. The result can be a looser fit.

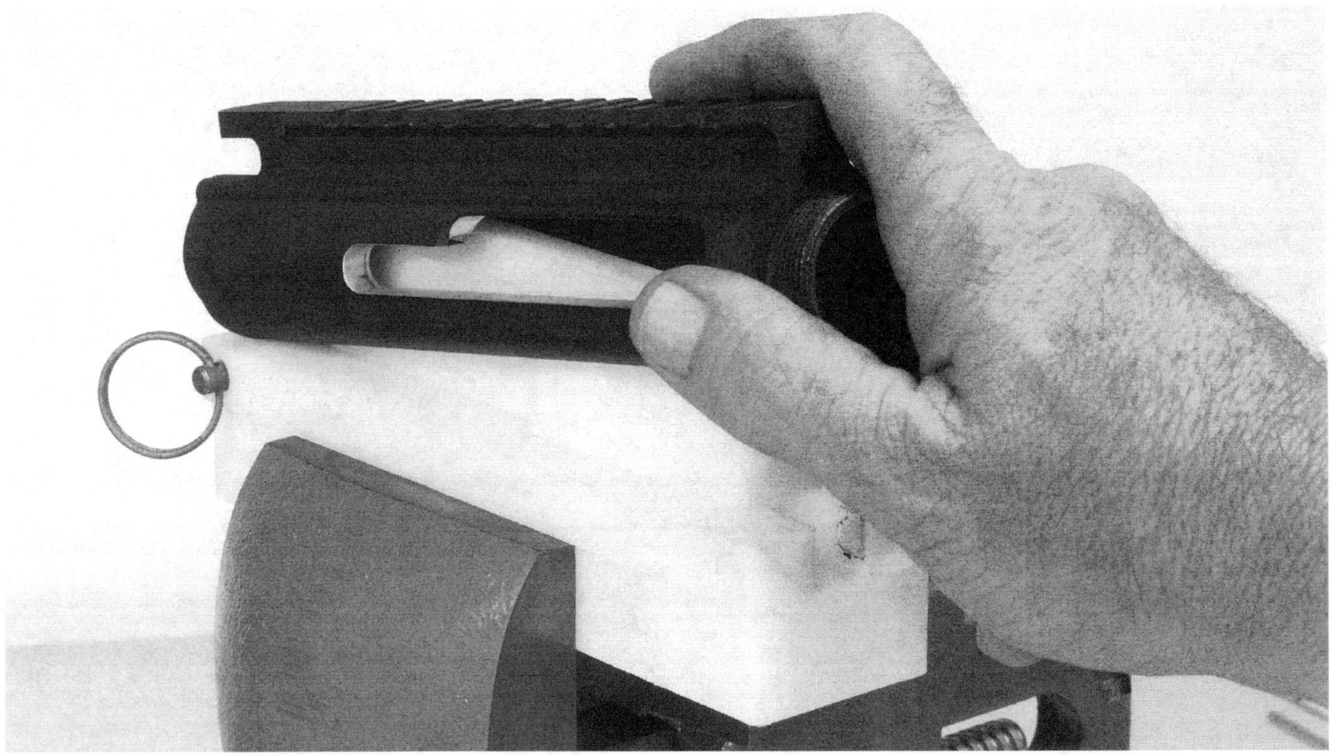

If you're only going to do one, get the clamp you'll need for the receiver type you will install it on. Conversely, if you're only going to get one clamp, get the "white" one. It fits into the upper receiver — any upper receiver — and provides universal utility. As mentioned in the tools segment, it doesn't look as sturdy as the clamshell-style clamps, but rest assured it is, and I'm not entirely sure it's not better in that regard. The fit into the inside of the upper receiver is flush and there is no possibility of clamp-induced damage. All the clam-shell clamps I've seen only work on a standard-form upper receiver, which means it won't work on any of the slab-sided or differently contoured and dimensioned uppers out there.

Wrenches, same sort of thing. If you're going to do nothing but mil-spec barrel installations, which here means that you'll want to install a barrel that already has the front sight housing attached, then you'll have to have a barrel nut wrench that's open so it will fit over the barrel. A "pin" style-wrench is the one to get if you only get one. I prefer "closed" barrel wrenches (as shown on right) because I find they are more secure since they fit into all the scallops on the barrel nut, but those can only be used on a stripped barrel. Not all free-float tube hardware allows their use either.

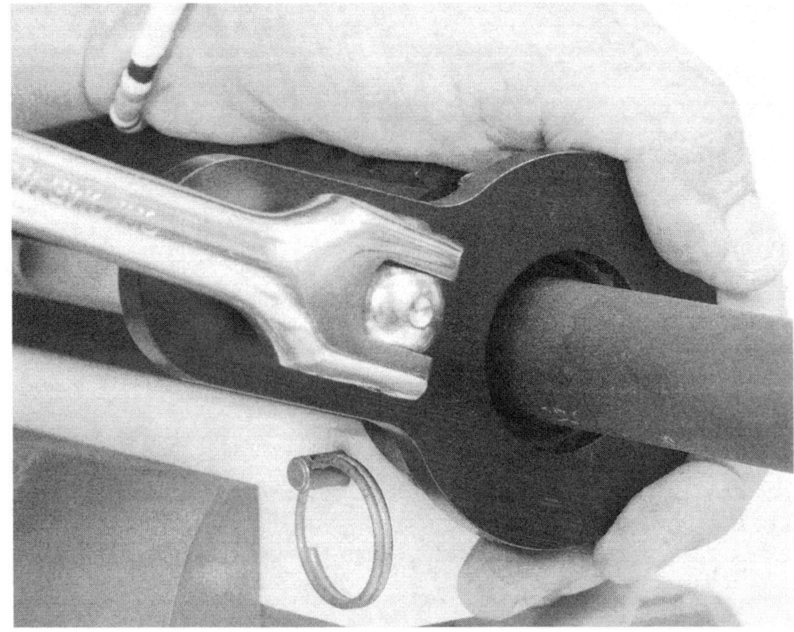

Secure and support the barrel wrench flush into its engagement area with the barrel nut. That means keep the free hand pushing the wrench to keep it from slipping.

PARTS & TOOLS

Most of us will be taking many of these parts off rather than putting them on. We'll get to that...

PREPARATION

Vise, very securely mounted

Upper receiver clamp

Tap hammer

Big hammer (for taper pins)

Roll pin punch and roll pin starter punch (#1 for gas tube roll pin)

Concave-tip punch (for taper pins)

Torque wrench, 1/2-inch drive

Breaker bar, 1/2-inch drive

Barrel nut wrench attachment for the above

Snap-ring pliers (for Delta assembly)

Gas tube alignment tool

Options (good ideas)

   Permatex "red" adhesive

   Contact cleaner

   Anti-seize compound (I like Loctite C5A)

   Small screwdriver (helps move Delta pieces into alignment)

   Gas tube wrench

   Front sight adjustment tool

   2x4 wood block (for taper pins)

PRECAUTIONS

Don't scrimp on alignment, and don't get frustrated!

Have to say it: Make sure the ejection port cover is installed!

PARTS

Upper receiver (with ejection port cover installed)

Barrel with barrel extension and indexing pin

Barrel nut

Front sight base and front sight base taper pins

Front sight (post)

Front sight detent

Front sight detent spring

Front sling swivel

Front sling swivel rivet

Flash suppressor (if applicable)

Flash suppressor crush washer or peel washer (A2)

Flash suppressor lock washer (A1)

Handguard cap (triangular for 20-inch barrels; round for 16-inch barrels)

Handguard slip ring (Delta ring)

Handguard weld spring

Handguard snap ring

Gas tube

Gas tube roll pin

Handguards (pair)

Again, since the focus of this book is about building your own custom AR15, I want to start off by showing how to install an assembled, standard-form barrel. Putting the pieces onto a standard-configuration barrel isn't hard, neither is taking them off. We'll get to both of those jobs next.

Front sight assembly is shown on page 171.

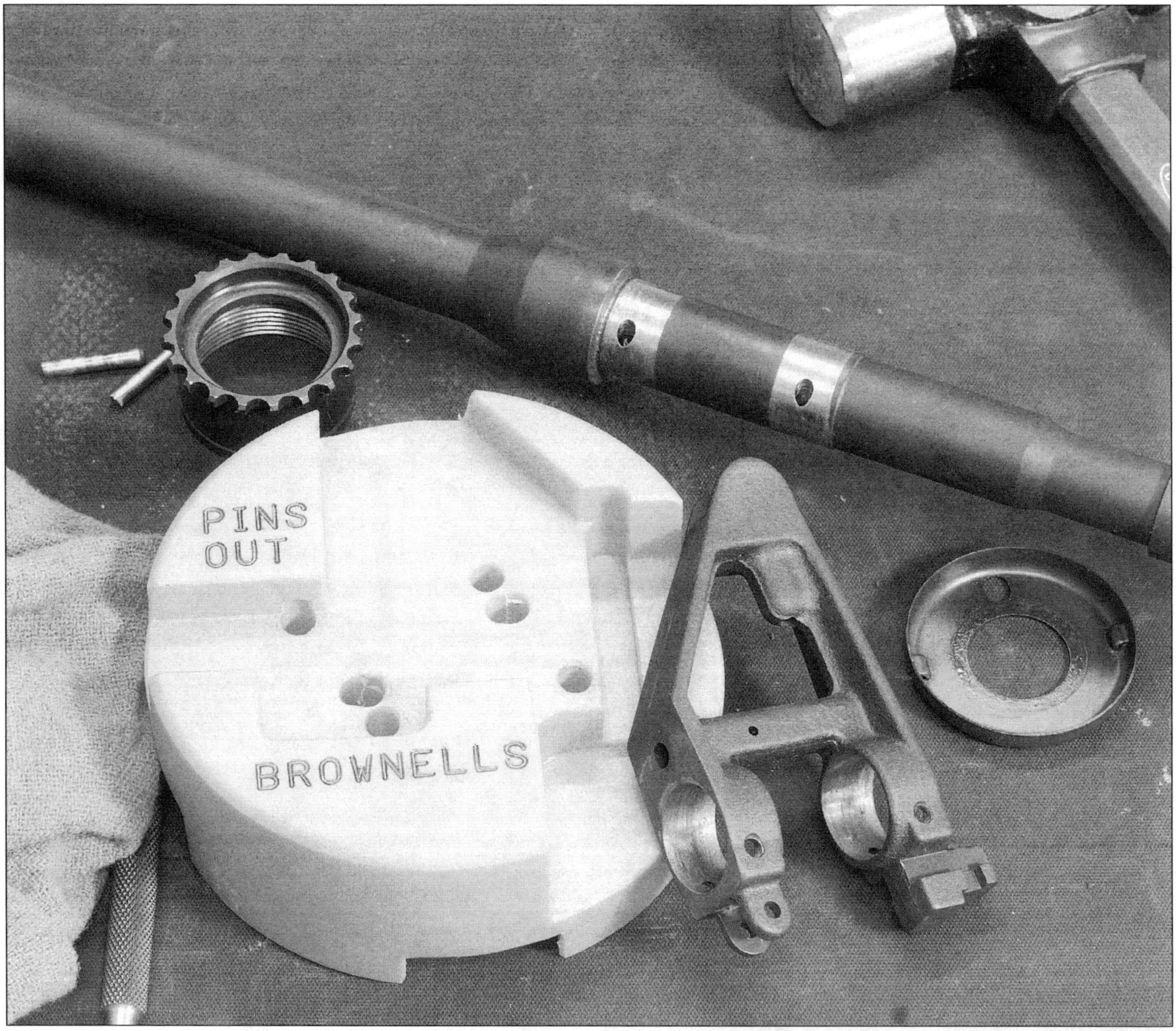

TAKE IT APART...

This is more than likely what most of you will be doing to get a project started. Whether it's a competition rifle, super-duty carbine, or most any other configuration we'll be interested in, the rifle will not reassemble with a barrel nut, handguard cap, or taper pins.

The fixture shown from Brownell's makes this job foolproof, well, almost. Still have to pay attention to how the pins are tapered for reinstallation. It normally takes a good deal of force to break the pins loose. Factory finish is most often applied after the front sight housing has been installed so there can be a little adhesive-effect to deal with. After the pins are freed, it may take a rap or two (plastic hammer) on the front sight housing to get it to move freely. The handguard cap and barrel nut come right off.

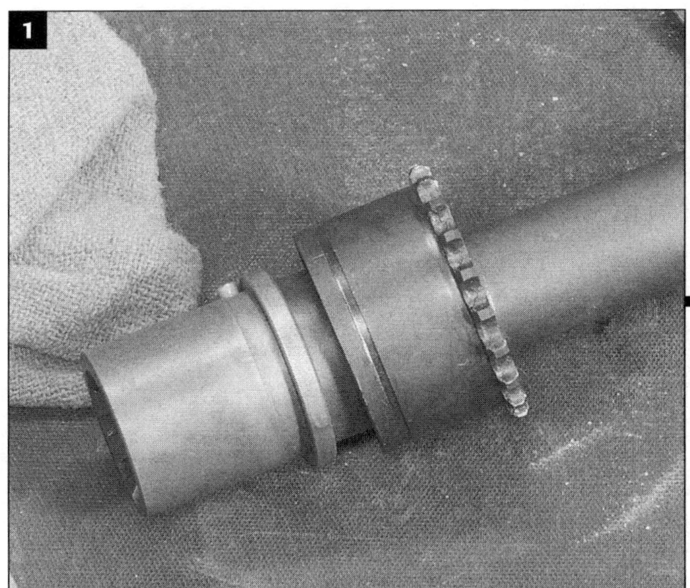

PUT IT ALL BACK...

It really is about that easy. Starting from the ground up, **1.** install the barrel nut first (scallops toward the muzzle), **2.** then the handguard cap (triangular for A2 rifle and round for a CAR, and make sure the gas hole cutout is aligned on the round cap), then the front sight housing itself. A couple of taps will seat the front sight housing securely, **3.** and then install the taper pins.

Taper pins don't seem to take as much force to seat as they do to unseat, and there will be roughly an equal amount of pin protruding on either side of the housing. Just make sure the taper is correct! Small end first from the right side of the sight housing. There are alternative to taper pins, certainly, for a conventional front sight housing. It's easy to use set screws and glue.

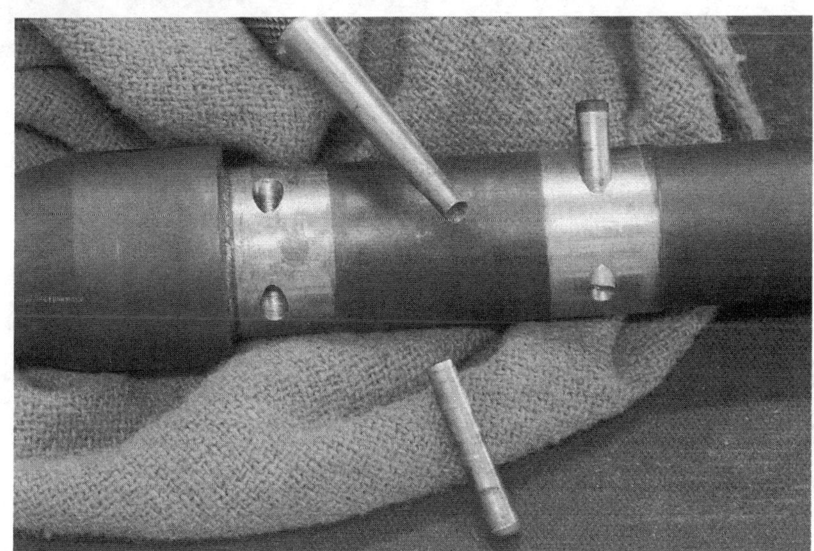

Use a punch with a concave face on taper pins. *This helps keep the punch from slipping and marring the finish. Most accessory barrels, especially those intended for competitive or precision use, won't even have the necessary holes drilled for taper pin installation. Those require some form of slip-on gas manifold.*

When you're pulling the front sight housing off the barrel, go slow. It's a tight fit where it seats, and is supposed to be. Grease can help protect barrel finish from incidental contact.

If you don't get the Brownell's fixture shown, and most won't, a piece of 2x4 wood with a hole drilled through it (to let the pin out) will suffice. Just remember: In from the right, out from the left.

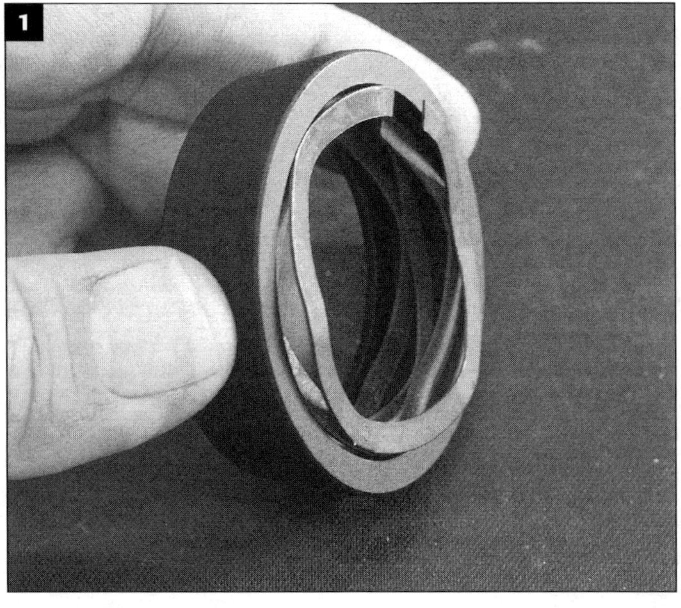

DELTA ASSEMBLY

The Delta assembly is a spring-loaded collar affixed to the barrel nut that serves to hold the handguard halves together. Any rifle that uses conventional-form handguards has to have one, including one built as an NRA Service Rifle with a free-float tube.

This is easy. It's just two pieces that install over the barrel nut and then a circlip or snap-ring that holds them in place by snapping into the groove on the barrel nut. Just get things oriented correctly, and it's done before you know it.

These three parts are rarely included in a barrel assembly, including a Service Rifle float tube. Make sure your order a set.

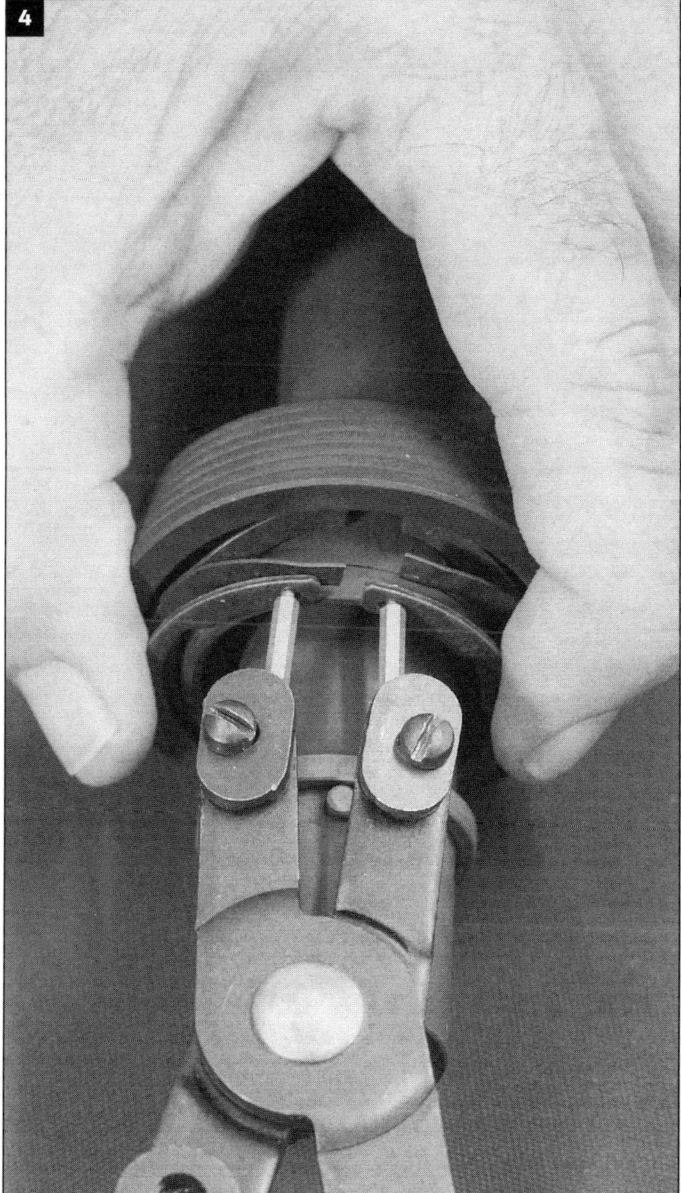

1./2. Place the spring into the cap. It doesn't matter which way the spring goes. Orient its opening with that in the cap.

3. Slip this over the barrel nut from the back of the nut.

4./5. Use circlip pliers (any auto parts store has them) to spread open the circlip ring and snap it into the groove on the barrel nut.

Make sure the ring is fully seated, all the way around.

Done. Well, all the gas tube openings have to line up but that's going to get fiddled with during barrel installation.

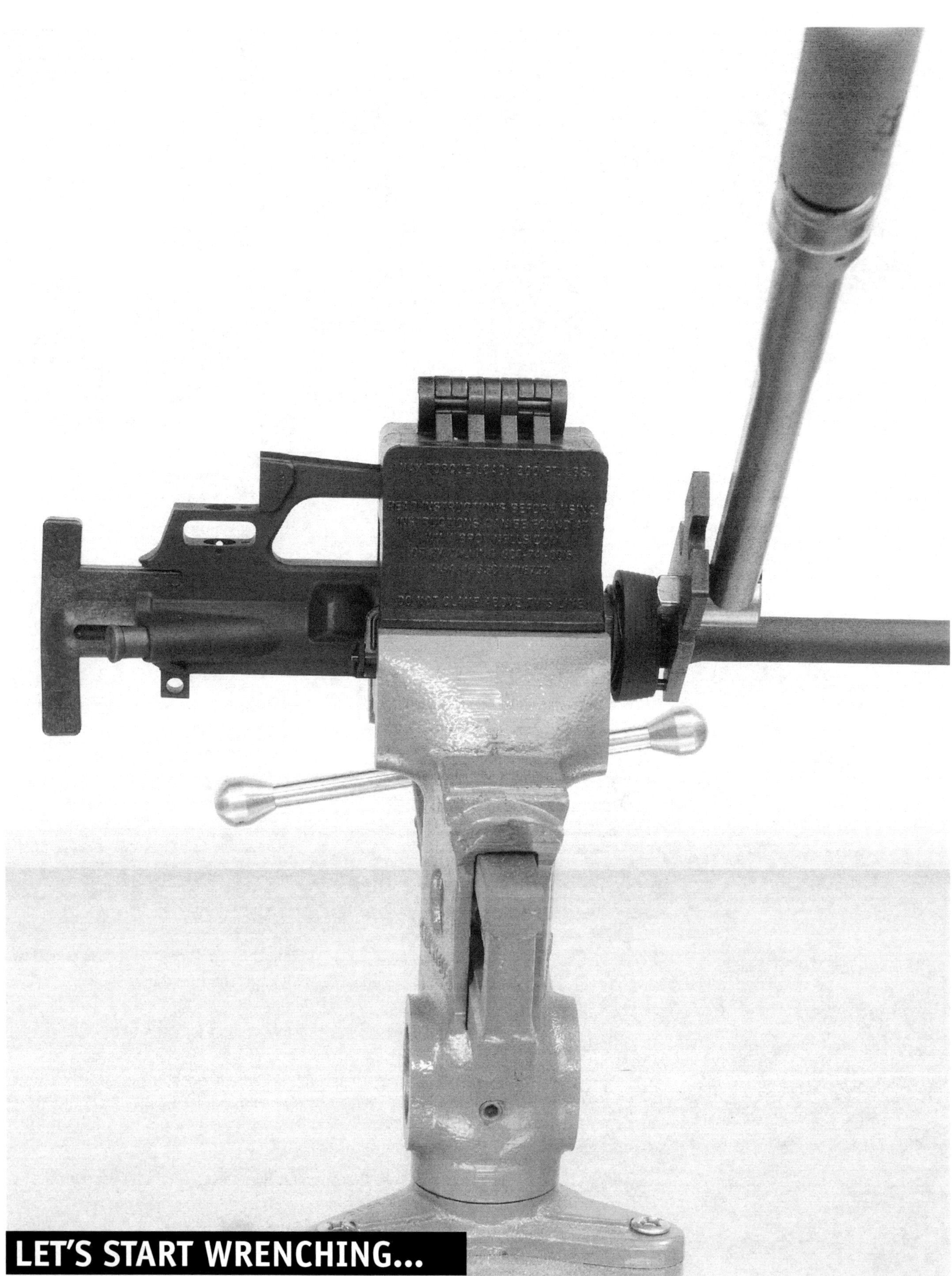

LET'S START WRENCHING...

7.0 BARREL TWO

BASIC INSTALLATION

You'll read about this a few times... We are getting ready to do some things that cannot easily be undone. Measure twice, cut once... Check this work six times, and then glue once...

OVERVIEW

An AR15 barrel is a slip-fit into the upper receiver. Specifically, the barrel extension fits into the opening in the upper and there's a pin that indexes it in place. A barrel nut then gets turned onto the threaded area that surrounds the opening in the upper receiver and this nut is tightened to secure the works. The nut bears against a flange or shoulder on the barrel extension. The barrel nut has 20 dished cutouts or holes that align with the gas tube hole in the upper receiver. Here's a trouble. One cutout has to align exactly. The smallest amount of contact between the tube and the nut can diminish accuracy.

Any and all ready-to-go barrels I've seen or heard of will unpack with the barrel extension in place. If it wasn't there then the barrel would be in no way ready to go. Installing an extension on a barrel blank is well beyond the scope of this book. Strictly machine shop.

Armorer's manuals call for a torque specification that makes it sound like it's all easy like. It's not always. Most call for 35 foot-pounds of torque. That is a *minimum* necessary for retention, and then one of the holes has to line up. That's the trick, and the trouble. Nothing is perfect, and among the least among these imperfect things are upper receiver threads and barrel nuts. It's rare to have a torque wrench click over at 35 foot-pounds and see perfect alignment. So rare that I've actually had it happen once.

SEGMENT CONTENT

99	**Overview**
102	**Assembly Process**
104	**Alignment Checks**
105	**Gas Tube**
108	**Handguard Assembly**
109	**Headspace Check**

The torque specification, then, is honestly a minimum because it's often, and I will venture to say most often, going to take more than that to get the job done right. The best thing that can happen, aside from perfection, is having the hole be just a little off such that just a little more tight gets it perfect. The worst thing is that it's sitting about halfway on the other side of aligned and will take dang near another whole hole's worth of rotation to align with the next available. If that's what it is, then that is also what has to get done. It will take some force. Double the "ideal" torque is common enough.

"Seating"

Before starting on the finish, it's a good idea to use a breaker bar to take the barrel nut to tight, then loosen, then tighten... Repeat a few times to help "square" the shoulders.

So don't put the barrel on at less than 35 foot pounds of torque. And don't be surprised if it takes 60 or better to get alignment. I've had some tell me they've approached 90 to get alignment copasetic.

I use high-strength glue to fix the barrel extension into its sleeve in the upper. Right. That means it's not going to come back off without heat. It also, however, usually improves the fit in this area, which, keep in mind, is a slip fit. Some uppers and extensions will combine to create a gap greater than others. The glue is not at all necessary to have a fully-functional AR15

that probably will shoot just fine, but it's a little step that someone looking for the best results really should take. Degrease both parts before applying the glue, and don't do it until after test fitting. You'll have plenty of time before the glue sets.

Apply a good anti-seize compound to the receiver threads. This is a product engineered to ease the stress on such an operation. It's not oil but a copper-based lubricant. It will in no way encourage a barrel to loosen and, quite the contrary, if its full function comes into play that barrel is not coming off very easily at all. It's a little assurance or insurance against galling, which can lead to cracking. And, yes, upper receivers can crack during this operation. Evermore reason to select a quality go-between to secure the upper into a vise.

After applying the anti-seize, use the breaker bar or torque wrench on the barrel nut to fasten down the barrel to either tight snugness or the recommended torque. If you're using a torque wrench, switch to the breaker bar to loosen the barrel. Ticky folks say you're not supposed to loosen anything with a torque wrench; they

are one-directional tools. Loosening can create inaccuracy in the tool. I don't know but it's easy advice to follow Ticky folks usually know. Do this two or three times to seat the barrel in the receiver extension with the idea of overriding and hopefully eliminating irregularities in this area. Torque it down, back it off. Repeat two or three more times.

To get the barrel nut aligned with the gas tube hole in the upper, and ultimately provide a clear slip fit for the gas tube itself, there's a tool required. All it is is a gas tube piece. Some make their own and those for sale as specialty tools work well. The ready-made tools fit the shank of gas tube in correct alignment with the carrier. Making the works work requires checking with the alignment fixture, moving the barrel nut, checking again, and so on.

If the barrel nut gets over-tightened out of alignment, bring it back where it needed to be without worry. As long as minimum torque was attained prior to, there's no harm in backing it off a tad, and there's also going to be no other option. Go carefully though.

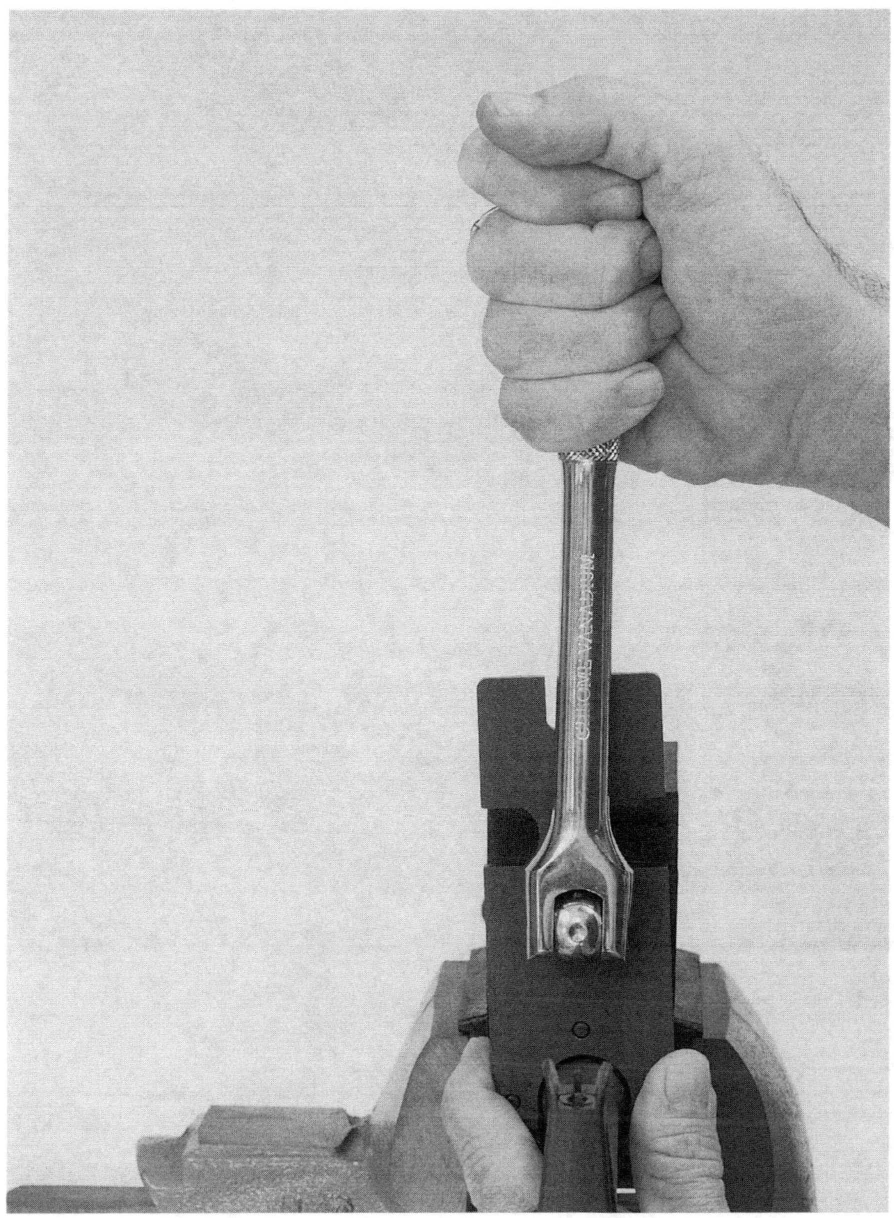

Ultimately, it's the gas tube that's installed in the rifle that needs attention to the quality of its fit. The alignment tool just gets the holes to share centers.

Point is, and this is just it, there has to be correct, stress-free alignment. Has to be. Work at it as long as it takes. If you don't the rifle is likely not to shoot too well. The gas tube should rattle when the bolt is in battery, that means fully forward. There should be no displacement of the tube in any direction when the carrier key is engaged and disengaged, back and forth. Feel, listen, look. I place my index finger on the gas tube and then run the carrier in and out. You can usually feel the tube shift as the carrier key engages and disengages the gas tube. If the gas tube is binding it will displace the carrier in the upper. Even just a little has a big influence. I'm convinced that the root reason for the accuracy of the AR15 is from its "floating" lock up. I further think that's why I've sure never seen "oversized" carriers shoot any better.

Bending a gas tube isn't easy. It's stainless steel tubing and will snap with very little warning. Rubbing an area of the tube back and forth over a wooden dowel while applying steady pressure helps result in kink-free displacement. Don't get greedy!

If the gas tube rattles, you are good to go.

Why alignment matters?

Look at it this way and here's why I think "it" happens. "It" is the accuracy level of an AR15. All the lock-up in an AR15 is in the bolt and the barrel. Unlike any other firearm I know of, when the bolt is locked into the barrel, all that's behind it floats. If the gas tube is pressing against the carrier key in a way that creates displacement of the carrier, it's not "free-floating," in effect, as it should be. Funny, well not ha-ha funny, but it's really the quality of the "loose" fit of action pieces that may well define this rifle's potentials. It is possible to tweak and bend the tube to observed perfection after installation, and sometimes is the only remaining "fix" for a tube that's centered but not "straight." That's probably

a bad gas tube. Try another one if alignment is correct and there's misdirected contact. Some say that bending is a crutch, others don't seem to mind it. Those who fret say that heat may relax, and therefore reorient, the bent tube to its original orientation, and that was what we didn't want. The target usually tells. A binding gas tube will, not can, ruin accuracy.

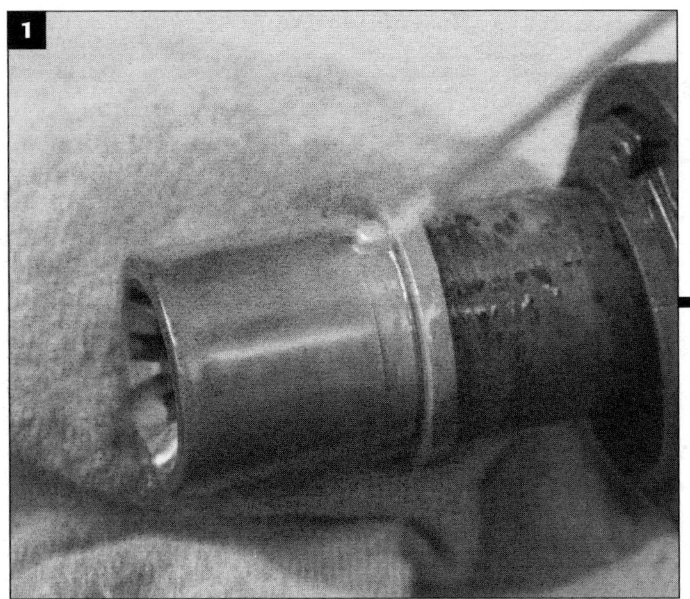

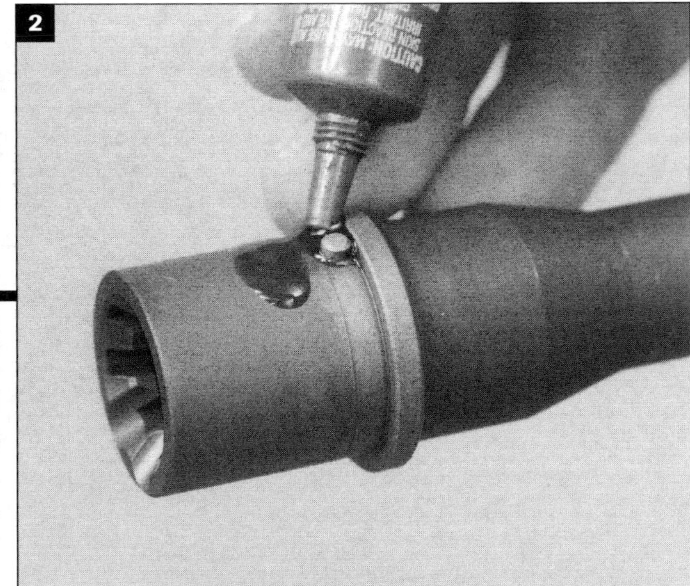

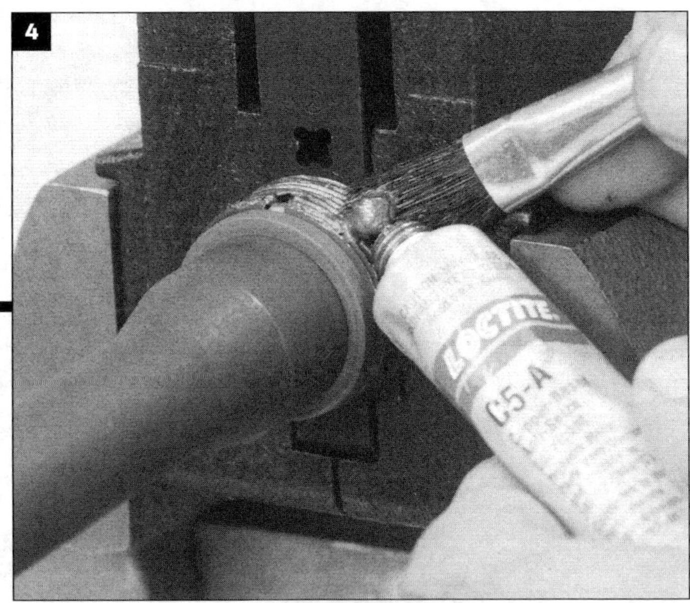

FINALLY...

Place the protected upper into a sturdy vise and snug it down. For this installation I'm using a "clamshell" style from Brownell's. Works great on any standard-form upper. Note the pin-style wrench. This type is necessary because the sight housing is on.

1. Clean and degrease the exterior of the barrel extension and interior of its receptacle in the upper receiver. Use contact cleaner. Do this even if you decide against glue.

2./3. Apply Permatex "red" adhesive to the exterior of the barrel extension. Push the barrel straight into the upper receiver, lining up the pin on the barrel extension and the slot in the upper receiver. Don't wiggle it in, just push it in, and then **4.** apply anti-seize to the exterior of the receiver threads.

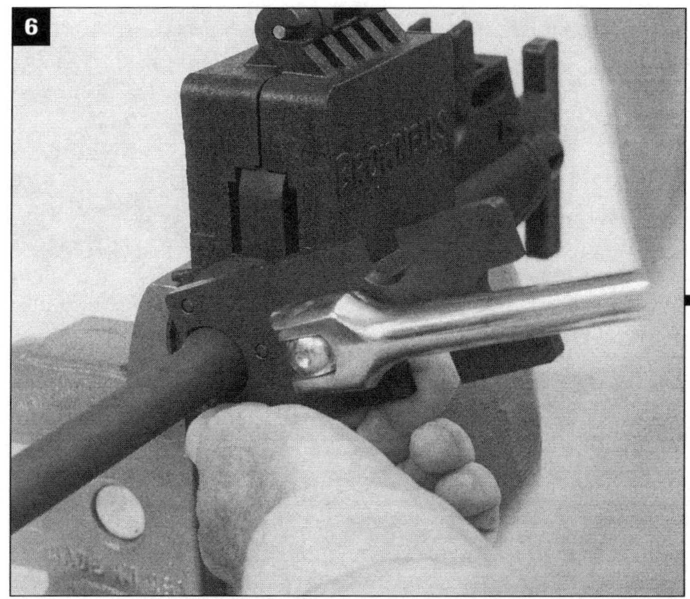

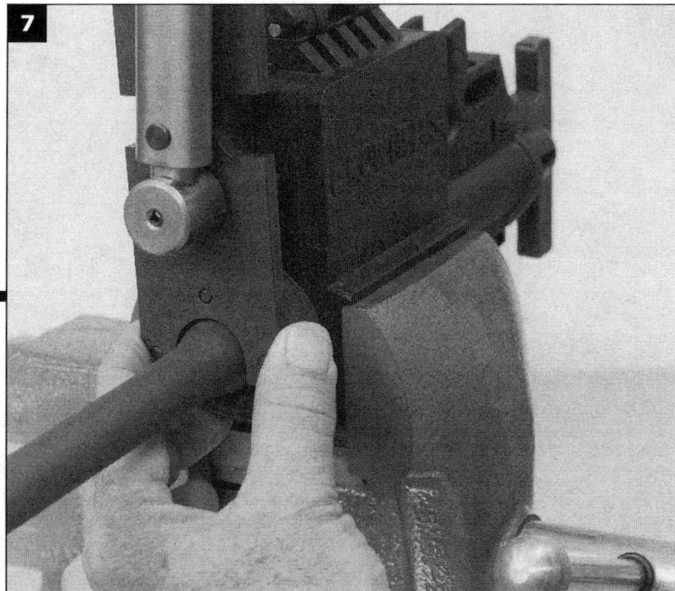

5. Thread the barrel nut fully onto the receiver.

Always hold the wrench firmly against the nut to keep the Delta spring compressed and the wrench engaged fully.

6. Follow the tighten/loosen process shown earlier using your barrel wrench and breaker bar. **7.** Switch to your torque wrench and pull until torque is attained.

Hold your breath, **8.** use a small flat-blade screwdriver to align all the Delta assembly parts to reveal the opening, and... **9./10.** insert the gas tube alignment tool. See where you are. Pray to the deity of your choice that one of the scallops is either just on or just a little off, and will be easily aligned heading in the "tighten" direction. If it's a lot off, you have no choice but to tighten the nut until alignment is attained. Be critical. Remember, torque specs are a minimum, so do not loosen the nut to get alignment.

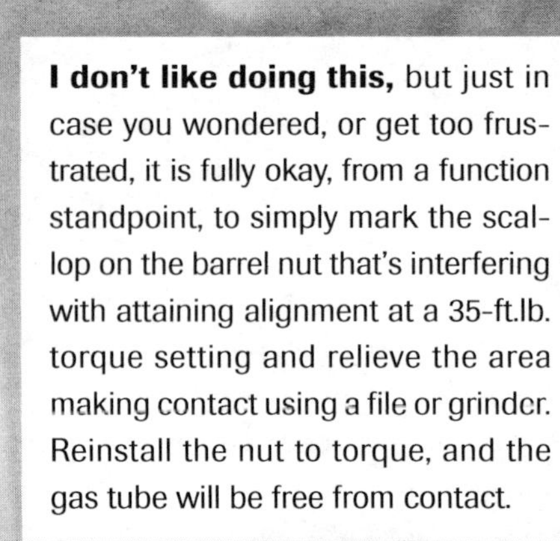

I don't like doing this, but just in case you wondered, or get too frustrated, it is fully okay, from a function standpoint, to simply mark the scallop on the barrel nut that's interfering with attaining alignment at a 35-ft.lb. torque setting and relieve the area making contact using a file or grinder. Reinstall the nut to torque, and the gas tube will be free from contact.

ALIGNMENT CHECKS

1./2. Keep at it until the tool is showing perfect alignment. Continue to double-check that the Delta assembly pieces are themselves not interfering, and move them accordingly.

3. One way I check for alignment is a little by feel. Moving the tool left and right should "feel" the same. Otherwise, just look closely.

4. When you think it's there, pull the works from the vise and clamp and insert the alignment tool into a bolt carrier. Check again. If it passes that test, what's left is installing the gas tube and triple-checking. If there's a problem, chances are now good that it's the gas tube itself. A precision-made alignment tool, used with care, is pretty accurate.

8.0 GAS TUBE

BASIC INSTALLATION

OVERVIEW

Installing the gas tube will complete this op, at least the functional end of it. Get the tube oriented correctly, meaning that the gas port or admittance hole end is toward the muzzle. Of course, make sure the handguard cap is oriented correctly to receive the tube, and also the

*A **Mark Brown gas tube wrench** from Brownell's makes this job way on easier. This clamps around the tube and lets you rotate and position it. It's a valuable tool.*

Delta or slip ring assembly pieces. Insert the gas tube, through the barrel nut assembly and into the upper receiver (sans bolt carrier).

Push it back into the upper far enough to clear the front sight housing to then allow installation into the housing. This is easier if the gas tube is rotated upside down. You'll still have to flex the tube upward to get it where it needs to go, but not as much. Take care here. I won't say it's fragile but will say don't get heavy-handed with it. Insert and orient the port-end of the tube into the front sight housing. Line up the pin holes, take care here to get them exact, start the pin with a roll pin starter punch and then tap it on home using a roll pin punch. Set the pin to the same on each side and that's that.

Check alignment again, check it with the carrier, check, check, check. This is a critical component connection.

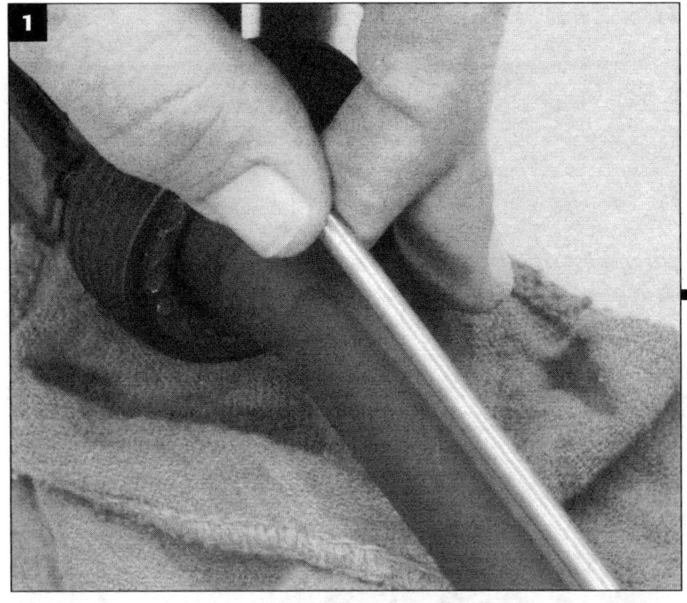

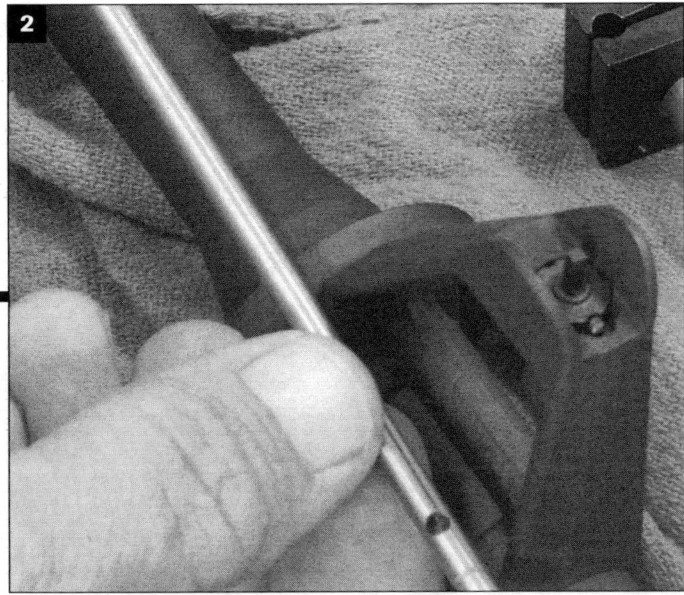

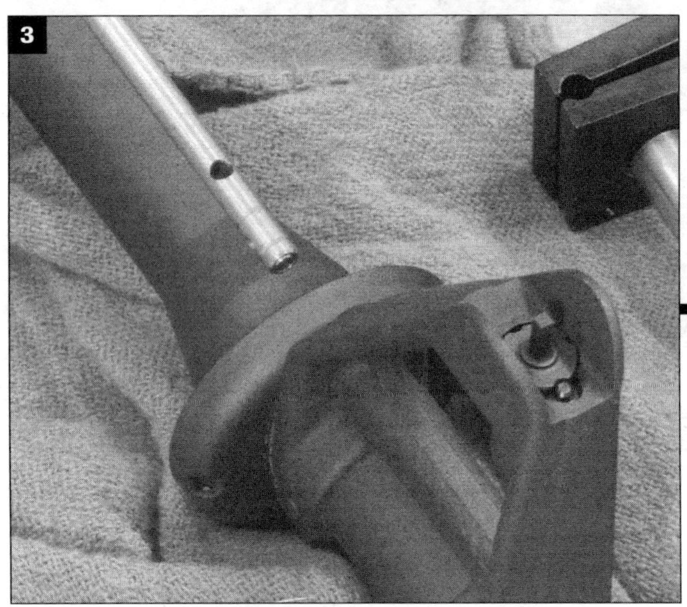

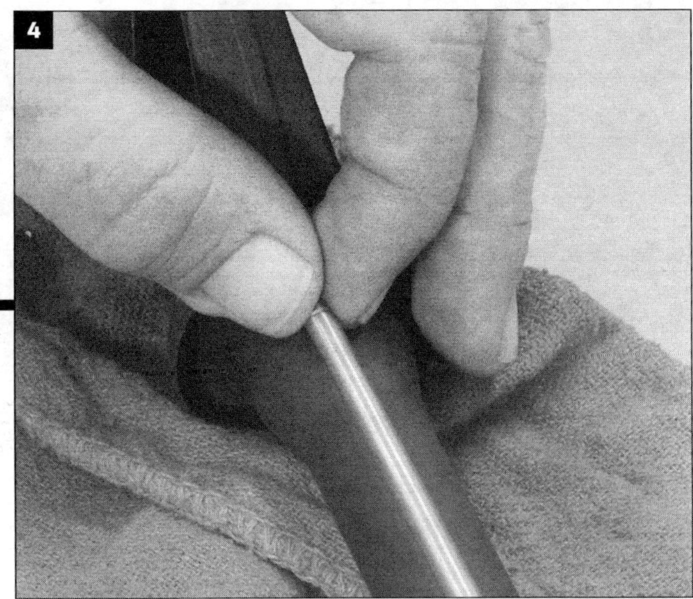

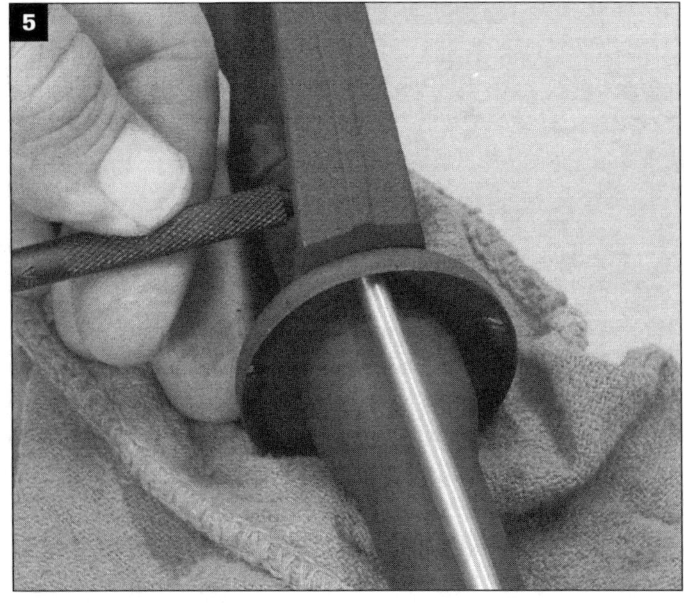

GAS TUBE

This installation shows a carbine-length gas tube. This little part takes a surprising amount of force to get situated. Rifle-length tubes are easier; there's more room to work.

Always test-fit the tube into the sight housing. Make sure it will install fully. There's often debris in the way. Clean it out.

(No bolt carrier installed.) **1.** Insert the tube into and through the openings in the Delta assembly and upper receiver. **2.** It helps, a lot, to invert the tube so the gas admittance hole is facing upward for this operation.

Once adequate clearance is obtained, **3./4.** rotate the tube so the admittance hole is oriented correctly (down). Flex the tube

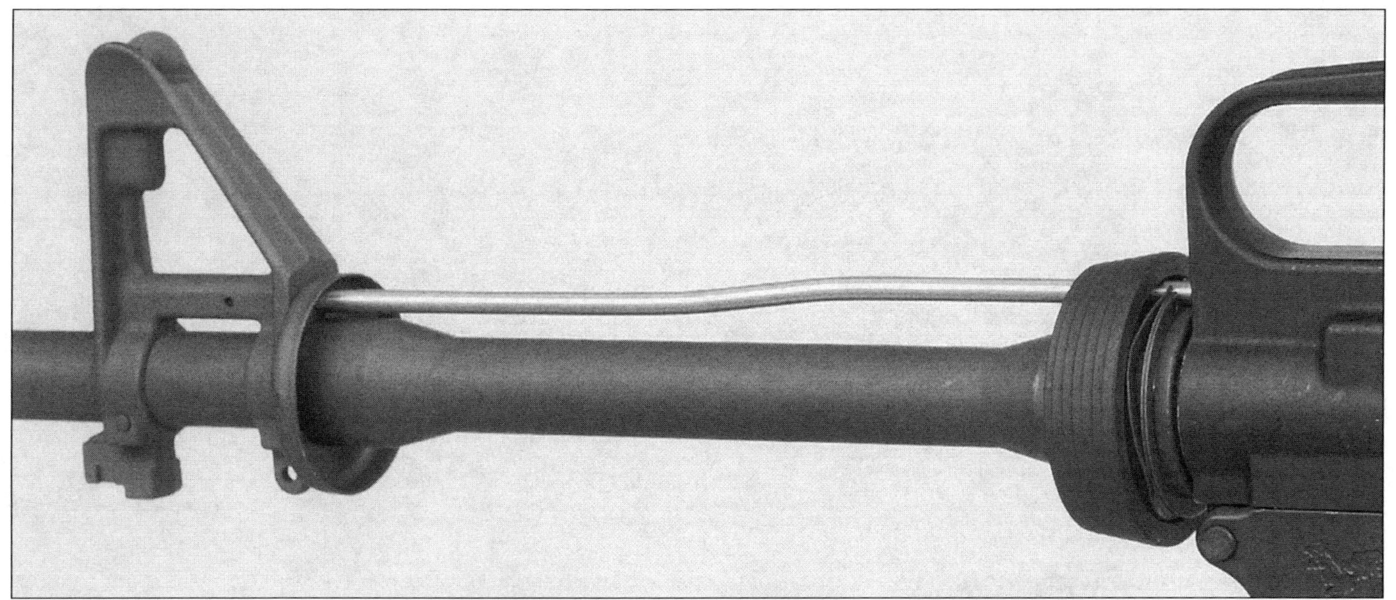

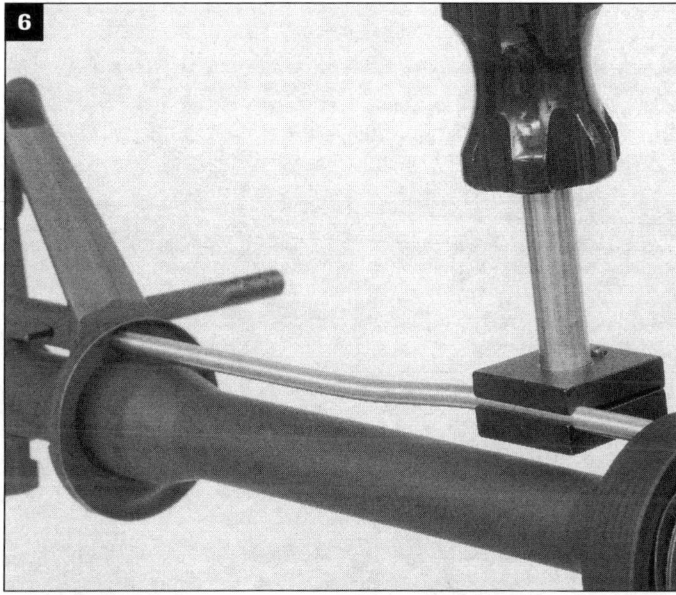

upward and back so it can enter the hole through the hand-guard cap and into the sight housing. **Make sure the roll pin hole in the gas tube is in alignment with its hole in the sight housing.** Get down and look at it. The gas tube is free to go farther into the housing than necessary. Line up the roll pin holes and sneak a capture punch in there. **6./7.** Here is where the gas tube wrench really comes in handy.

8. Start the pin with a roll pin starter punch and then finish it up with a roll pin punch. Both sized **#1** (1/16-inch). Make both sides equal, dab a dot of touch-up paint on either end, and you're done.

For what it's worth, gas tubes in most of the project rifles upcoming are easier to install because they can be installed into the gas block or manifold (or sight housing for that matter) prior to the tube actually entering the upper receiver.

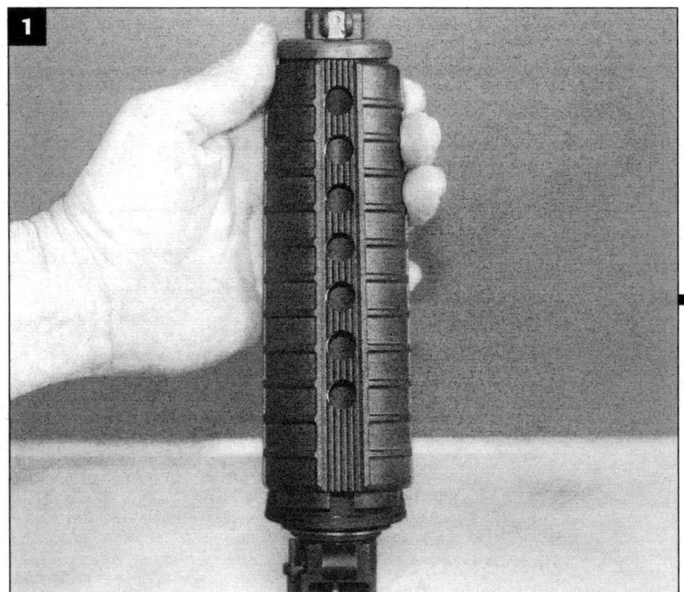

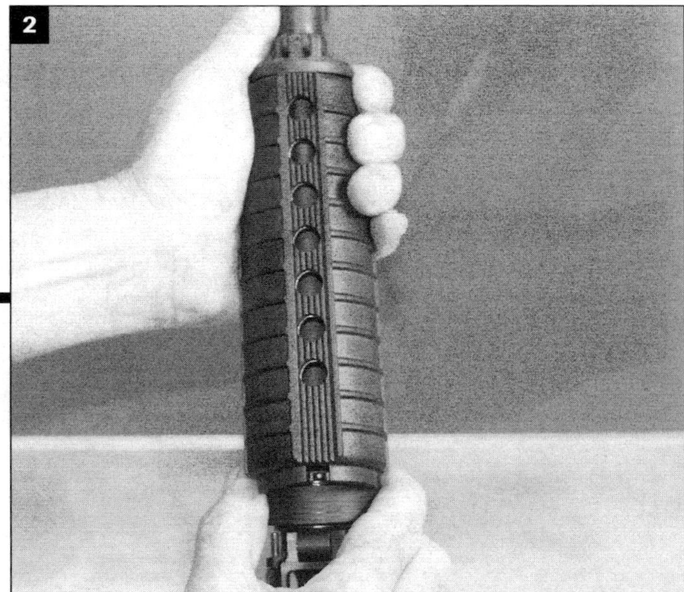

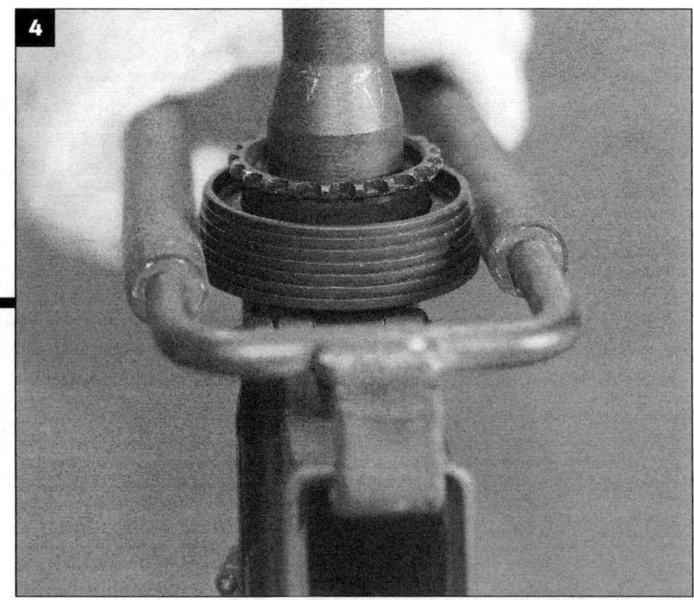

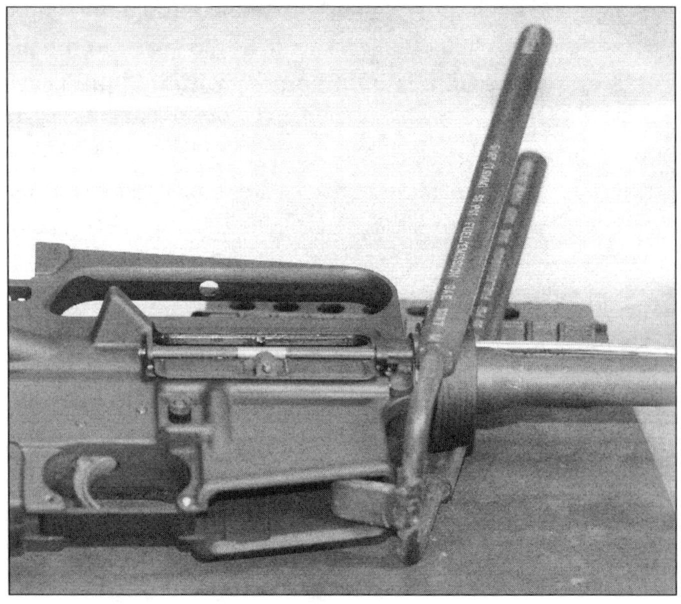

HANDGUARDS (STANDARD)

This is a really simple job. **1.** Insert the front portion of the handguard into the handguard cap, and **2.** compress the Delta assembly far enough to snap the back part of the handguard into place. Repeat the procedure for the other half.

It takes a lot of grip and force to get the Delta assembly compressed enough to slip the guard in. So, get your handy dandy handguard tool and breeze right through it. **3./4.** This is one of the best investments I ever made to make a routine job easy, especially for the Service Rifle shooter, who *will* use this. It hooks into the magazine well and offers plenty of leverage.

Not all handguards are equal in quality. Some are flimsy and result in a "creaky" installation. **Thermold** is what you want.

9.0 HEADSPACE CHECK

ALWAYS DO THIS

OVERVIEW

Chamber headspace checks are mandatory. We get pretty complacent after enough builds because it is very unusual to encounter a problem using mil-spec and commercially-produced bolts and barrels. It's very easy to check headspace and requires only a moment or two, and, of course, gages.

Headspace, in essence, is the distance from the bolt face (when the bolt is in battery or fully forward, rifle ready to fire) to the location of the "datum line" along the case shoulder. On .223 Remington, this line is a point 0.330-inches diameter. Height to this line will be in the vicinity of plus 1.4636 inches (SAAMI headspace spec). When the chamber reamer is run, the builder or manufacturer stops forward progress of the cutting tool with respect to this. Headspace can vary a little from maker to maker depending, among other things, on their beliefs as to what "playing it safe" means.

If the reamer was run too far then the datum line measure would be farther into the chamber, and that would allow too much room ahead of the cartridge case shoulder, and that is excessive headspace. If the reamer wasn't run far enough, then the height to the datum line would be shorter, and that is insufficient headspace. Both are bad.

*Maybe you can borrow, but you sorely need to have a gage set available. These are from **Forster**. For a new barrel installation, you'll need a "GO" and a NO-GO." A "Field" gage is a specialty tool to check suspected well-worn rifles.*

Excessive headspace opens up the potential for a case failure, what we can call "blowed-up." Insufficient headspace means the bolt might not close fully on a chambered round, and that opens up the potential for an out-of-battery discharge, what we can also call "blowed-up."

We want nothing to blow up! Headspace gages work just as intoned by gauging whether there is, one, enough, and, two, not too much, space. You will need two, can't run with just one. A "NO-GO" gage is too long to replicate excessive headspace. If the bolt closes on this one, then don't fire the rifle. A "GO" gage replicates correct headspace. The bolt is supposed to close on this one. If the bolt does not close on a "GO" gage, then the chamber is too short, then don't fire the rifle. My gages range 0.003 inches difference. Not much.

If excessive headspace is indicated, try another bolt. If that fails, there's really nothing that can be done save for a new barrel. If insufficient headspace is indicated, that can be remedied by a gunsmith, who can cut the chamber to the correct depth.

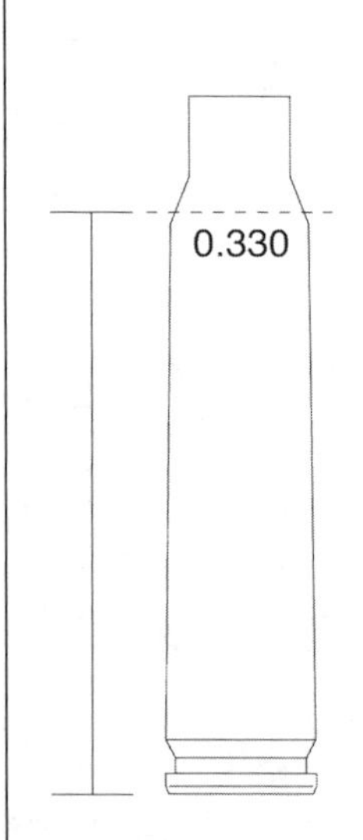

Here is what we're working with, and for. The case datum line on a .223 Remington is 0.330 inches. From the point of that diameter (dotted line) back to the bolt is headspace in the chamber.

Likewise, from that line to the base of the cartridge case is our concern in sizing cartridge cases for reuse in a rifle. Get resized case headspace to read 0.003 inches under the actual chamber height to this line and we're good to go for handloads.

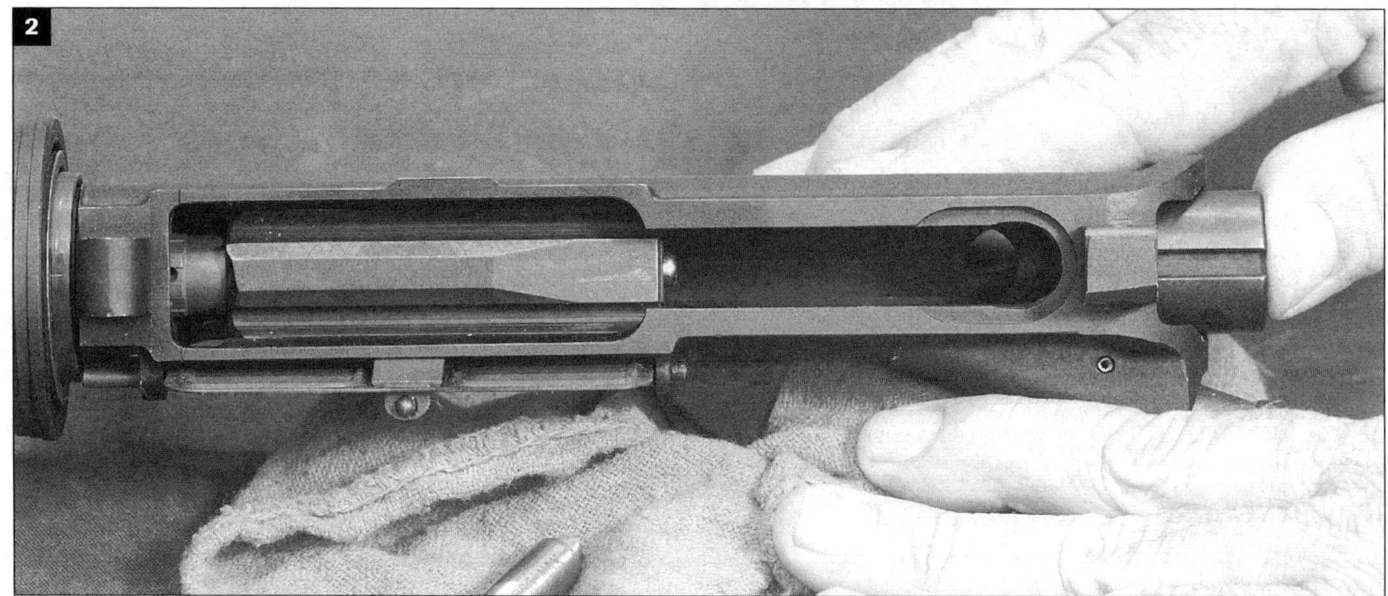

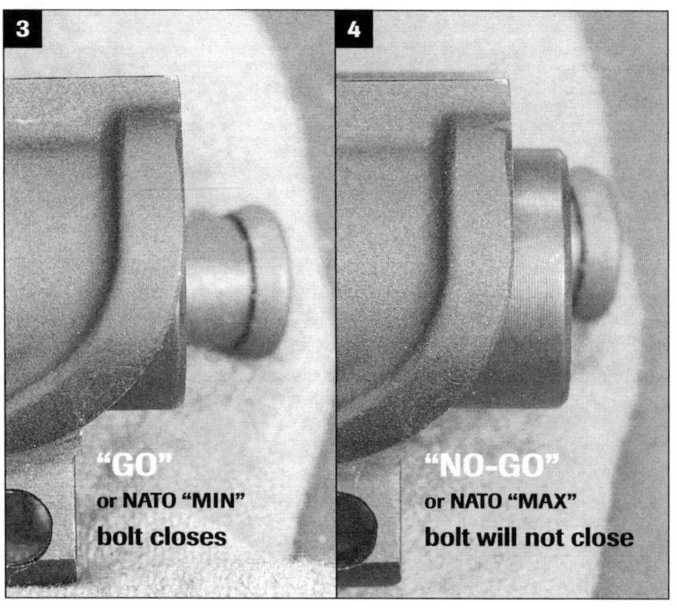

GAGE SELECTION/OPERATION

Make sure the chamber and the gages are clean. It doesn't take much at all to impede and corrupt gage operation.

Get your GO and NO-GO gages ready. **1.** Insert one fully into the chamber and, by hand, **2.** move the fully assembled bolt carrier forward toward battery as far as you can. **If the back of the bolt carrier comes to flush with the receiver, that's "closed."**

3. Has to close on a GO. (Forster gage is 1.4636 inches)
4. Cannot close on a NO-GO. (Forster gage is 1.4666 inches)

If it's a 5.56 chamber, use a NATO gage! 5.56 MIN is 1.4636 inches (same as .223 Rem. GO). 5.56 MAX is 1.4736 inches. Big difference! That's 0.007 longer than .223 Remington NO-GO.

10.0 BOLT & CARRIER
THE HEART OF THE RIFLE

COMPONENT FORMAT

The "big" book, *The Competitive AR15: the ultimate technical guide* covered this pretty much to death and ruin, and it's clear from reading about some of the specific choices I made in outfitting these project rifles that I actually do follow my own advice. Both ways. There's what I do and what I say to do, and I admit to having a penchant for trickery.

SEGMENT CONTENT

112	**Component Concerns** (getting the right parts)
113	**Bolt & Carrier**
115	**Carrier Key**
116	**Parts & Tools**
118	**Ejector Installation** (and tuning)
122	**Extractor Installation**
125	**Gas Rings**
126	**Carrier Key Installation**
129	**Component Assembly**

> ***Pay close attention to component quality here.*** *Format may not really matter much to rifle performance, but parts that are what they should be sure will. Of special note is the firing pin hole size.*

When I have the choice, and when I think the rifle really "matters" I like to have a carrier set that's all it can be. That means proprietary carrier design, select bolt, and on down the list of pieces, each representing what I think is the best of its kind, or just makes me feel good. All of this is in the "big" book.

Otherwise, "good" is good enough, as long as good is not a conjecture. I honestly don't think a "premium" carrier can make a rifle shoot any better. They are usually well conceived, however, and good to go. Quality concerns are not concerning.

Coated or plated carriers I do think are an asset, not an advantage. The coating makes them clean up easier and is "slicker" than Parkerizing or oxide. Oil, of course, is slicker than any finish and if either is lubricated like it should be then there's little to no point in reality for this notion. Mostly they clean up easier. Again, they don't perform better meaning that your rifle won't shoot any better with a plated or coated carrier.

Bolts? When I can find them I use Colt's. I've yet to see a bad

COMPONENT CONCERNS

one. Machining is nice, especially on the faces, and firing pin hole sizes are consistently smallish. Firing pin hole size should be checked and critiqued.

Blueprints call for a 0.058-inch diameter pin hole. That's

Firing Pin Hole Size

There's no precisely-dimensioned drill bit to use as a gage, although a 1/16-inch (0.0625) pin punch can be turned down to suit. Mostly we're guarding against a hole that's too large. To that end, 1/16 (0.0625) or #53 (0.0595) drill bits can be used as checks. If the first fits the pin hole, get another bolt. If the #53 won't fit, by all means use it. Measure your bits with a micrometer. Bits have a tolerance also.

ideal. If the hole is too large then primer problems will, not can, show. How big is too big? I say 0.062. Measure with a caliper, which is not precisely accurate, but that will show if the hole is in the smaller or larger range.

Firing Pin Protrusion.

Firing pin protrusion is blueprinted for 0.029 inches. If it's more than that, it's probably not going to hurt, and the only solution is to try a different firing pin. Same if it's less, but, again, it probably won't hurt. This is a firing pin protrusion gage. The gage shows minimum and maximum, and this can also be measured with a caliper (use the "depth" end to measure from the rim of the bolt to the bolt face, zero the caliper, push the pin fully forward, and then measure from the bolt rim to it).

BOLT

Starting with a bolt body, we'll add the ejector assembly, gas rings, and extractor assembly. As mentioned elsewhere, I really think it's a good idea to disassemble a complete bolt to do a little fit and finish on the ejector and extractor anyhow.

Gas rings are easy. Just hook one end into the groove on the bolt body and rotate/snap the ring into place. When all three are in and on, rotate them so none of the gaps provide a tunnel. Technically they should be 120-degrees apart, but just as long as each is sealing against the next, that's going to work.

Ejector assembly. Not so easy. Small parts in small spaces under high spring tension are never easy. Huge help comes from a specialty tool shown shortly. It's a three-handed job without this tool.

Extractor next. No aftermarket toolage help here but it's easier to work with than the ejector, or so I say. A standard spring installation starts with putting the polymer insert inside the spring, big end to big end. A better installation starts with a chrome silicon spring all by itself, and that is addressed elsewhere. The spring then fits into the extractor, big end down into the recess in the extractor. This spring is a tight fit into its recess and needs to be fully seated. Do this before installing onto the bolt. Push it down (quite firmly) and it will snap into place.

Installing the extractor itself is a simple matter of fitting it onto the bolt body, lining up the holes, compressing the works adequately to line up the holes, and then pressing the pin through. No punch. This pin should go in easily. Test fit, of course, before installation. Just make sure the pin is centered and equal on each side.

CARRIER

Bolt carrier designs are essentially limited to two formats for the back end of the carrier to take. Specifically, how much back end it has. M-16 bolt carriers and most aftermarket specialty carriers have a longer portion of full diameter. That makes them heavier. Everyone knows I like heavier carriers. Higher-mass parts move more slowly, and that means the bolt will stay locked up a little longer. If you're shooting a carbine or higher-pressure loads through a rifle, that's some insurance against pressure-induced extraction difficulties, as well as against case failures from the same cause.

Heavy-weight carriers also have a shrouded firing pin area. That provides smoother operation but does require the use of a "large-collar" firing pin. Again, none of this will make or break your rifle, but better is better. For more, check *The Competitive AR15: the ultimate technical guide.*

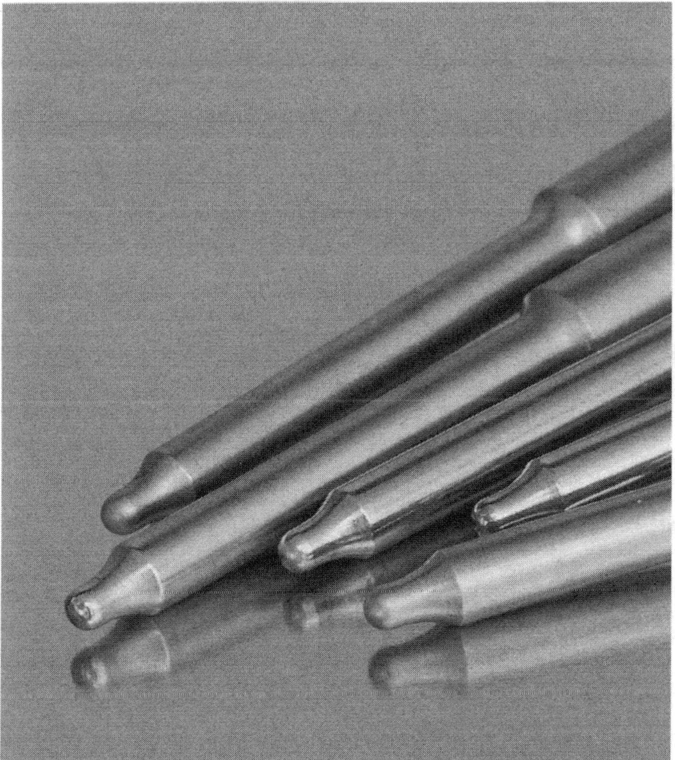

Pins. *I do look over and sometimes measure firing pins. Close inspection (use magnification) will often show, if you have enough to compare, differences in lengths and pin tip "shapes." Checks are noteworthy but results are likely unnoticeable.*

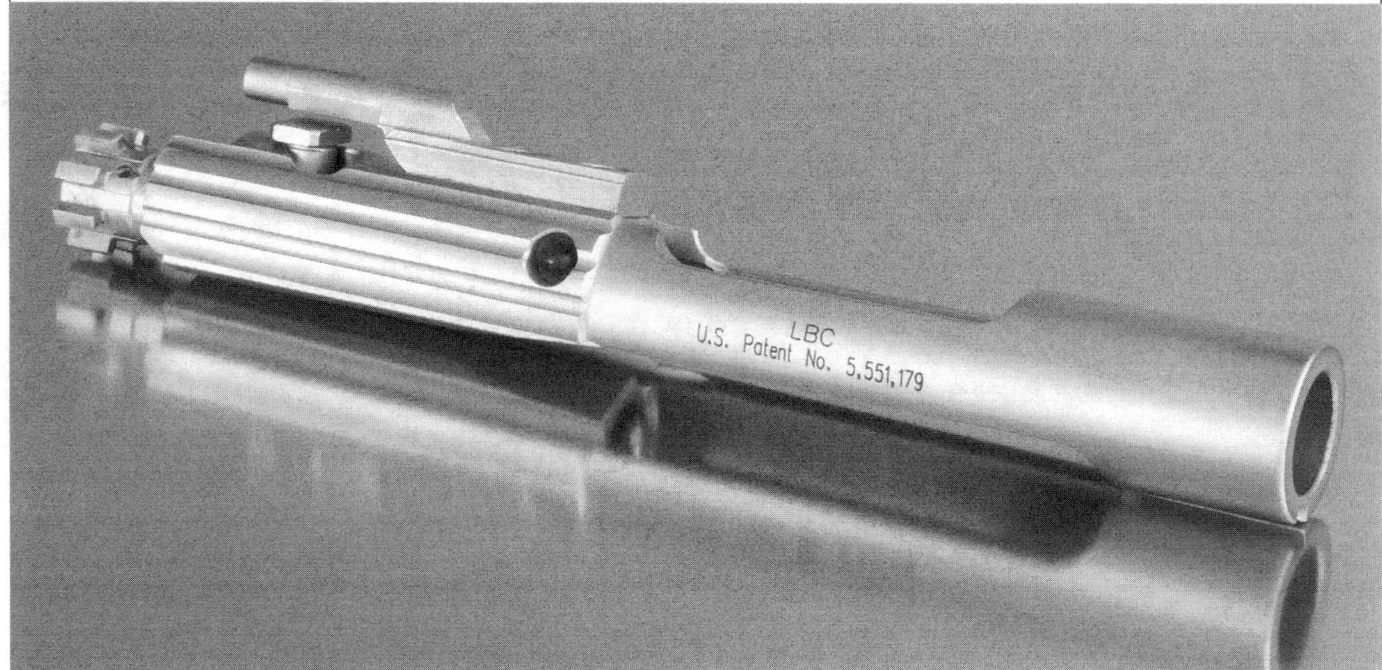

Carrier Selection. *Thing of beauty, joy forever. Probably. A premium carrier like this from Les Baer is money well spent. Expect correctness. Funny though that the carrier key still needs to be staked and that's harder on chrome.*

The differences here make a difference, for me. Up front is a standard AR15 carrier, behind it is an M-16-style carrier. The greater length of the full diameter at the back of the carrier makes the M16-style carrier about one ounce heavier, which delays bolt unlocking a little longer. Good thing. The firing pin area is also shrouded on the heavier carrier and open on the other. Shrouded runs smoother, but requires the use of a large-collar firing pin. The pins in the foreground are respectively small-collar and large-collar.

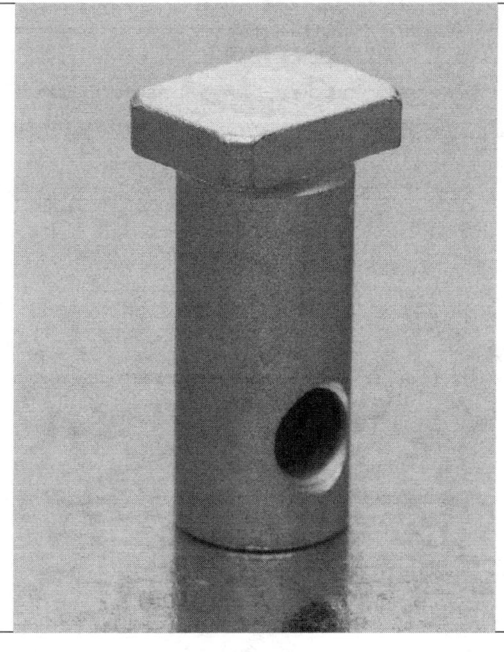

If it looks bad it just may be...

I've seen some rough looking cam pins. On left is one fished out from a baggie-full of GI parts; on right is a chrome version from Smith Ent.

There's never anything wrong with smoothing a part, just as long as it's not dimensionally changed. Parts will eventually smooth out from being worked through their operational cycles, but short-cutting the "break-in" may save a little unnecessary wear and tear.

It is very important, considering this, to very thoroughly lube a new part.

Firing pin retainer

I have developed a serious peeve about firing pin retainers. I got one batch that were borderline unusable. They just didn't want to fit and the problem was poor quality: one leg longer, very rough ends, and the like. I like the DPMS solid pin retainer on right. It's the ArmaLite original design and works for the life of the rifle.

After making sure it's a worthy part and replacing it if it's not, I polish and "round" the ends of a cotter-style firing pin retainer to make sure it's as easy as it should be to remove and reinstall, and it should be easy.

CARRIER KEY

Usually this is installed on most complete assemblies, and one reason it's a complete assembly. Otherwise, it's not a simple matter of just screwing it into place.

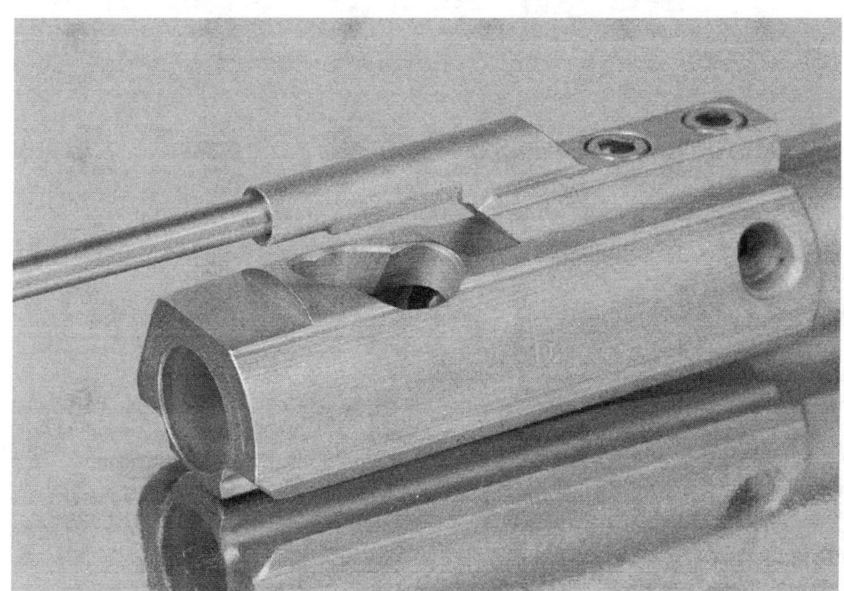

Thoroughly clean and degrease the area surrounding the installation. Sometimes there's preservative that has to get gone. The first start is easy: position the key and tighten the screws. That's the start. See elsewhere later on for more. It gets more involved...

PARTS & TOOLS

This may not happen often because it's easy to purchase assembled bolts and carriers, including the bolt, of course, and certainly doesn't have to ever, but at the least it's wise to learn how to completely strip an assembly.

PREPARATION

Roll pin starter punch (#1)

Roll pin punch (#1, 1/16-inch)

Prick punch (for staking carrier screws)

Hex wrench (for carrier key screws)

Options (good ideas)

 3/8-inch diameter wooden dowel (fits the bolt face)

 Permatex "red" adhesive

 Contact cleaner

 Small screwdriver (helps move gas rings into alignment)

 Scrap wood piece (to back up bolt)

 Torque wrench (reads in inch-pounds)

PARTS

Bolt body

Extractor
Extractor spring
Extractor pin

Ejector
Ejector spring
Ejector roll pin

Bolt gas rings (3)

Bolt cam pin

Firing pin
Firing pin retainer

Bolt carrier
Carrier key
Carrier key screws (2)

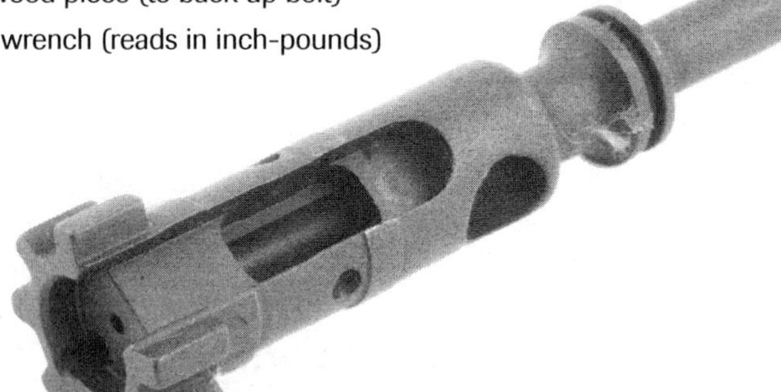

PRECAUTIONS

The only "take-a-really-deep-breath" part of this assembly work is staking the carrier key screws. The rest is easy.

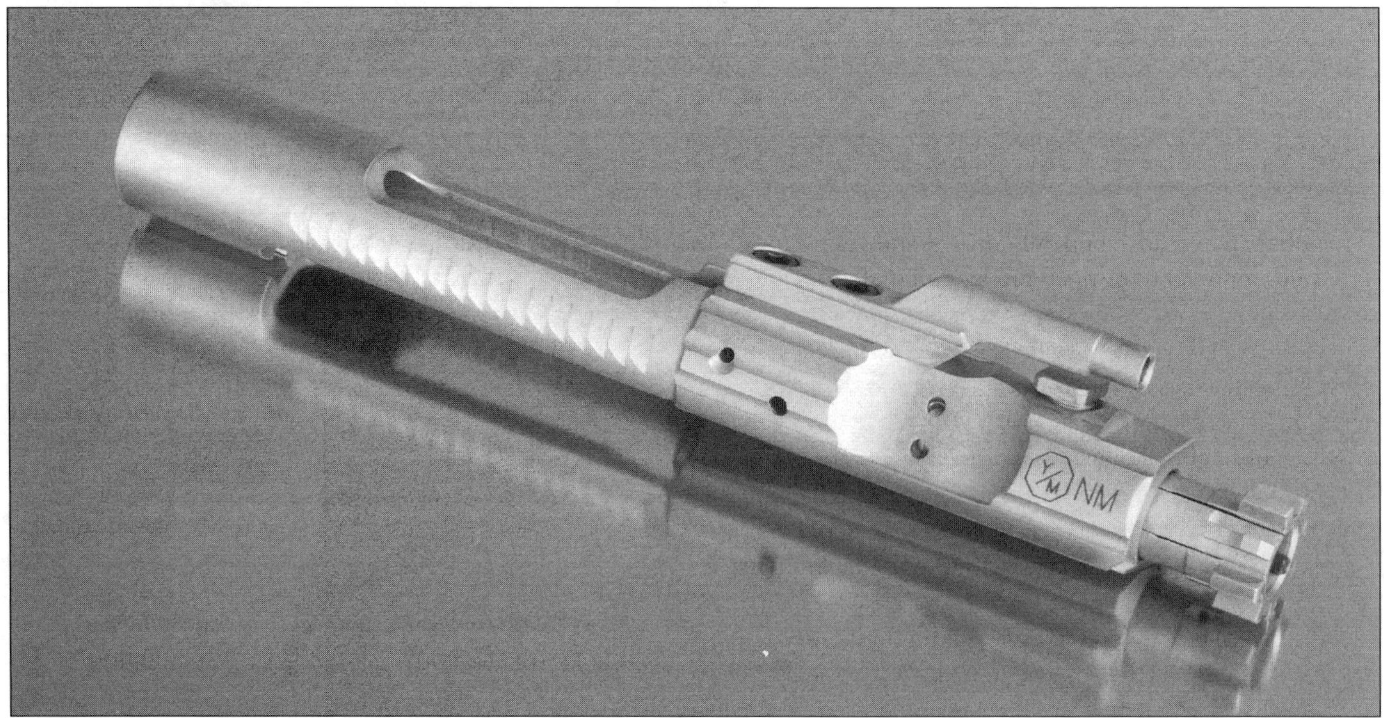

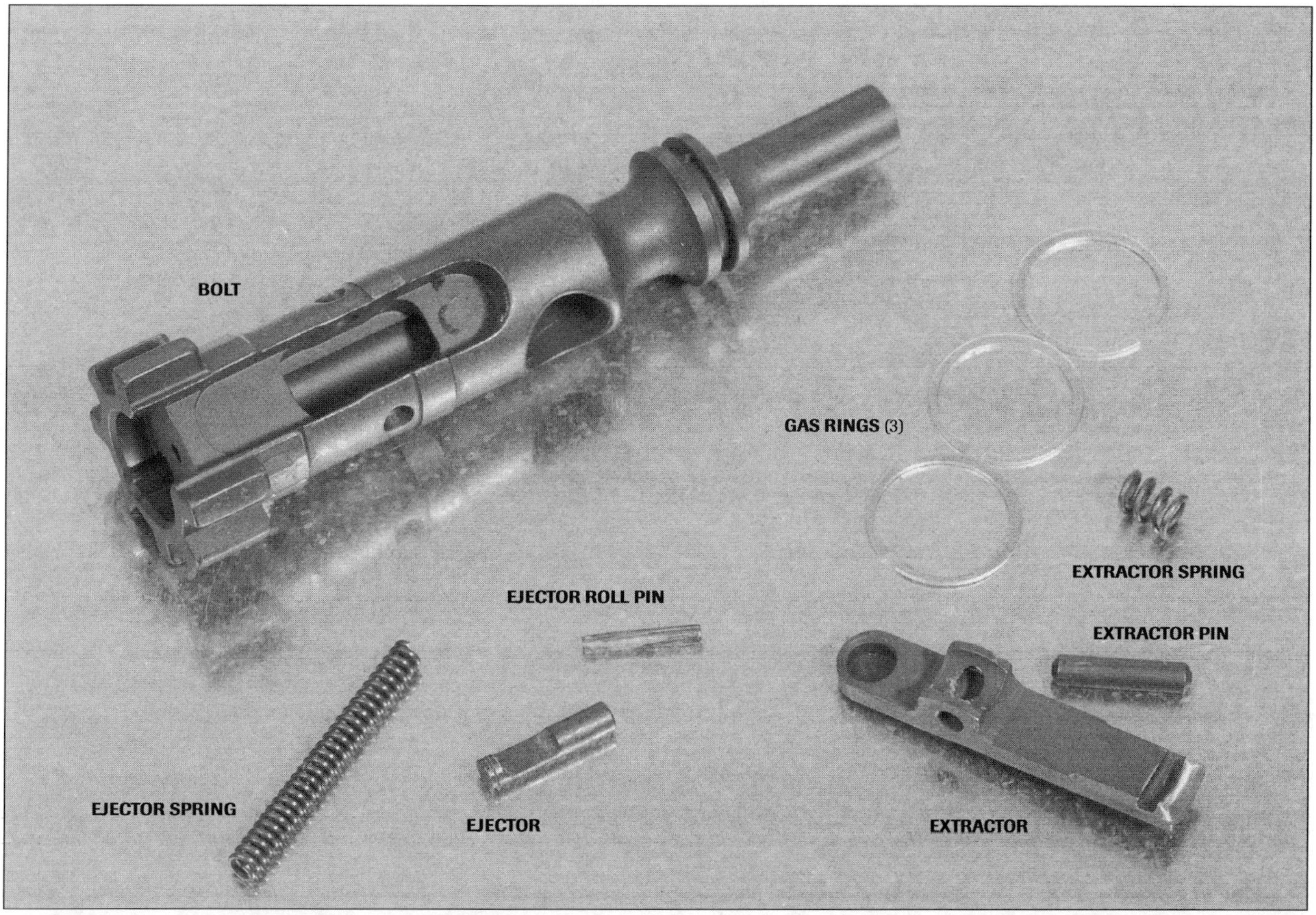

BOLT
GAS RINGS (3)
EXTRACTOR SPRING
EJECTOR ROLL PIN
EXTRACTOR PIN
EJECTOR SPRING
EJECTOR
EXTRACTOR

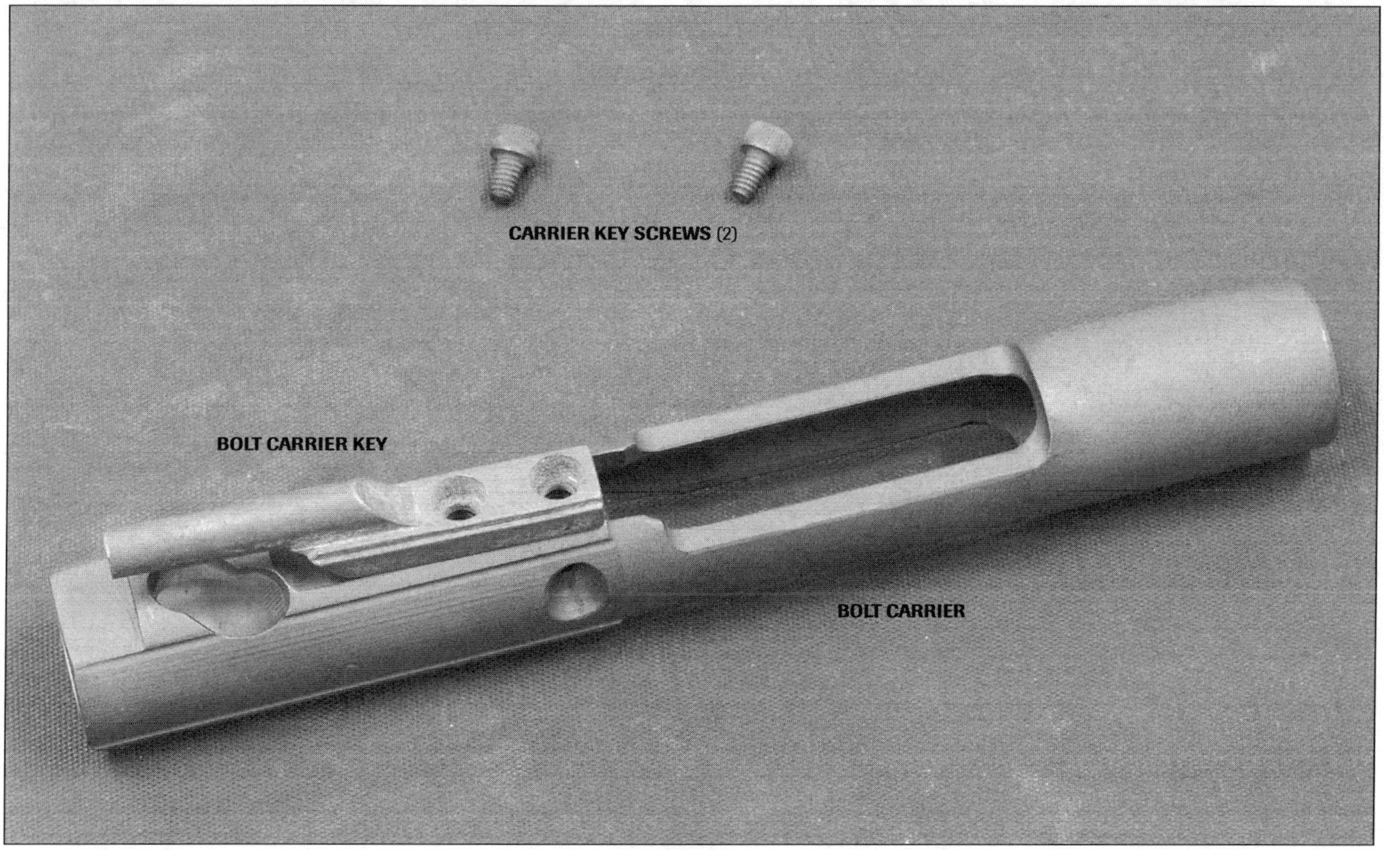

CARRIER KEY SCREWS (2)
BOLT CARRIER KEY
BOLT CARRIER

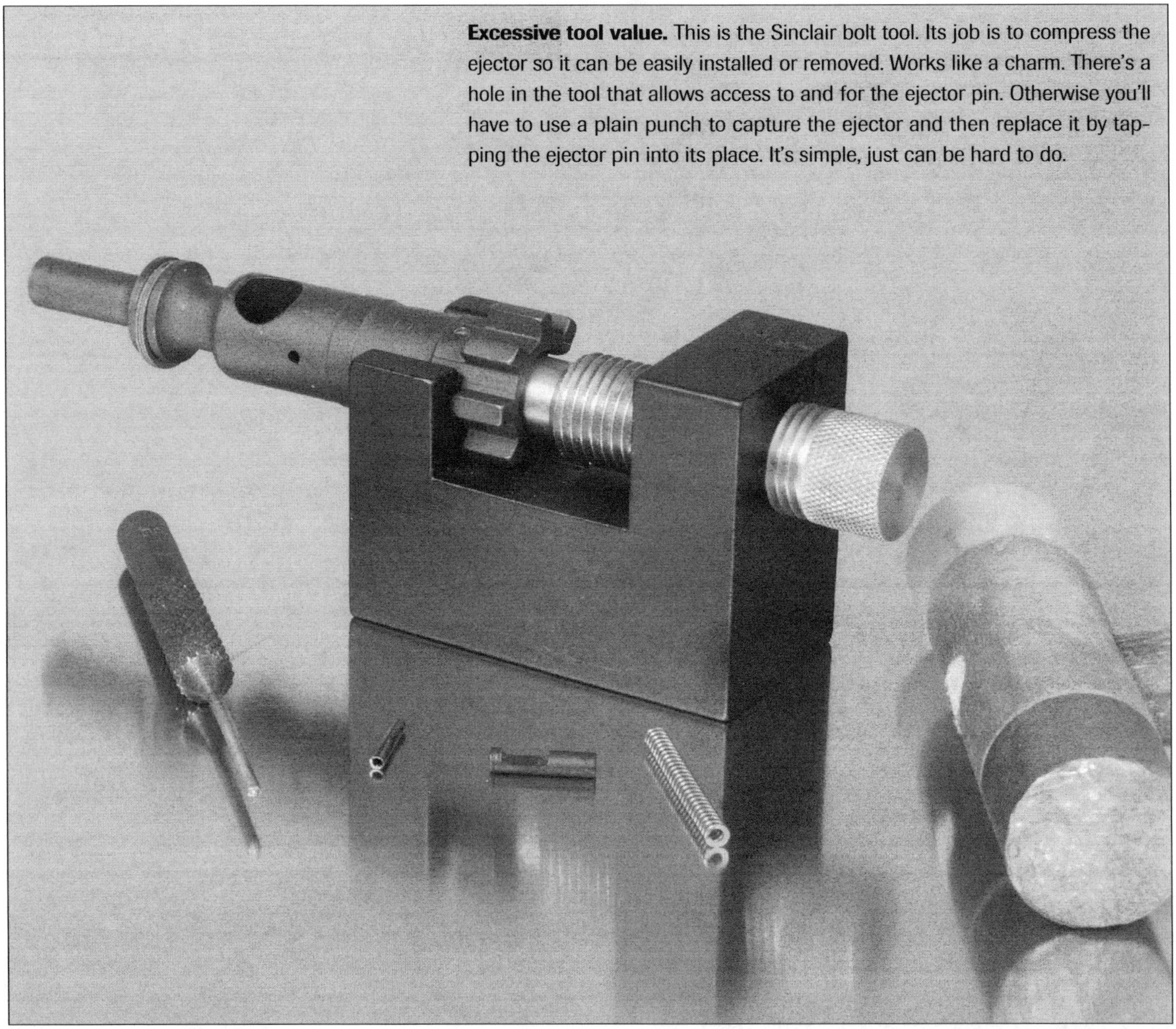

Excessive tool value. This is the Sinclair bolt tool. Its job is to compress the ejector so it can be easily installed or removed. Works like a charm. There's a hole in the tool that allows access to and for the ejector pin. Otherwise you'll have to use a plain punch to capture the ejector and then replace it by tapping the ejector pin into its place. It's simple, just can be hard to do.

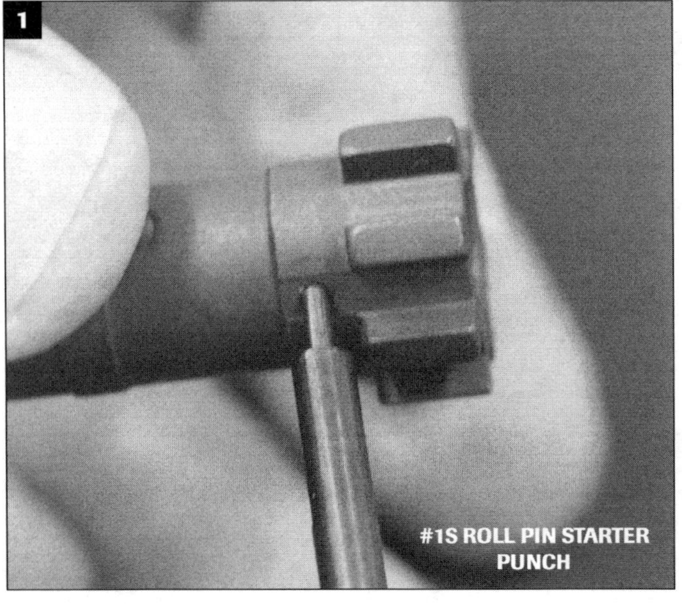

EJECTOR

1. Start the ejector pin with a starter punch. **2.** Drop the spring into the bolt body, **3.** and then position the ejector plunger (note its orientation), **4.** push it down, hold it down, line up the holes, and **5.** insert a #1 (1/16-inch) punch to capture the plunger. A 3/8-inch dowel or similar helps keep the ejector compressed. Back up the bolt with a piece of wood. Make double-sure the plunger is oriented correctly so the clearance cut on the ejector shaft is oriented towards the pin. This is where the capture pin really helps. Tap the pin on through, displacing the punch, and **6./7.** finish off by seating the pin so it's below flush on each side.

It's not hard, just tedious sometimes getting the plunger oriented correctly. Round side faces the outside, flat side faces inside.

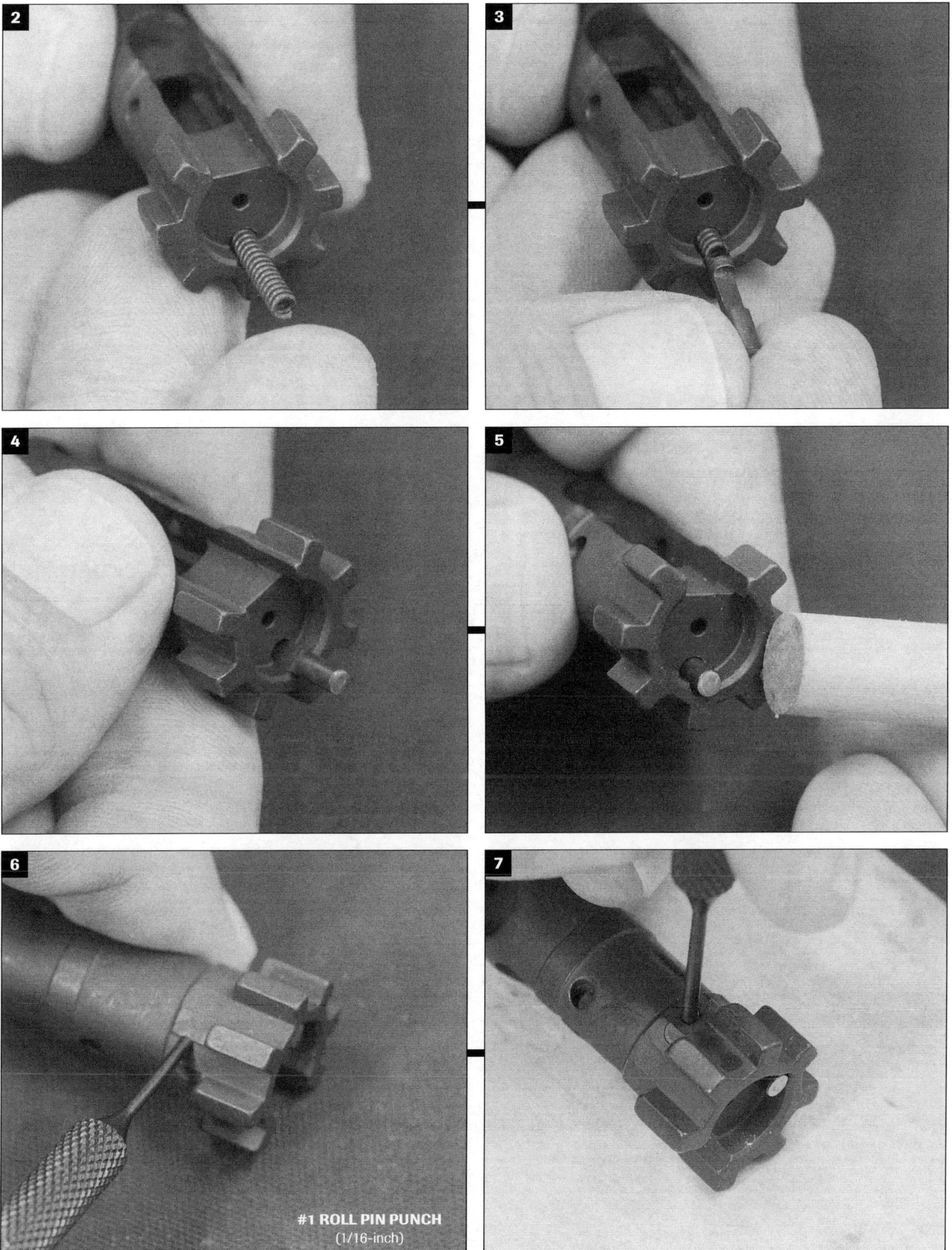
#1 ROLL PIN PUNCH
(1/16-inch)

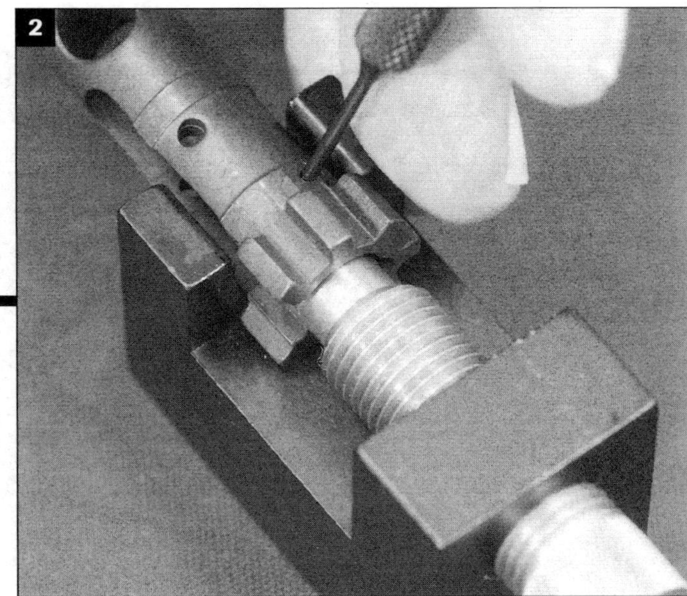

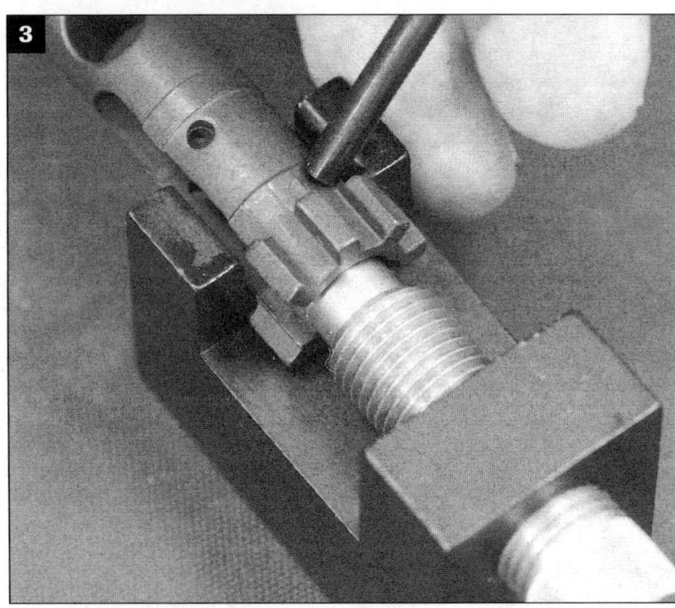

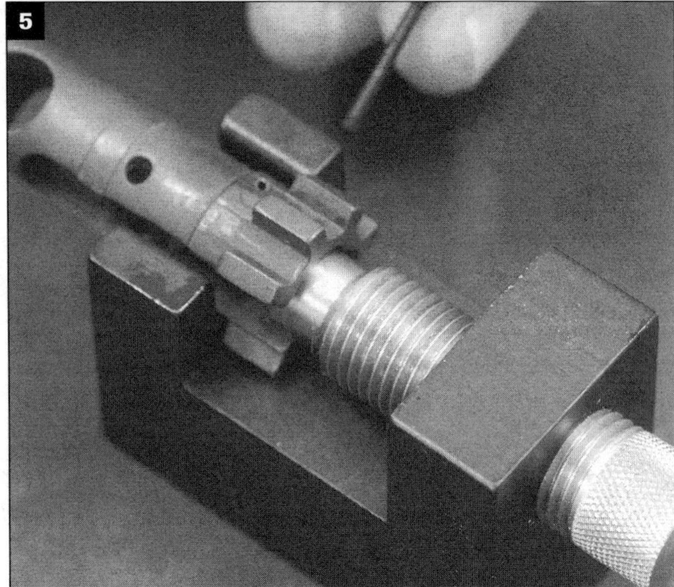

EJECTOR (EASY WAY)

The Sinclair tool makes this far easier. Position the ejector spring and plunger into the bolt in the correct orientation and turn in the knob on the tool until the ejector is fully compressed. Check orientation and alignment with a capture punch, start the ejector roll pin, and take it on home.

Disassembly is likewise easy using this tool. There is a hole that allows the pin to be driven fully out while the bolt is contained by the tool.

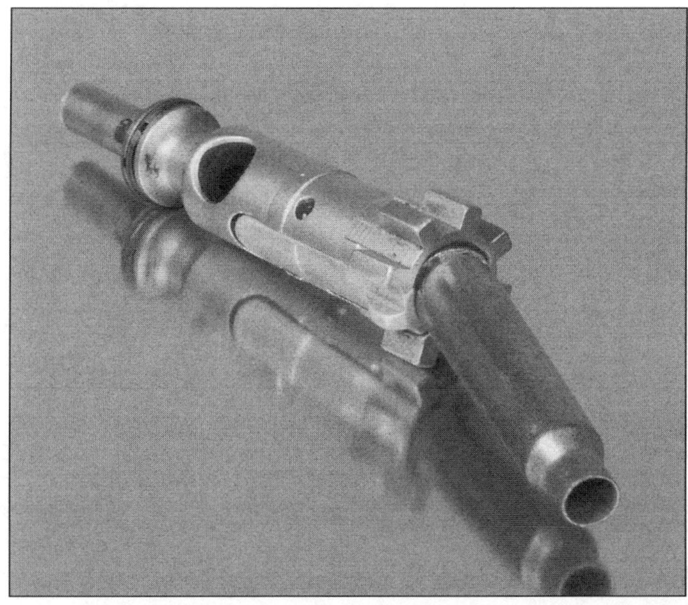

TRICKS & TUNING

The ejector pushes forward against the case head to tilt the case toward the right (shooter perspective) so the rearward-traveling, and rotating, bolt can let the empty whiz on out the receiver port. The ejector itself doesn't send the case out; that's from the case rim pivoting on the extractor hook.

When it's working right, there should be a small pile of undamaged brass about 3-o'clock (shooter's perspective) and no more than three feet away. Dings, dents, and creases on spent cases result from contact with the rifle on the way out. A wildly spinning case usually results from the empty banging hard off the rear edge of the ejection port, and such a condition usually is remedied only, or at least in a large part, by slowing carrier velocity.

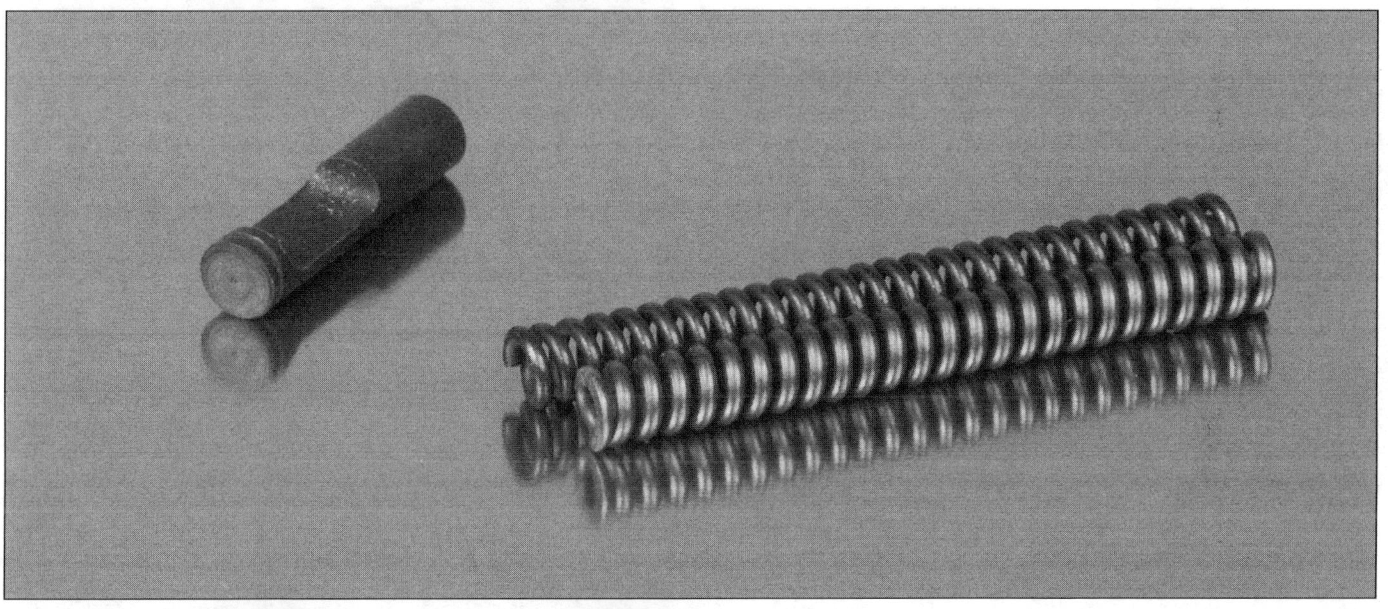

Tuning extraction/ejection can involve adjusting the height of the ejector spring. Generally, there is more pressure on the ejector than necessary. Reducing the pressure means cutting coils. Use a cutting disc attachment for a hand-held grinder, like a Dremel. On a new rifle I usually start at 0.800 inches total length. That's for a competition gun. Essentially, that is a "no-tension" installation that puts the ejector sitting flush with the bolt rim with no spring pressure pushing it above that height prior to installing the roll pin. For a carbine I start at full spring height.

Polishing the face of the ejector reduces friction as the case is rotating on its way out.

The biggest helps, still, are a polished chamber (that is also kept clean), and reducing bolt carrier velocity (there are several means to that end).

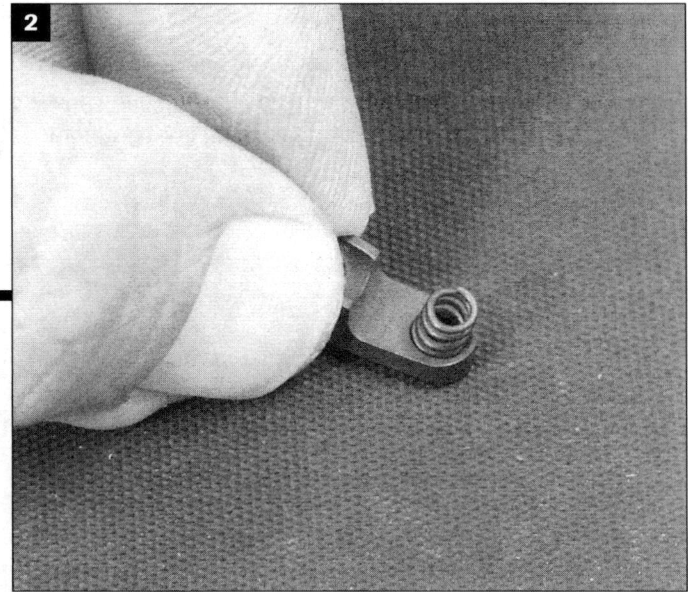

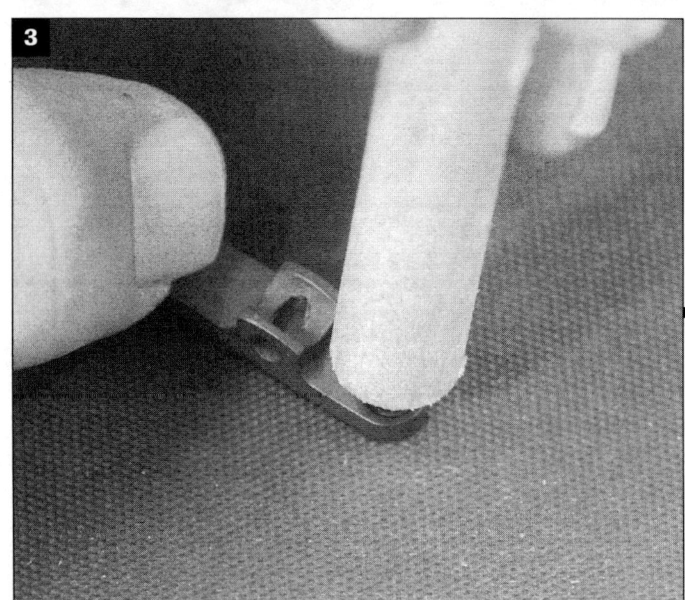

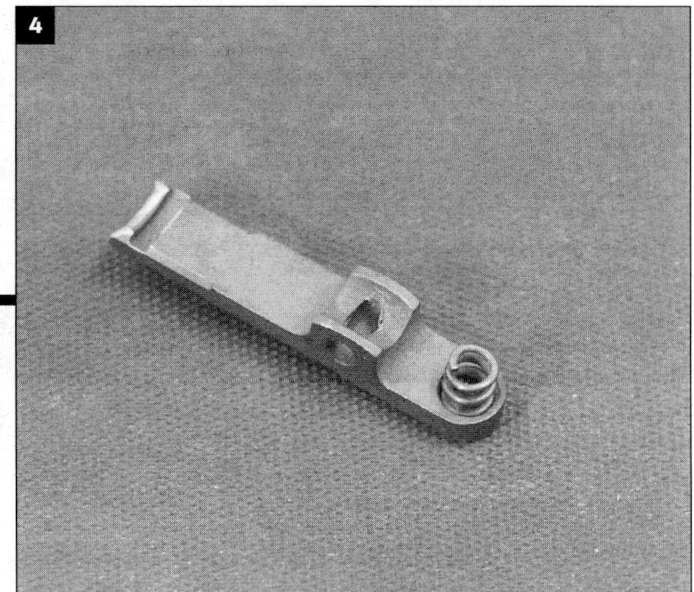

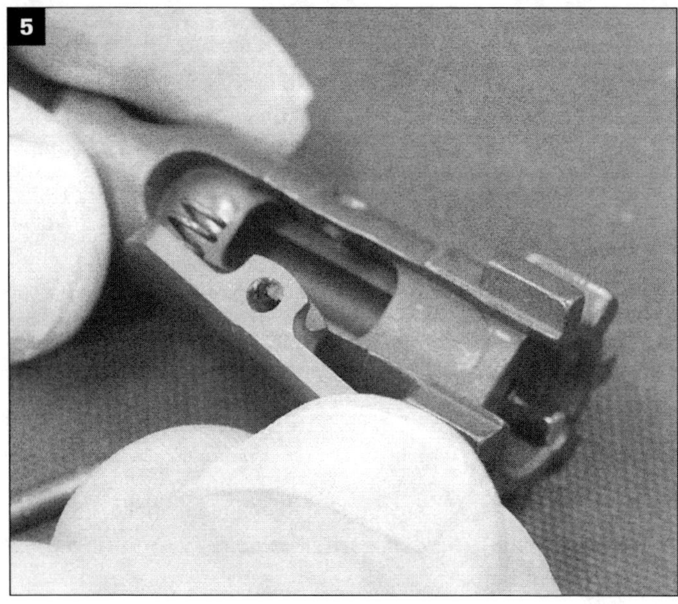

EXTRACTOR

1./2. Place the extractor spring, big end down, into its recess in the extractor.

3./4. Push that bad boy in until it is fully seated. Use a dowel or similar.

5./6. Position the extractor into its cutout in the bolt body.

7./8./9. Push it down, hold it down, line up the holes, and insert its pin. Make sure the pin is fully contained and below flush on either side of the bolt body.

That's it.

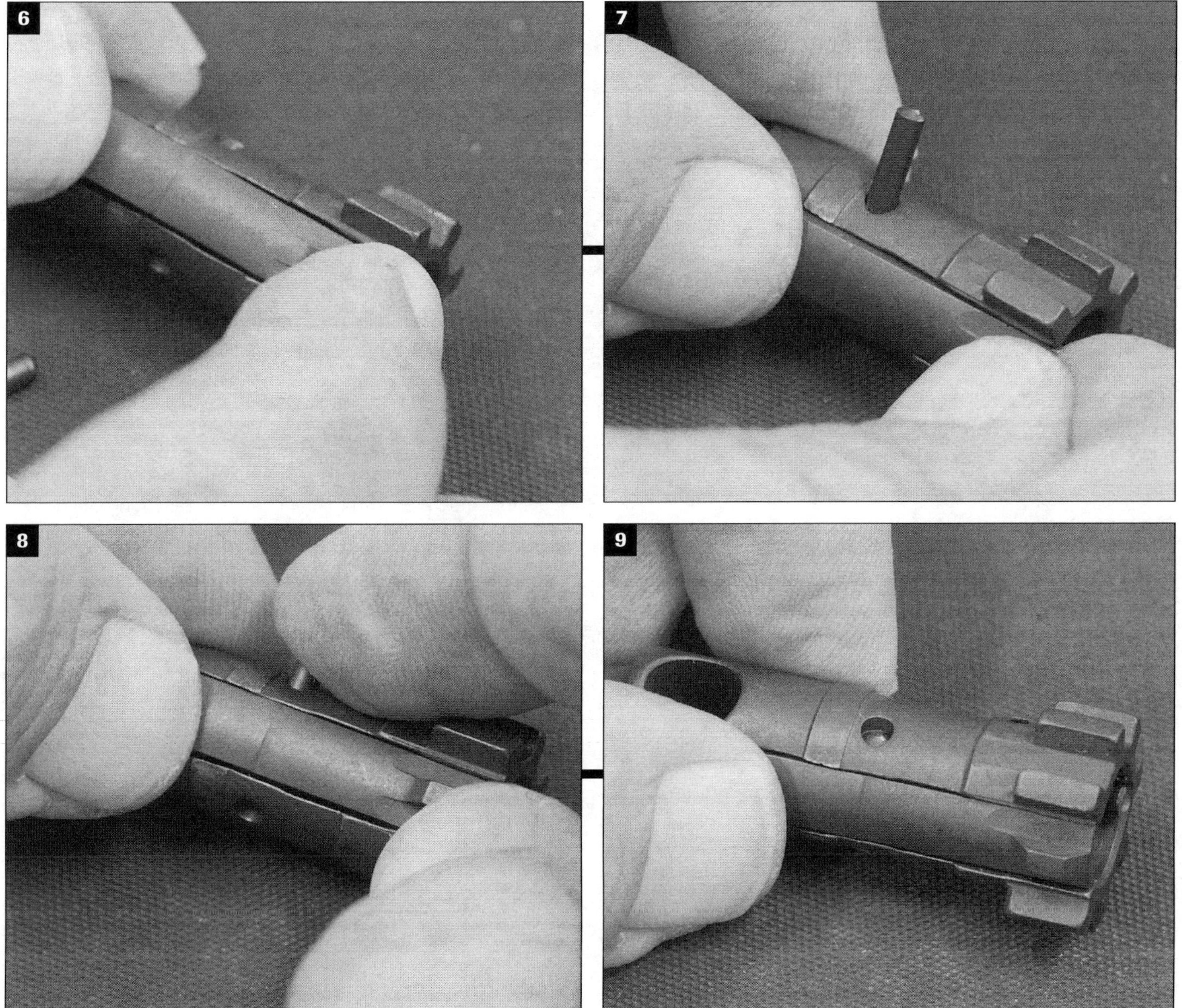

[CONTINUED NEXT PAGE]

Extractor polish

Polishing the extractor is easy and does make a difference. Use emery paper. Don't touch the lip or "inside" area, just smooth the outside. This lets the extractor slip more easily over a chambered case head. I polish the ejector face also, and in the same way. Don't reshape anything, just shine it up.

Fixes, not cures...

Shown left to right: standard spring with phenolic insert (routine configuration), a D-Fender phenolic insert engineered to accompany a standard spring and boost tension, extra-power chrome silicon spring from Superior Shooting Systems Inc.

Of these, my preference is running with the SSS Inc. spring, but with a coil cut off the top. The spring material lasts the life of the rifle. This spring is never to be used with an insert.

EXTRACTION HELPS

I really don't like to see anyone up the pressure on the extractor. It's not the ultimate cure for extraction problems. Poor extraction is usually the result of too much gas too soon. That makes the carrier move too early, the bolt unlock too early, and the extractor takes a yank at the case rim while the case is still swelled up and gripping the chamber walls. Sometimes the extractor can lose its grip. Increasing the tension on the extractor through the addition of phenolic inserts and extra power springs is done to keep the extractor more firmly engaged in the case rim. Polishing the chamber and slowing bolt unlocking are more preferred means to the same end, which is trouble-free extraction. Increasing tension means that the extractor will push more forcibly against the case head to slip over the head and engage the case rim when a round is chambered, and that can push the case that much farther into the chamber, damaging the round.

Inset shows the effect of the premature bolt unlocking. Look carefully at the case rim. You can see where the extractor tried to yank it off.

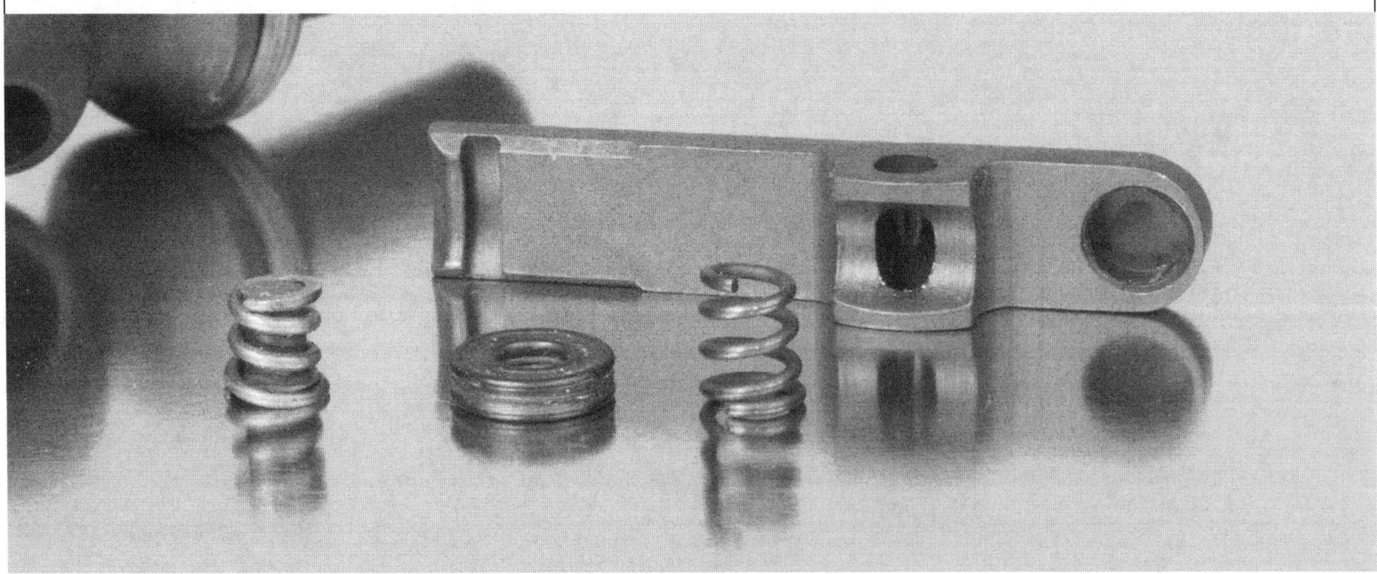

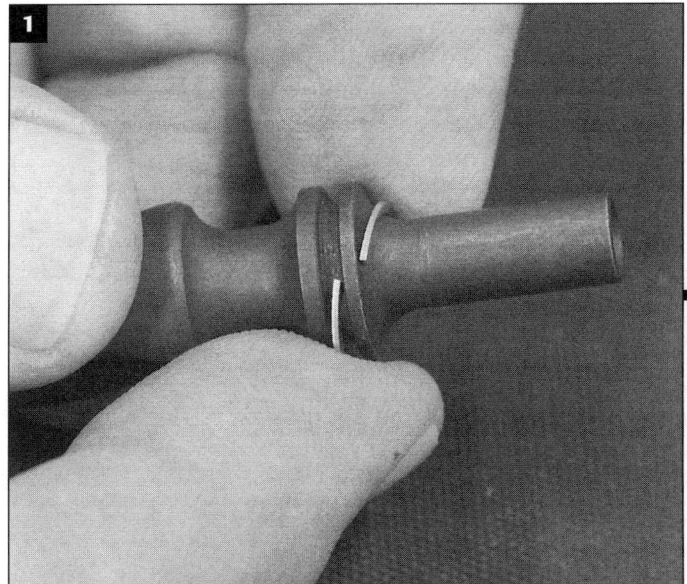

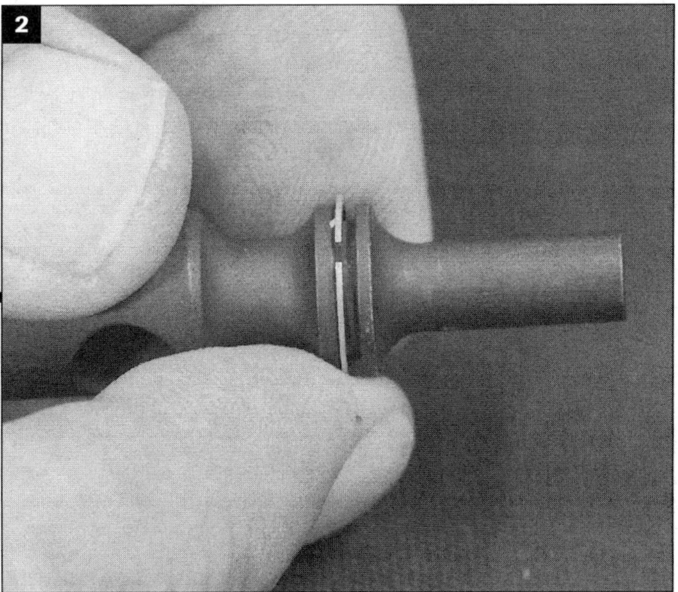

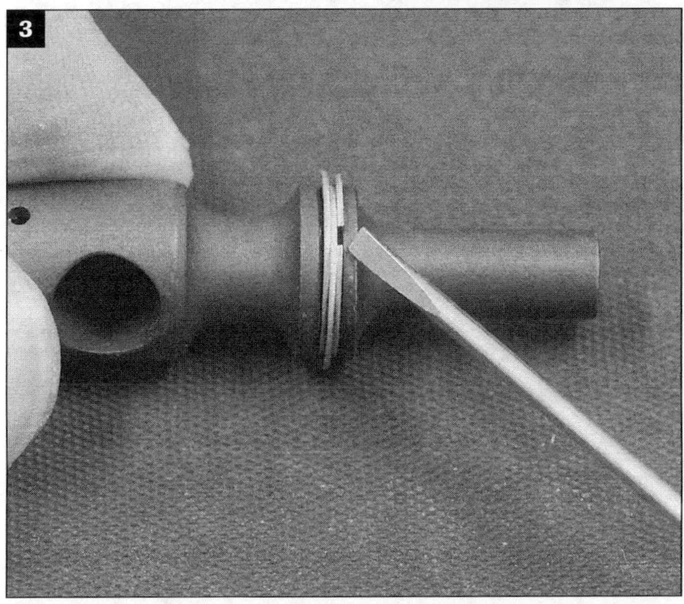

GAS RINGS

One note on gas rings. They don't last forever. Replace them every 2500 rounds or so and always keep spares.

Installation is pretty simple. **1.** Slip a ring over the tail of the bolt. **2.** Get one end into its groove and then the other end. A little twisting motion to get the end to spiral and snap over will prevent having to spread the ring too much. They can bend. **3.** Repeat for the other two rings and then orient them so no gaps are in proximity to each other. A small screwdriver helps.

Shown below on left is a **McFarland** one-piece ring. These last a long time and install the same way. It just spirals or "threads" into place.

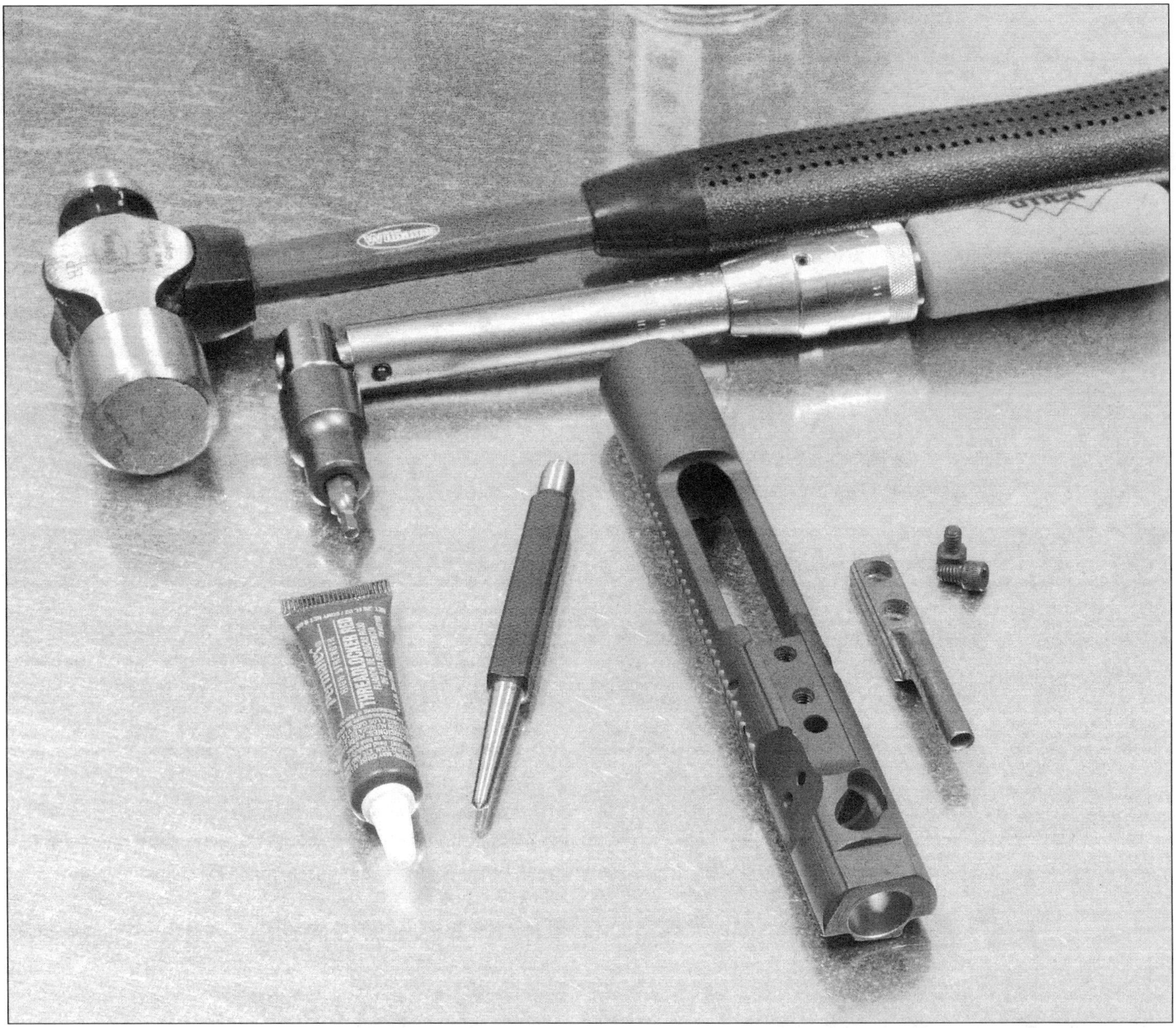

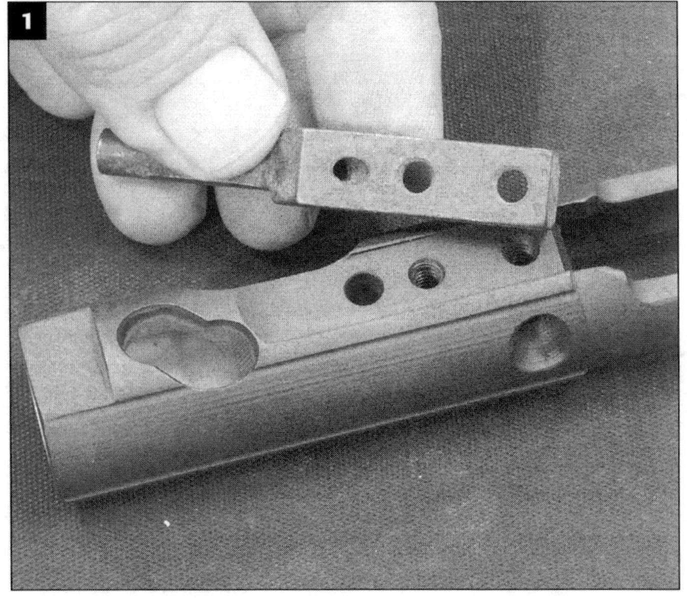

CARRIER KEY

Dang. I hate to disagree with so much published information, but the 30-40 inch pounds suggested for the carrier key screws is not enough. Note that is *inch*-pounds, not foot-pounds. They need to be daggone tight and I don't think a torque wrench is even necessary. Tighten them equally by alternating tugs on each screw; don't tighten one fully before having the other roughly equally tight. A loose carrier key is a sure malfunction.

1./2. If you want to use glue, choose "red" and degrease with contact cleaner prior to installation. **3.** Position the key in place and thread in and tighten the screws. **That was easy.** But we're not nearly yet done. We have to make sure that those screws stay tight.

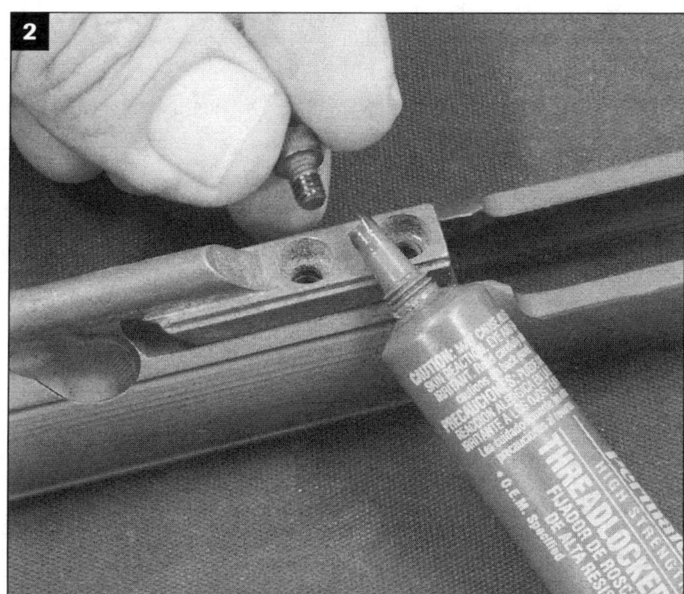

STAKING THE SCREWS

The screws have to be staked in place. Glue won't always do it. Some do both and that's fine, but this part gets hammered continually and will loosen. Not may. Use a "prick" punch and a real hammer. You'll see them **1.** staked from the side of the key or **2.** staked atop the key in two or three places. Staking from the side I think is easier because the little ridge gives something to help position the punch. The idea is to move metal from the key into the screw heads, which are always knurled for this reason. Position the punch and give it one good whack with the hammer. It's not hard to do. It may also not be the prettiest work you've done either. There is a specialty tool you'll see soon enough that are worth the money. (Tool shown is a **Starrett 5/32 center punch**.)

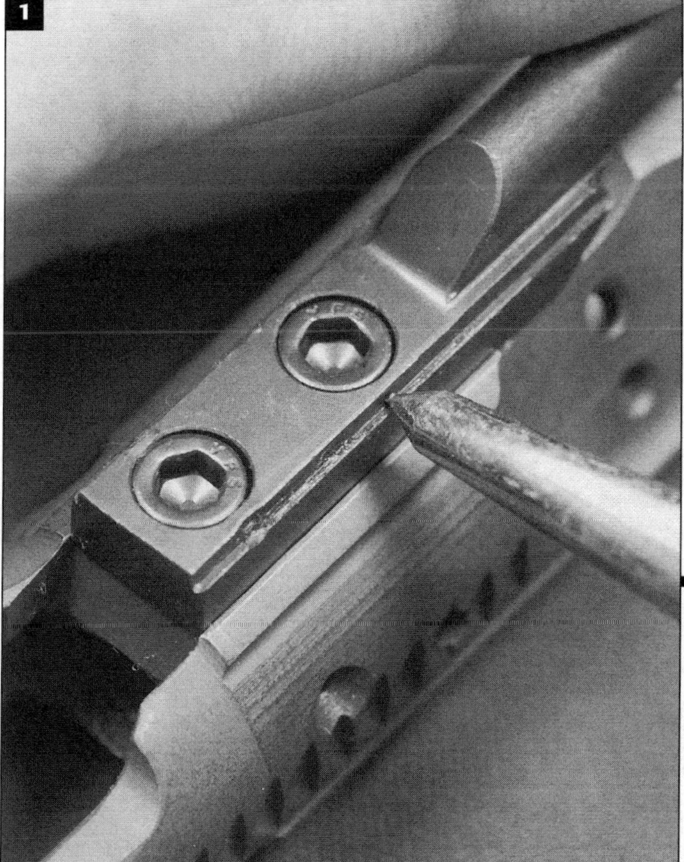

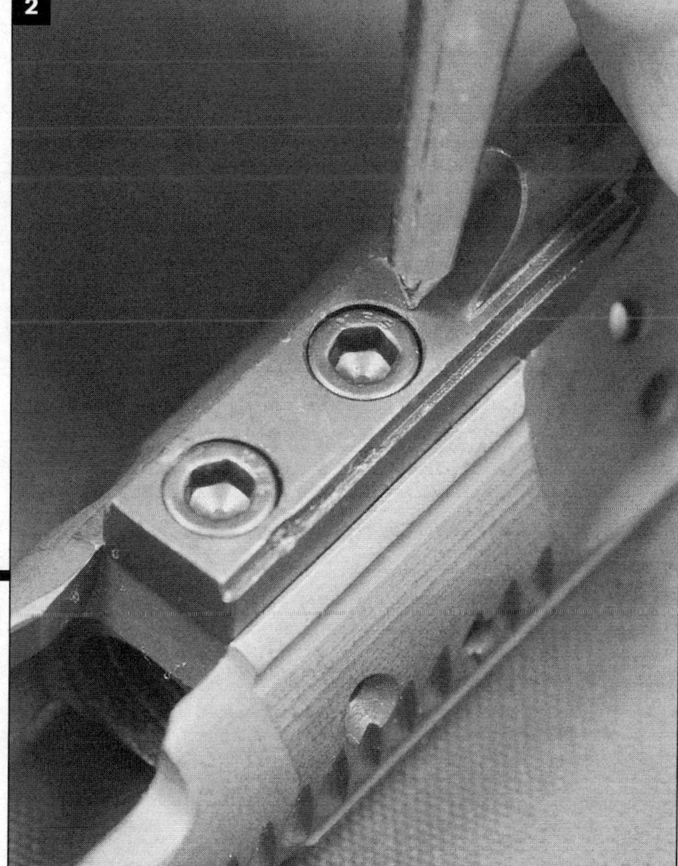

Moacks Tool

This tool puts a sano stake on the carrier screws and is pretty foolproof. Carrier key screws have one of two different hex sizes in their heads, so figure out which you need and that wrench will position the bolt carrier in place. Turn the screws in from the side to get the amount of displacement you want. Job well done.

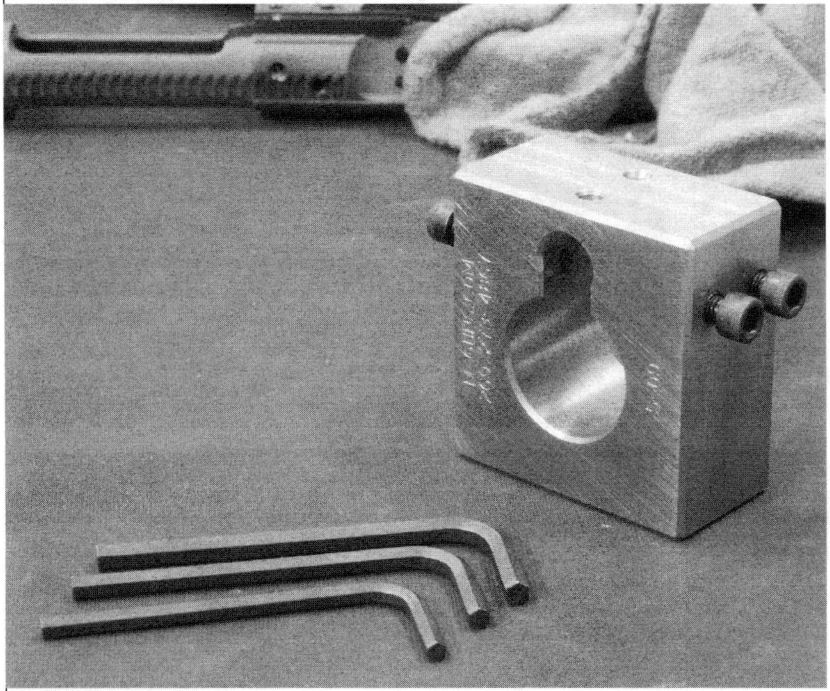

This is a Moacks "Plain." There are fancier and somewhat more plain versions, but this one works great. Moacks aren't cheap. If you do much of this, however, it's more and more valuable, as with many tools. On the other hand, I have seen so many carrier keys that didn't have staked screws that there's a pretty good chance you'll end up using yours even if you never build a rifle.

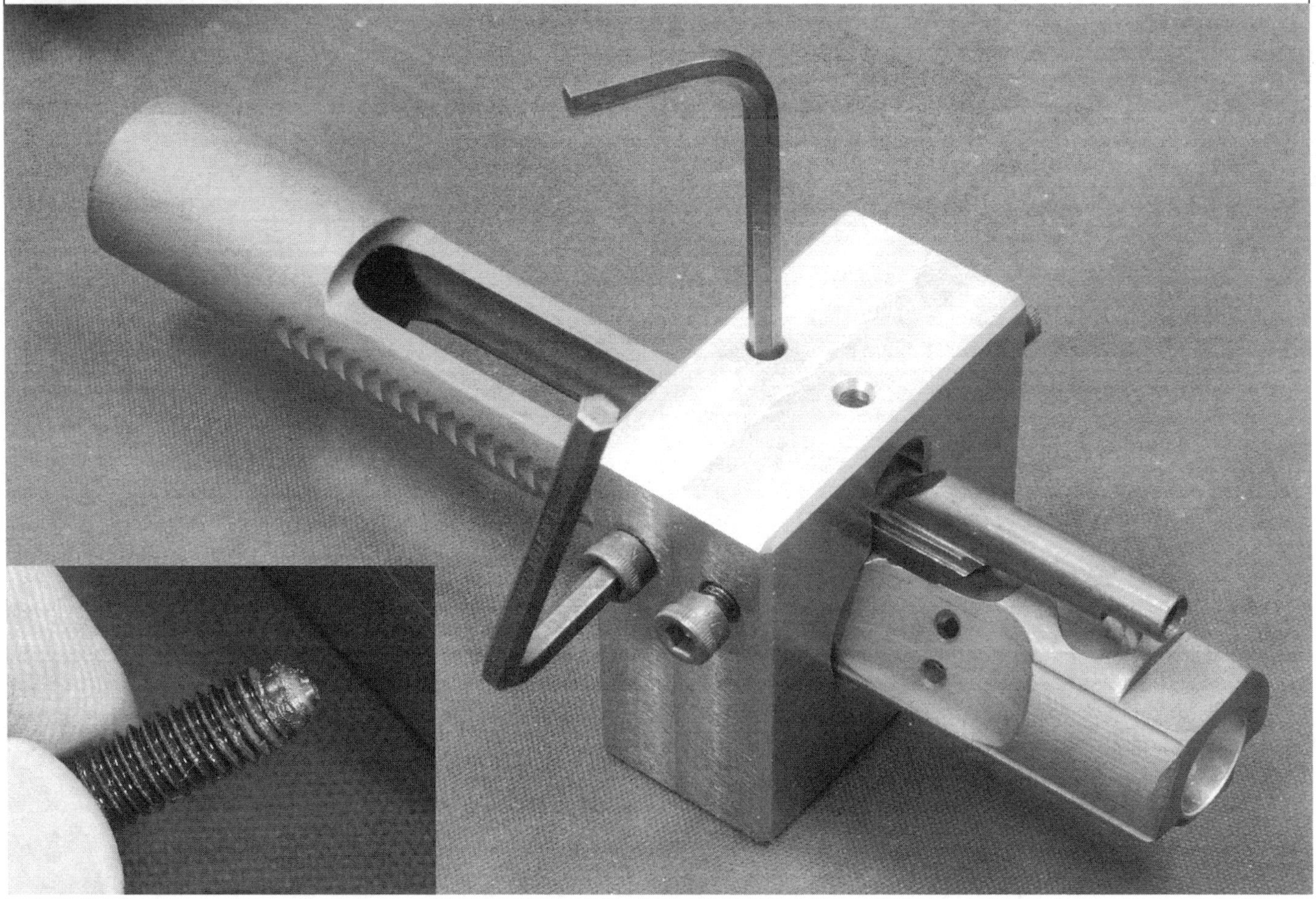

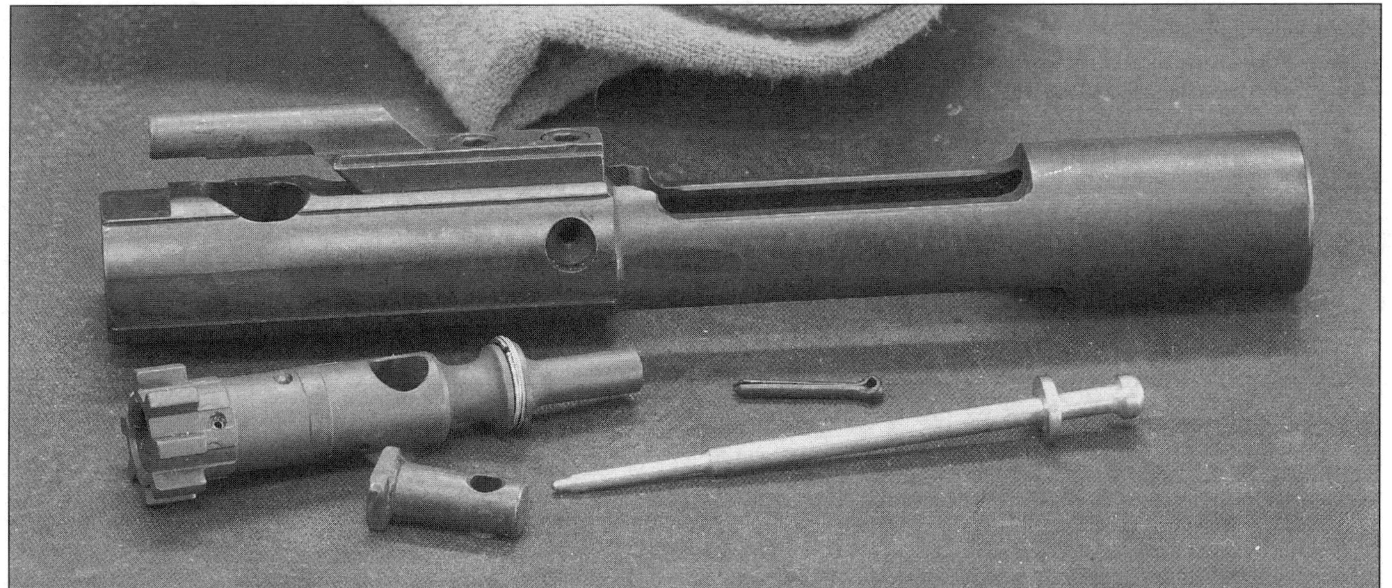

PUTTING IT TOGETHER

1. Insert the bolt into the carrier and push it fully in. Ejector on the left; extractor on the right (cam pin won't fit any other way). Wiggle the bolt just a little forward and orient the cam pin hole in the bolt so it can accept the cam pin, which is **2.** inserted in a lengthwise orientation and then **3.** rotated into position. This lines up the holes to accept the firing pin. **4./5.** Insert the firing pin and push it fully forward. **The bolt is pushed fully to the rear during these operations. 6./7.** Insert the firing pin retainer from left to right (the head of the retainer fits into the dished out area on the left side of the bolt carrier body). Seat it fully. Take a look at the last photo of the AR15 carrier on the next page to see that **the retaining pin has to be behind the collar on the firing pin. 8.** To install the bolt carrier assembly into the rifle, the bolt must be pulled fully forward. **There you go.**

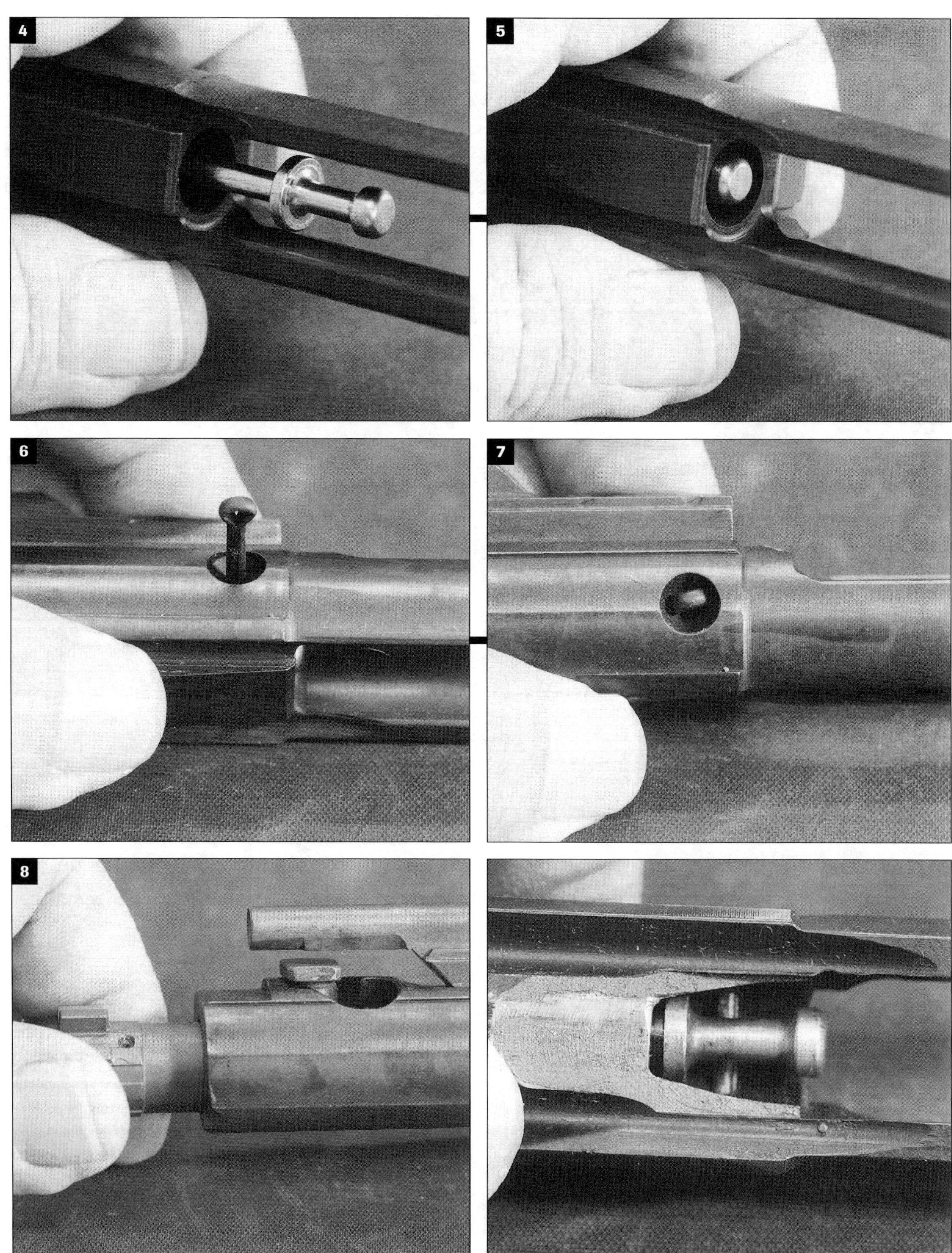

11.0 SERVICE RIFLE

NRA/CMP COMPETITION RIFLE

[There is a lot of material in this segment, which is actually broken into a few sections. The reason is that I will use this project opportunity to fully complete a build of an A2 style rifle, including front and rear sights. I didn't do those separately elsewhere because this is an opportunity to explore the "match" options, while still thoroughly covering the basics. The only real difference in installing a GI rear sight and a competition-use rear sight, for instance, is the parts specifications.]

SEGMENT CONTENT

132   **Barrel** (choices and chambers)

134   **Float Tube** (and other parts selection)

139   **Parts & Tools**

140   **Barrel Nut & Barrel Installation**

143   **Float Tube Installation**

146   **Gas Tube & Sight Housing Installation**

149   **Flash Suppressor Installation**

152   **Handguards & Sling Swivel Installation**

154   **Disassembly**

I have a suspicion this part of the book is one reason most bought it. The NRA/CMP Service Rifle is the heart of High Power Rifle competition, and therefore squarely on center-stage in this book. It's the most complex and time-consuming project in here.

[If you're refurbishing an existing rifle, or building up an already assembled stock gun, that's addressed later in this segment.]

PARTS SELECTION

It's the NASCAR of competitive shooting. Looks like, but isn't. Because of that, one of these is actually harder to do up than an NRA Match Rifle. There are literally more parts and that's because it has to fit the standard A2 form. That means a rear sight installation onto the upper, forward assist and trapdoor on the upper, and a more complex free-float tube system and installation. There's also a front sight housing that can, and should, be modified for better function (or replaced with a modified part). There may be a few more tools to order too. No steps for a stepper.

It's not hard to find target parts for smoke poles. It's also not hard to find good ones. The trick is knowing whiches are which. This book segment shows a few options, and they're things you'll be happy with. The better things usually come from builders who have been motivated by the lure of fair profit and offer their proprietary parts for sale singly. In other words, and this involves sights mostly, builders like Derrick Martin developed their takes on systems originally engineered for the custom rifles they build, and then decided to offer the same for sale in a poly bag. Those are also going to be the more expensive items, but the cost difference is patently insignificant compared to the quality and function difference.

Choosing the parts for this project respected quality, cost, and availability. Three things most see as important. I didn't spend the least I could have, and didn't nearly spend the most, but I did chose pieces I have confidence in, with a goal of producing a rifle capable of shooting to anyone's capacity.

BARREL

First, always, barrel. There are a good number of ready-to-go "HBAR" configuration barrels available. "Drop-in" barrels from Wilson, Lothar Walther, Olympic Arms, and more similar price-point units are available from several outlets. Taking a step up, Pac Nor will sell a barrel ready to install too, and offer a choice of rifling styles, materials, twist rates, and chambers. Krieger can box and ship finished barrels too, and that choice will be a star in the crown of any Service Rifle.

I know most of this has already been discussed previously in this book, but refreshing anyone's memory never hurts. I know I can't remember why I went into the kitchen anymore…

The short course on barrels for these guns is that we're looking for a heavy-weight under-handguards contour, usually just under an inch diameter, that then steps down successively toward the muzzle to A2 exterior standards. That was and is the essential configuration for a Service Rifle competition barrel. Bigger differences exist then in barrel quality, as mentioned, chamber specs, and twist rates.

Since it's a Service Rifle, the gas port portion will be parallel at 0.750 inches, and all A2-style sight housings will fit the same, and are the same in this dimension.

My choice for this one

An outstanding barrel comes from Satern Custom. It's stainless steel, nicely finished, and has a Wylde chamber. It's cut rifled and there are several twist rate options. This one has a flash suppressor installed but that's an option too. The front sight housing comes with the package and is modified for set screws that go into flats already machined into the sides of the barrel. A complete bolt is included so no headspace worries. It's hand-lapped and perforates as well as any. Recommended. Chosen. Coincidence?

Brownell's offers a Shilen-brand stainless steel, complete with bolt and Wylde chamber, in a Service Rifle configuration. This is a good barrel. Things like this just didn't exist a few years ago, and it's great for us now that they do. It's one more reason to do it yourself because it's one less reason you have to envy anyone's custom barrel.

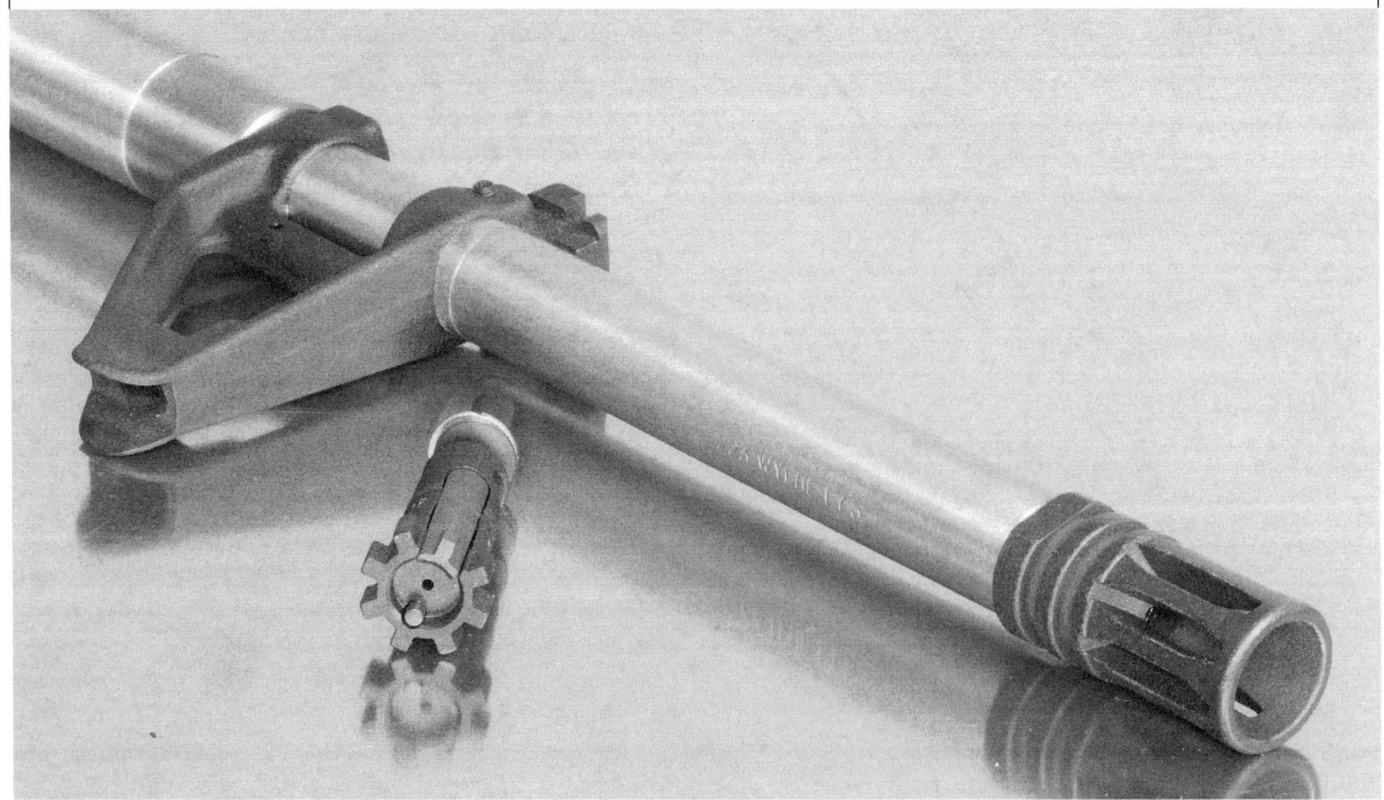

Understanding Chambers

There are a good number of .223 Remington chambers. These, for the most part, came about over the past decade to accommodate the needs of High Power shooters using longer bullets. Therein lies the glaring difference in them: they produce varying amounts of leade, or throat length, which is shown in an illustration at page bottom. For more, way more, about this check The Competitive AR15: the ultimate technical guide or Handloading For Competition.

Essentially, best accuracy almost always comes when a bullet in a loaded round is seated to engage the lands. Since 80-

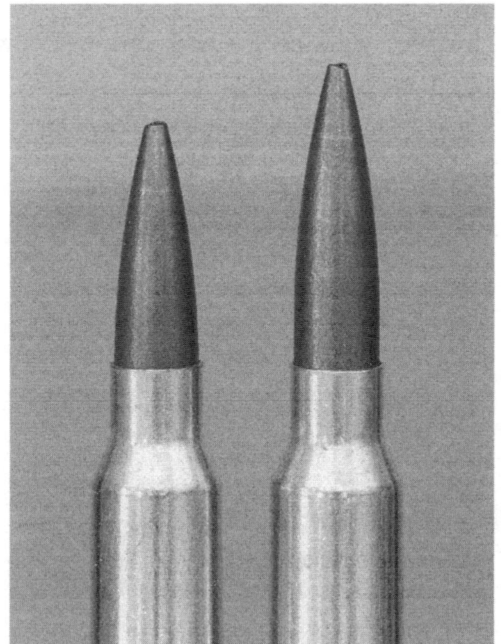

grain-vicinity bullets are pretty long, that means the option is to scoot them way back in the case on a short-throated rifle or let them scoot out into a longer throat. A shorter throat keeps the first point of contact with the rifling closer to the bullet for those seated in to fit into the magazine. Many believe this makes them shoot better, but a High Power rifle, in my estimation, needs to perform its best at 600 yards where the longer bullets are used. And! Based on the scores and groups, related, that we get making short bullets jump what anyone would say is an excessive amount, it doesn't really matter if the chamber throat is longer or shorter.

Although it's not "really" done this way, we do it, so the commonly expressed measure of the differing chambers in use is based around a cartridge overall length with a Sierra 80-grain MatchKing seated to engage the lands. For instance, a Wylde chamber, which is my overall recommendation, is "2.475" inches, meaning just as said that that is what a Sierra 80 will measure just touching the lands.

Many available barrels in Service Rifle configuration, or other forms for that matter, carry chambers that are way too short for best results across-the-course. Many have a SAAMI minimum chamber which usually measures 2.420 (or so) inches using our 80-grain yardstick. Keep in mind that there are many chamber reamers in use and what constitutes a SAAMI minimum may vary from maker to maker. However, they are always too short. On left shows the difference in seating depths that touch the land between a SAAMI minimum and a NATO chamber. Big difference.

Two of the recommended barrels mentioned, the Brownell's-Shilen and the Satern, have Wylde chambers. These were the only ones I could find that did, as a stock chambering. It's possible to get that chamber, or another, from a barrel maker that will do up a custom drop-in for you. Wylde is a safe and well proven bet. Its other dimensions are generous, and that's good. No .223 should have a "tight" chamber in any dimension. Function first. By the way, a NATO chamber, if you can find one with a decent rifled section extending from it, is even longer throated than a Wylde and won't handicap a High Power shooter from any yard line.

Always check the throat on a barrel, even if you know it has such and such chamber. If it's not something you know, such as purchasing a nothing-special HBAR configuration from a supplier (whether or not advertised as a "match" barrel), there can be a wide variation in the chamber specs. As said, it usually will be shorter than we'd like, and often it's SAAMI-minimum. If it's NATO that's fine, and borderline dandy. Use one of the Hornady LNL gages shown on page 88. It's the only way to know what you have to work with in your barrel.

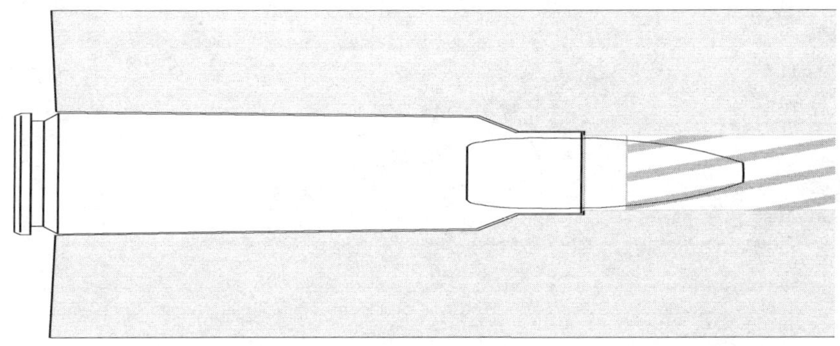

The illustration on the left shows throat length, which determines the amount of jump a bullet will have to span to engage the rifling. Its esentially the distance between the cartridge case mouth and the first point of "major diameter" in the rifled section of the barrel. If it's a 0.219 land diameter, then that's the same point on a bullet that will make first contact. Farther forward is good for longer bullets seated out; shorter is better for shorter bullets coming from rounds fed out of the magazine.

Get at least a 1-8 twist for a competition rifle, no slower. A 1-9 will not shoot an 80-grain Sierra, or similar. The barrel I chose had options ranging from 1-8 to 1-6.5. That last will rotate a 90-grain bullet enough to stabilize it. I chose a 1-7.5 for honestly no reason beyond believing a little faster is a little better, and the longest bullet I planned on running through this one was an 80.

Alternatives are mentioned and many are shown, and this topic got beaten to ruin in *The Competitive AR15: the ultimate technical guide*. Accessorizing a Service Rifle largely revolves around trigger upgrades, weighting, and sight pieces. Those will all be addressed later in this material.

FLOAT TUBE

I chose a Rock River free-float tube for this example because it's probably the most commonly used on the market. It's also easy to install. Others are not overly difficult but this one is simple and easy to index, and that's why it's easy. A few are more complex or require extra steps, and, again, none are easier to set in place correctly.

Aligning a float tube that doesn't have some means built in is easy, if imperfect. Custom builders may end up creating their own alignment fixtures, but for the most of us either visual or (my preference) self-made indexing marks help. Eyeball the front handguard cap and also check gas tube alignment through its hole in this cap. That last, of course, is done in test fitting. Make a pair of marks from cap underside to barrel when you have it sitting straight and then return it there on final assembly. Don't worry, you're going to check it many, many times before pronouncing it "good." It's important.

Service Rifle float tubes are small diameter, much smaller than Match Rifle tubes. They are very close to the barrel, and shims can and should be used to ensure that the barrel is centered in the tube. Strips of cardboard work fine. Keep them in place while the glue is setting on the collar. You want to see an equal gap around the front handguard cap before you call it good to go.

GETTING READY TO START

Assemble the upper receiver. Goes pretty much without saying.

I used a good bolt carrier for this bad boy, and chose what I chose because it is the heavier M-16-style configuration. I got this one from a good GI parts supplier and it's nothing tricky, just correct. Since the Satern barrel came with a completely assembled bolt, I used it, after disassembly to polish the extractor and work over the ejector.

Since it's a Service Rifle, nothing changed in the routine-function pieces parts and the only attention paid was to their correctness, deburring, and functional capacity.

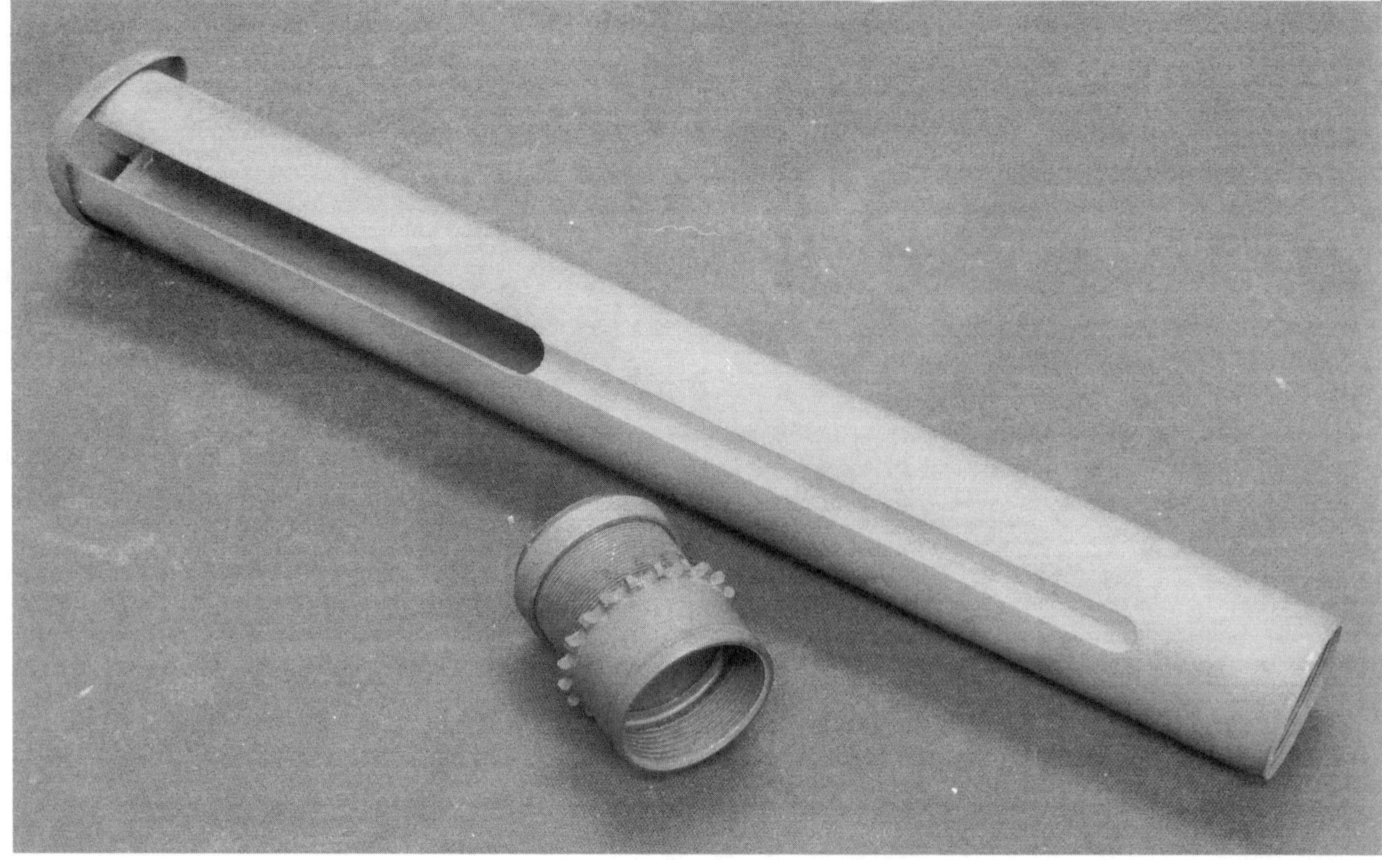

My choice for this one. The Rock River free-float tube set comes complete with a gas tube, modified handguards, and the tube hisself. Nothing else needed. It's a complete package. Shown here is just the tube and barrel nut. The tube is plenty strong. I chose it for this project because of all that, and also because it's very popular. Make sure you order a Delta assembly and sling swivel with rivet.

Bushmaster

This is a Bushmaster tube. It's the same thing as on all the zillion "DCM" rifles sold by this company, and there's enough of them in service with nothing approaching a failure record.

DPMS

I've gone with the DPMS tubes on some of my Service Rifles. DPMS has a few of what some builders see as advantages. One is that the threaded engagement between the tube and barrel nut is greater and the threads themselves are finer that most others. Security may be improved. Next is that DPMS gives the most "room" for gas tube installation, and configuration, and that's

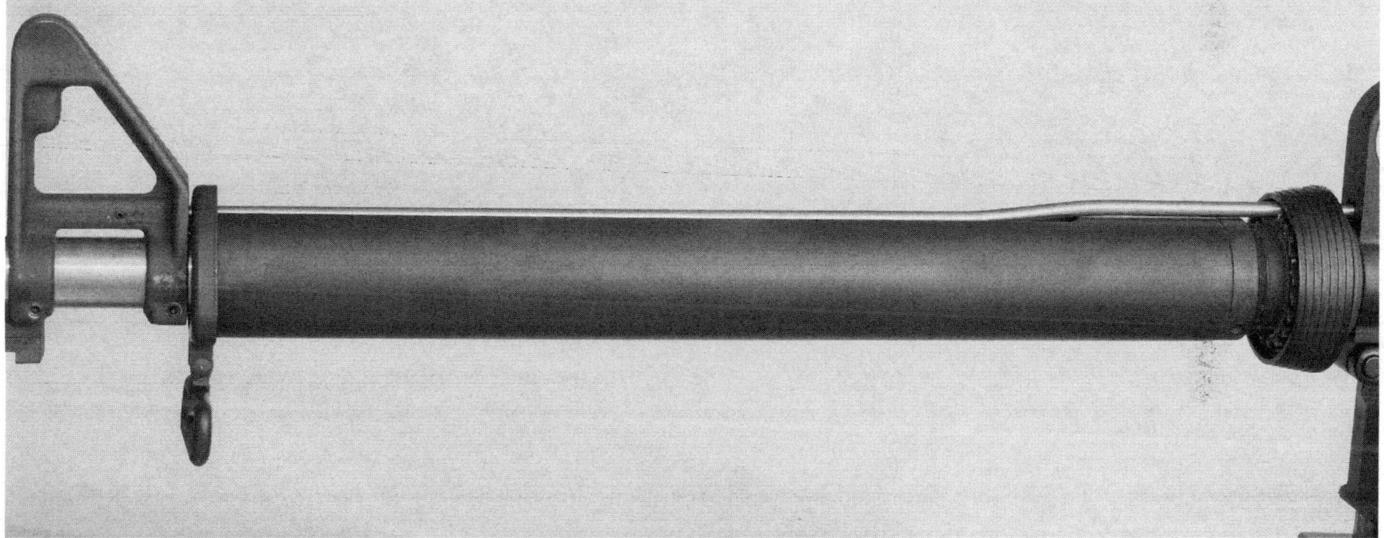

minor but it's been mentioned as a plus to me by more than one professional builder. The gas tube cutout is full-length. Another I favor is that the sling swivel attachment point is under the welded on handguard cap, not sticking out ahead of it as on the Rock River. If your front sight housing has a set screw on the underside of its back flat the swivel mount will interfere with securing this screw.

Again, the more complete the package is the better it will be for the home builder. The DPMS comes more complete than any of the others, and it's honestly my recommendation. I went with the Rock River, as said, primarily because of its popularity.

I glue the float tube collar. *I use very high sling tension. Permatex Blue is "enough" glue for this job due primarily to the thread pitch and area the threads occupy, but Red makes for a permanent installation. The Blue can be more easily released so the tube can be used again for the next barrel, along with the upper receiver. Oh, and on that topic, there's been talk for years now that an upper shouldn't be reused. Not true. That's the answer. It may be a chore to get all the threads and fitments loosened and the upper re-readied for a new barrel and parts installation, but all those parts can be reused. The only exception is if the upper receiver has a crack, and this crack will usually be near the barrel receptacle.*

Make sure you have all parts as part of the package. Gas tube, gas tube roll pin; Delta cap, weld spring, retaining clip; sling swivel, sling swivel rivet. And if a stripped bolt was supplied you'll need all the pieces to assemble it: ejector, ejector spring, ejector roll pin; extractor, extractor spring, extractor pin.

Upper and lower are from DPMS and nothing more (or less) than a quality 7075 forged pair. Routine drill-bit-checks of all holes and spring cavities, grip screw tap, and overall assessment, was all it got prior to fitting up the hangers-on.

All buffer parts (receiver extension tube, buffer, and spring), buttstock, and pistol grip are stock-stocks from DPMS. The Rock River free-float tube was unboxed with a pair of already modified handguard halves, which were common Thermold items. Some manufacturers require us to modify the handguards to work with their tube and that step is shown. It's easy.

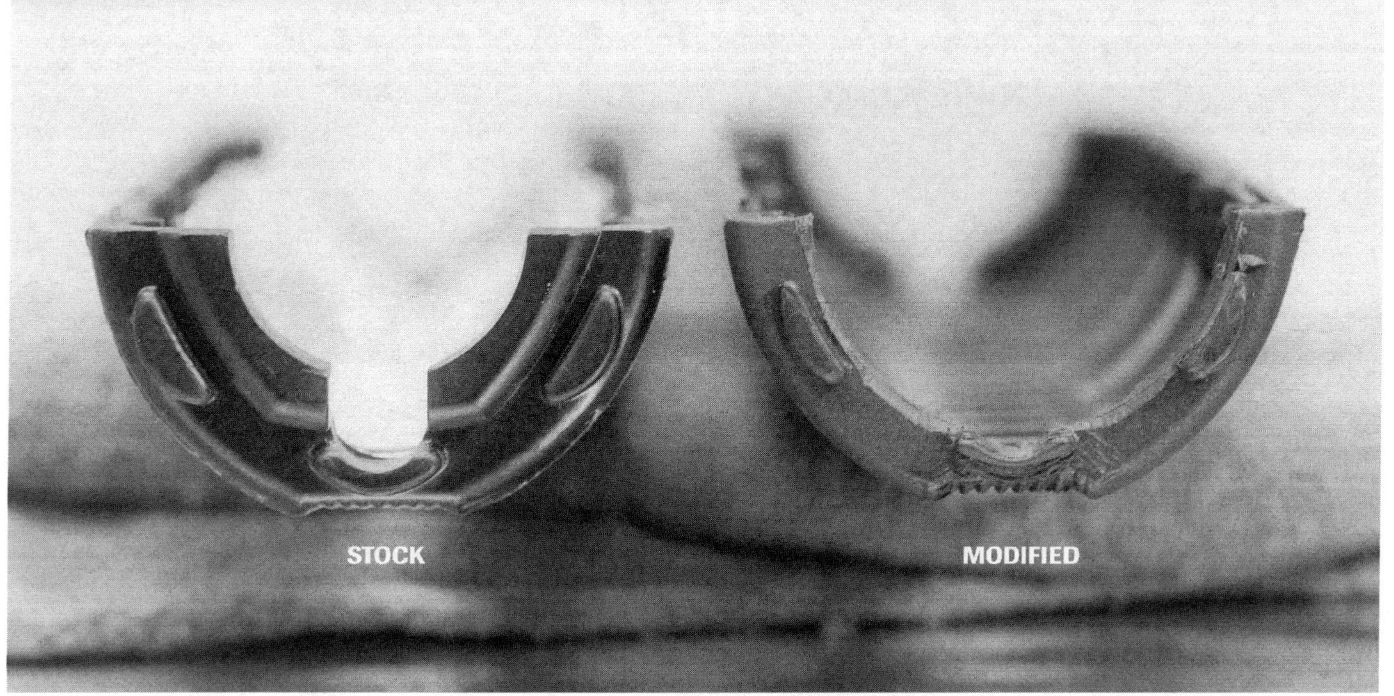

*This one is from **Accuracy Speaks** and it's a thing of beauty (shown also lower photo on preceeding page). It's also more tedious to install, which is not at all the same as saying it's difficult. Derrick Martin provides fine instructions. This tube is stainless steel and extremely well made and strong. You will feel a little more like a gunsmith tackling this project because you'll have to modify some stock parts. The gas tube (maybe, depending on barrel contour) and handguards (surely) both have to be fitted to work within the system. Modifying the handguards is easy. Remove the metal liner (the one that says "do not remove") and essentially pare away plastic as shown. A hand-held grinder works great for this process.*

Receivers Jump step way ahead here and I'll say "get your assembled lower…." A Service Rifle lower receiver assembly has all the same parts as an A2 except the trigger. Alls you need to get that from there to here is a stripped lower and a baggied lower receiver parts kit. Follow the step-step on lower assembly and then come back to this page. I have recommended building the lower for any rifle from pieces because it's easy and certain, but here can now be an exception. If someone wants a really easy time of Service Rifle assembly, he can purchase a lower from DPMS, or another maker, with a two-stage trigger already installed. It is not a preferred two-stage but it's not a lost cause either. I've used two-stage triggers from

DPMS and Bushmaster and they're not bad. They're just not outstanding. A spring change can work wonders. The Superior Shooting Systems set in particular is very good. Neither trigger will be as good as a Geiselle but most wouldn't notice unless they had one to compare it to. That's essentially the difference, if that makes any sense. It's a matter of degrees, and better is always better!

Here's a nice set from Rock River. It's advertised as a "matched set" and indeed the upper and lower have a great fit. Notice that it ships with take-down and pivot pins. I can't honestly tell you that your rifle will shoot better with this pair in place, but it's a nice assurance for those who wonder if it matters. That's the closest to an honest answer as I can give. Sometimes quality matters more as a perception than as a reality. I of all people appreciate that!

PARTS & TOOLS

I'm going to skip the routine list of parts and tools to assemble the receivers. Aside from the trigger, it's all A2. This first section in this book segment will address installing the barrel and float tube, with associated other parts. We'll look at triggers and sights in more detail soon enough. It's essentially the same list as for a basic barrel installation.

PREPARATION

Vise, very securely mounted

Upper receiver clamp

Tap hammer

Roll pin punch and roll pin starter punch (for gas tube roll pin)

Torque wrench, 1/2-inch drive

Breaker bar, 1/2-inch drive

Barrel nut wrench attachment for the above

Gas tube alignment tool

Lower receiver block (for vise mounting)

Options (good ideas)

    Permatex "red" adhesive

    Permatex "blue" adhesive

    Contact cleaner

    Anti-seize compound (I like Loctite C5A)

    Gas tube wrench

    Cardboard shims (I use notebook backs)

PRECAUTIONS

Test fit everything and take your time for alignment checks on the float tube. Let glue set completely before handling the rifle and certainly before doing any more work on it. Give a minimum of 24 hours.

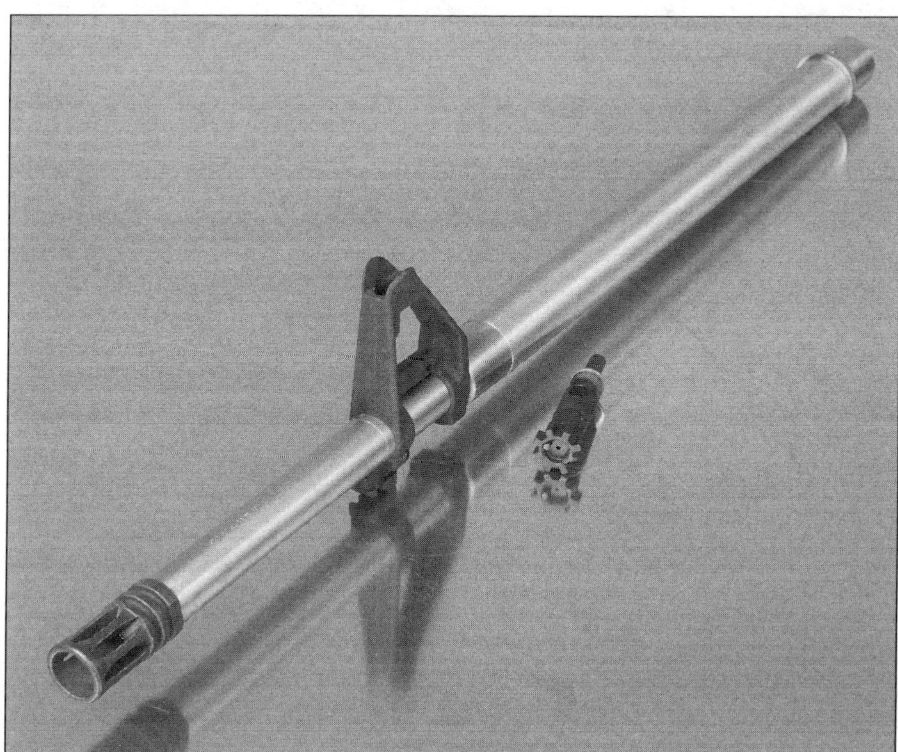

> **Port cover!** The only part that needs to be installed onto the upper receiver before doing up the barrel and float tube is the trapdoor cover, and make sure you did that! There's no undoing anything if you forgot.

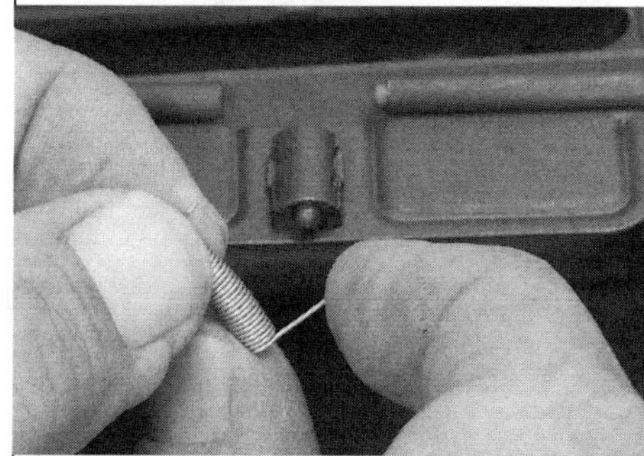

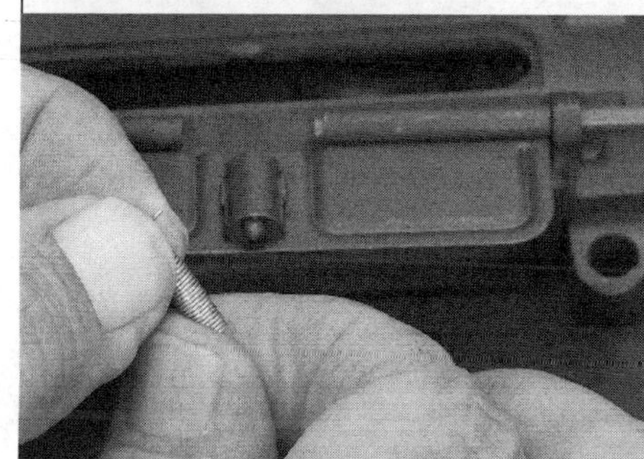

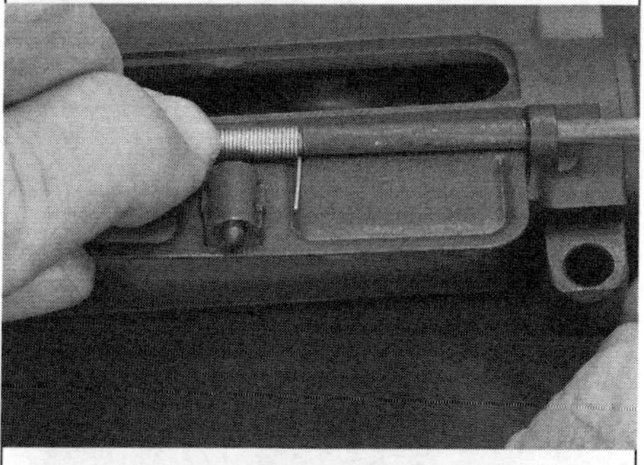

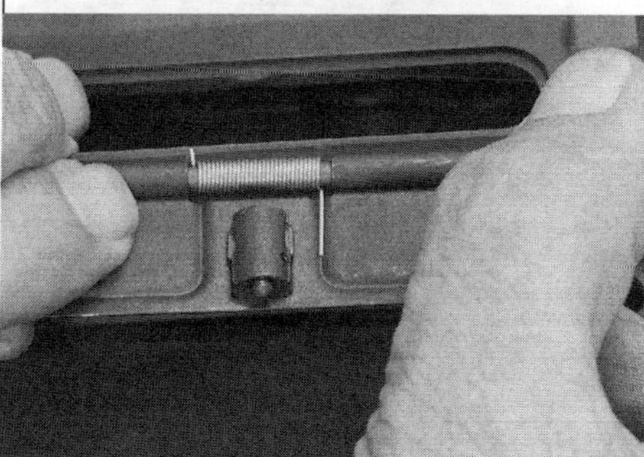

12.0 THE BUILD

FLOAT TUBE AND BARREL

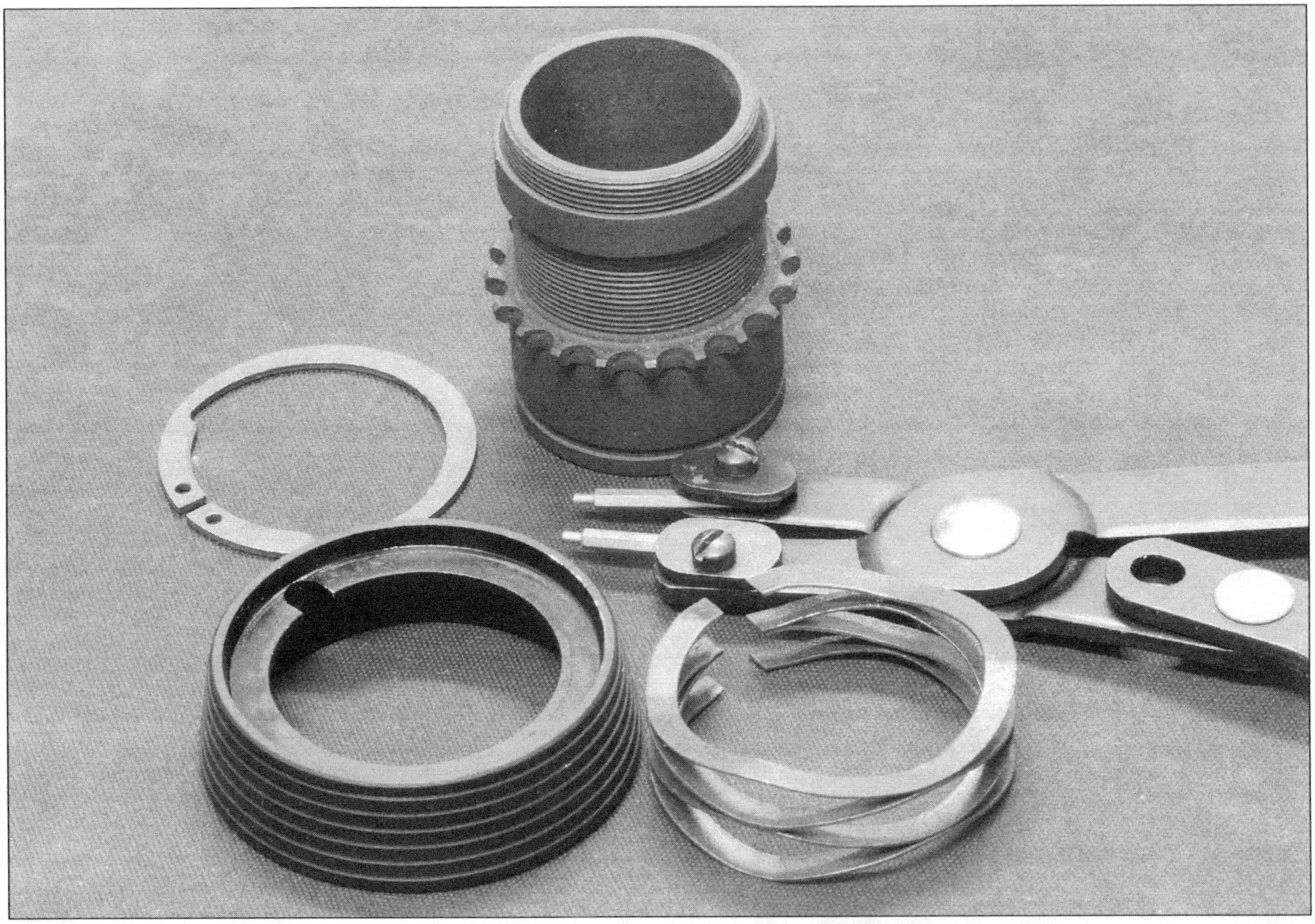

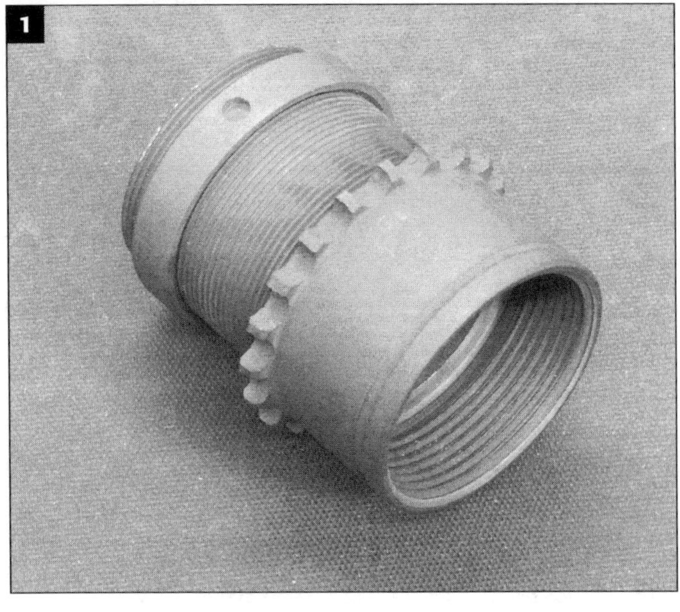

BARREL NUT ASSEMBLY

The first step is assembling the barrel nut. 1. Thread the float tube collar on. All we have to do now is attach a Delta assembly.

2. Put the weld spring inside the cap, line up the openings, and slide the works over the back of the barrel nut. **3.** Get your circlip pliers out, spread the retaining clip, and snap it into the groove. Make sure it's seated fully in its groove, all around.

Get the upper receiver, barrel, and nut ready for installation. 4. Decrease the barrel nut threads that retain the tube, the barrel extension, and inside the upper. **5.** Apply glue to the barrel extension (Permatex "red"), **6.** position the barrel into the receiver, **7.** apply anti-seize to the upper receiver threads.

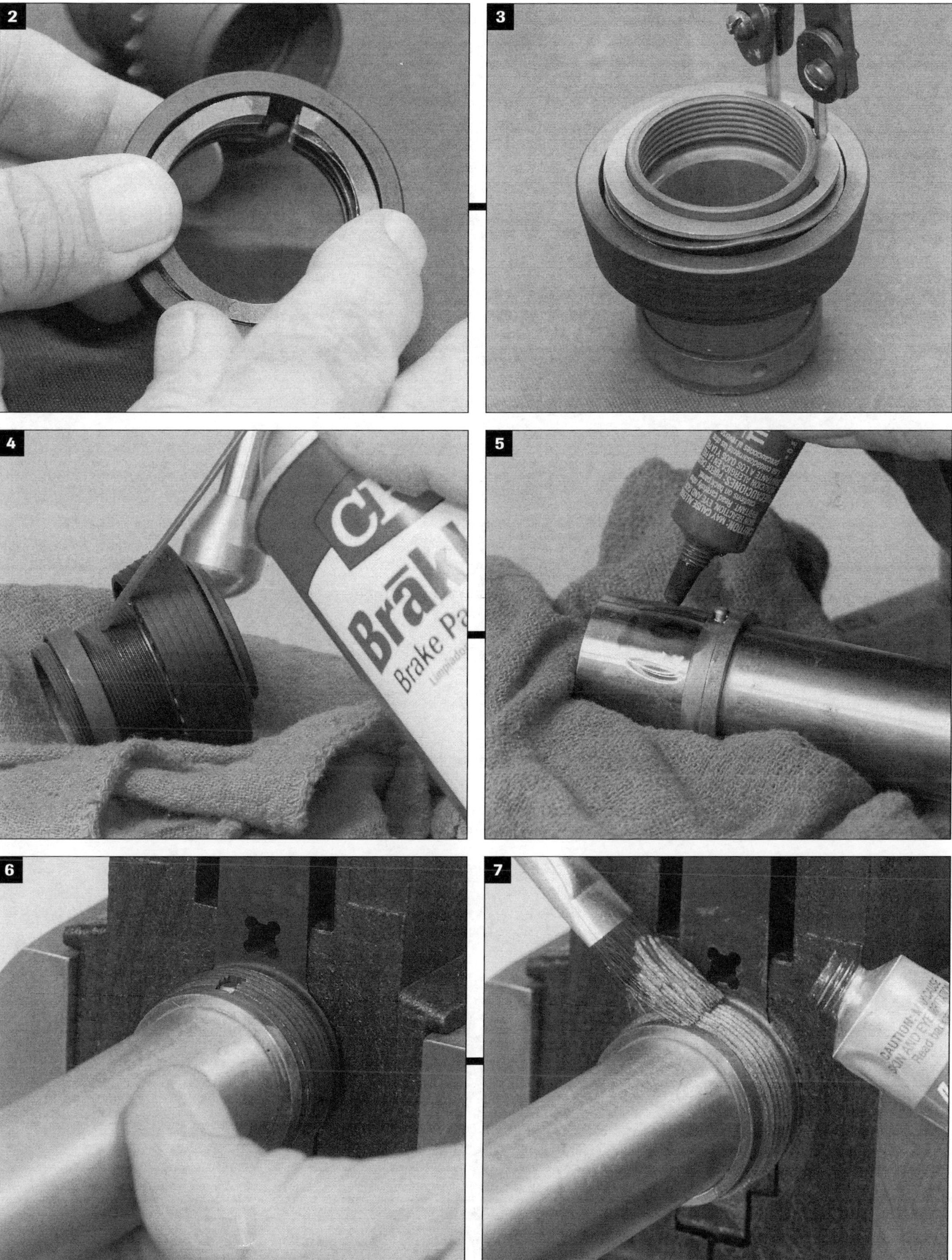

BARREL NUT & BARREL

8. Remove the retaing ring from the barrel nut (it might interfere with the barrel nut wrench.

9./10. Do the tighten-loosen procedure a few times with a breaker bar to seat the parts. **11.** Take it to torque.

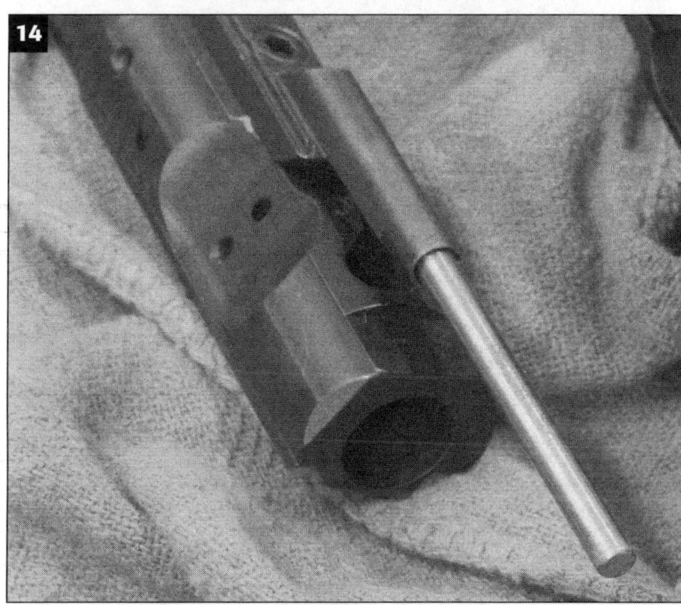

GAS TUBE ALIGNMENT

12. Use a flat-blade screwdriver to position the Delta assembly pieces so the gas tube opening is unobstructed, and **13./14.** check alignment with the gas tube alignment tool (and always then with the tube in the carrier.) Always take the barrel nut in the "tighten" direction to get it lined up and don't be surprised by how much force it takes. Get it right.

TUBE INSTALLATION

The tube now has to be installed and aligned. 15./16. Install the retaing nut and run it all the way to the back of the threads. Thread on the forend tube itself.

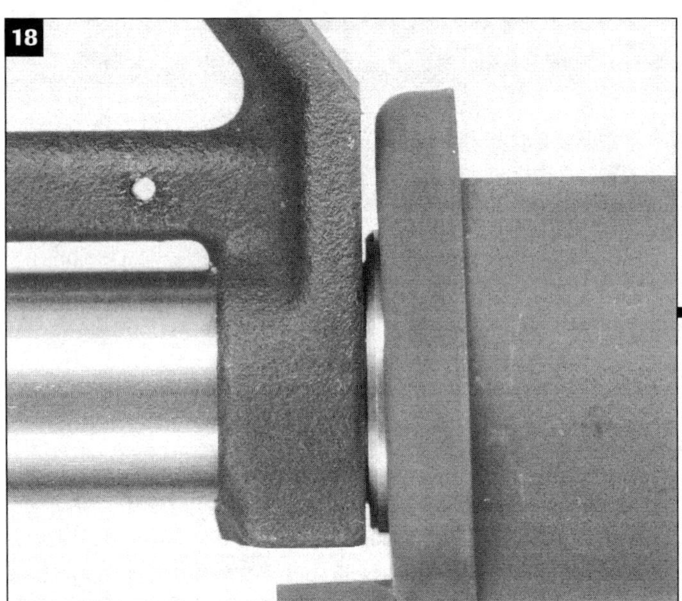

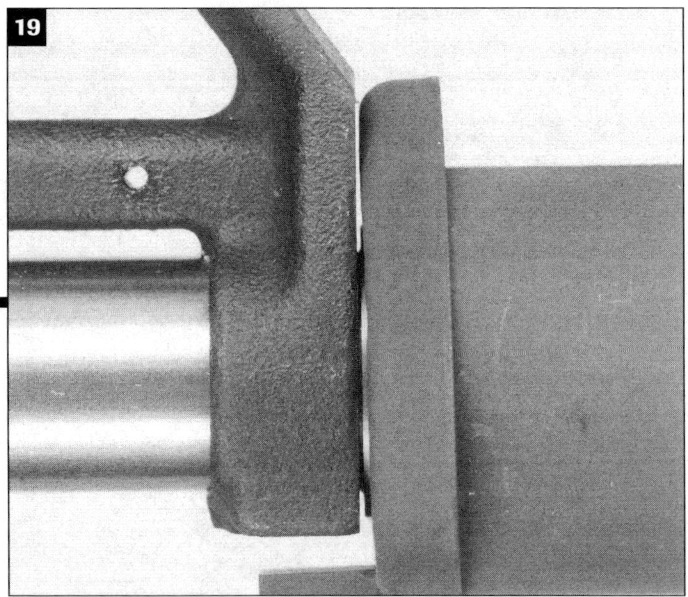

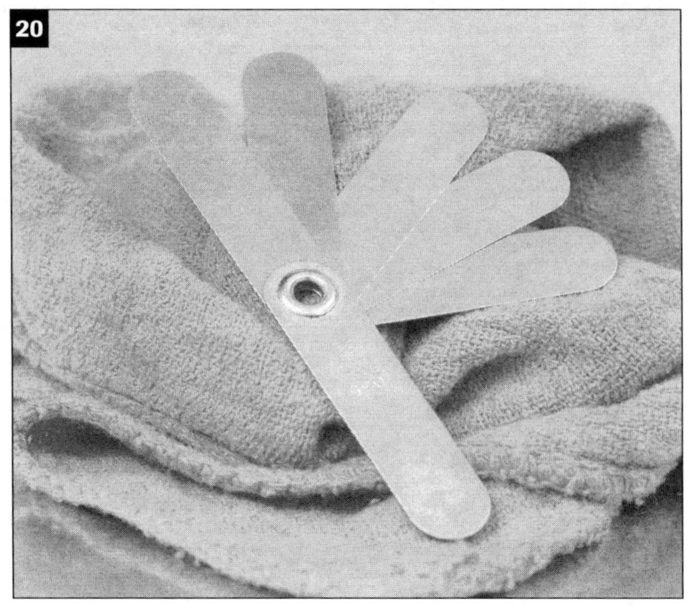

SIGHT HOUSING CLEARANCE

I test fit all this before applying glue mostly for the sake of clarity in the photos. This is a little side step for detail.

17. Run the tube back until its front edge is a little behind the barrel shoulder behind the gas port.

18./19. Slip on the front sight housing and run it back fully flush against its shoulder on the barrel. Run the tube forward to establish its forwardmost position. Remember, it takes a full turn at a time for the tube to be aligned. There has to be daylight showing between housing and handguard cap. These photos show the difference one turn makes. Finer threads are better!

20. No need for these really, but at least 0.020 inches is required.

TUBE ALIGNMENT

It's okay to start from here, by the way. Again, the sequence on the preceding page was mostly done for clarity. It all goes together either way and you'll find you do, as well you should, many, many alignment checks all during the process...

21./22. Apply Permatex "blue" to the degreased threads. Run the tube on paying attention to front sight housing clearance as shown on the opposite page. Alignment may not have to be perfect to make it work, but you'll want it as close as you can get it! It's primarily an eyeball issue.

23./24. I use a Sharpie and make a tic mark on the barrel shoulder in line with the gas port hole in the barrel and one on the front sight housing in line with the gas tube opening.

GAS TUBE & SIGHT HOUSING

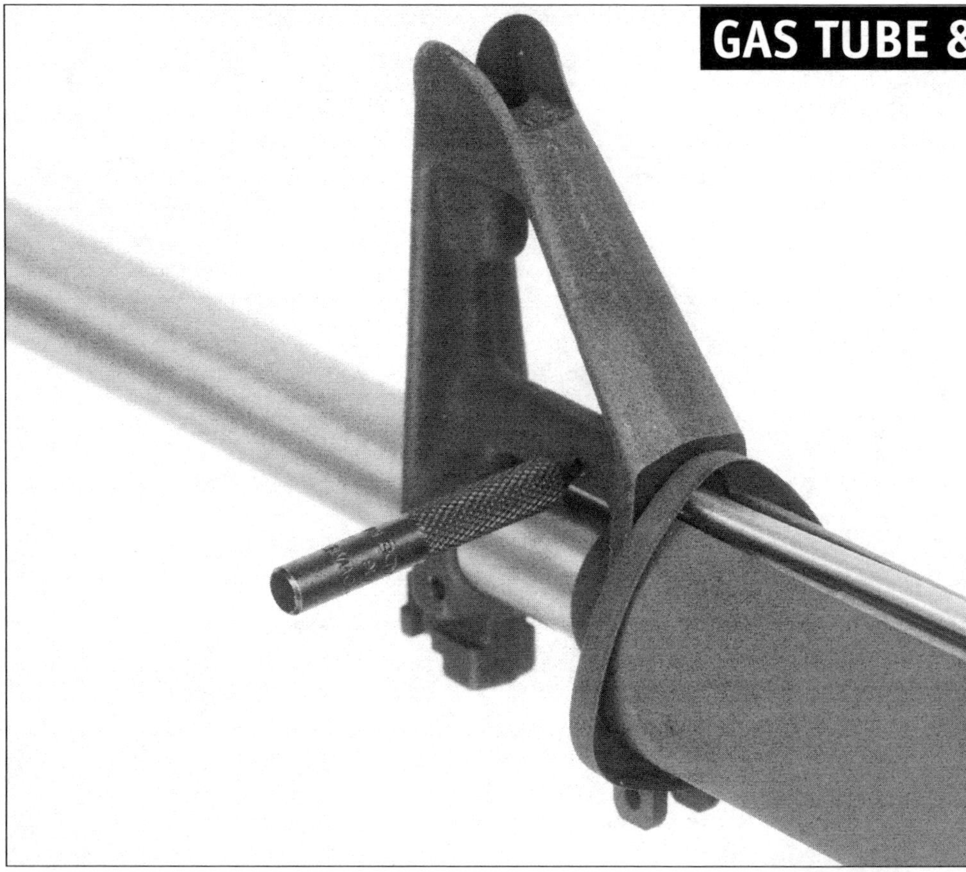

There are two ways to handle gas tube installation. Either mount the front sight housing on the barrel and then install and pin the tube in place, or, go ahead and install the tube into the housing and then slide the whole works in place as a unit. I prefer to assemble the housing and tube first. It's just easier and no banging on the housing afterwards.

FINISHING UP

Once it's all together, snug the retaining ring against the forend tube. If you don't have a spanner that fits it, use a flat-ended punch. A small strap wrench can help keep the tube in place while the retainer is being tightened. The glue will hold it.

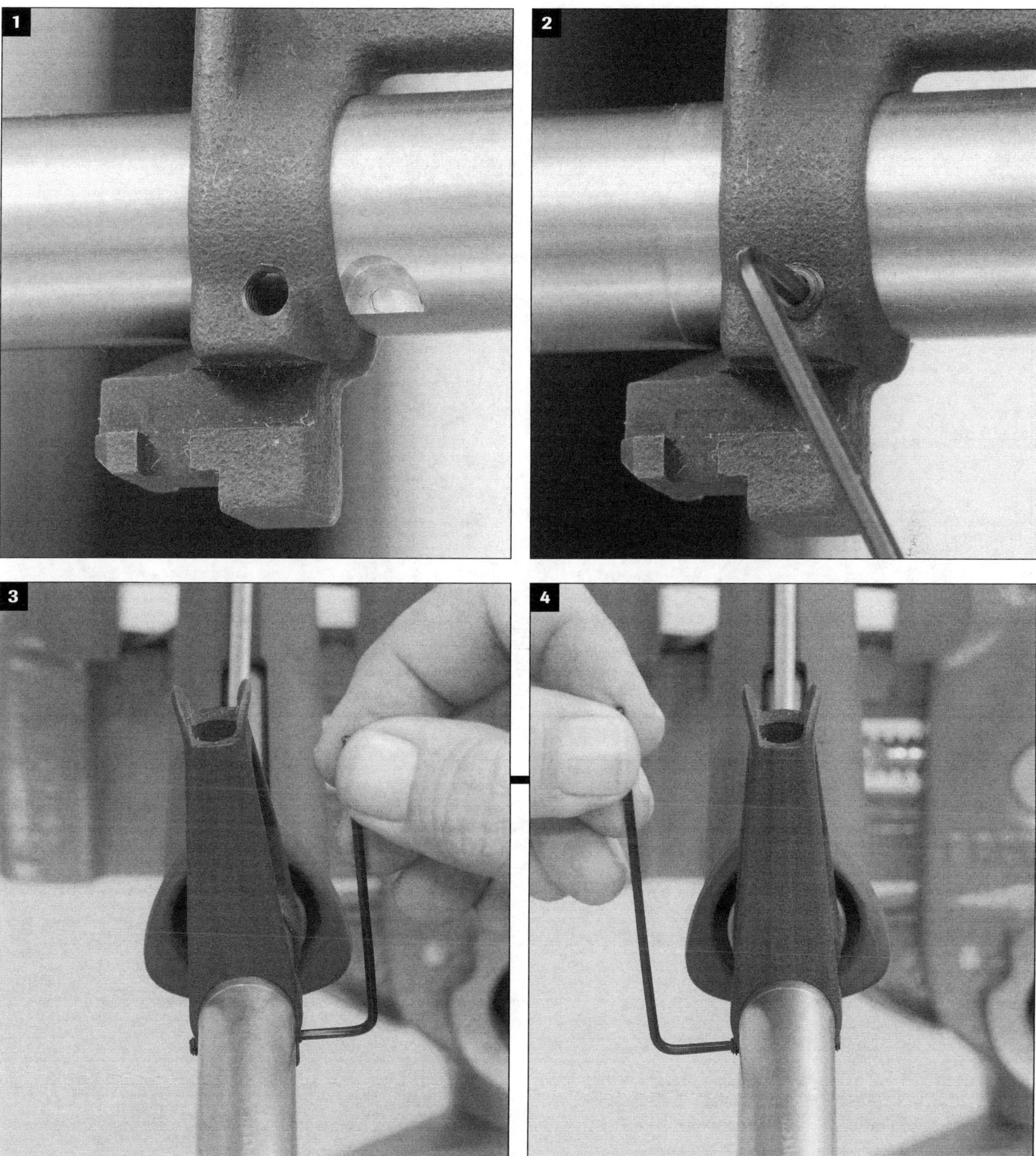

1./2. The Saterrn Custom front sight housing uses screws that fit into flats machined into either side of the barrel. It also has a set screw located under the gas-port-end lug but I couldn't find a way to use that because of how the Rock River sling swivel stud eclipsed access to that hole. It's not really necessary for retention, but if you're using another type of clamp-on housing, make sure there's a screw hole in the front lug on the sight housing to use this tube.

3./4. Here's how that works by the way. Tightening and loosening the screws on either side moves the housing so it's windage adjustable. Pretty spiffy. Effective too.

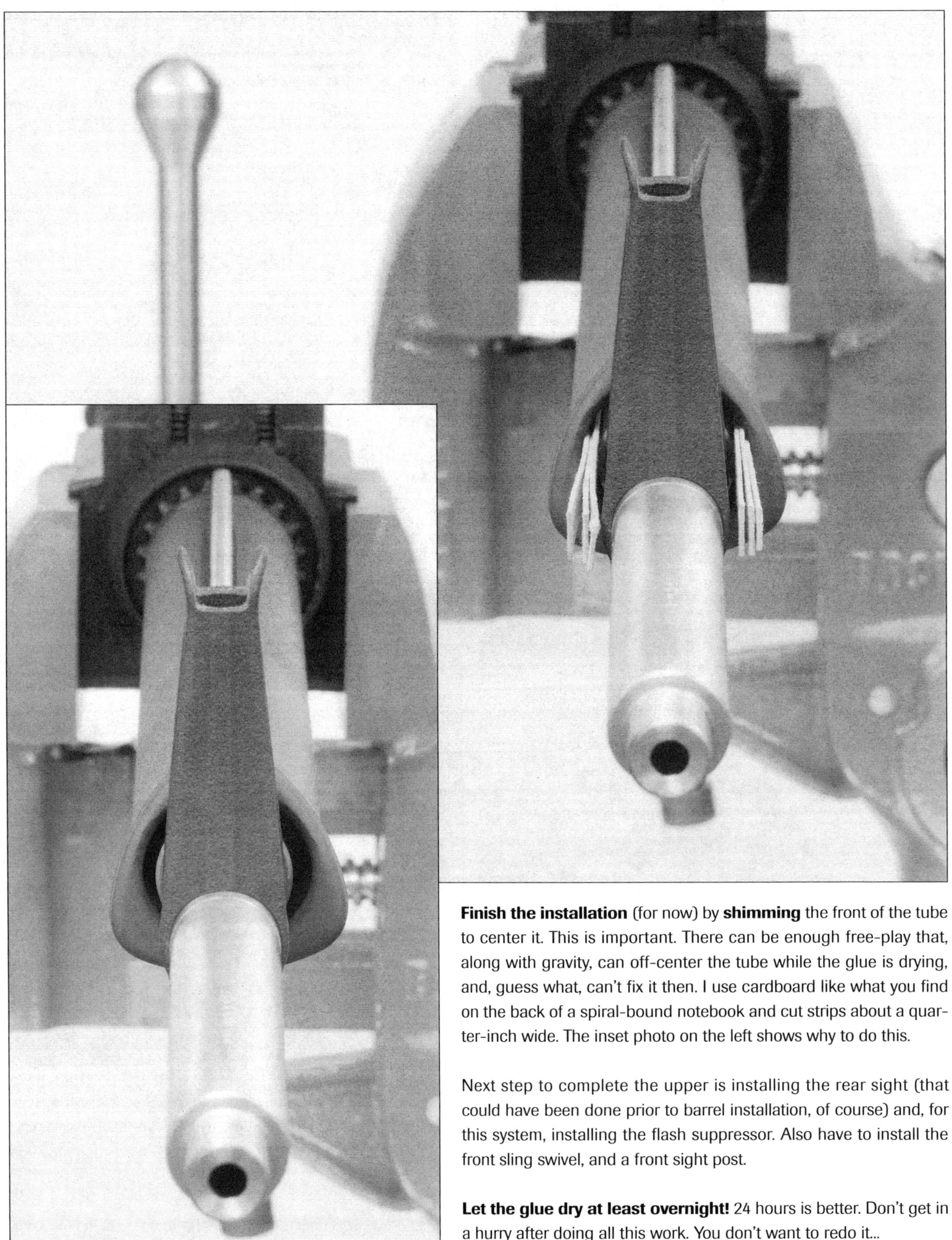

Finish the installation (for now) by **shimming** the front of the tube to center it. This is important. There can be enough free-play that, along with gravity, can off-center the tube while the glue is drying, and, guess what, can't fix it then. I use cardboard like what you find on the back of a spiral-bound notebook and cut strips about a quarter-inch wide. The inset photo on the left shows why to do this.

Next step to complete the upper is installing the rear sight (that could have been done prior to barrel installation, of course) and, for this system, installing the flash suppressor. Also have to install the front sling swivel, and a front sight post.

Let the glue dry at least overnight! 24 hours is better. Don't get in a hurry after doing all this work. You don't want to redo it...

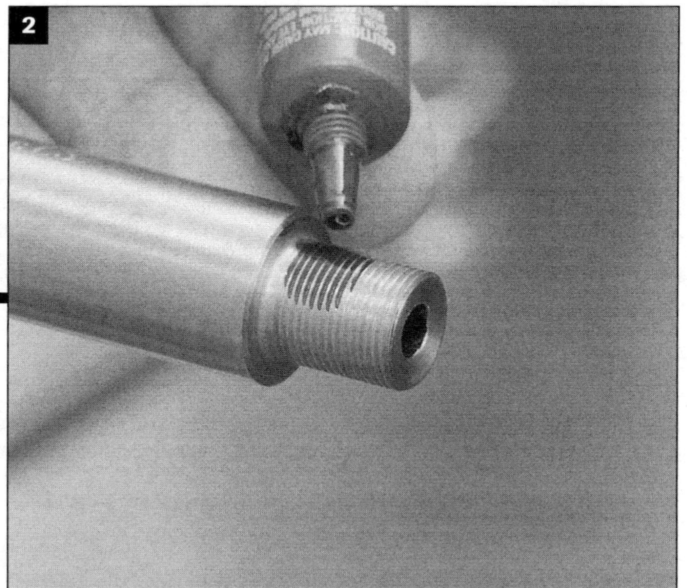

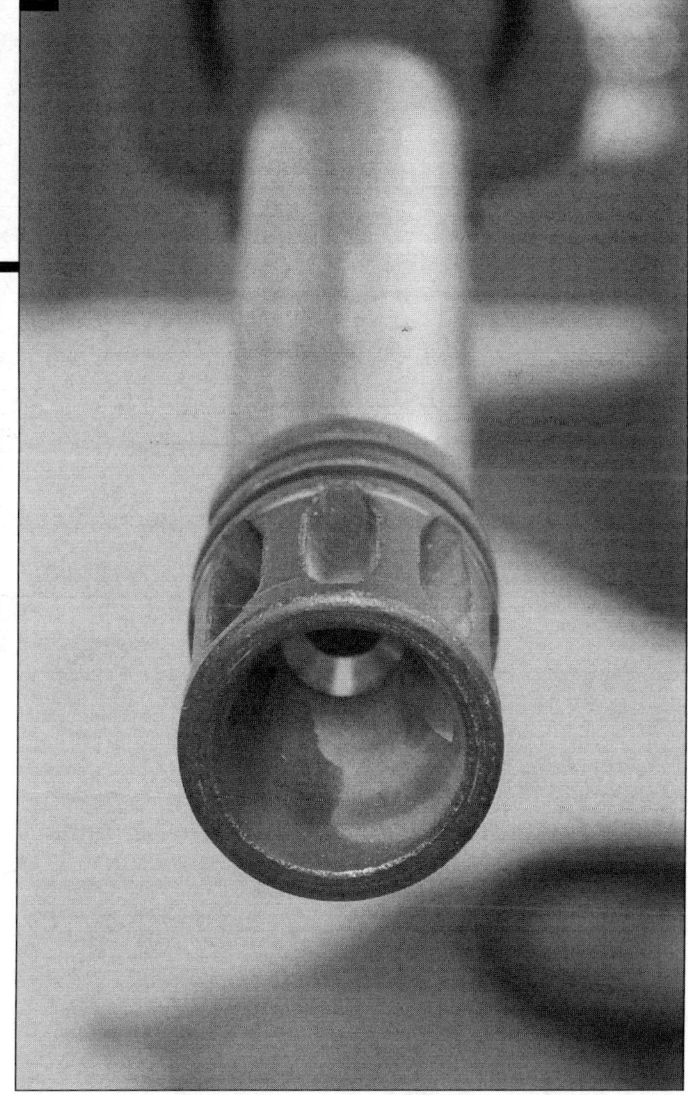

FLASH SUPPRESSOR

This barrel was threaded 1/2-28 for an A2-style flash Suppressor, and that part was supplied by Satern Custom. This one installed with no washers needed or intended because of the machining precision and intent of same. That's a bonus. In other words, when it threaded down and then snugged up, it was straight. You'll see another take on installation using washers on the next couple of pages.

1. Decrease the threads on both parts using contact cleaner.

2. Apply Permatex "red."

3. Thread it on and snug it down (take it to tight and then one sharp tug on the wrench is plenty). Check for alignment. It's not nearly always this easy... The glue overcomes the need to attain perfect alignment at any certain torque. Once it's cured, it's on for good.

Threads or Plain

There's no benefit from a flash suppressor on a target arm. Although as of this writing you can get a 1/2-28tpi threaded section on a barrel, it's honestly another expense and complexity best done without. Most will have an 11-degree target crown, often recessed, so there's little increased assurance against crown damage by having a suppressor.

However! If you want to run a flash suppressor, know that peel or crush washers and glue (Permatex "red") are the right aids to assembly. With the glue, just snug is all it takes, and alignment is important, and that means getting the solid portion centered right on the bottom. Some believe that orienting the open slots at the top slightly toward 1- or 2-o'clock helps the muzzle jump straight up for a right-handed shooter, but I honestly have never seen any difference given the low-level effectiveness of an A2-style suppressor and also weight of a competition Service Rifle. Such tuning does make a difference running a highly effective brake on a lighterweight rifle. Next page shows how.

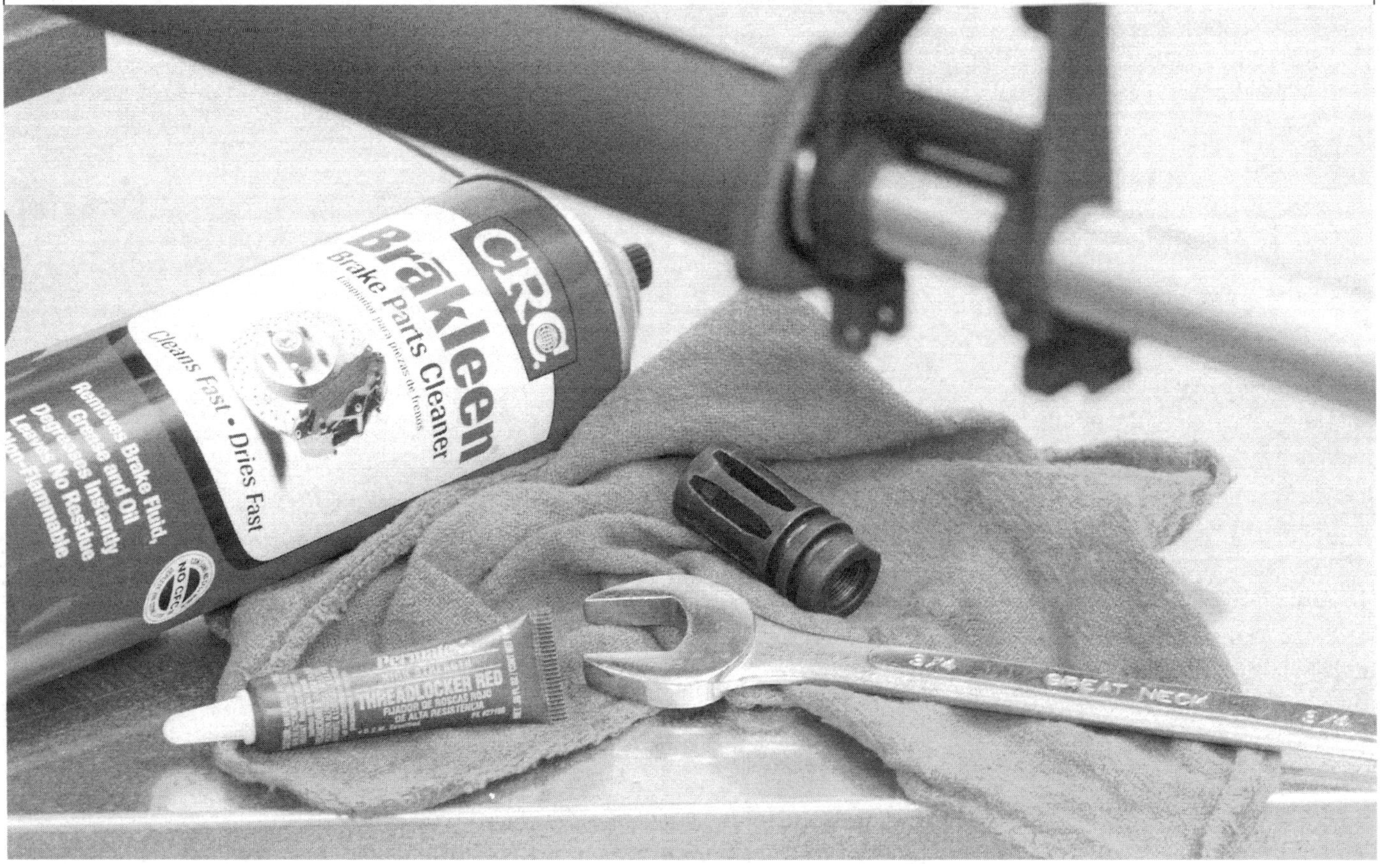

Warshers (well that's how they say it in Indiana)

I tend to use a **crush washer** (left) to install a flash suppressor. The idea with a crush washer is that the dished-out area will flatten as torque is applied, and it will, and that will let you achieve alignment. However! It can take a lot of torque. Since we can simply use glue that's often way on more torque than necessary. After it's "snug" about 30-percent of the torque applied to a flash suppressor goes toward constricting the muzzle. I dress a crush washer down using #220 grit aluminum-oxide paper until I get alignment at a minimum torque. It's a little tedious but also easy and straightforward. The washer installs big end toward the flash suppressor and small end toward the barrel shoulder.

This is a **peel washer** (bottom). It's laminated. It starts out approximately 0.725- to 0.0750-inches thick and each layer is approximately 0.004- 0.005-inches thick, so, do the math and see that's a lot of layers, so precision alignment is very easily possible. Each layer equals about 10-degrees of rotation. Remove layers until it lines up at the snugness you want to apply (15-20 ft.lbs. is plenty, no glue). Some like to add a lock washer if there's no glue used. Peel washers can be a challenge to "peel." The layers are secured with glue. It takes heat to delaminate the layers and a heat gun or even lighter will do it, along with something to initiate separation, such as a razor-knife point. Crush washers have become decidedly more popular.

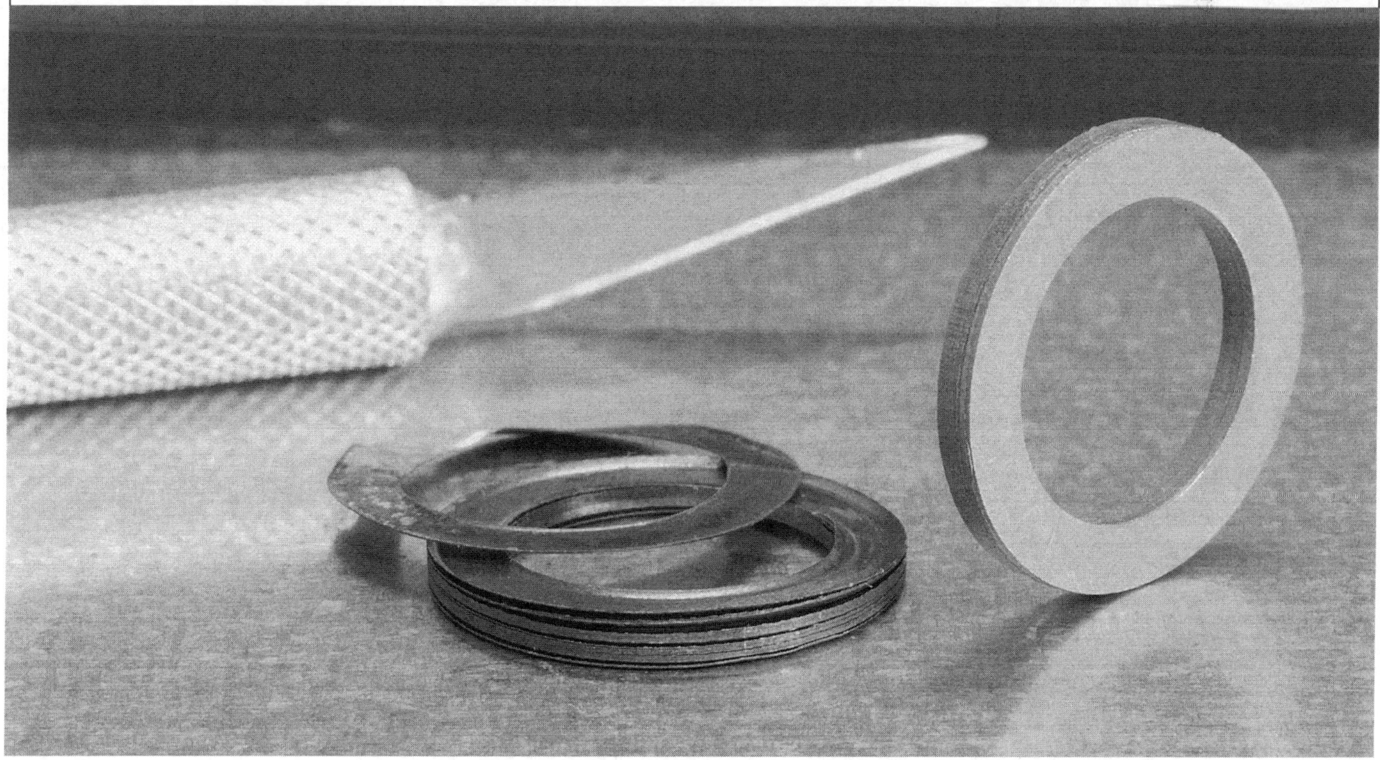

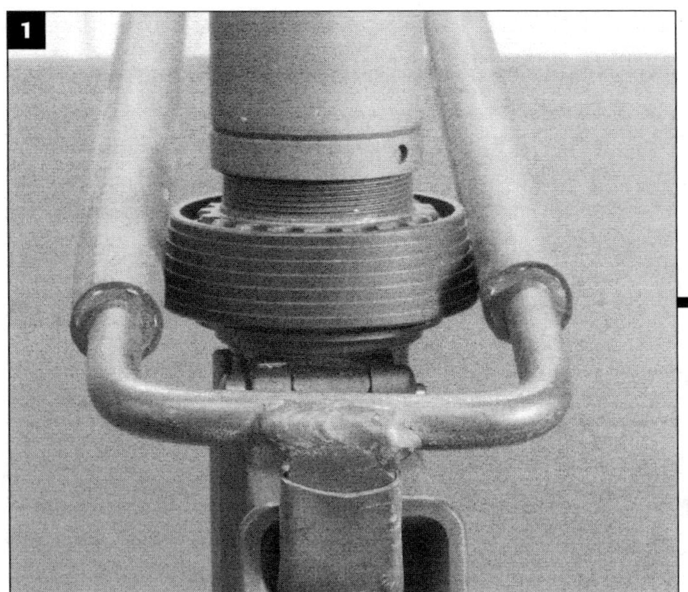

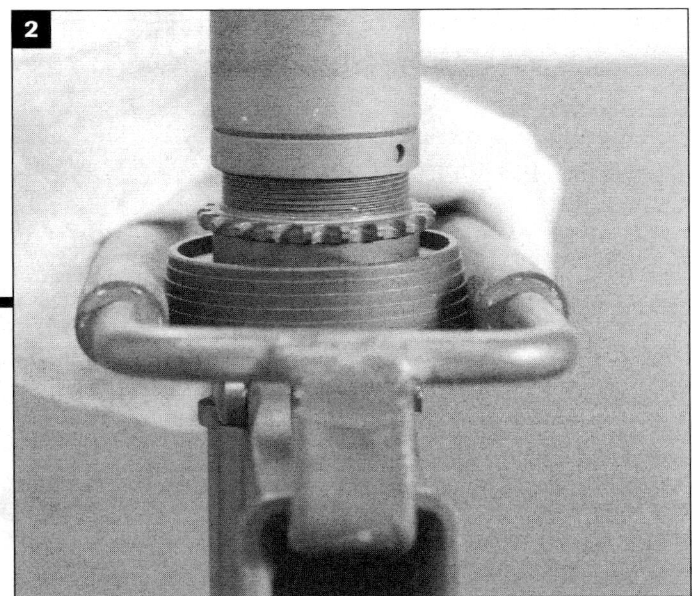

HANDGUARDS

Handguards install just in the same as always. The only potential quirk that they can be a little more difficult because there can be differences in the space they have fore-aft to fit. Reason is attaining clearance with the front sight housing can sometimes shorten this space. I have seen some before that had to be sanded a little on their back ends (part that fits into the Delta assembly).

I love this handguard tool. It makes really easy work out of what can be an otherwise somewhat painful job.

To use the tool, position its hook into the magazine well and its arms over the Delta ring, squeeze and lever it down. **Easy.**

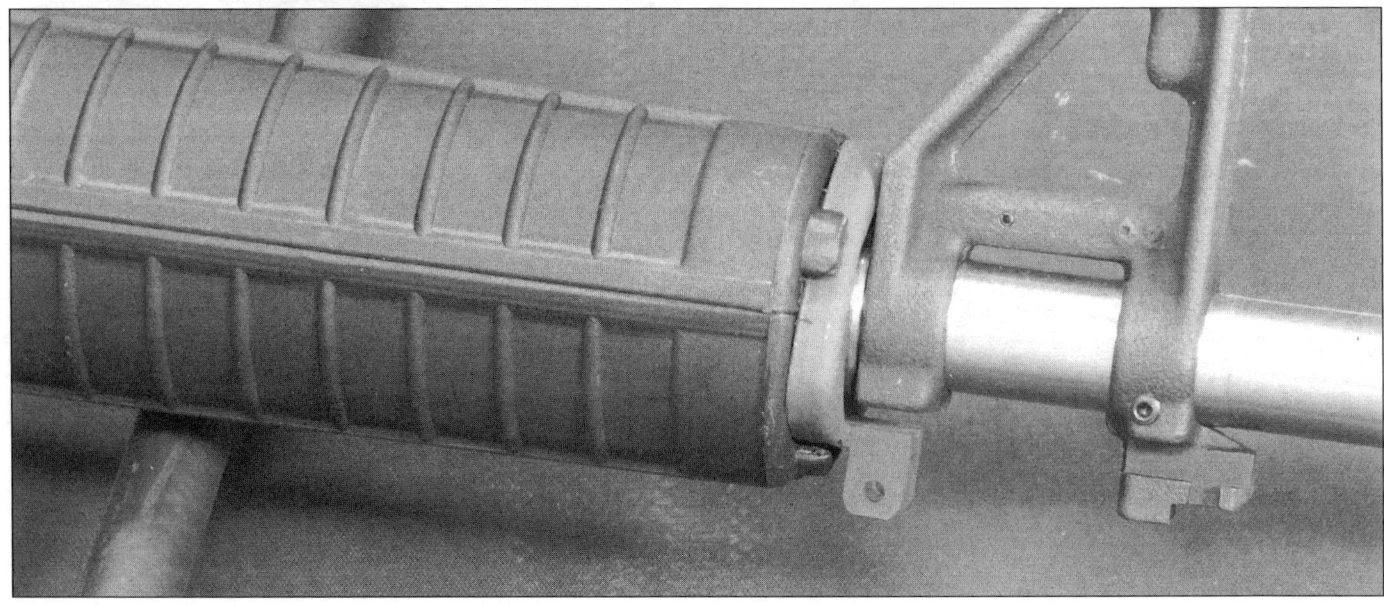

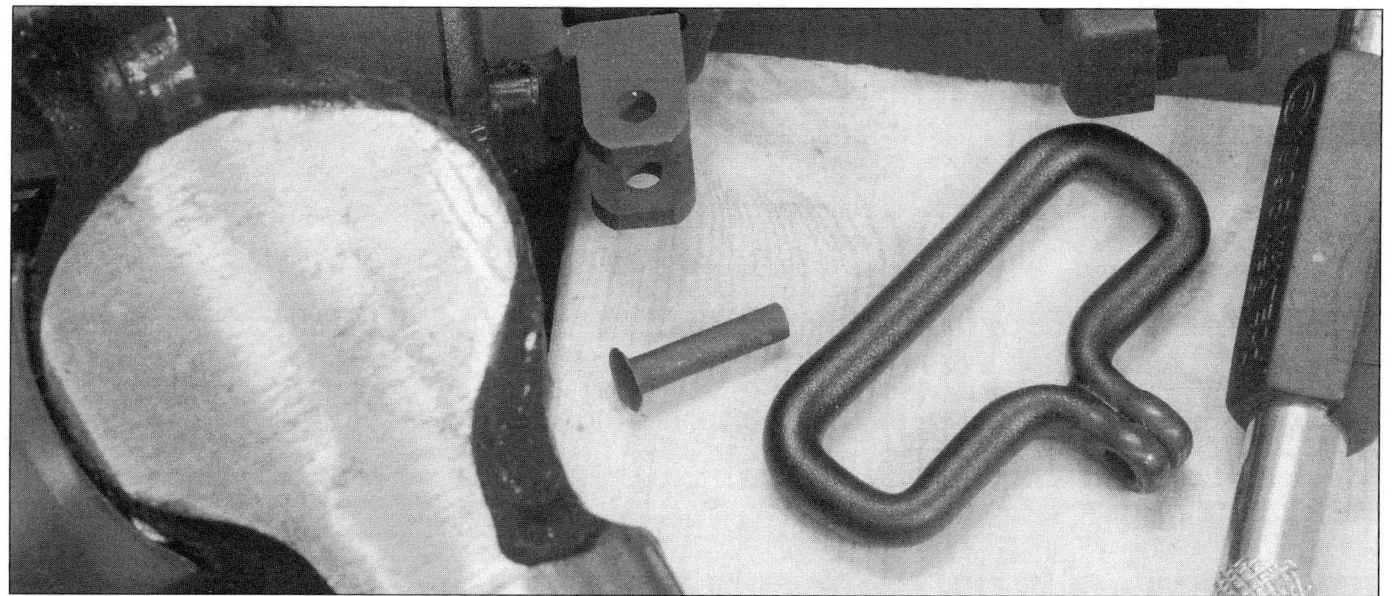

SLING SWIVEL

Some tubes come with a swivel that's either attached (bonus) or must be attached. Either way, you'll need to have one. A standard installation on an A2 handguard cap uses a rivet, and if the swivel must be reused on the new tube, then the rivet must come out. The rivet cannot be reused. There's a rounded end and an open end on this rivet. Either file down the opened end until the "ears" are gone

or, and this will work, just punch it from that end with a flat punch that fits into the opened area. A solid whack or two will shuck the ears so it will release the swivel. Back it with wood.

1./2. To install a new rivet, first, don't get too critical of your work. There is a tool that does this job but I searched high and low and could not find one. I ground on a punch until it was the right contour to spread the rivet fully open and right depth to seat it against the bracket on the float tube or handguard cap. It's hard not to crack the edges, but if it's seated fully, touch it up and it will work to do its job. On left is a hardware-store solution.

Making Good Better

If all you want to do is install a free-float tube over the barrel you already have on your rifle, and that's a possibility if you have a Colt HBAR (now the #6601), or anything else that already had a heavy configuration 20-inch, then alls we have to do is get it shed of its extraneous parts.

There are options in the order these pieces are removed, so read through this next to decide the best way for you to proceed.

Pull the handguards.

Vise up the upper, get your gas tube pin punch and tap that out. Pull the gas tube. Now, I will note that it's not really necessary to remove the gas tube from the sight housing just to release the assembly; you could reinstall this as-was with the tube in place, but let's assume you're going to lose reliance on the sight housing taper pins and either replace or modify the housing. Likewise, there's likely to be either another gas tube or possibly required bends made to yours to fit with the new float tube.

Remove the taper pins from the front sight housing. This will free the housing and the handguard cap.

Now the barrel nut is exposed so get your wrench and break that deal loose, thread it off, and pull the barrel.

Unclip the Delta assembly from the barrel nut. Remove the spring clip and the rest slips right off. Your new float tube most likely will not have this assembly included, so keep it handy.

If the barrel has a flash suppressor you will have to remove that also. The sight housing and handguard cap can't otherwise be removed. Flash suppressors are often, and are supposed to be, installed with a good deal of cinch. It can be removed while the upper is vised-in or afterward by clamping the barrel. It's easy to argue, and for me to accept, that it's best done with the barrel clamped after being pulled from the upper to avoid stress on the upper.

The flash suppressor, front sight housing, handguard cap, and barrel nut should be in a shoebox or your choice of container.

Disassembly of a Modified Rifle

If you're refurbishing a competition NRA Service Rifle that's had all this trickery and accepted assembly procedures performed previously, then we're going to need heat. It's the glue. It's also now when you might be questioning the wisdom of glue. Folks, glue is all about security and performance, not convenience. Not nearly convenience. It's going to take heat to get the glue to release its grip. Heat comes from different sources and, depending on the surface area and glue type, it's going to either be an industrial heat gun or a torch.

Permatex Blue Threadlocker isn't supposed to require heat. It's supposed to be removable with hand tools. However, It's good to 300 degrees F, so some heat from a heat gun might help if this is covering a larger surface. Permatex Red takes a lot of heat to break, about 450 degrees F. Even though it's supposed to hold only to 300 degrees, it takes that much more to actually free the assembled parts. The only way I've done it is with a torch. Permatex Green (Penetrating Grade) takes heat too, and sometimes the same approximate 450 degrees F to actually break down the bond. Epoxy requires heat to free also, but not usually as much.

I can tell you that some patience is required. Even after it's at the temperature where it should have broken lose, struggles can ensue. Even when its bond has broken, glue can still retain pieces, especially threaded pieces, much in the same as a threadlocking nut insert. It's heat penetration that's key. Keep it hot for a while and disassemble when its hot. Protect your fingers!

Pay attention before and during cranking up flame in your work area. I cover parts with heavy-weight aluminum foil to keep them from discoloring and otherwise exhibiting finish damage. I can't tell you how to know something has attained 450 degrees (there are ways, but they're not necessary to what we're doing), but I can tell you not to get too greedy too soon, and also that heat does need to penetrate to do its job on many parts. Keep the torch moving within the area. One trick to release glued screws is to heat the suitably sized wrench and place it into the screw head.

Permatex Gasket Remover is good at dissolving residue from most of these products and, along with a brush, can clear residual goop from disassembled parts. Take it easy with the brushing on threads, though. Don't use a wire brush.

13.0 SERVICE SIGHTS

A2 CONFIGURATION, MATCH OR STANDARD

[I decided to separate "sights" into another segment because there's a lot to talk about. It really doesn't matter whether you're installing match or issue sights because they all go together in the same way. The differences will be in the configuration or specs on the parts themselves. Keep in mind, of course, that the following material follows along with the NRA/CMP Service Rifle project underway now in the book. This segment also has all the information on assembling a front sight housing and modifying it for better performance.]

SEGMENT CONTENT

155   **Match Defined** (component differences)

156   **A2 Rear Sight Assembly Overview**

160   **Parts & Tools**

162   **Windage Assembly** (assembly and installation)

165   **Elevation Wheel**

166   **Sight Base** (assembly and installation)

168   **Elevation Spring Installation**

171   **A2 Front Sight** (assembly)

173   **A2 Front Sight** (housing modifications)

Aftermarket or GI, the sight components are the same, as is the work involved, tools, and process for assembly and installation. If you can do a match sight, you can do a standard sight, and vice versa.

MATCH DEFINED

It's the truly high-zoot sights that are going to be harder to reproduce if we're working up a rifle at home, and mostly the rear sight. The Grade A rear sight configuration is one having pillars incorporated into it to improve both stability and tracking, and that is a machine-shop modification. Some commercial builders have hinted at producing upper receivers with the pinned or pillared tracking arrangement but as I finished this little ditty here there were none cataloging any such. If it's going to cost you sleep, you might want to talk to one of the builders performing this modification and see if they'll do your upper for you prior to your assembly of the rest of the components.

The barrel chosen for this project came, fortunately, with what I think is the best way to affix a front sight housing. Otherwise, you won't be losing much going with the modification soon shown.

REAR SIGHT

A target-grade rear Service Rifle sight, any of them, will have at least the following two things in common. The elevation base will have finer threads, as will the included elevation wheel. The rear aperture will be smaller. That's it. Beyond that, there is a long list of component specs and manufacturing tolerances that might or might not be part of your package. For instance, the click value of elevation and windage stops, along with the means used to attain finer correction values. Wind screw tolerance. Aperture styles. Overall, and specific, machining tolerances.

I do think extra-quality sight parts are worth the money, especially if you have the pillar mod done on the base and upper. Taking a look at what the pillars do for the rear sight will show that even the tightest tolerances in a rear sight

Match Specs

Here are basic rear sight mod parts to make up a "match" sight. A new base with a finer-threaded elevation post and a rear aperture with a smaller hole. The base is necessary to get shed of the coarse 1.4 moa elevation click stops and provide something closer to 1/2 moa increments. I say "something closer" because choices include 1/4, and also 1/3 from some. The wind stops are already approximately 1/2 moa on the standard A2.

This is from Northern Competition and has 1/4 stops each direction. Tolerances are radically tightened and finish quality is considerably ahead of what the major makers offer. Cost is also higher, as well it should be. This is machined from bar stock and is sold with aperture options.

are going to fall short of the improvement from the pillar pins, but I can tell you that better rear sights will track better than the issue or routine "match" assemblies sold by the big outfits. Better is better even if it's a step or more shy of best. Best takes about one hundred dollars in addition to the cost of the sight.

Installing any back sight takes some patience and tools. Of course it takes tools. One that's indispensible is something that can hold the rear sight spring compressed to allow you to tap its retaining roll pin through. This is a tricky and often frustrating job.

Grease everything! Well, everything that threads, clicks, or rubs.

Grease is the right lubricant for sight parts, not oil. You'll want something that will hang in (which means hang on). Reducing metal to metal contact means better precision for a longer time.

ASSEMBLY OVERVIEW

The rear sight is an assembly itself and it's assembled onto the rifle in segments. Different sights will be more or less assembled on unpacking, but this will assume all parts are loose.

As mentioned, I chose the Accuracy Speaks rear base because it offers a means to "tighten" the tolerance in the elevation tracking and also square the base so it sits straight. Its treatment is discussed and shown in the assembly steps. If you picked another base, skip that fitting process, of course, and move ahead to here.

Some find wind assembly installation easier with the sight base off the rifle and others like to do it after the sight has otherwise been attached to the upper. We'll do it with the base standing alone.

First is installing the (flat) aperture spring into the sight base. This spring pushes up against the bottom of the aperture and allows it to "flip." It's spring steel and just sets in its place. That was easy. Too easy. There's more. If this spring isn't seated right, and some don't want to fit their confines, the pressure bearing up against the bottom of the aperture piece can be inconsistent. They sometimes need to be fitted to the opening. The spring should sit, and stay, flat at all four corners when the aperture is flipped. Fit it by gently bending away any warp that's preventing it from sitting perfectly flush across its bottom.

The other pieces-parts can now go into place. There's no difference in installation of the wind screw in a match or standard sight. Parts tolerance and specifications are the only differences. Put the aperture into place, push it down to compress the aperture spring, and thread the screw into the sight base. Make sure the aperture is correctly oriented, and make sure to thread the screw into it so there will be enough protrusion to attach the wind knob.

Speaking of apertures, I can't tell you what to choose no more than I can know what you're seeing when you look through your system. I might help you understand what you should see through the system and that's detailed in other books. Know there are options, and when we throw lenses into the equation, I can tell you that the sight picture can get radically better with effort, and, unfortunately, some expense. The effort is in trying different combinations, and that's also where the expense is.

Place a heavily greased spring and ball bearing into their receptacle on the flat side of the knob. Line up the hole in the end of the wind screw with the hole in the wind knob. Use a capture punch to

Best Base

This is the Accuracy Speaks rear sight base. It's probably the best setup next to a pillared modification as made available from John Holliger, Clint McKee, and others. Can't do that one with a hand drill, and even with the proper equipment it's stepping into adventure in metal-working. Put it this way: if you can do a pinned A2 rear sight then you probably don't need this book. Derrick Martin's base has a little weld-tack build up that can be dressed down to improve tracking and also take the "twist" out of the base. The sight assembles the same as always around the Accuracy Speaks base, but does work better, so says my dial indicator. The elevation wheel has two sets of detent stops, one for 1/4 and one for 1/2 clicks.

Accuracy Speaks rear base provides means for establishing lower tolerance and abating unwanted sight movement. To use the fitting pad, get a wide flat file (1 inch or more and fine) and keep it handy. Put the threaded elevation post down into place in the upper and thread the elevation adjustment ring onto the post. The fitting pad is on the opposite side of the spring that ultimately will push back against the base. It's this spring that makes the base sit cockeyed. The weld spot essentially closes the gap that makes it possible for the base to rotate under spring pressure. Pressure the side with the weld fitting pad forward against the upper receiver. The pad should be making it sit cockeyed the other direction. Remove the works, secure the sight base, and dress the pad. Pay attention to making a dead square cut. That's the reason for the wide file (it's easier to align). Go one stroke at a time and test fit again. When the base is sitting straight with pressure applied as described, then it can't cock over any more after the final assembly and the spring is installed. Now it will work better. Again, go slowly and be critical, and not overeager. Don't go for a flush or interference fit. The base has to move easily and freely (related) through its full range of elevation. Make sure you have tested this prior to calling it done. Better is better, and this base makes it possible to do way on better with respect to fit than any other. ArmaLite, by the way, offers a similar approach on its match sight via making that edge itself on the base oversized and thus requiring it to be treated to the same fitting process. It's not nearly as overt as what Accuracy Speaks does, however. Notice also the little bushing pressed into the wind screw hole. That's done to take up play. Very necessary.

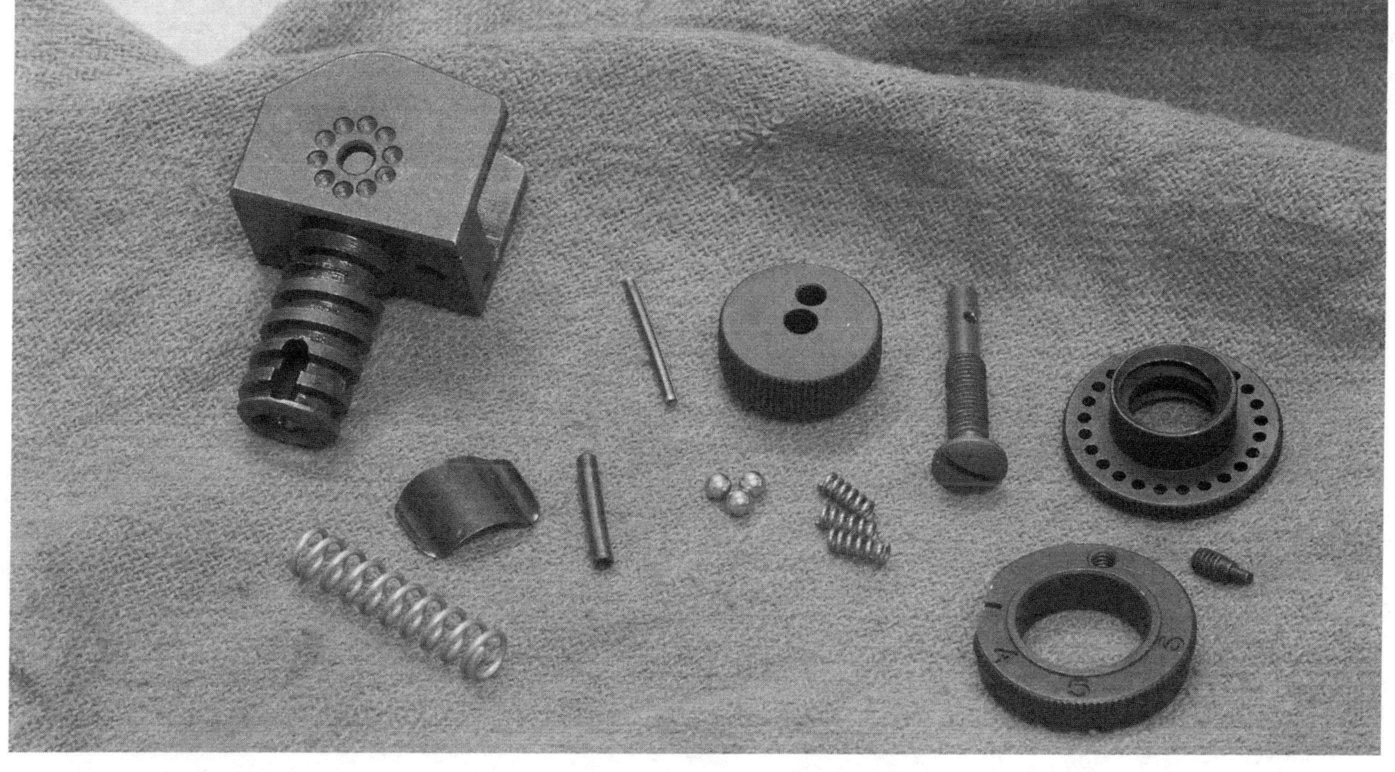

align the assembly and maintain the knob in position so you can set the roll pin. (You can also pre-assemble by setting the pin into the knob a fair ways prior to attaching the knob to the screw so there's not as far to go to finish.) Tap the pin into place. Make sure the pin is seated the same on each side.

When you install the wind assembly, make sure the parts are greased with a high-viscosity grease. No oil. The main caution is to make sure the parts are backed up prior to any hammering. The screw is easy to bend. All that's necessary is a little block of wood placed against the wind knob backside (side opposite pin insertion point). That also makes it easier to install. I put a drop of oil on this pin prior to installing it, preceded of course, by smoothing its ends on emery or a stone. If the wind screw gets bent then there goes your tracking.

The next piece that gets pinned in place is the big spring that rides up inside the elevation post, and the first step in that process is threading the sight base down into the receiver.

Put the greased ball detent spring into its hole in the receiver and top it with a heavily greased ball bearing. Grease the underside of the elevation wheel heavily and slide it into place. A small flat-bladed screwdriver helps to compress the ball bearing. Make sure the wheel "clicks" into place, ensuring the detent is captured.

Install the spring and ball bearing that pushes against the sight base. Grease it heavily. Insert the sight base and thread it down into the receiver by running the elevation wheel. Push down on the base as you're doing this to make it easier, or just turn the works upside down and keep a little pressure against the base against a padded surface. Turn it in far enough to see the hole in the threaded post on the base align with the hole in the receiver that will accept the pin.

To install the elevation spring you really have to get hold of a specialty tool. It's not hard to make one and it's even easier to buy one from Brownell's. It's difficult otherwise to compress this large spring and then be able to drive its roll pin retainer in place. The pin has to be able to go under the spring, and has to go under the spring, which means there has to be clearance in the tool to get it to go together properly.

The first time I needed one I made it from a quarter-inch diameter metal tube that I drilled an eighth-inch hole through. Not that you should try to duplicate that, but it's not hard to rig up something that will suit. I like the one I bought better.

Insert the spring into its recess. Start the roll pin, locate in place and then push the spring up using the tool, keeping the tool recesses aligned with the receiver holes. (Masking tape is a very good idea to protect the receiver.) Compress the spring so it's holding above those holes, and capture it with a punch. Then tap the pin home. You have to push the spring up a good ways to get it above the receiver pin holes, and make sure you do.

This isn't easy.

It is a stout spring and a relatively small diameter, lengthy pin and it will be necessary to maintain the spring under compression until the roll pin enters the far hole. It will also be necessary to keep the spring compressed throughout this operation. Back off its tension just a little bit and the pin will never align with the hole so it can be driven through and home. The pin can bend and at the least get its end buggered to the point where installation may require starting over with a fresh pin, and likely a fresh perspective... One tip is to really whet down the end of the pin to help ease it into the far hole, and add a drop of oil to it.

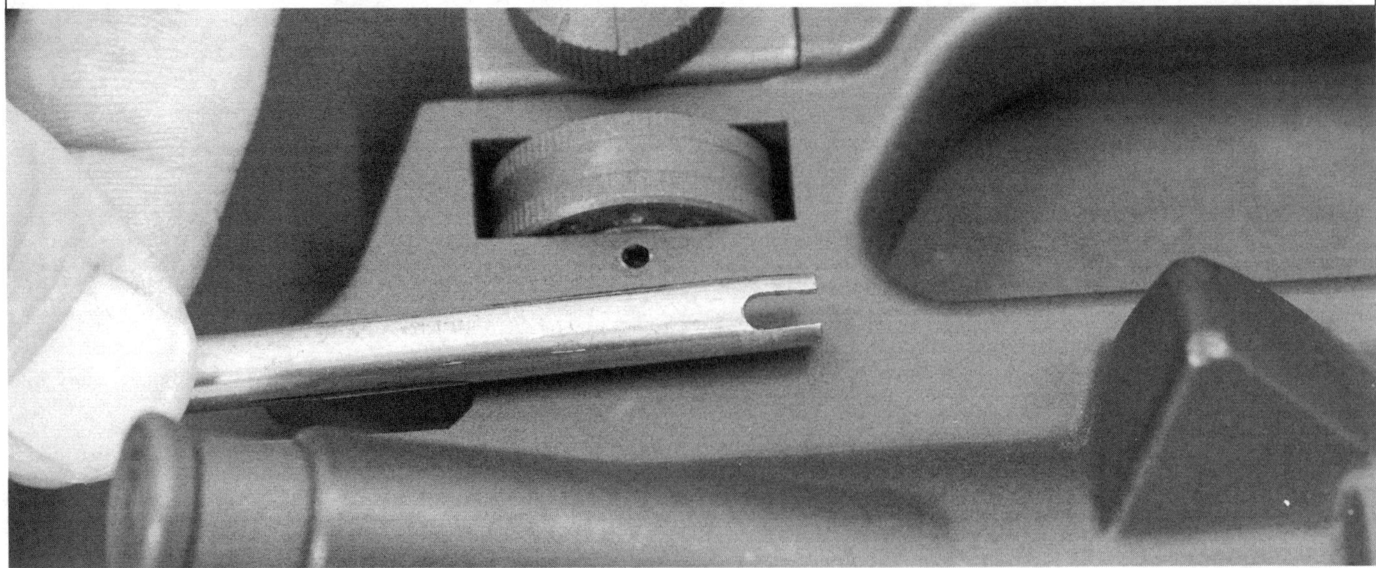

Necessary, make no mistake. *This is a sight installation tool from Brownell's. The roll pin has to go under the compressed spring, and this tool provides a means to compress the spring and the open area gives room for the roll pin to move through. Use this tool to remove the pin too. If you don't the pin will be virtually impossible to drive out. There are other tools available, including one that essentially assembles into a stand-alone fixture. That's very nice because it gives more free hands, but, as with so many tools, the frequency of use and value must both be high to warrant the investment.*

Aperture Options

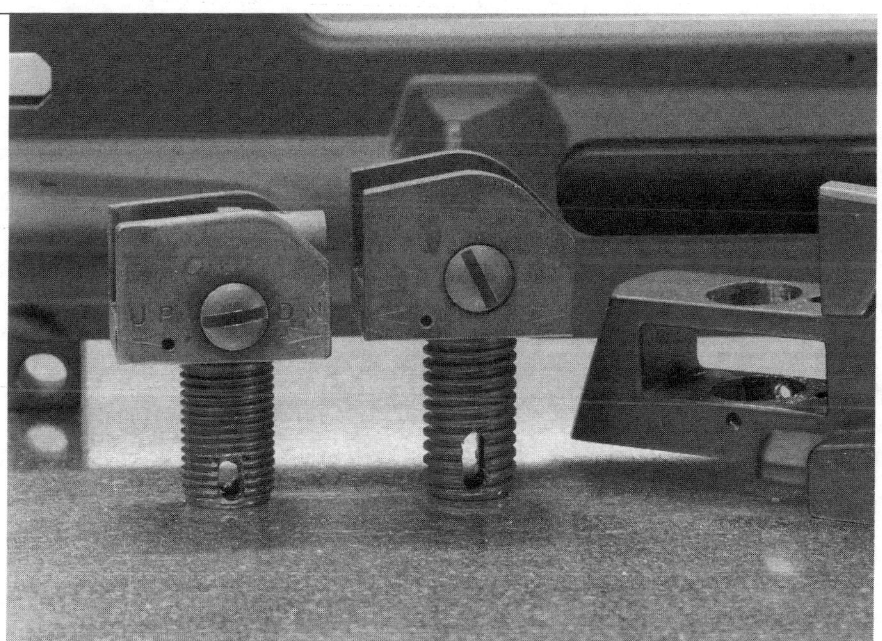

Rear sight options include apertures and lenses. There are several formats or formulas to follow and only experience knows. If you are swapping aperture assemblies, make double sure you're getting a thread match to your wind screw. It's usually 64 tpi (turns per inch) pitch for 1/4 MOA clicks and 32 tpi for 1/2s. Some aperture sets come with a screw, and most better pieces have a little additional diameter added to the screw shoulder (that's the part just ahead of the head) to reduce wiggle.

This is an essential hooded aperture from DPMS. This one has an 0.042-inch aperture, fixed, and a fairly narrow hood. Narrower hoods are good because they allow nearer to maximum wind range. Watch out for those with excessive outside diameters. There are some that restrict the amount of available wind movement a good deal, and enough that you might notice. Hood length also has options. Shorter is, well, shorter, and since we're very close to the rear sight that's good for some. I like short ones. There's little to no difference in the appearance of the look-through.

Also, the best is when the rear aperture sits directly over the centerline of the wind screw. Some don't. If it's ahead or behind the screw, and the base isn't pinned in place using one of the pillar arrangements, then there can be a zero change. Fixing the base to sit square, such as is possible using the Accuracy Speaks part, eliminates this effect.

Thread-style interchangeable apertures are popular, but do not trust them. Get a zero confirmation for each one you might use. Derrick Martin has a good idea with his Accuracy Speaks is "flip" sight. You get a choice of two aperture sizes for different light conditions. The one shown in this project has a 0.040 and a 0.042. Caution generally follows against flipping a sight aperture, and that's warranted, so make sure you know zeros for each aperture "side."

The Competitive AR15: the ultimate technical guide *thoroughly covers these options and how to know what to expect from each.*

Good Idea?

Maybe. It's not a bad idea. Rules permit CMP and NRA Service Rifle conformation when outfitted with a detachable carry handle. That means, of course, that the rifle can lose the carry handle and then have the full Picatinny rail surface on the upper for optical sight mounting. Similar conventional or competition A2-style rear sights can install in the carry handle portion. The problem is that there is a loss of elevation range due to the shortened travel of the threaded elevation post, and it's a significant amount. That's because there's less room. Some really like this idea, others, like me, really think a Service Rifle ought to have a conventional A2 upper for best utility on the firing line.

Very important. Always, always, always back up sight parts prior, and during, pin installation. Just use a piece of wood and make sure it's flush-fitting the part against all movement. Wind screws are easy to bend.

Any and all sight installation steps require grease! Now is the time to add it. I use Plastilube although a lower-viscosity product will substitute. Plastilube stays put and does its job, and that's why it does its job. Grease the elevation post and underside of the elevation wheel, wind screw and underside of its knob, and the receiver-side of the base. Grease is the "glue" that holds the ball bearings in place where they are installed as well.

PARTS & TOOLS

Once again, whether it's a match-quality assembly or a standard issue package, the A2-style rear sight assembles in the same way and all the parts are shared in common. The only differences are in the parts details.

PREPARATION

Tap hammer

Roll pin punches and roll pin starter punches

Flat-head screwdriver (for windage screw head)

Spring tool (to compress elevation spring)

Grease (fairly thick, like GI grease)

Piece of wood (to back up parts during assembly)

Hex wrench (for elevation indexing knob set screw)

Front sight wrench (4 prongs for A2, 5 for A1)

Options (good ideas)

    Masking tape (to hold parts, also to protect receiver)

    Small flat-blade screwdriver (helps install windage wheel)

PRECAUTIONS

It helps to have extra punches to use as capture punches.

Grease everything!

Make sure the rear aperture is correctly oriented.

PARTS

Rear sight base

Rear sight base positioning ball bearing and spring

Rear sight aperture

Rear sight aperture spring (flat spring)

Rear sight windage knob screw

Rear sight windage knob

Rear sight windage knob detent ball bearing and spring

Rear sight windage knob pin

Rear sight elevation knob

Rear sight elevation indexing knob set screw

Rear sight elevation spring

Rear sight elevation spring pin

Rear sight elevation detent ball bearing and spring

There are three ball bearings and three springs used in the rear sight assembly and each is the same

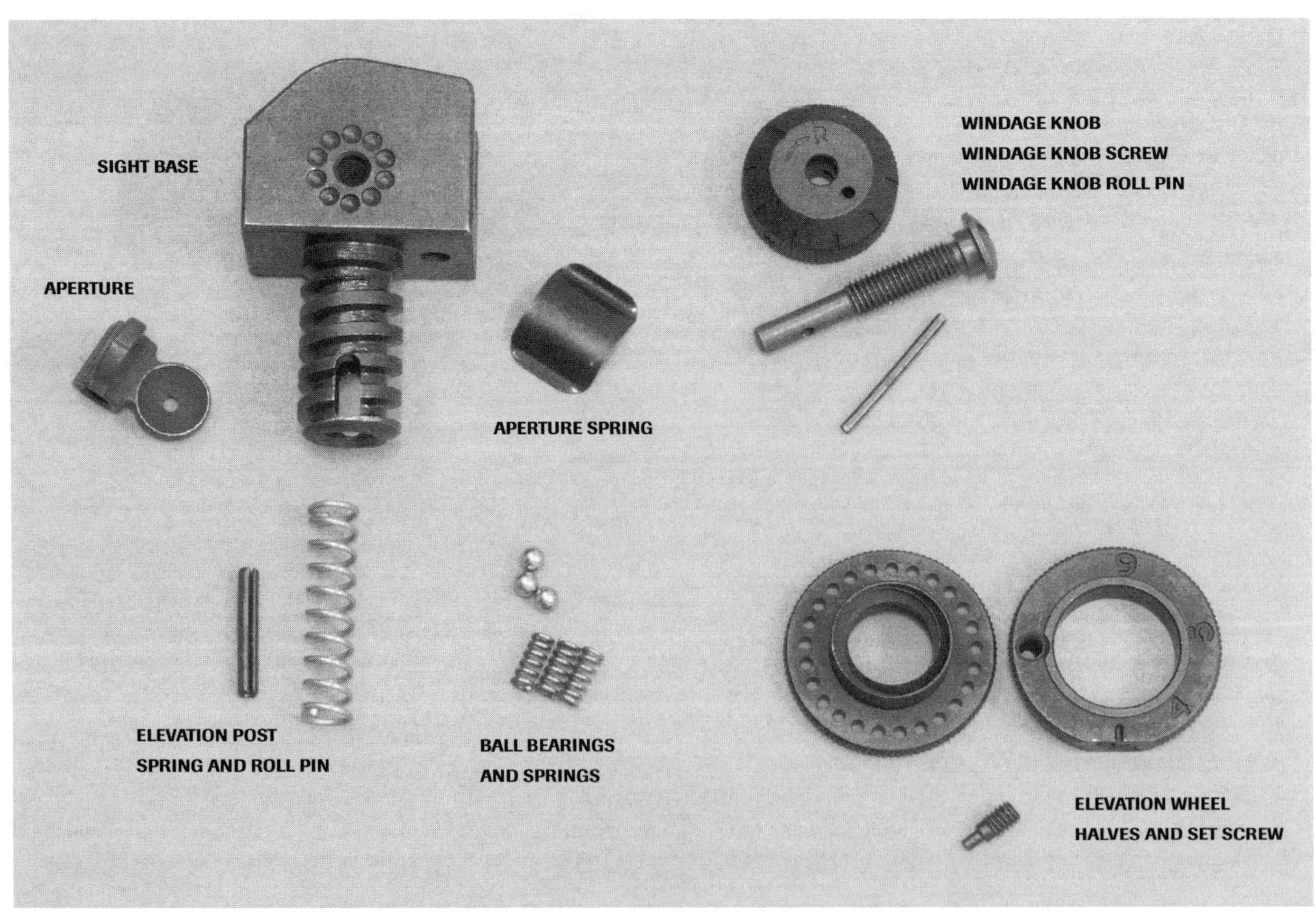

14.0 A2 REAR

ESSENTIAL MATCH OR STANDARD INSTALL

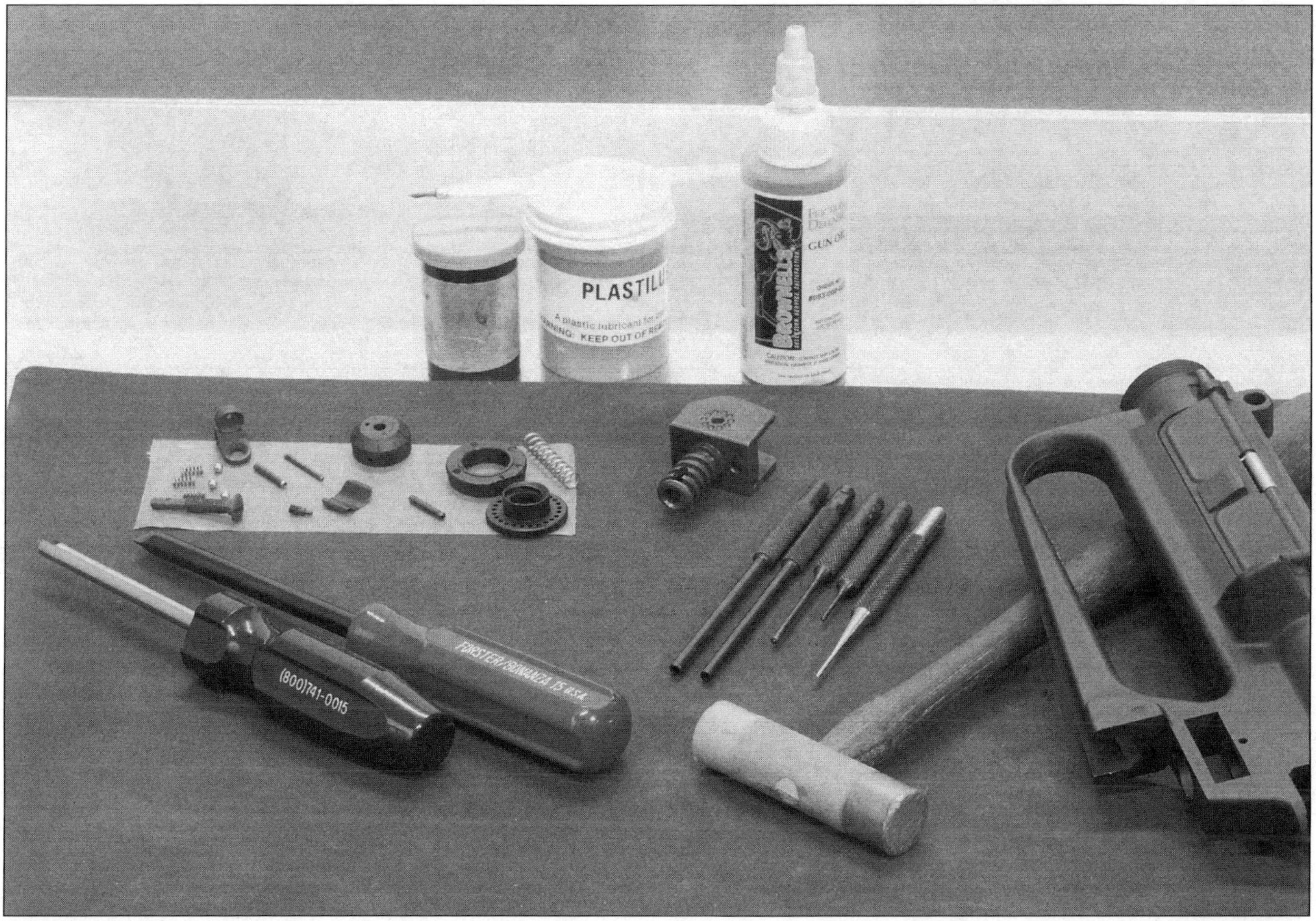

SIGHT BASE PREPARATION

It doesn't matter whether you install all the parts onto the base first or install the base and then add the parts. I did the parts first for the sake of photos.

Since the Accuracy Speaks base has a fitting pad, the first step is to mount it into a vise (with protected jaws), get a fine-cut file, and dress down the pad. This process was explained four pages back, and the essence of that is to go slowly, cut squarely, and test fit continually. It's worth the effort though.

1. The weld spot essentially removes the gap that makes it possible for the base to rotate and sit cockeyed when it's installed and under spring pressure.

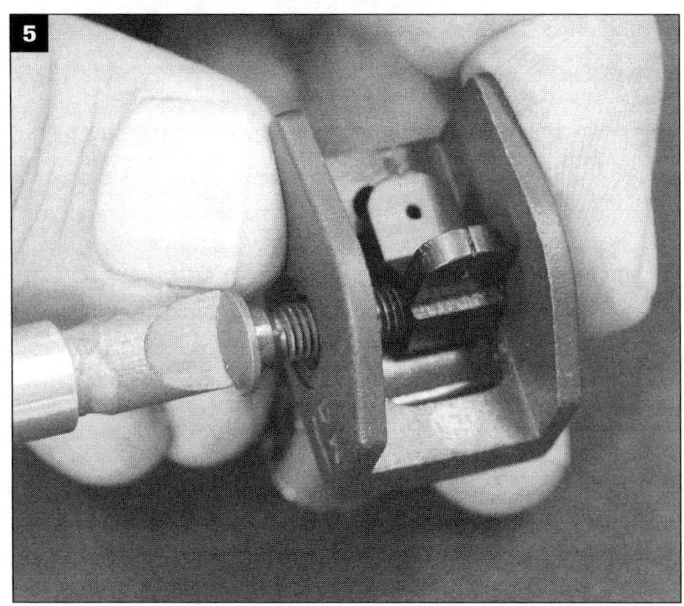

WINDAGE ASSEMBLY

1./2. Position the aperture spring into its recess in the sight base. Make sure the spring is sitting flush at all four corners. If it's not tweak it until it is.

3./4. Position the rear aperture (making sure it's correctly oriented) and insert the windage screw. You'll need to push down on the aperture to get the screw located into its hole on the right side of the sight base.

5./6. Using a correctly fitting flat-head screwdriver, thread the windage screw into the aperture and then on through the right side of the base. Make sure you start the threads through the aperture or the screw won't reach across to the other side of the base, but don't screw too far or you won't be able to align the

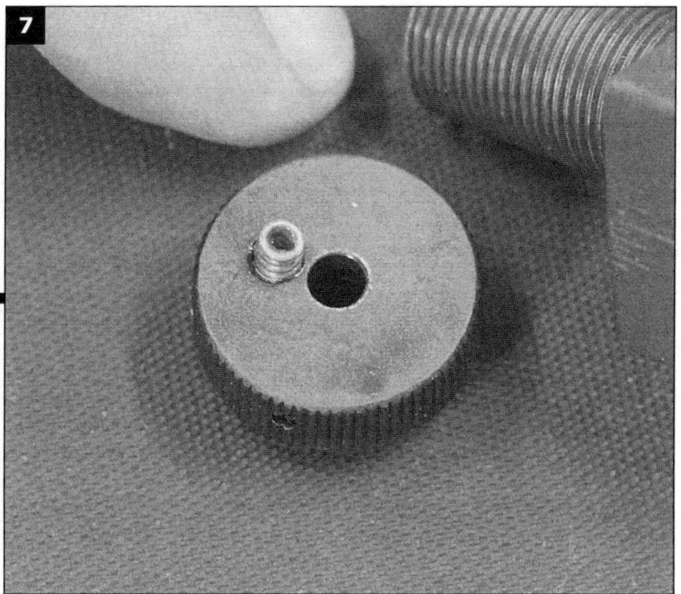

end of the screw with the hole on the other side. It's most easily done the first time by paying close attention to progress. The screw should be flush on the left and protruding on the right far enough to accept the wind knob.

5./6. Grease and place one of the three detent springs into its recess on the flat side of the windage knob, and then cap it with an even more heavily greased ball bearing. The grease is necessary for lubrication but also serves to hold the ball bearing in place.

The photos below show the correct orientation for a standard issue aperture. If you're using something from the aftermarket, take a good look at it and understand its installation needs.

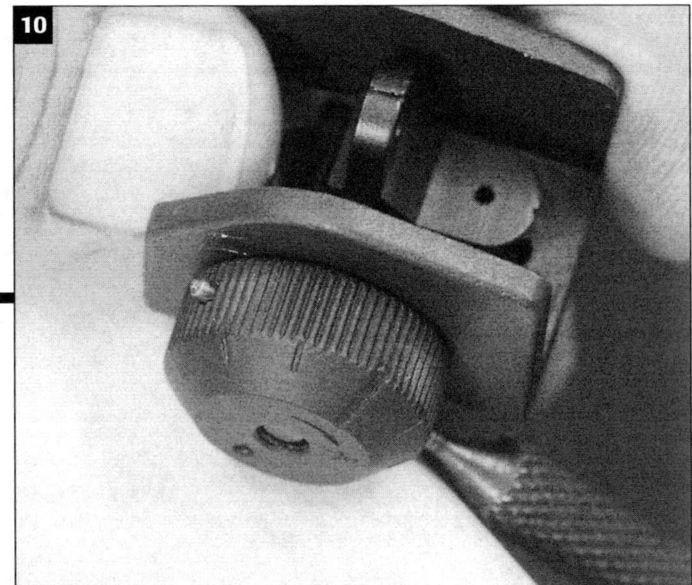

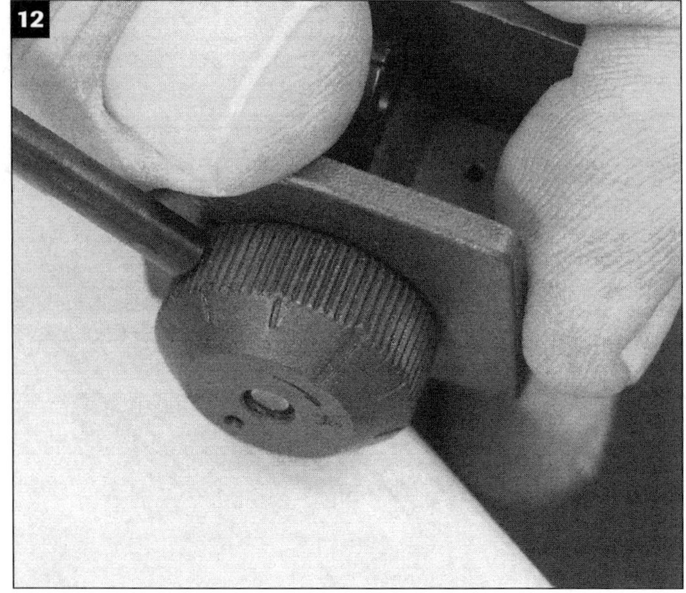

WINDAGE ASSEMBLY

5./6. Position the windage knob against the sight base, aligning respective holes in the knob and screw, then push it in to compress the spring and insert a capture punch (#1, or 1/16-inch). The capture punch is a huge help. Small holes and a relatively long pin make it very hard to eyeball alignment.

10. Use the punch to rotate the knob to a favorable position to now insert and drive the roll pin.

11./12./13./14. Back up the knob with a piece of wood, keep the capture punch in place, and then start and drive the roll pin. Let the pin displace the punch. Make sure the pin is driven in the same on either side so you can't feel it with your fingers.

ELEVATION WHEEL

15./16. Not much to this, but it does have to be done. So, locate the halves together, and thread in the set screw. Done.

Some set screws have thread locking compound on them. If yours does, remove it and then don't put it back. It's not necessary and, due to the tiny size of the screw, can create installation and removal problems.

It doesn't matter the orientation of the halves for now because the wheel gets indexed after the sight assembly is complete.

The Accuracy Speaks wheel is one piece, so it's ready to go. It has 1/4-minute detent stops on one side and 1/2 on the other. I used 1/2s.

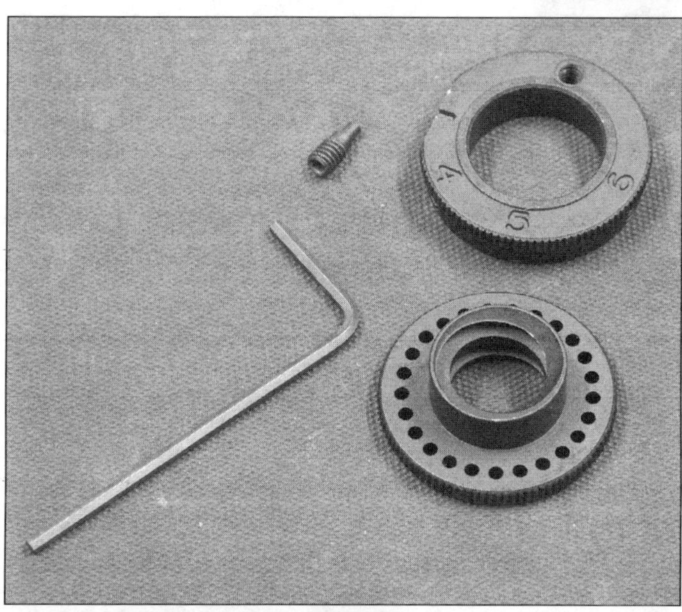

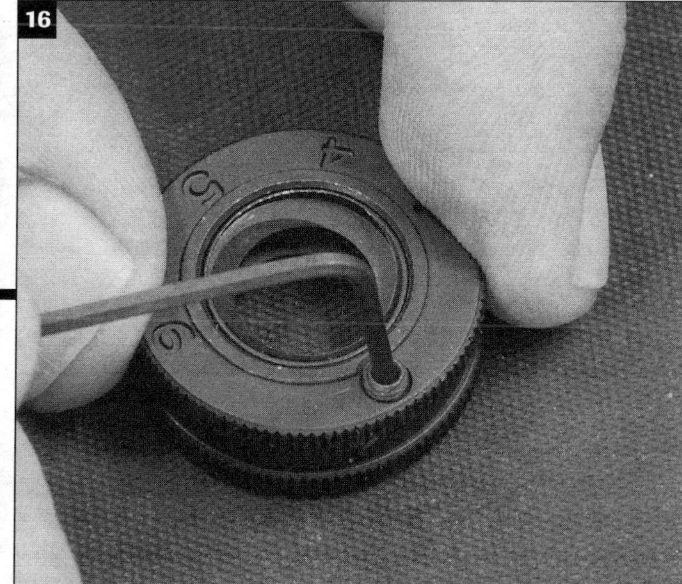

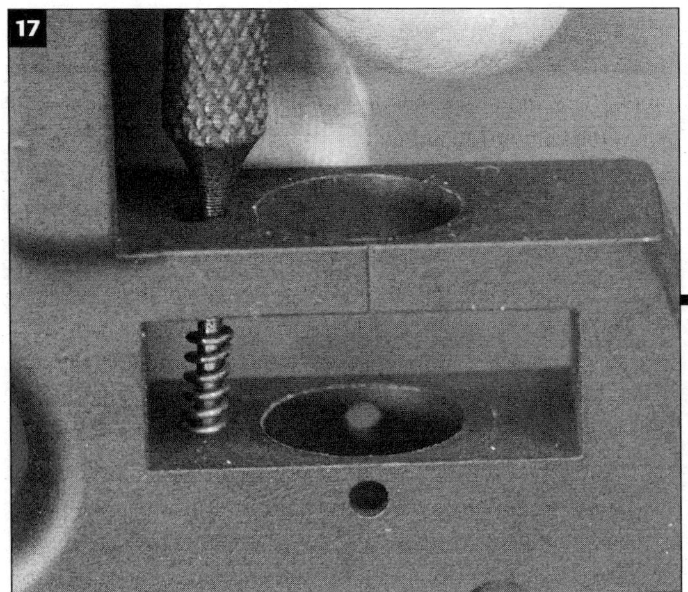

SIGHT BASE INSTALLATION

17./18. Place a greased spring into its spot in the upper receiver, capped by a greased ball bearing. A small punch helps get the spring in place.

19. Slide the elevation wheel into its spot, making sure, of course, that the ball bearing stays put. It will click into place. A small screwdriver helps to compress the ball bearing so the wheel can slip over it.

20./21. Place the third spring and detent into the front of the sight base. Grease heavily! **22.** Then position the base and start threading it down into the receiver by running the elevation wheel while pushing down on the base.

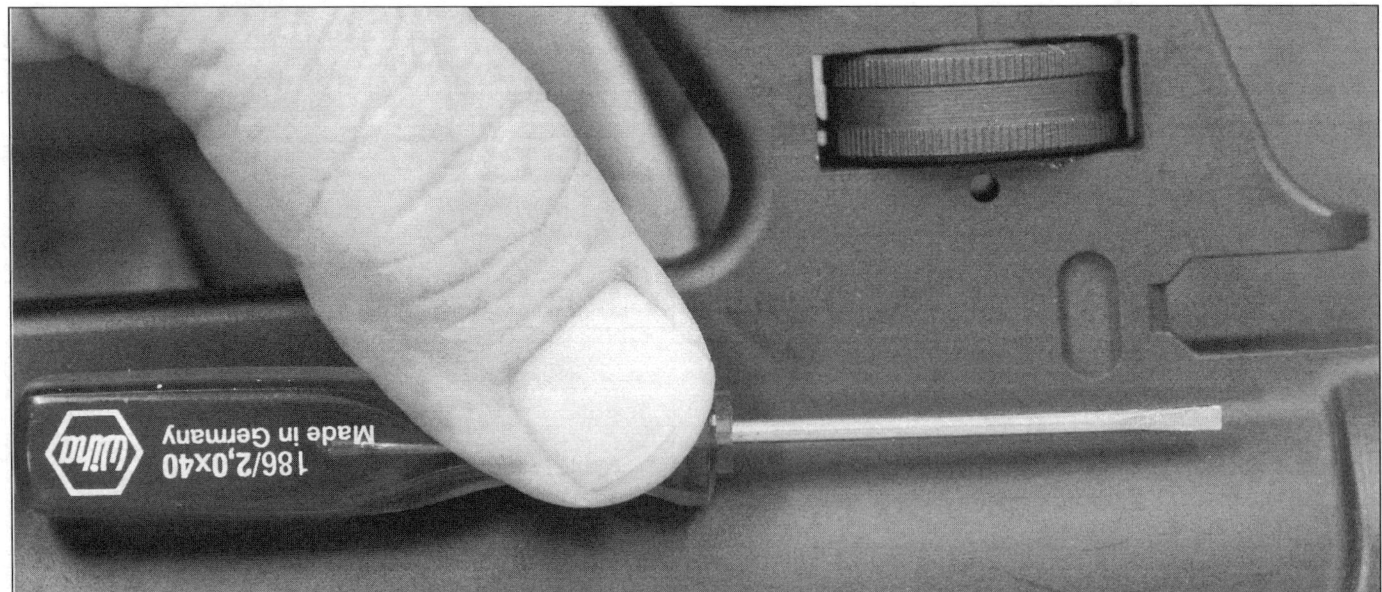

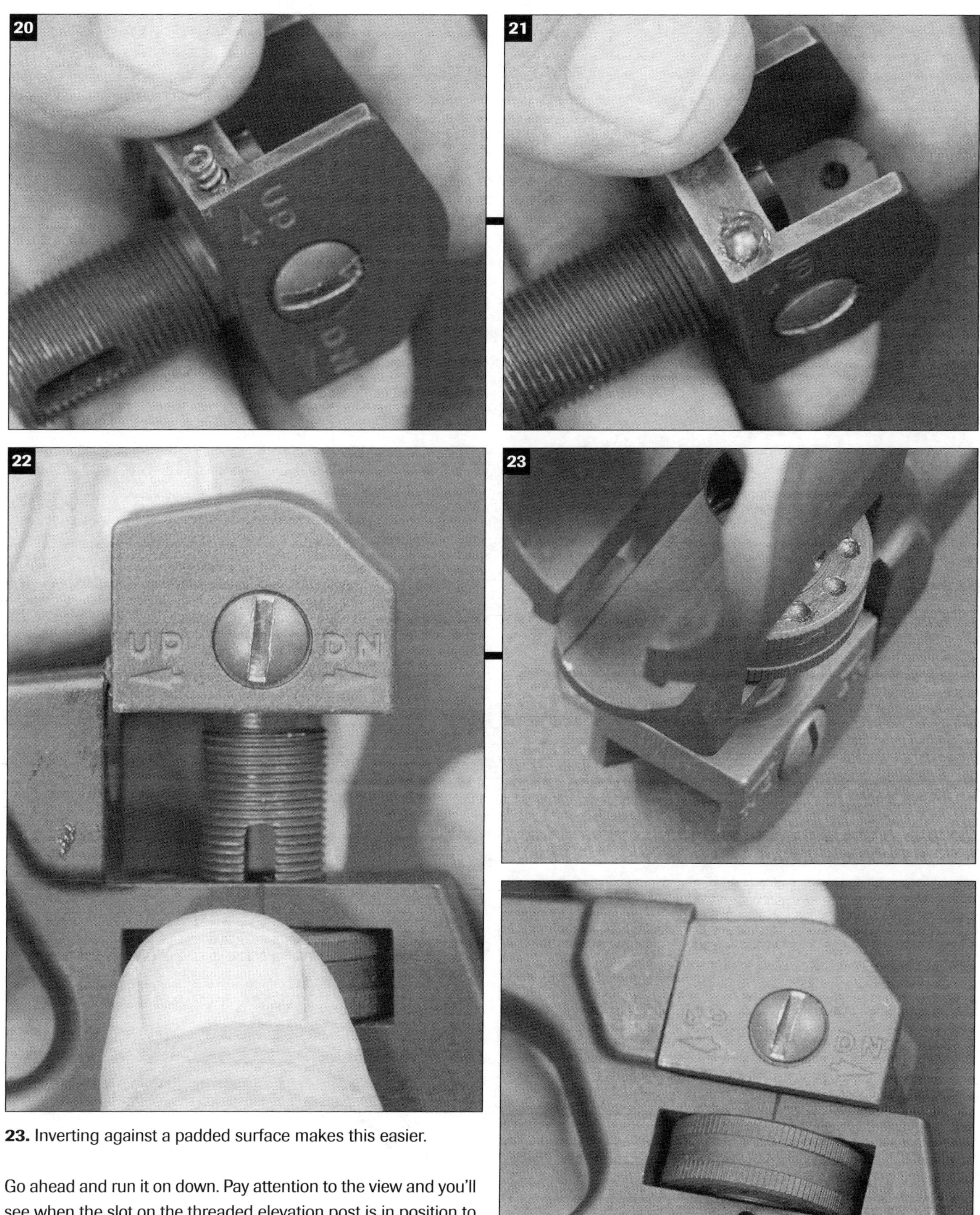

23. Inverting against a padded surface makes this easier.

Go ahead and run it on down. Pay attention to the view and you'll see when the slot on the threaded elevation post is in position to take the roll pin.

Take a short break and let's get ready to grumble!

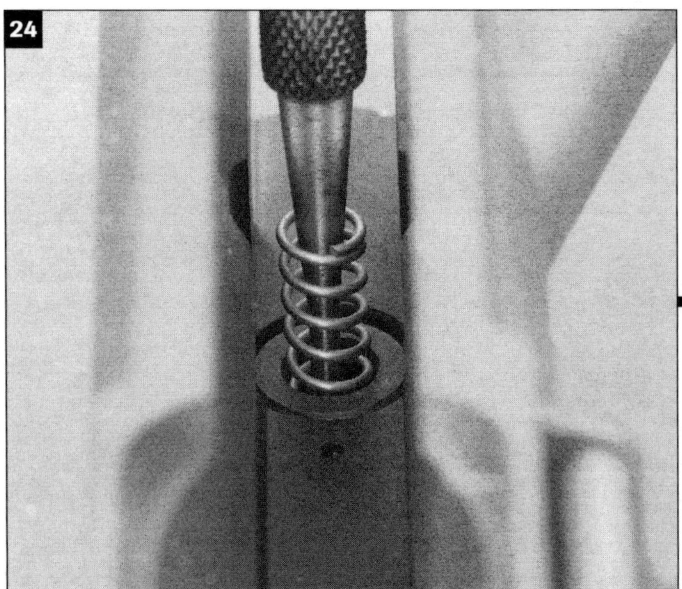

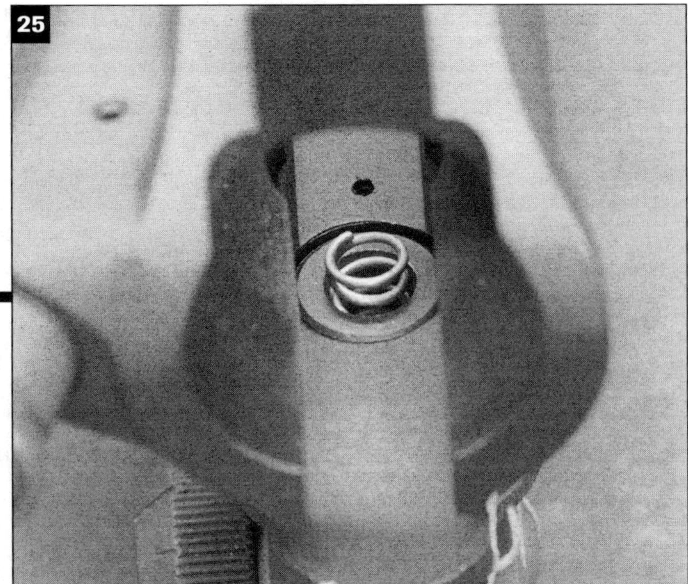

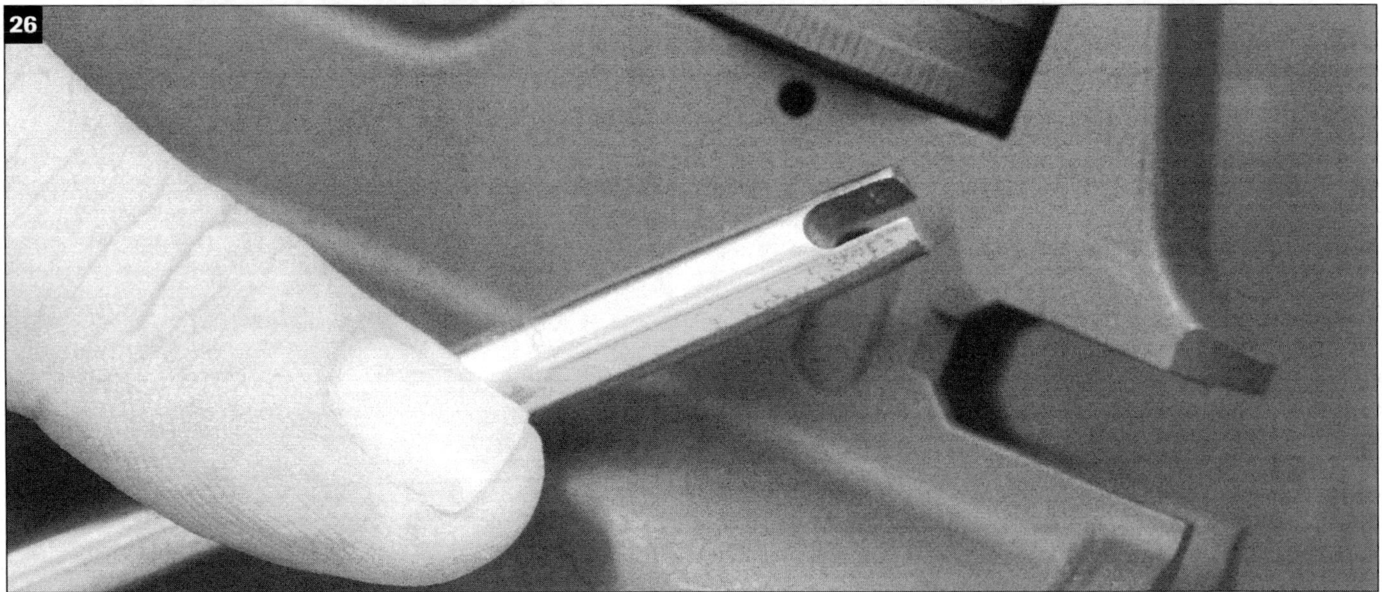

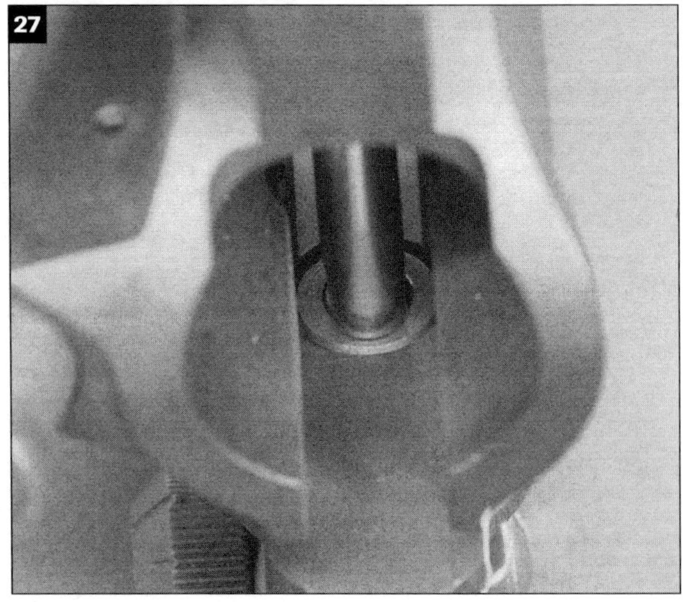

ELEVATION SPRING

24./25. Insert the elevation spring into the threaded post on the sight base. A punch helps serve as a guide for the spring.

26. Get your tool!

27. Push that bad boy to fully compressed and then **28. get a capture punch involved.**

You'll be able to feel how much pressure is at work here. It is very important to keep this spring compressed all the time the roll pin is being driven in. There's enough pressure that even if the roll pin barely misses the far hole it's not going to happen.

29./30. Start the pin, drive the pin using respective punches.

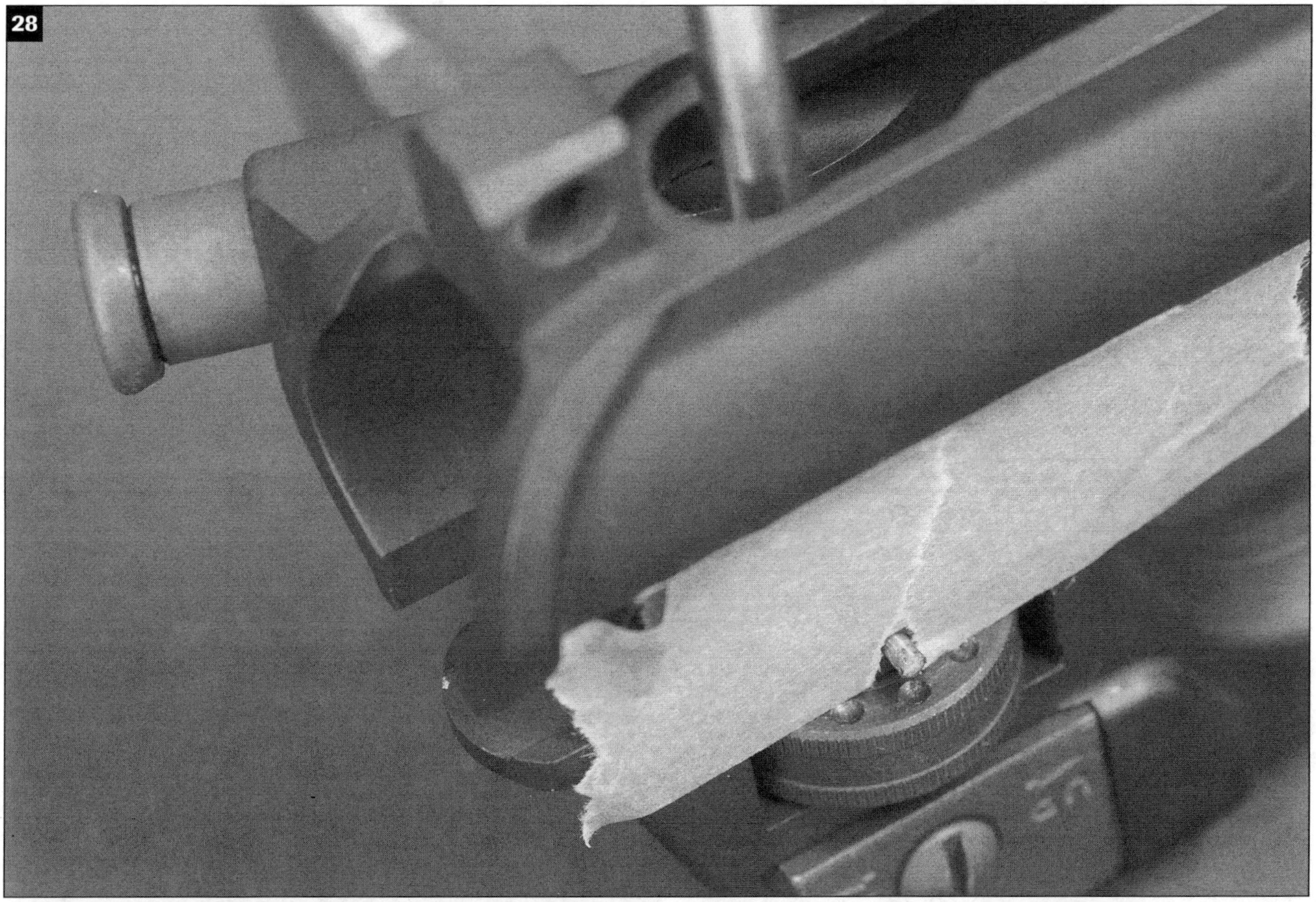

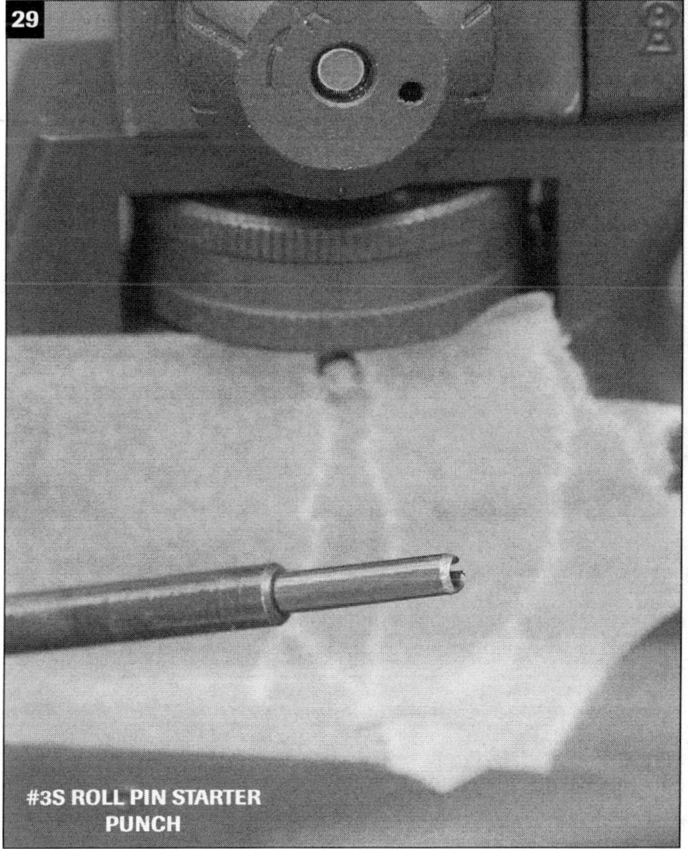

28

29
#3S ROLL PIN STARTER PUNCH

30
#3 ROLL PIN PUNCH
(3/32-inch)

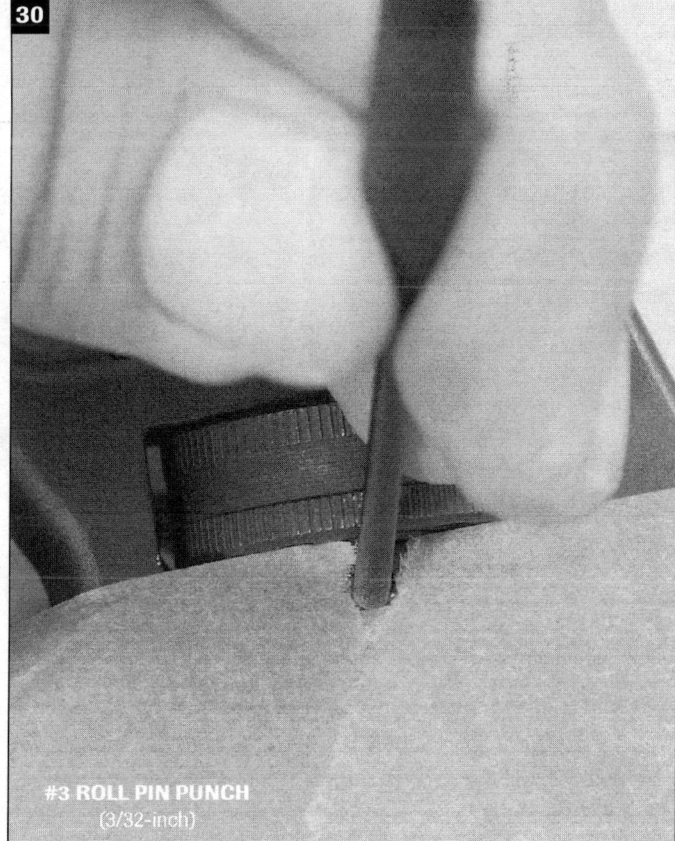

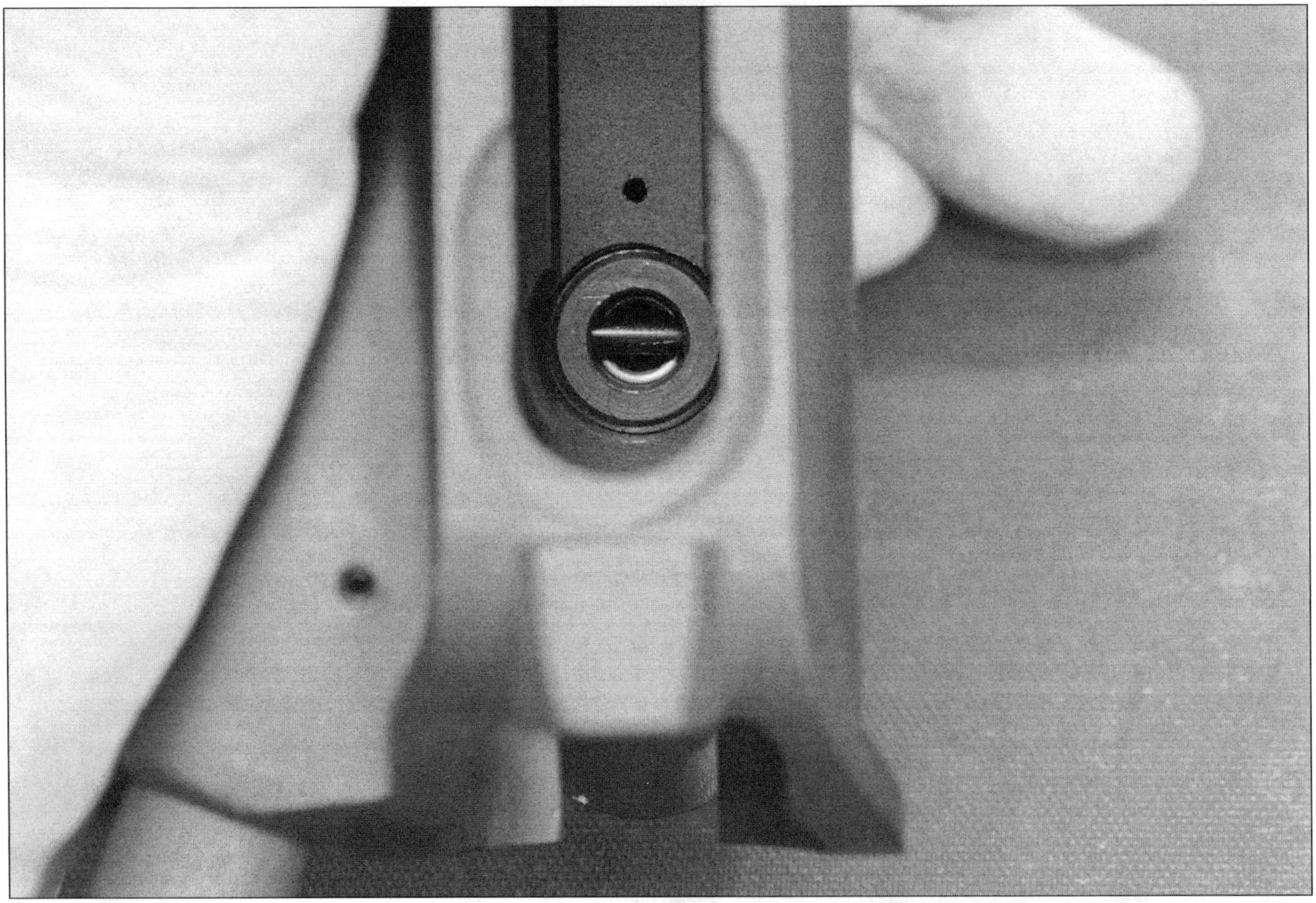

FINALLY...

The last step is to index the elevation wheel. (The Accuracy Speaks sight doesn't use one, but in case yours does...) Turn the wheel until "8/3" is showing on the left side of the rifle. That aligns the set screw in the wheel with the hole in the receiver that allows access to it. Access and loosen the screw using a 1/16-inch hex wrench so the halves separate enough to rotate the bottom half until the rear sight is bottomed out. Bring it back up a couple of clicks and, keeping the "8/3" showing out, tighten the set screw. **Done!**

14.0 A2 FRONT
PLUS MODS

ESSENTIAL INSTALLATION

1. Place the spring into the detent body.

2. Position that in place in the front sight housing.

3. Secure the front post using a specialty wrench.

4. Thread it in. Careful starting it and make sure one of the little tabs on the wrench is depressing the detent.

5./6. How far? A good starting point is to have the top of the post 0.200 inches above the sight housing. Use the depth measuring utility on a caliper to make it easy.

Pretty simple.

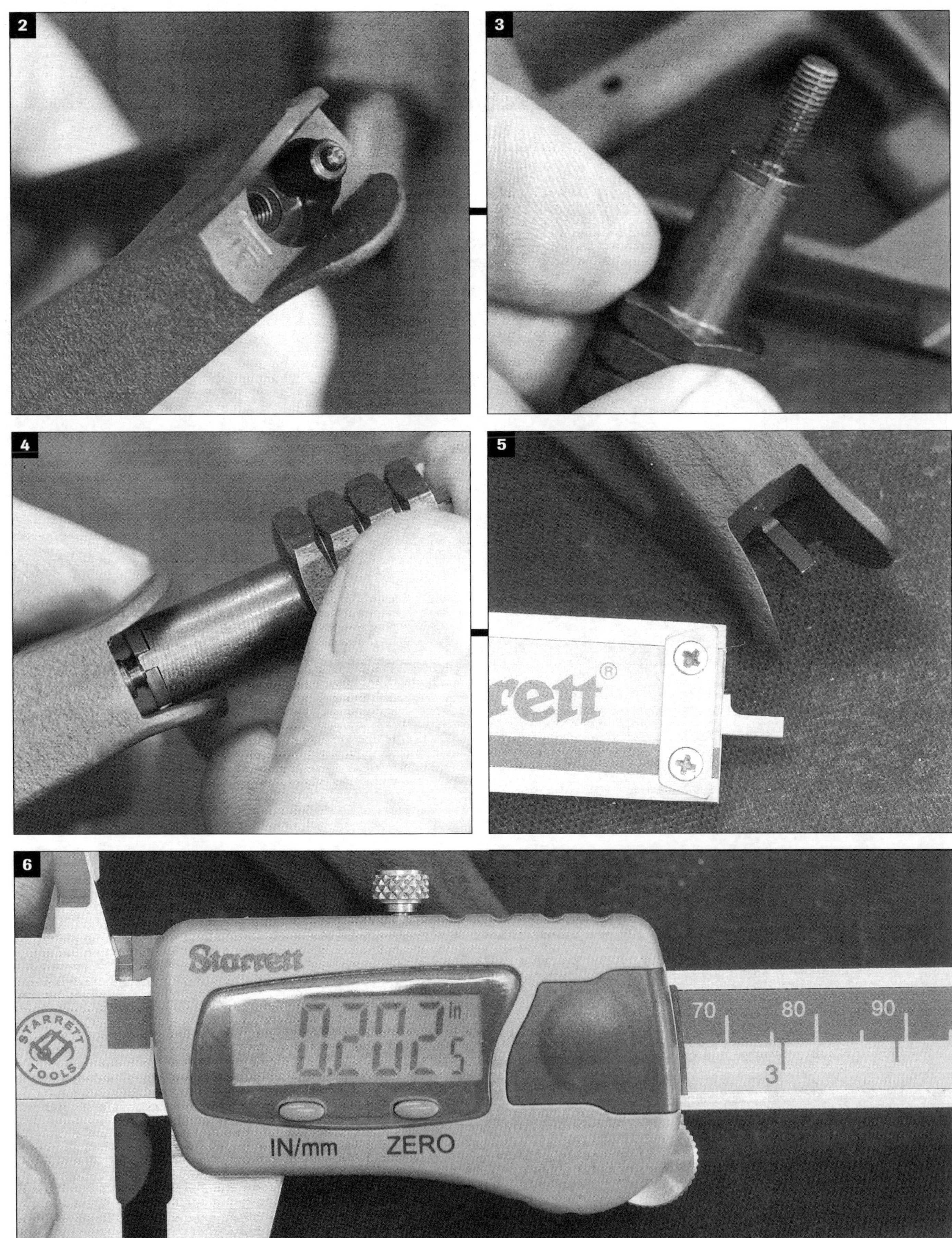

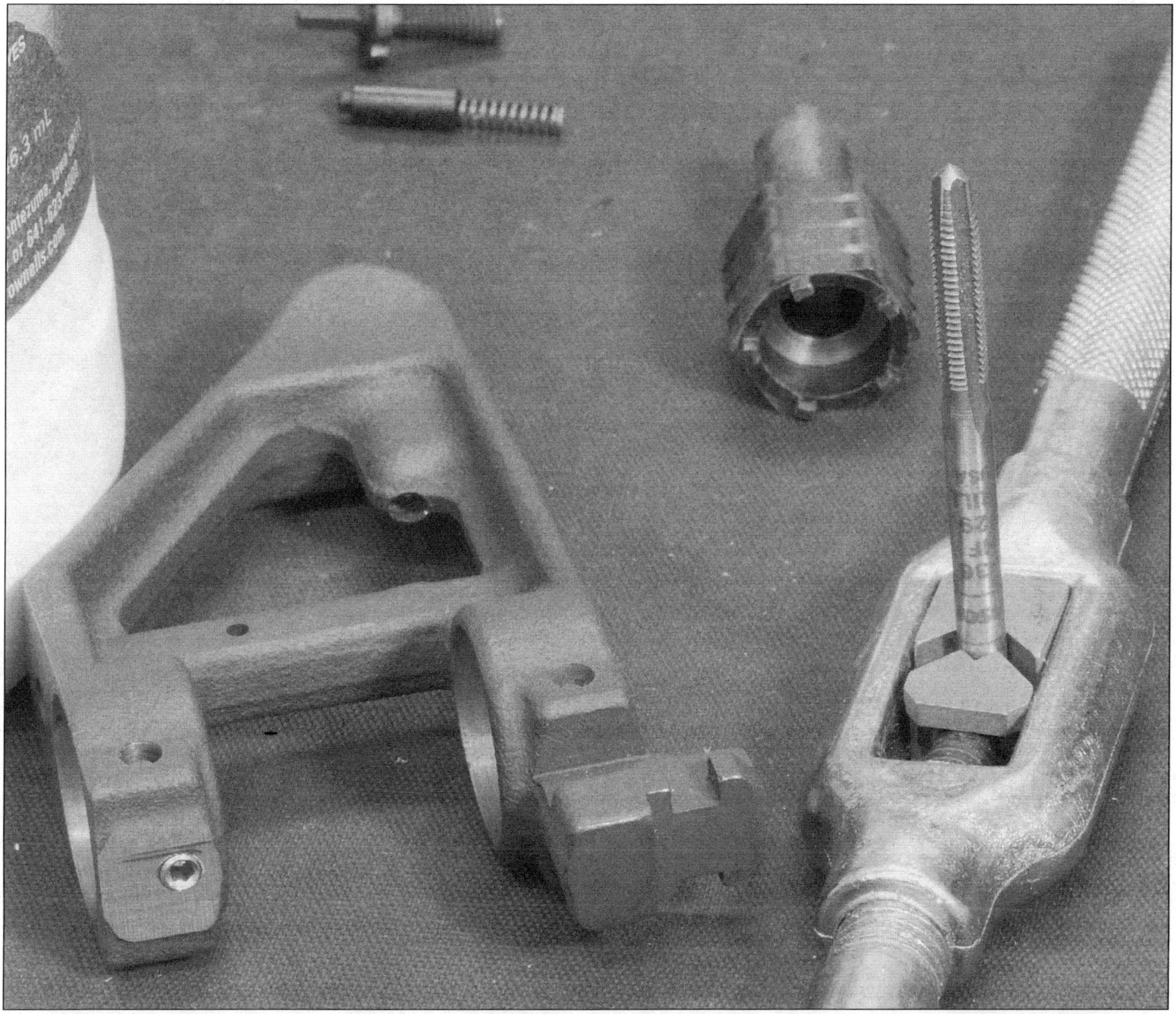

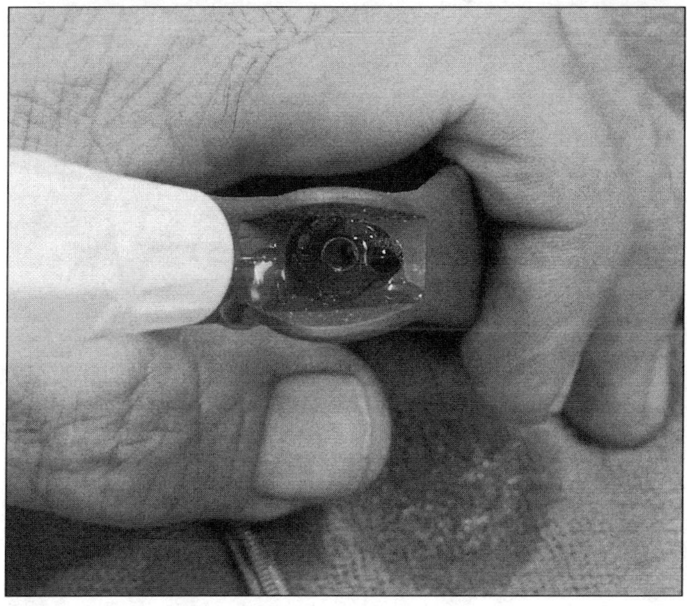

POST HOLE TAP

The idea is to thread the front post hole in the front sight housing all the way through. They're drilled through but not threaded through. Doing this then lets you run a small 8-36 set screw up under the bottom of the front sight post to lock it absolutely against any movement. Machinist supply for the parts.

Brownell's sells an 8-36 tap. Beware lesser taps. Brownell's are good and this is no place you want a broken tap. I know from bad experience with a lesser tool. Use plenty of cutting oil, and Brownell's brand is likewise recommended, promise yourself slow progress, and this job is easy.

Go slow, no more than a 90-degree turn at a time (turn, back off, turn a little more, back off, and so on), clear chips frequently.

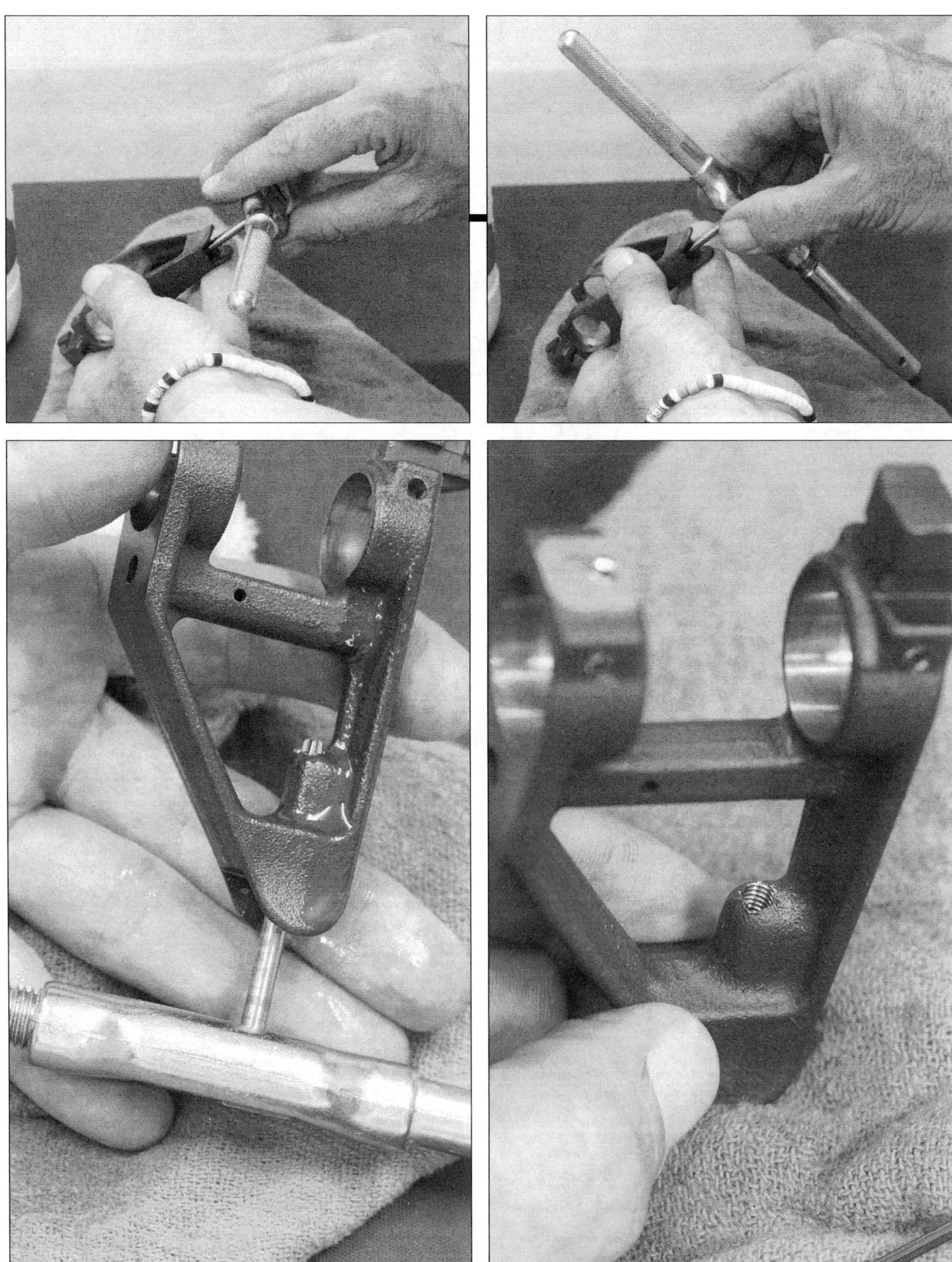

Modded Housings

Options were discussed, and this project already had one, and one done very right. The Satern barrel needs no more than what it has.

Clamp-on sight housings are available aftermarket.

It's no step, or shouldn't be, to get a local machinist to drill and tap the centers of the undersides of the housing to accept set screws. 8-32 is fine, and common. Now you don't have to buy a clamp-on housing. One screw per "ring" is fine, especially since we eventually are going to glue down the front ring. It's perfectly okay to have it tapped 8-36 if you already have an 8-36 tap, but 8-32 is hardware store shopping.

We're doing this so the front sight housing, one, can be more easily installed (no taper pins); two, so there's less pressure on the barrel (no taper pins); and, three, so the sight housing can then be moved a little left or right to get the rear sight smack in the middle of its wind range (no taper pins). That's important.

The Olympic Arms front housing shown on the right needs a post and parts and, I say, post hole modification to be complete.

The lower photo is one I had done at my local machine shop. The hole is already there, just needs to be "fixed." I fixed this one for an 8-32. A little cold blue or paint touch up and it's good to go.

Finishing This Package

Since this here is fairly close to a full-race Service Rifle, it can get some tricks to be complete, and here a few are.

Someone might want to add a forend cuff weight to the bottom handguard and lead wedge to the buttstock. Those, of course, just make the rifle heavier. I like very heavy Service Rifles. Either of these can be trimmed, or substituted, for less weight.

A CWS in the carrier and a CS spring in the stock both keep the bolt locked longer, and that, if nothing else, improves spent case condition. See these parts on page 192.

The extractor and ejector treatments and tricks thoroughly explained and also

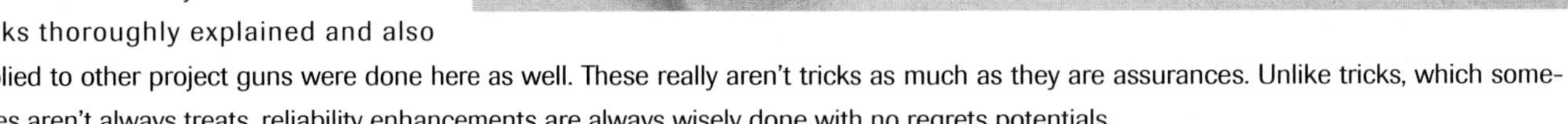

applied to other project guns were done here as well. These really aren't tricks as much as they are assurances. Unlike tricks, which sometimes aren't always treats, reliability enhancements are always wisely done with no regrets potentials.

I also suggest sealing the charging handle using RTV and this modification decidedly is more treat than trick. This is done in case a case decides to leak. Service Rifle shooters tend to get greedy on bullet speed.

Oh yeah, and you'll need one of those too… The Competitor Plus from Brownell's is a good choice, as is a Turner. Get a 54-inch length for an AR15. Oh, and a trigger! And that's coming next…

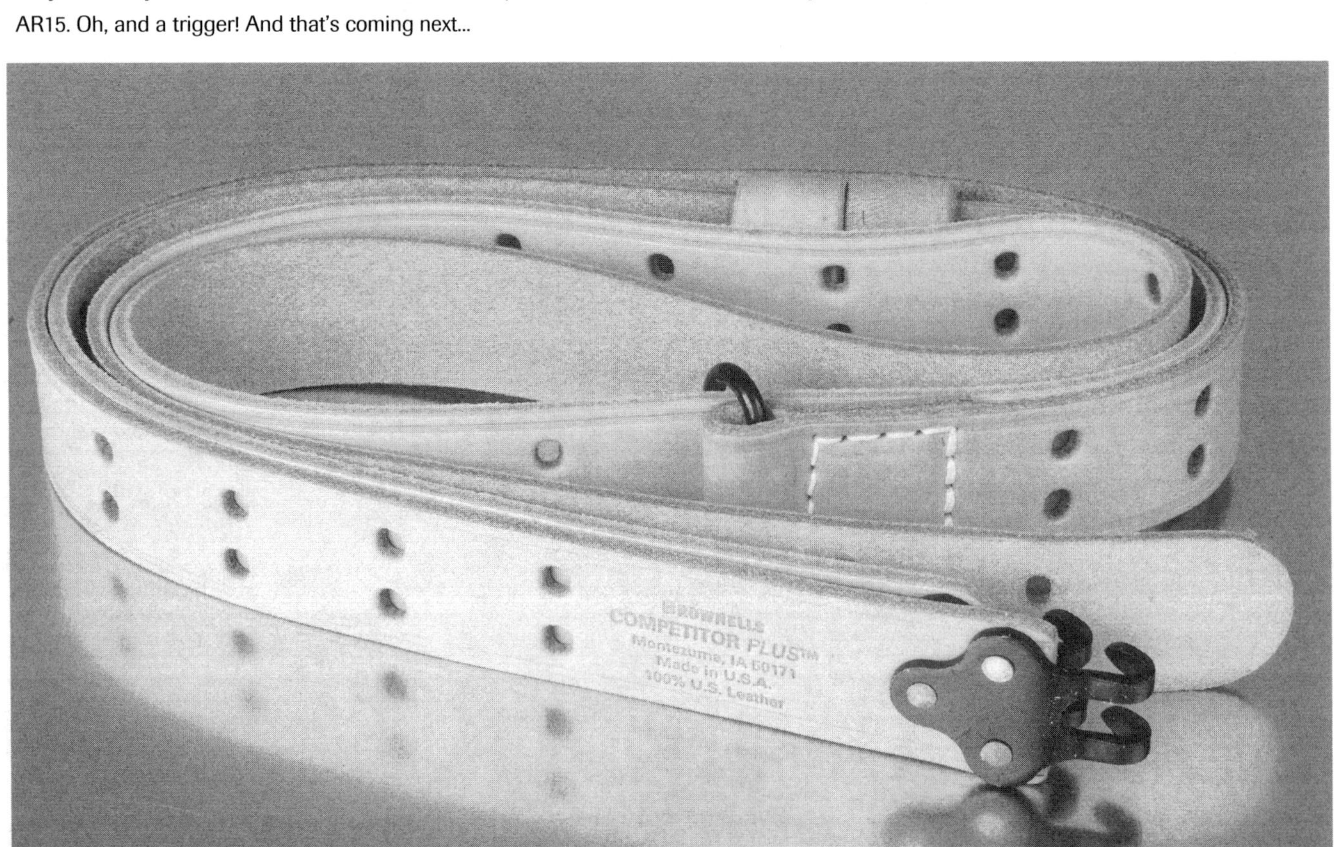

15.0 JEWELL TRIGGER
FINISHING THIS PROJECT

[There are several competition two-stage triggers available. Check The Competitive AR15: the ultimate technical guide for more about what's on the market, and keep your eyes open because there are sure to be more. Generally, the more you pay for a trigger the better it will be. That's true with many things. What makes a trigger "good" in my opinion is how easily it can be adjusted after installation, and, of course, how it functions. Since this is a Service Rifle, the overall weight of the trigger pull cannot exceed 4-1/2 pounds.]

SEGMENT CONTENT

177   **Overview**

178   **Jewell Package**

179   **Installation**

184   **Adjustment & Tuning**

This rifle got a Jewell two-stage. Pay attention! *Jewell triggers are easy to install but won't seem so the first time this is tackled. They are different. It's the appearance of the parts that creates quizzical looks the first time they're laid out. You'll easily see how they work.*

DIFFERENT LOOK

Jewell makes a good trigger that's easy to adjust. The only mark against these is that, compared to more recently available others, lock-time is a little slower.

The first step is getting out all the parts and taking a hard look at how they function together. It's the first-stage adjustment that really looks different. That's the big buzz-saw looking thing. The springs are different too. As you'll see, the pieces in a Jewell set look different but function in essentially the same ways as those of any other two-stage trigger. Of particular note are the roller arrangement on the hammer spring and the spring pin retainer. There are also a couple of specialty tools part of the package. Don't lose them.

Jewell instructions could be (way) better. Hope that you'll find those in this book are.

JEWELL PACKAGE

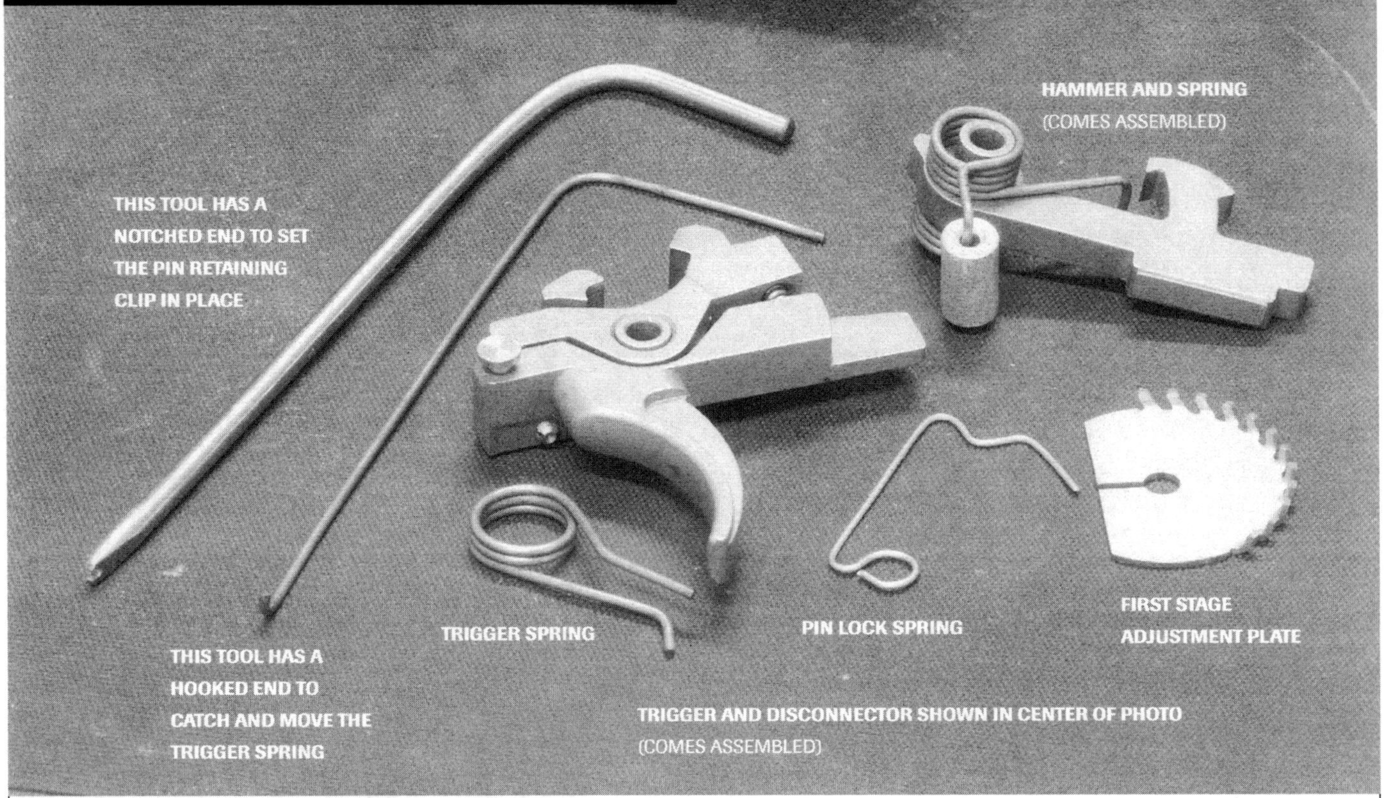

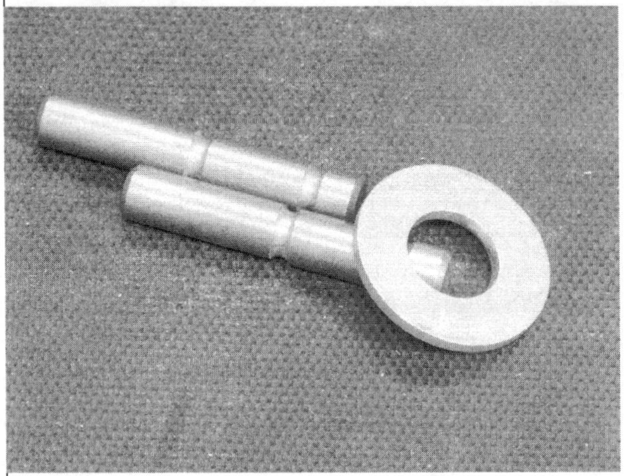

Helps

This has been said many times many places but I don't think any Jewell trigger installation should commence without first collecting a set of KNS trigger pins. Jewell doesn't supply its own pins and that's a mistake. Standard trigger pins are frequently very "loose" in their fit and this can, no joke, negatively influence trigger performance. KNS pins are effectively oversized but correctly sized, if that makes any sense, and will make your Jewell a jewel.

Some suggest adding a washer to shim between the the first-stage adjustment plate and receiver wall. This reduces the lateral movement that sometimes accompanies a Jewell installation. Find a washer about a half-inch diameter and approximately 0.035 thick. I bought a small baggie-full at a hardware store. Size is #14S.

Making weight. It's not precisely easy with a two-stage to get reliable weight reads using a display-type trigger pull gage. This is a Lyman tool purchased from Brownell's. It's accurate but it can be hard to settle on a confident assessment of second-stage influence. Fixed weights (the kind that hang from the trigger) are a better indication of legality, and so always use one, or rig something similar, to check off on having attained accepted weight minimum prior to showing up to your first tournament with your new rifle.

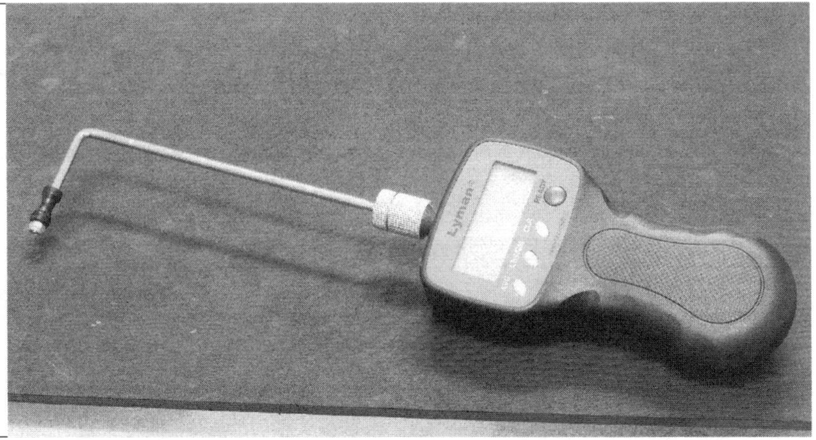

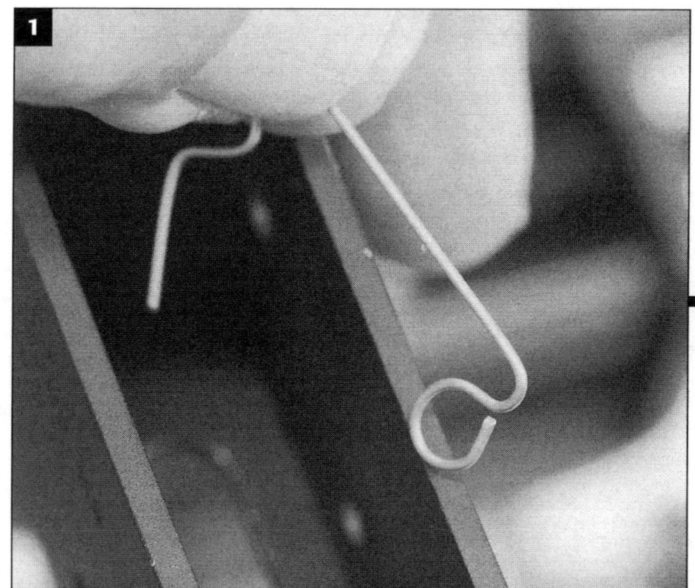

JEWELL INSTALLATION

I wish this sort of thing was easier to photograph. It's a dark area and the parts are small. Details are tough to show, and some things are easier said and done than shown in the first place. Read the process descriptions carefully and I think you'll manage this installation just fine.

Also, for reasons of best clarity, I didn't grease all the parts prior to installation. Of course (of course) you should grease them.

The Jewell trigger installs just fine, by the way, with the safety lever in place as normal. Some aftermarket triggers won't.

1./2. Start by locating the pin lock spring and a trigger pin. Start the trigger pin into the lower receiver from the right side, which is the right side. Trigger pin (and hammer pin) install from right to left, smooth (ungrooved) end first. Slip the pin lock spring over the pin.

The pin lock spring serves ultimately to fit, under tension, into the first groove on the hammer pin. It will be the last step in the installation, but first thing that goes into the rifle. We'll take another look at this later.

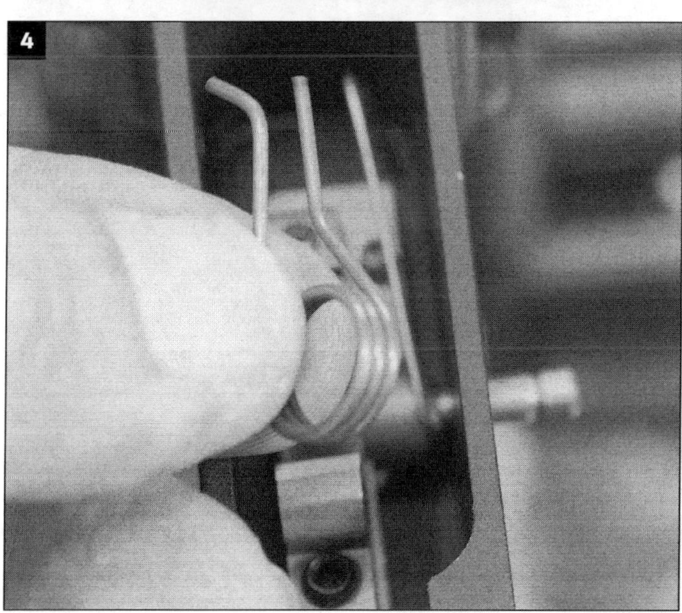

As you're installing the hammer assembly, rotate the free end of the pin lock spring up over the hammer pin hole. This will keep you from having to fish it out later.

3. Drop the trigger into place and push the pin into the trigger body, but no farther.

4. Get hold of the trigger spring and turn the page...

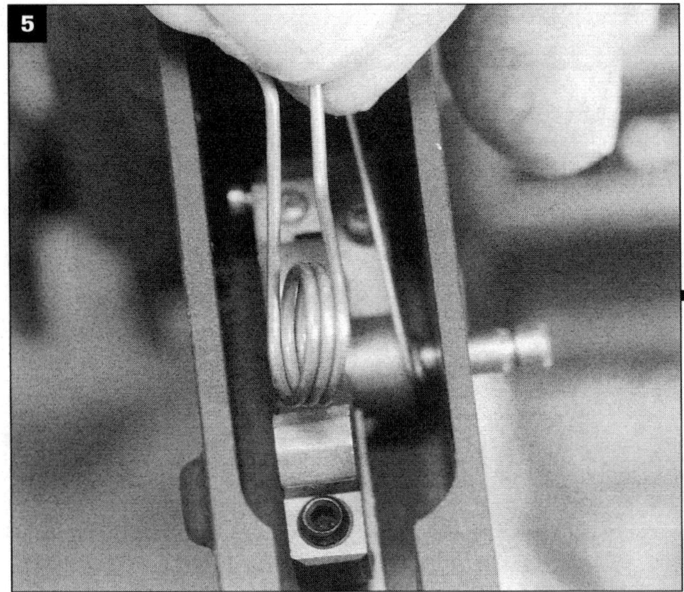

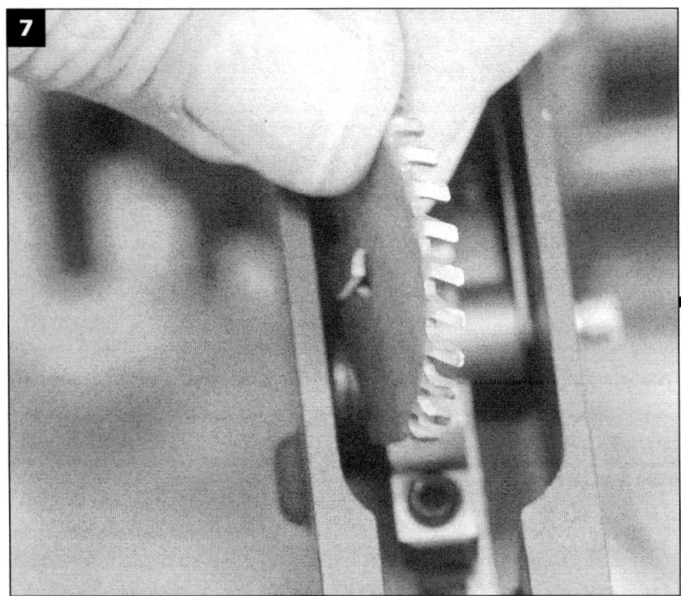

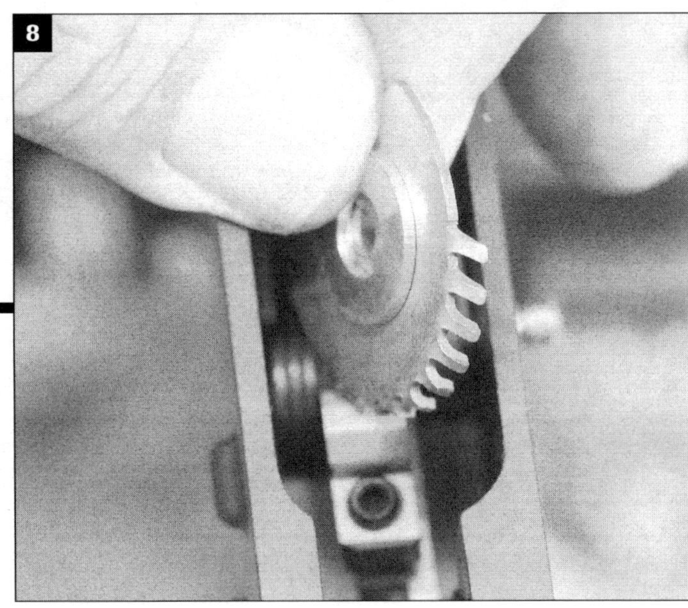

5./6. Place the trigger spring to the left of the trigger body. Pay close attention to the orientation of the spring. Make sure the straight leg is under the stud toward the front of the trigger. That's the part inside the white circle.

7. Place the first-stage adjustment plate (the buzz-saw-looking thing) to the left of the trigger with its "teeth" facing inward or toward the right.

8. Now is the time to add the washer talked about earlier, if you want. Grease it and it will stay put better.

9. Get all the holes lined up and go ahead and seat the trigger pin fully through and into the left-hand receiver hole.

That's how it looks, or should.

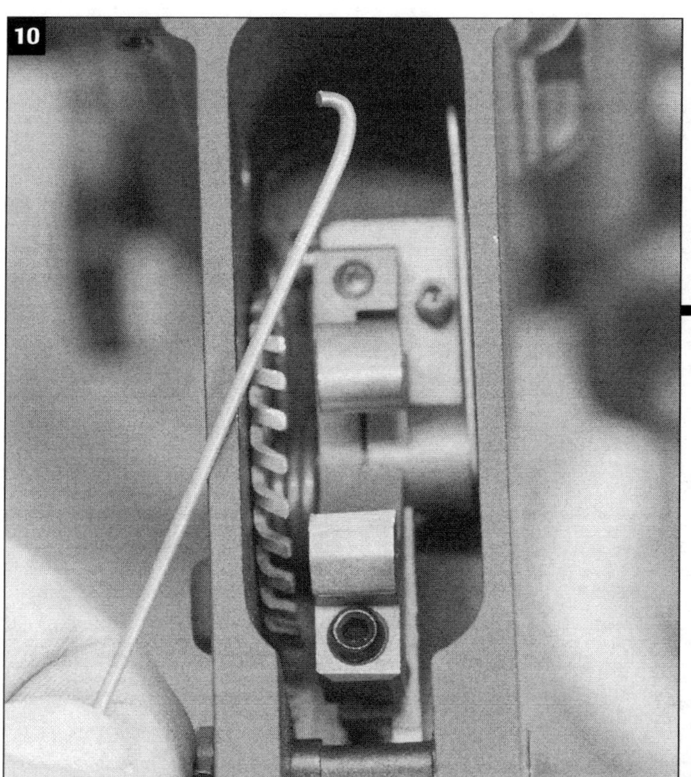

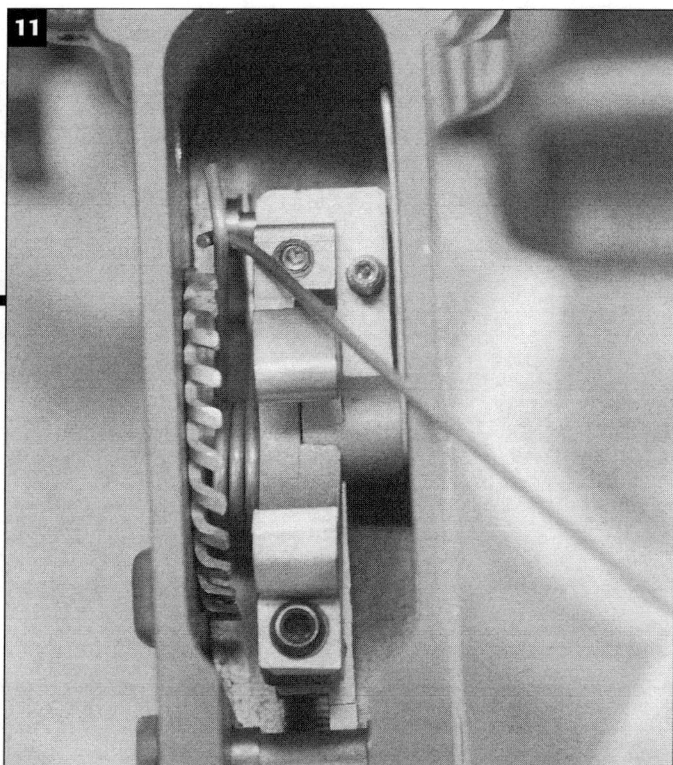

10. Get the hook-end tool, reach it in, and pull the free trigger spring leg (the one with the bend in it) up.

11./12. Locate it into a notch. I would suggest about the fourth or fifth notch to start. This can be adjusted later. The farther back the spring goes around the wheel the heavier the first stage pull will get. Not at all a bad thing since it's a Service Rifle. First stage added to second stage gives overall weight, and higher first stage can mean lower second stage break weight.

I went way too far around the wheel here with this one simply because it placed the end of the trigger spring leg centered in the photo.

Get the hammer, hammer pin, and a capture punch handy.

13. The hammer spring has a roller on it. This roller ends up resting on the bottom of the lower at the juncture of the inside receiver wall and the bottom of the receiver.

14. Push the spring roller down fully into its place in the lower and then line up the receiver holes with the hammer body hole.

15. I use a capture punch inserted from the left side of the receiver. It takes a hard push to attain and then maintain alignment to install the pin. Mechanics gloves help push down the hammer.

16. Insert the hammer pin all the way.

To finish, we need to engage the pin lock spring into the groove on the hammer pin. First make sure the free end of the spring is above the hammer pin.

17. The bend indicated inside the white circle is what engages the hammer pin groove.

18. Take the fork tool and push the spring down and back. When the free end clears the hammer spring **19./20.** the lock snaps into place in the groove on the hammer pin. **21.** The hammer spring can interfere with this so sometimes has to be pushed toward the left to provide clearance (use a small screwdriver).

Careful with that pin lock spring on subsequent removals and reinstalls. It's easy to bend out of shape. Make sure this spring is engaged into the groove on the hammer pin. It is positive when it engages, and is easier to feel than to tell about or show.

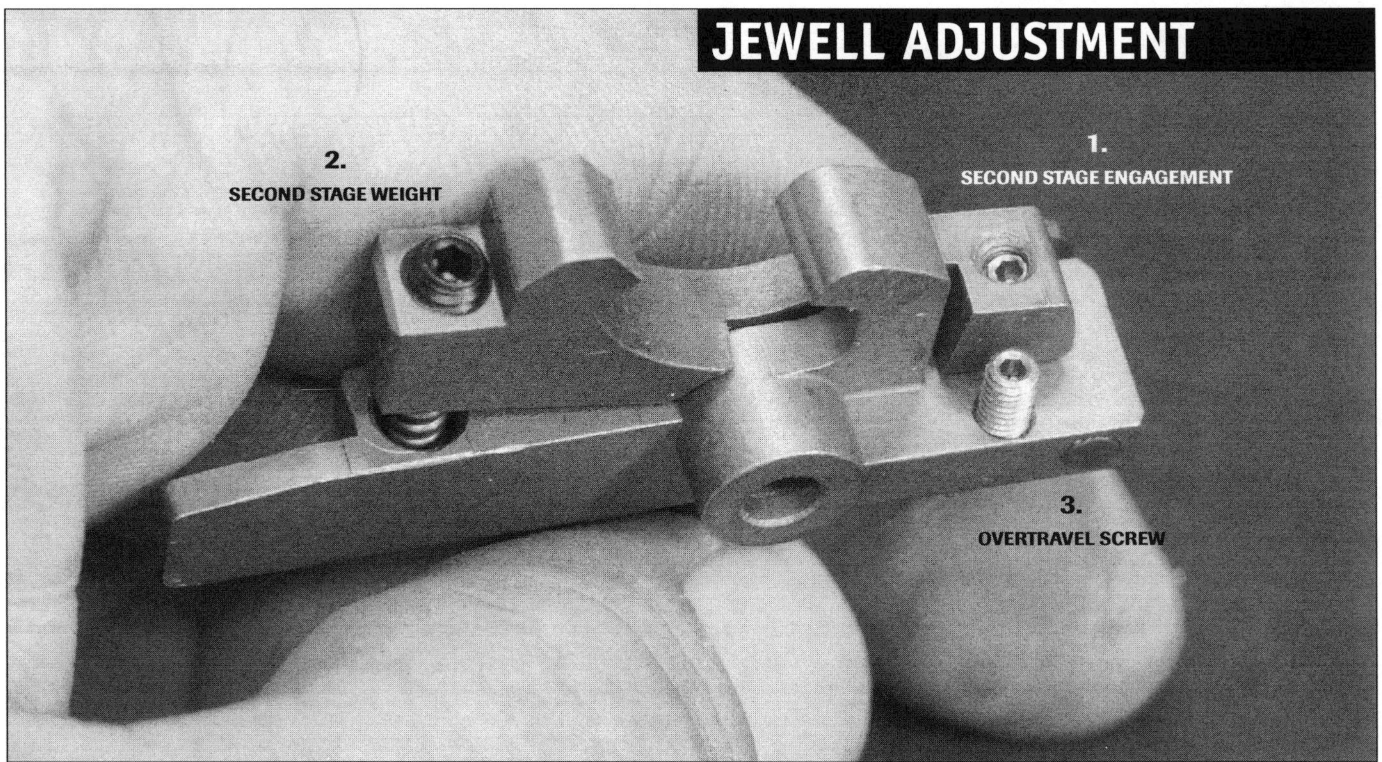

There you go. Now we can adjust it. First, make sure it's lubed before adjusting, and as always, make sure the lower receiver wall is backed up by something that will cushion it. Never let a hammer fall freely onto this wall.

First-stage pull weight is the solely determined by the notch selected for the trigger spring on the adjustment plate. That's easy.

1. Second-stage sear engagement determines the "creep" in the trigger, the added travel of the second-stage. Usually we want the least we can get. Try it and see what you have. If you feel no second-stage present, then turn the screw out until you get what you want. To decrease engagement turn the screw in. This is a preference or "feel" characteristic. The only thing is that the more engagement you can accept the longer the adjustment will last.

2. Second-stage pull weight is established using, no shock, the second-stage pull weight screw. Turn the screw in to increase weight and out to decrease. This is really what you feel as the trigger break. First stage plus second stage equals overall weight.

3. Overtravel is the amount of movement in the trigger after sear release and hammer fall. It's hard to miss from reading anything related that I've ever written that I like all the overtravel I can get, meaning I want the trigger to continue to "swing" after the hammer falls. Most don't like that, and the majority want the least amount of movement they can get. There has to be some. To get the minimum, cock the hammer, turn the overtravel adjustment screw in until pulling the trigger will not release the hammer. The trigger is essentially "frozen." At this point, keeping the trigger pulled, turn the screw out until the hammer falls. Take another quarter turn out for a cushion, check its feel again, and that has established minimum overtravel.

Safety checks on a Jewell are the same for any, and that is to ensure that the safety works freely and does its job. **That means the hammer won't fall when the safety is engaged no matter how hard you pull the trigger.**

Disconnector checks are also the same here as for any. Cock the hammer, pull the trigger to release the hammer, and keep the trigger held to the rear. Still keeping the trigger pulled fully to the rear, cock the hammer again and then very slowly release the trigger. The hammer is captured by the disconnector when the trigger is pulled to the rear. When the trigger is released the hammer should be handed off from the disconnector to the trigger sear. Do this several times and even try pulling up on the hammer with your other hand while releasing the trigger. This test must again be conducted at the range. If it's not working right, go back to the second-stage sear engagement screw. There's rarely a problem as long as you maintain a reasonable amount of engagement.

16.0 CARBINE PROJECT
SHORT GUN

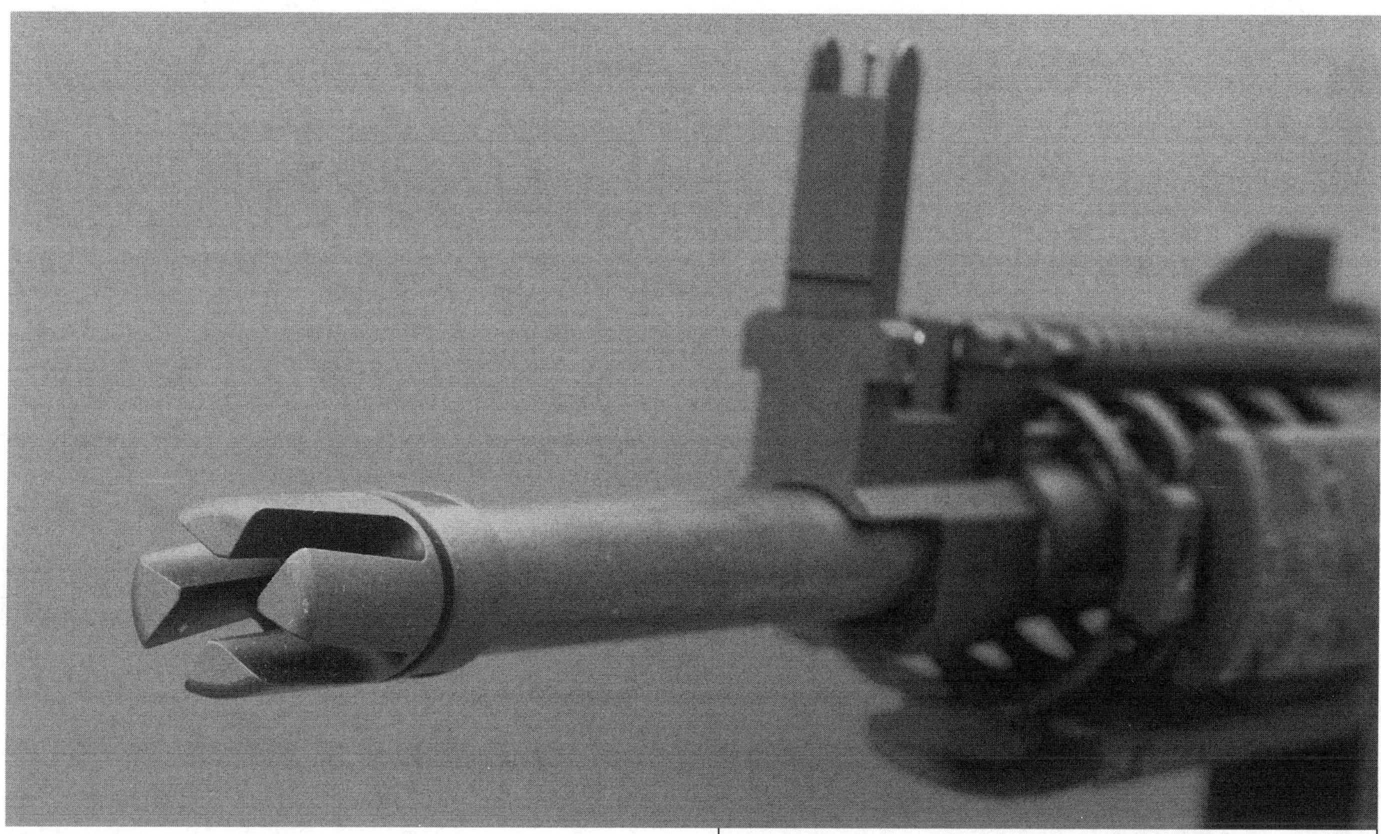

This one is about as streamlined as they get. *But, hey, there's four Picatinny rails that will accept anything most any heart could desire.*

CARBINES CONSIDERED

The short gun. As with all rifles done for this book, component choices alone separate and define this project. The assembly work is no different. This gun got a float tube and aftermarket trigger and stock. Three things that change the face of a carbine, any carbine.

Beyond the basic build, it's the hangers on that seem to catch the attention and occupy the focus of producing a suitable carbine.

SEGMENT CONTENT

186	**Component Selection**
187	**Float Tubes**
189	**Barrel Selection**
192	**Tricks** (system and accessory options)
195	**Parts & Tools**
196	**Barrel & Float Tube Installation** (plus manifold)
201	**Vortex Flash Hider Installation**
202	**SOCOM Stock Installation**
205	**Trigger** (Accuracy Speaks)

That's all about latching on and screwing things down onto Picatinny rails after the fact.

There's way on more about this in *The Competitive AR15: the ultimate technical guide*, and my interpretation of a civilian carbine is one that performs well on target and is streamlined of accessories. That's, of course, where I differ with most folks on these guns. Again, virtually no difference in the build itself, though, so all these choices will be only matter-of-factly shown with respect to my opinions.

I chose a receiver set from TKS for this build. These are very robust machined-from-billet receivers. Quality is outstanding as is the finish. The upper is slab-sided, considerably thicker walled than a common-form receiver, and retains the mil-spec forward assist and trapdoor ejection port cover. The lower is likewise über-sturdy and has a solid trigger guard, that also has a more generous opening for gloved use.

Free-float forend tube installation on this one was very straight-

COMPONENT CONCERNS

forward and simple. You'll be done with it before you know what happened. I selected a Yankee Hill tube for the project outlined here, carbine length. These are high quality, I think reasonably priced, and easy to install. The most sano installation of a Yankee Hill tube requires the purchase of a Yankee Hill spanner wrench.

You'll see other float tube options in this segment.

Installing a float tube such as this one starts with a barrel nut, just like in a standard barrel. Big difference, of course, is that there's no Delta assembly and, at the start at least, not even a forend. The nut itself goes on, the gas tube hole gets aligned, and the rest of the assembly work is then done after the fact, including, if wanted, the gas manifold and gas tube. The forend itself threads on over the barrel nut.

One suggestion I will offer (and that I didn't do here) on a carbine is not put a carbine forend on it. Go with a longer float tube. This will in no way influence the length of the rifle, obviously not in reality and also not in deploy. It will, big time, influence how well anyone can handle one of these guns. The CAR-length forend is too short to do your best offhand shooting with. Some manufacturers, Yankee Hill for good instance, offer varying length tubes, including "in between" lengths.

Another thing a longer tube does is gets us more room for forend-mounted add-ons, and likewise positions them ahead for greater flexibility and utility. Say, for example, that you want to mount a flashlight, laser sight, auxiliary optic, or, and this is a big one, a front sight. The longer handguard allows positioning these items farther forward, well ahead of the hand, and, especially for a front post sight, greatly improves utility.

I went against that advice, however, on this build and installed an accepted length carbine tube. One reason is that this represents the vast majority of what most will choose. The one hiccup on installing a longer tube on a barrel with otherwise carbine-length barrel dimen-

---

Daniels Defense

The Daniels Defense float tube is the best of its type I've seen. It's got full-length rails at each 15 minutes on the clock face and a can't-miss alignment means. After the barrel nut is secured, and with that the gas tube hole is centered, there are two pins that fix into two recesses on the float tube. These pins are located into a pair of opposing remaining barrel nut holes. It just locks together and goes on straight (or crooked, your choice). This is so stupidly simple I don't know why all aren't done this way.

All Picatinny cut edges on the tube surface have been broken prior to finish. It has less "porcupine" feel to it than others similar. The addition of Brownell's rail covers, the quality of which are a match to the tube, makes for a comfortable handguard to handle. My hands are on the larger side so its girth might become too much for some, but it's entirely sano. The Daniels tube is expensive, about $400.00.

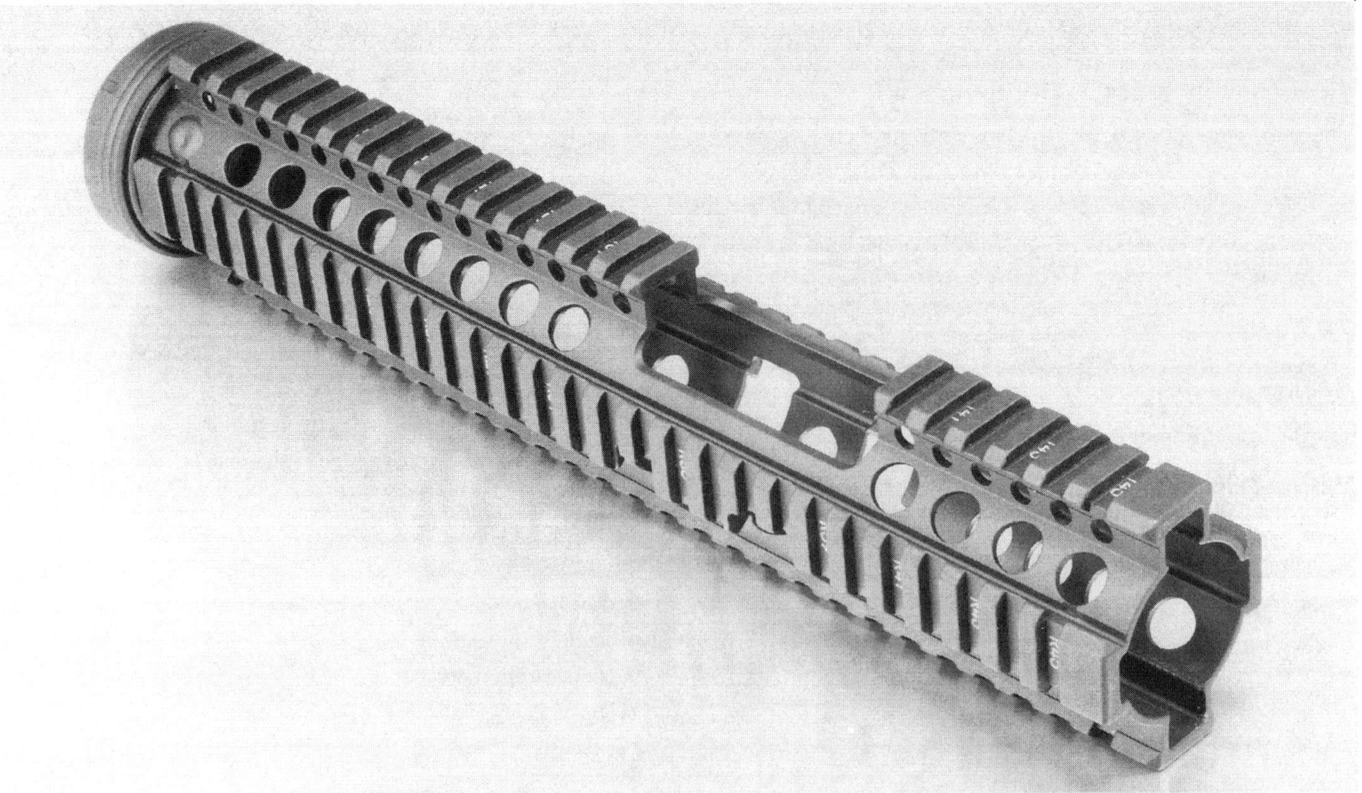

Length options.

This carbine float tube is from Medesha Firearms and is 12 inches overall with a cut for the front sight housing or gas manifold on a barrel with common carbine dimensions. The extra length makes it better to hold and handle and, therefore, connect with the targets. Shoot carbines back to back, one with this tube and one with a CAR-length (8-inch or so) and you'll see its superiority before as much as the first shot.

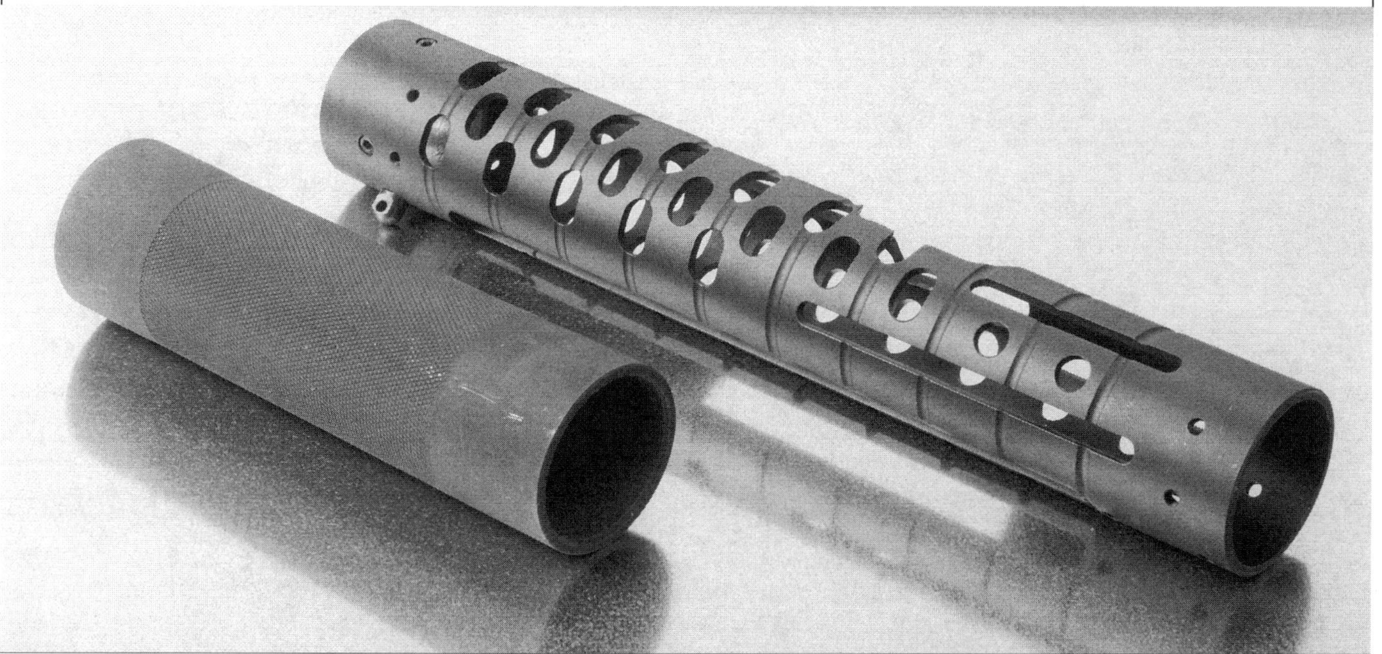

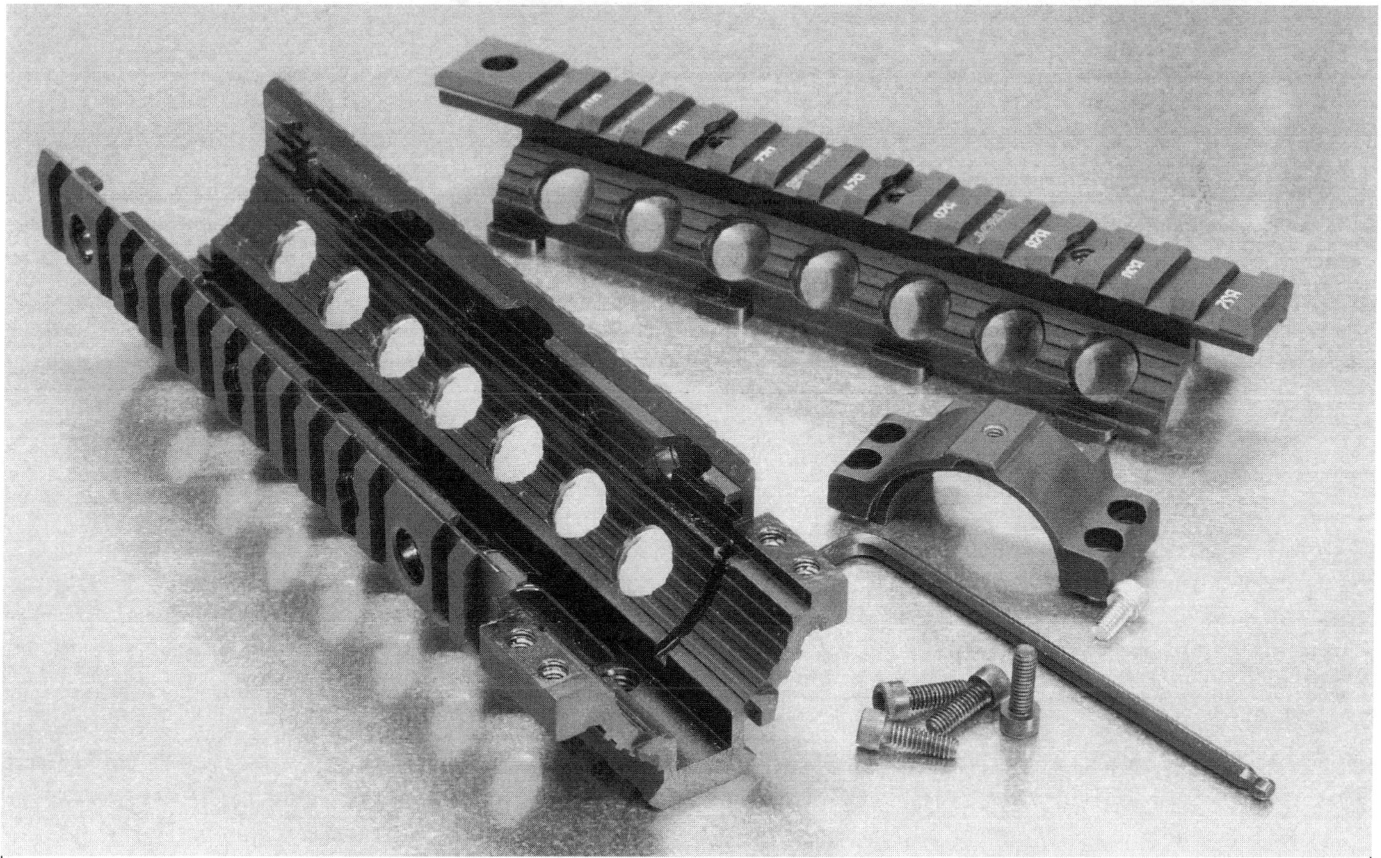

Easy. This is a two-piece 4-rail forend from Troy Ind. It fits securely over an OEM barrel nut. It can be retrofitted to a carbine having conventional handguards by simply cutting off the Delta assembly parts (after removing the spring clip). Front sight and gas tube can stay in place too. Good idea.

> ***My choice for this one.*** *I chose a Yankee Hill skeletonized free-float forend tube for this project because, I admit, it's just very cool looking. It's also very light weight and high quality. I've installed a number of YHM float tubes and have yet to encounter a problem. YHM makes good parts, that fit and function easily, and they don't carry huge price tags. Not cheap by any stretch but so many tactical parts are outrageous.*

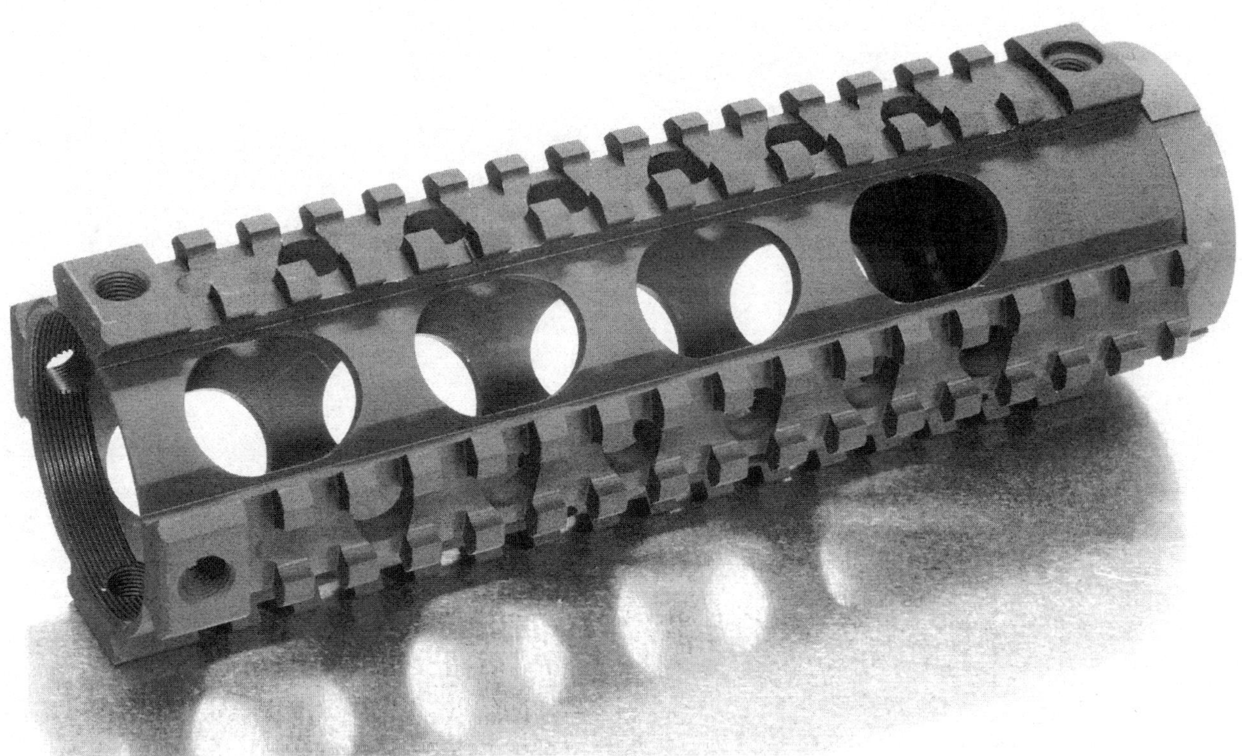

sions is finding a gas block to fit with it. I've had sessions in other build projects getting parts that worked together. There's not much room inside a tactical-style tube. The advice, again (and again) is to stick with the same manufacturer for gas manifold and forend tube, and let them know that your intention is for these parts to work together.

I don't always follow this advice for myself, but what I will give for everyone is to select a barrel with a NATO chamber. This particular barrel does indeed have a NATO. That will let it handle virtually any ammunition that can be purchased, borrowed, or handloaded. There's a side here nearby about the difference between 5.56x45mm NATO and .223 Remington, and it's a very real consideration, especially on a rifle built to actually serve as a professional's tool. I can also tell you that, given the realistic field-deployed ranges at which such a rifle might be aimed and fired, there will be zero difference in the operator's capacity to hit targets dead center. Honestly, not that there would be in the first place. The question is reliable function, and NATO is the answer.

Fluted barrels do have a place on these rifles, and, although it's an extra cost, fluting is probably the best way to save a good deal of barrel weight without compromising on the performance capacity of the barrel. The reason to flute is to maintain a larger diameter at a lighter weight. Cooling effect is real but insignificant. Again, one reason anyone might want to take steps to reduce barrel weight is because of the penchant tactical users have for hangers-on. A carbine should be a lightweight handy-to-handle firearm.

Buttstock furniture is, in my experience, a worthwhile place to consider options from the aftermarket. The common-form CAR collapsible stock is loved by none. It's uncomfortable. There are others available that can radically improve the shooter's relationship to the rifle. Of those I've had experience with, the SOCOM "Boom Tube" is the best, and that's what went on this rifle.

All parts inspections and preparations and initial concerns were addressed for these just as with all others: test fitting, deburring, and on down the now-familiar list.

Some attention was, and I think should be, paid to function. This one is easy: if a carbine won't work, it can't work. There are no alibi strings on a two-way firing range. As has been covered in far more detail in my other publications, a want for what I call "smooth" operation defines the aftermarket involvement. Carbine gas systems are overactive, which makes for overactive operation. That is the source for many failures to feed, extract, eject, and so on. Slowing down cycling cures many to most of these problems.

Every carbine should have a polished chamber. Refer to other

Barrel Selection

Barrel chosen was a DPMS chrome-moly threaded 1/2-28 for a muzzle attachment of choice. The barrel chosen was a heavy contour 1-9 twist and unboxed with front sight housing, handguard cap, and barrel nut attached. First step was getting it stripped as shown on page 94. And, by the way, the "M4" contour shown on right only matters if you actually have a grenade launcher. Otherwise, round is fine, and dandy. An M4 barrel, however, is also considerably lighter weight and that, as said, can matter a lot to the owner of a loaded-down carbine. Hope this makes sense: If you're going to mount a lot of accessories on one, go with a light barrel; if you're not then go with a heavy contour barrel. In my experience the gun is a little light for best shooting manners with an M4-type contour and gets too daggone heavy once the rails are loaded down with lights and sights built with a heavy contour barrel. This rifle finished right at eight pounds, and is very easy to shoot well.

material on the whole ejector, extractor bidness, and pay attention to it as well. Any functional reliability enhancements should be done.

Any rifle should have a good trigger. Any aftermarket trigger out there is going to be an improvement over stock. That's a given. What's not so given, though, is the suitability of aftermarket triggers for use in a "serious" rifle. I don't know that a competition-style two-stage is the right choice for a carbine. Depends a lot on the shooter. A good single-stage with a higher-than-attainable break weight is my choice and recommendation. In other words, get a single-stage trigger and just don't tune it "down" to its minimum weight. Keep it 3+ pounds. If there's no built-in means for adjustment, to make pull weight heavier, bend the trigger spring legs downward.

I'll not get too far into theory because I'd rather leave that to the speculative magazine writers, but will say that if someone's life (at either end of the rifle) depends on the operator's capacity to control

moment of firing, a two-stage had better be positive and, in my opinion, a single-stage is "safer."

I won't advise definitively on sighting and accessory choices because, again, not only are those personal but there's many that simply can't be answered definitively. I will say never to short change quality. Something as simple as a flashlight really, really has to work when it's needed. Selecting a cheap mount, for anything, is borderline irresponsible.

If you choose an electronic optical sight, same applies. Get a good one and keep its batteries fresh. Take time to perfect the mounting of any optical sight with respect to deploy. The aiming image should "be there" when you shoulder and look for it. That's one key to first-shot speed.

I say now the same for any and all control pieces. Before installing any "extended" controls, think (and test) hard before say-

NATO

Choose a mil-spec chamber for a carbine. This will, or should be, marked "5.56" and that's one way of defining a NATO chamber. The marking might be "NATO" or "5.56 NATO" or "5.56X45mm" or another similar, but it will not be ".223 Remington." Primarily, NATO chambers have more room ahead of the case neck than SAAMI-spec .223 Remington chambers. A NATO chamber will take a good deal more pressure. Some tests done in the industry have shown up to 15,000 psi difference in NATO ammunition compared to commercial .223 Remington ammo. Essentially, a NATO will handle virtually any ammo but **do not fire** NATO spec ammo in a .223 Remington chamber. Here's what will happen if you do. Ouch!

Accuracy Speaks trigger.

I think that Accuracy Speaks makes the best single-stage trigger. It's not dependent on screws or trickery. It can be tuned to very light weight, but I don't suggest doing that for a carbine installation. There are a few "tactical" triggers out there with operational characteristics suited to shooter-side safety, which also means target-end safety, and of these the Geissele is the best choice.

ing it's good to go. A magazine release, for good instance, should only work when you engage the button with your finger, not if you bump the gun against a doorway.

The rest is all up to you. Think it through and, by all means, test, test, test. Make sure it works and that you can work along with it.

Finally, as you read through this next segment, and then the next, you'll see there is no difference in assembling one of these rifles than the NRA Match Rifle project to come, or most any other custom-configuration AR15 for that matter. The float tube goes on the same way, as does the buttstock. The parts are the same, just different features and points of focus in design.

Iron Sights

Front. I picked a decidedly effective and efficient front sight assembly for this one. It's a Young Mfg. combination folding front sight and gas block assembly. This ends up mounting the same height as an A2 and its simplicity by virtue of consolidation is bonus for my tastes. I like this part a lot. There are other similar, and Yankee Hill makes a very good one. If you don't get a "unit" then the option is either a standard A2 front housing or a gas block with a rail-mount surface and then a clamp-on front. That's more expense and more clamps and neither is necessary. This Young Mfg. block uses a standard front sight post, spring, and detent. I replaced the post with one from KNS.

Back. The rear sight is just cool. It works well too, which is one reason it's cool. Brownell's-brand "360" went on this rifle. For its intent I liked the design of this better than an A2 for this carbine. The 360 is sturdy, reliable, precise in its movements, and relatively less snaggish than most I've seen. It's easily elevation-adjustable for indexed distances, and these are based on mil-spec ammunition.

Taming it down.

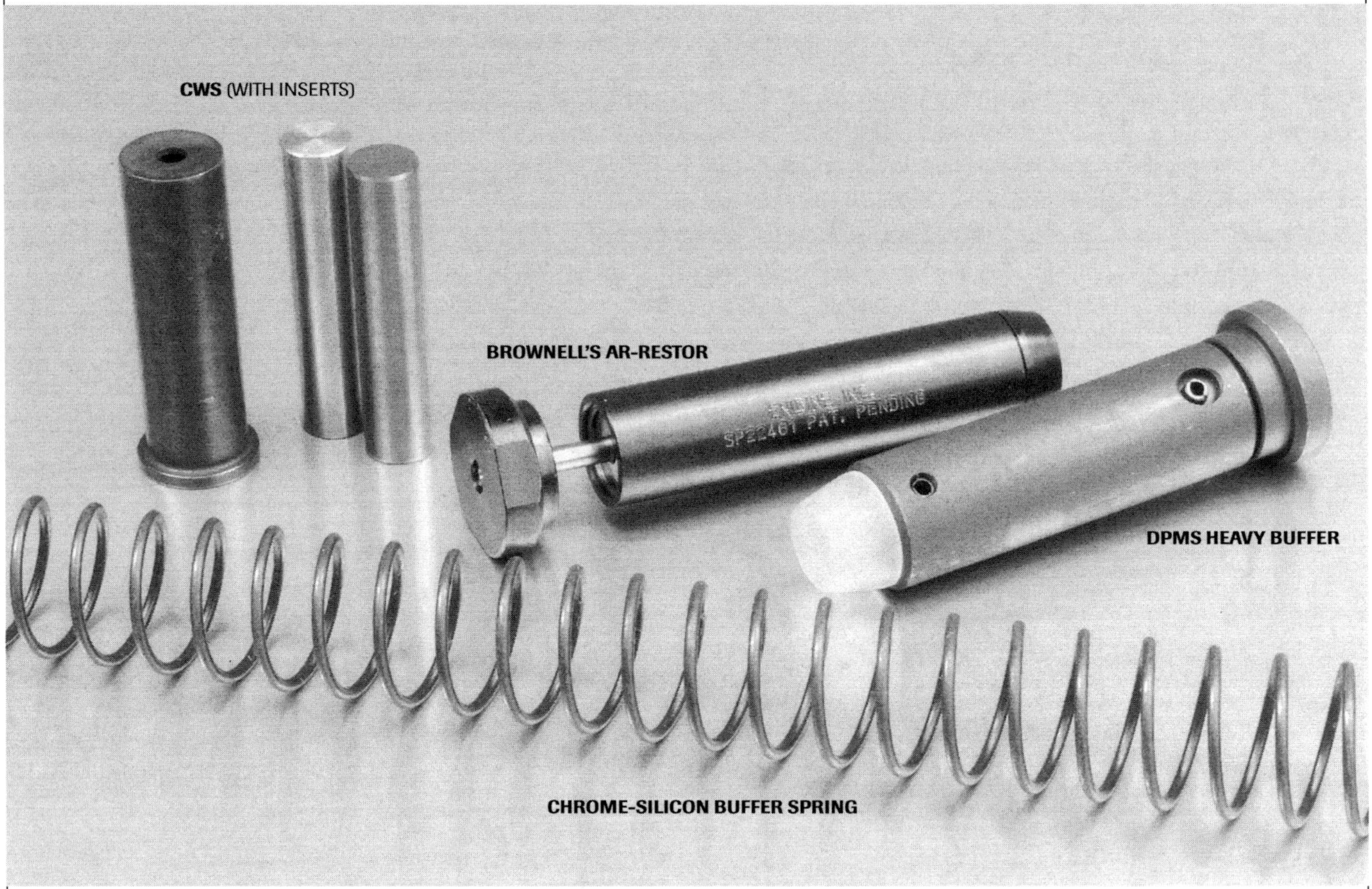

Here's a mess of help to calm a carbine's hyperactive operation system. I don't say throw the whole toolbox at this leaky faucet, but a better buffer spring and something near to it to increase the effective weight of the bolt carrier system is strongly recommended. I've had good results using the CWS carrier insert. I will caution, however, that when combined with an aftermarket buttstock, which might not have as much interior space, you may encounter bolt-stop overrun problems, which means inadequate space. I was, for instance, unable to use the AR-Restor buffer with a SOCOM stock. CWS likewise reduces stop overrun. A heavier-weight carrier is probably the easiest way to delay unlocking without running any risk of functional foibles. Left to right: SSS Inc. CWS (carrier weight system); Brownell's Ar-Restor; DPMS heavy-weight buffer; SSS Inc. CS buffer spring.

Well. You all know me. This gizmo fits up in place of a common gas tube, and it's a very uncommon gas tube indeed. It's from Armforte and the little block provides a means for gas flow adjustment. Ha! Something to fiddle with. Honestly, this works well, and the overt caution is to err on the side of reliable function. Think in terms of the maximum needed not the minimum. Point, however, that will become clear is that maximum needed is not the same as maximum available. This gadget installed fine on either the light or heavy barrel, but different float tube designs may not allow ready access to the adjustment screw. It's positioned to coincide with the first hole on an OEM handguard. Make sure the screw post is "floating" within the handguard hole.

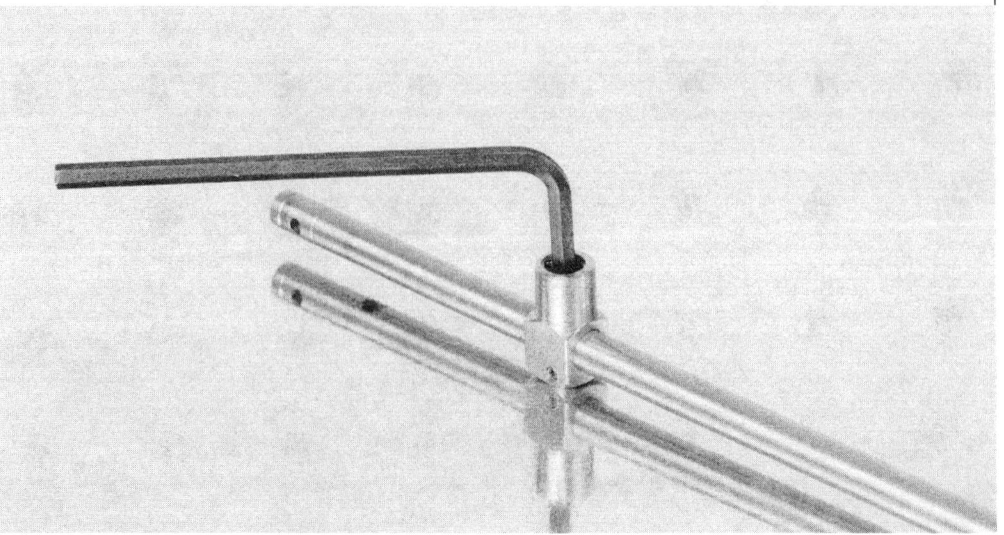

Buffers and springs* are different for carbines and rifles, and the springs are just shorter. There's nothing to keep someone from cutting down a rifle-length spring to suit a carbine application. Rifle springs usually have 43 coils and carbines have 37. One note on springs: Always count coils because you'll encounter buffer springs from different suppliers that are obviously different lengths, but they should have the same number of coils. If a spring is a little longer it's not necessarily stronger. The only thing to consider if you want a stouter spring for your carbine is that there can easily be space problems such that could cause insufficient bolt stop overrun. A flat-wire-style spring (inset) is one way to get just about as much spring punch anyone could want and not encounter solid-height problems. These compress fully to a much shorter package.*

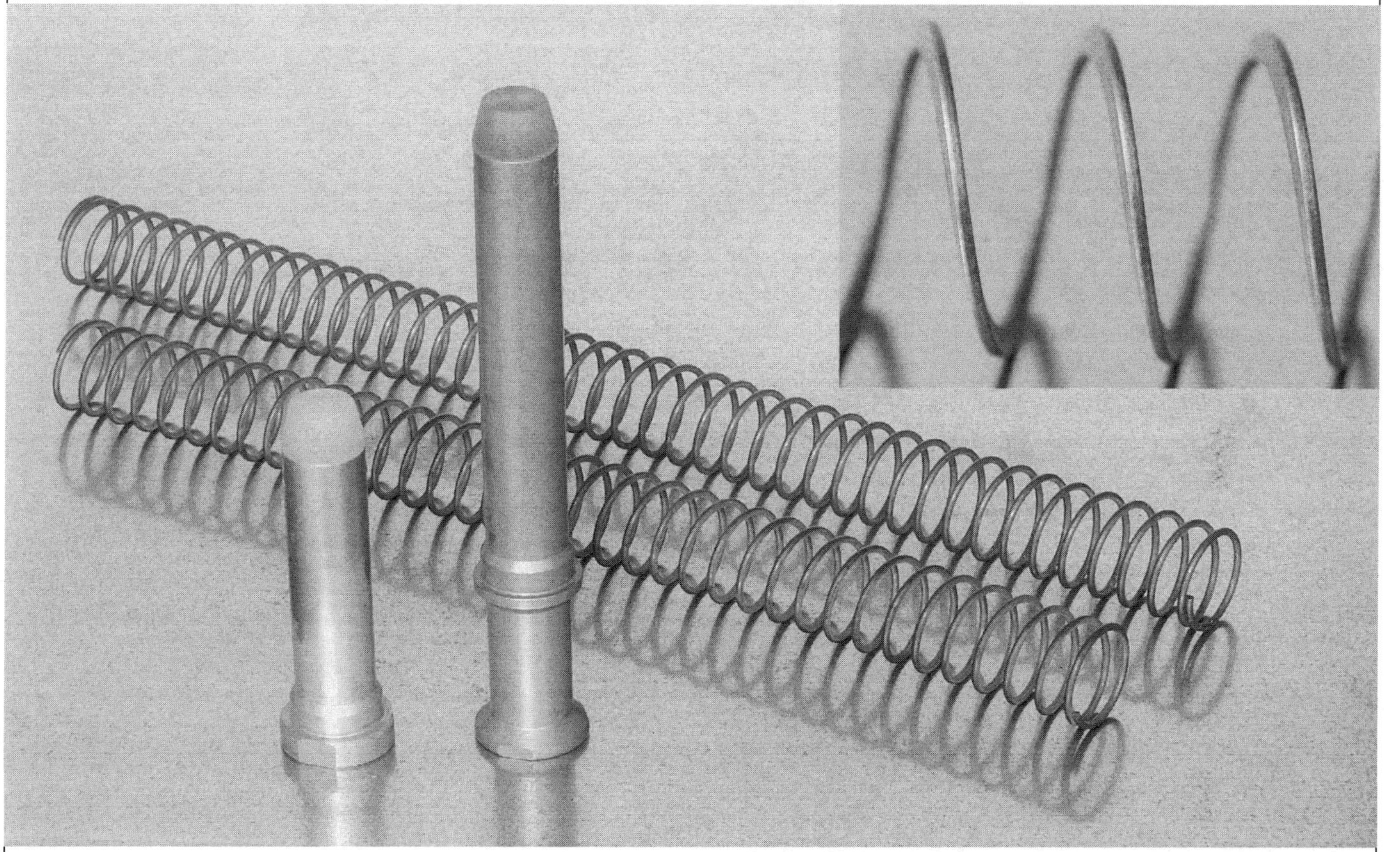

*****Muzzle-mounted devices**** are popular, and I caution against installing a truly effective brake. With a short barrel the blast can be atrocious and actually, in my experience, curb follow-up shot speeds under some conditions. Fired near a wall or in the dark, the bark and blinding effects are massive. This is a Vortex from Smith Ent. and is (effectively) designed to reduce flash. It's the sort of thing that can help more than hurt, and, therefore, is a wise choice. Otherwise, a plain old A2 serves most just fine (some, me included, prefer the overall performance of the very old style A1 three-prong). Save the compensator-style or multi-port brakes for another venue. Last, I really do think a carbine should have some sort of blast-direction control mounted up front, which means "something" threaded on; plain-ended fronts are likewise detrimental to shooter-side happiness with such a short barrel.*

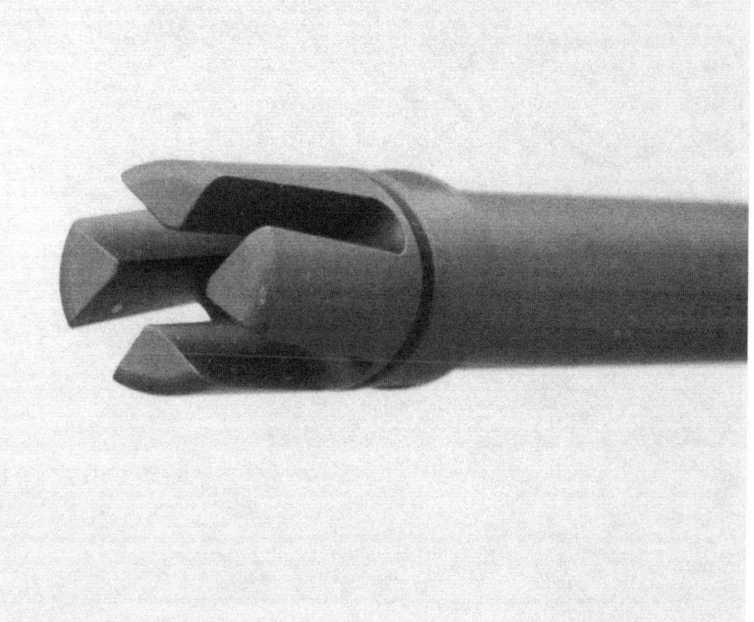

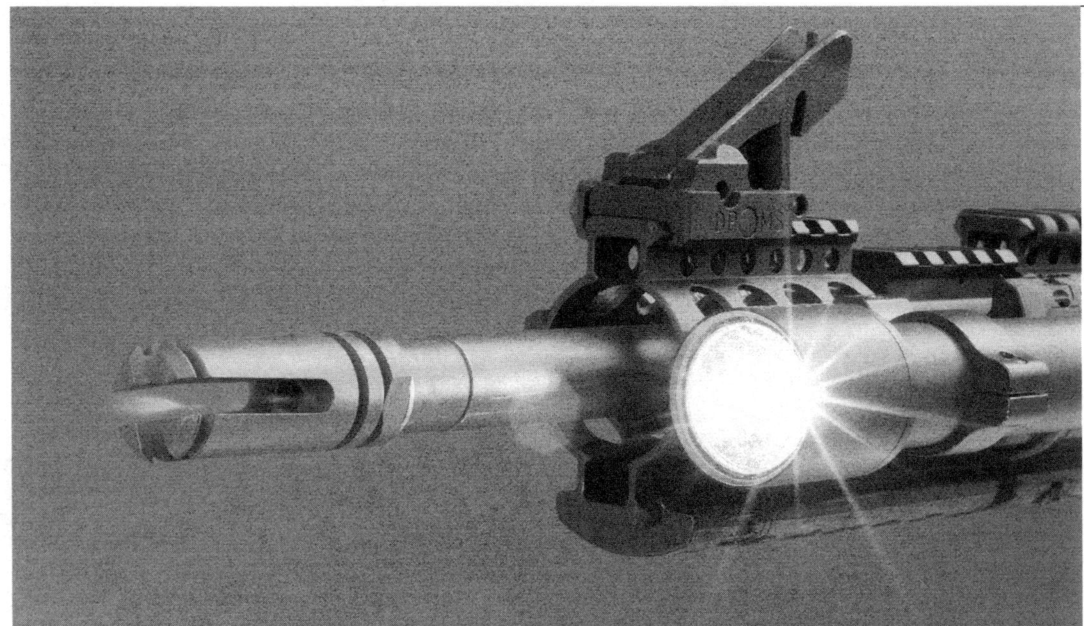

Easiest way to mount a flashlight, or at least most flashlights, is with a plain old 1-inch scope ring. Works well and usually stronger than those supplied with the light.

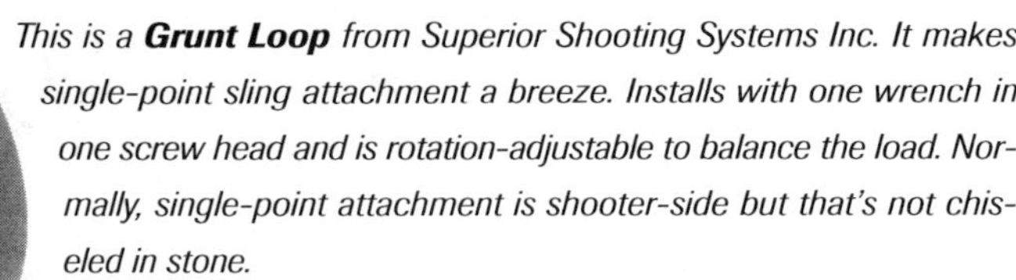

This is a **Grunt Loop** from Superior Shooting Systems Inc. It makes single-point sling attachment a breeze. Installs with one wrench in one screw head and is rotation-adjustable to balance the load. Normally, single-point attachment is shooter-side but that's not chiseled in stone.

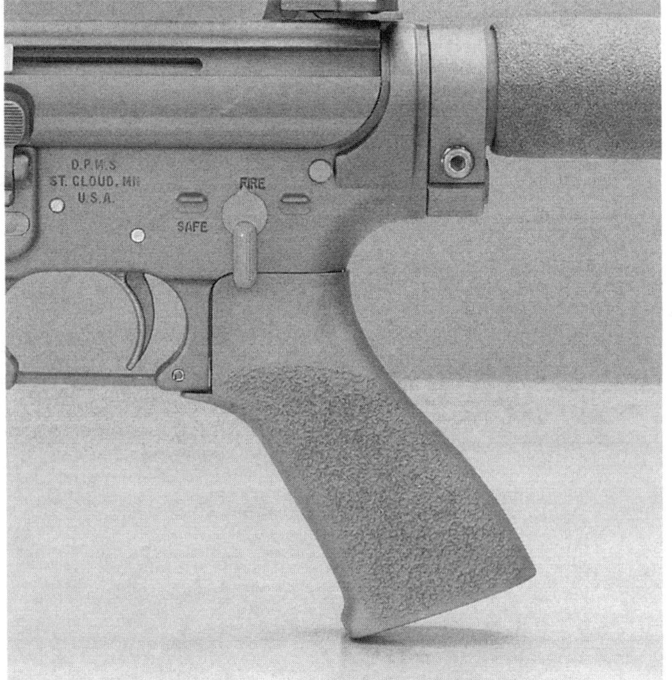

Grip choices abound and it's personal. The only objective consideration is not to go unusually form-fitting or oversized due to reasons of easy deploy and controls operation. Keep it simple and sleek. No "target" style grips. This one is from Tangodown and decidedly feels and fits better in the hand (most hands) than an A2 but there's no extended thumb or heel shelves or palm swells to interfere with safety or magazine release controls. This grip has a weather-tight base plate to cover a compartment for batteries. Spare batteries are wisely carried.

Uh, step one is to get all the upper and lower pieces-parts installed as first shown in this book.

I chose nothing different or tricky for this one. All parts installed onto and into these receivers just as normal. The only thing I encountered were a few areas where the (evidently) low tolerances on the machined lower required a little fitting to get smooth operation (the magazine catch in particular). I used GI parts.

PARTS & TOOLS

I'm going to skip the routine list of parts and tools to assemble the receivers, and, depending on components you choose, there could be differences certainly in what you'll need to finish the rifle.

PREPARATION

Vise, very securely mounted

Upper receiver clamp

Tap hammer

Roll pin punch and roll pin starter punch (for gas tube roll pin)

Torque wrench, 1/2-inch drive

Breaker bar, 1/2-inch drive

Barrel nut wrench attachment for the above (pin-style)

Gas tube alignment tool

Lower receiver block (for vise mounting)

Options (good ideas)

Permatex "red" adhesive

Permatex "blue" adhesive

Contact cleaner

Anti-seize compound (I like Loctite C5A)

Gas tube wrench

Brownell's rail alignment fixture

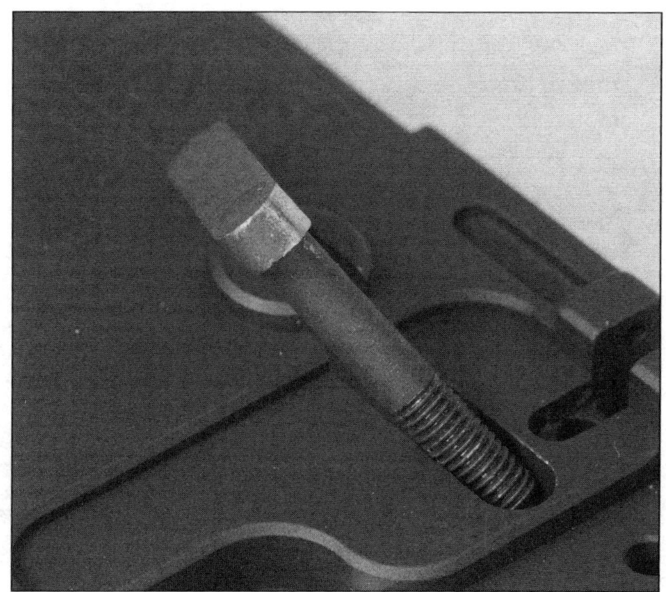

PRECAUTIONS

Test fit everything! Especially aftermarket parts. It's common to encounter a need for files...

Make sure all your parts work together. Forend tubes and gas manifolds in particular can be hard to match up.

Take your time with alignment, especially if you're using glue on anything.

17.0 THE BUILD

OPTIONS ABOUND, ESSENTIALS FOLLOW

The Yankee Hill float tube system uses a barrel nut that threads down onto the receiver. The tube itself is threaded on the inside so simply gets screwed on over the nut. The locking ring then secures the tube in position (along with glue). I don't trust only the lock ring.

Note the different style of upper receiver clamp used in this build. No option. The TKS upper won't fit into the clamshell-style because its lines are different than a conventional A2-style.

BARREL & FLOAT TUBE

Refer to the "basic barrel" installation material to get started with the barrel itself: 1. Degrease the extension and inside the upper, apply glue, position the barrel into the receiver, apply anti-seize to the upper receiver threads.

Use a pin-style barrel nut wrench attachment. **2./3.** Follow the same tighten-loosen procedure described in the basic barrel installation to seat the parts. **4.** Take it to torque (35 ft.lbs.) and hope you got lucky... **5.** Check for gas tube alignment using an alignment tool, and **6./7.** then (always) using the carrier. Always take the nut in the "tighten" direction to get it lined up and don't be surprised by how much force it takes. The large surface area to contain the tube in the YHM barrel nut makes it easy to judge alignment. The tube will rattle when it's right. **Get it right!**

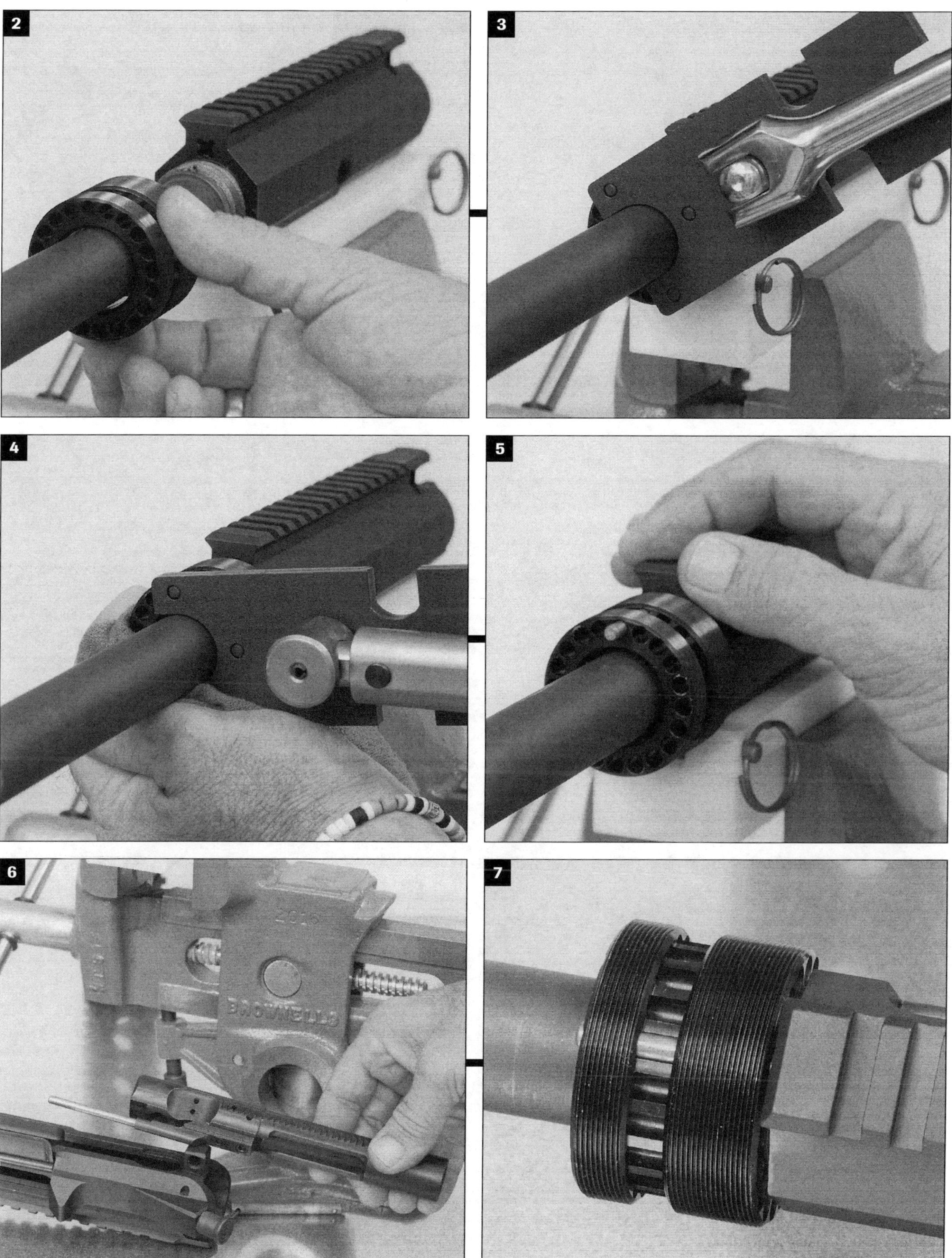

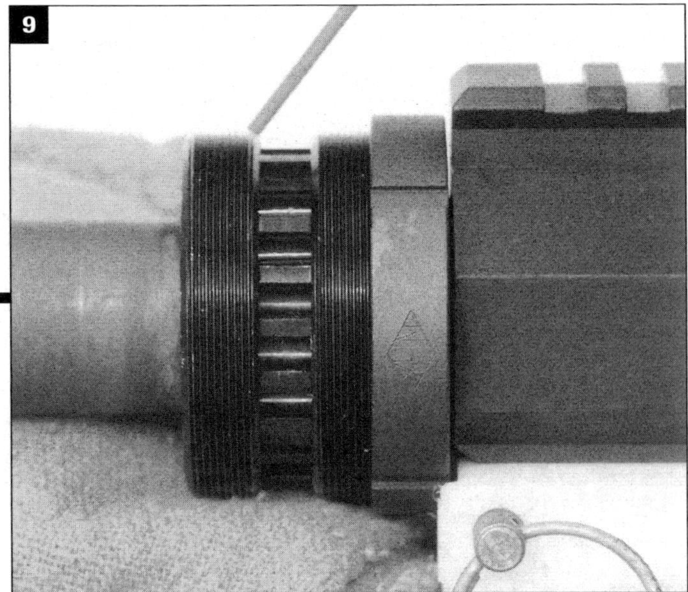

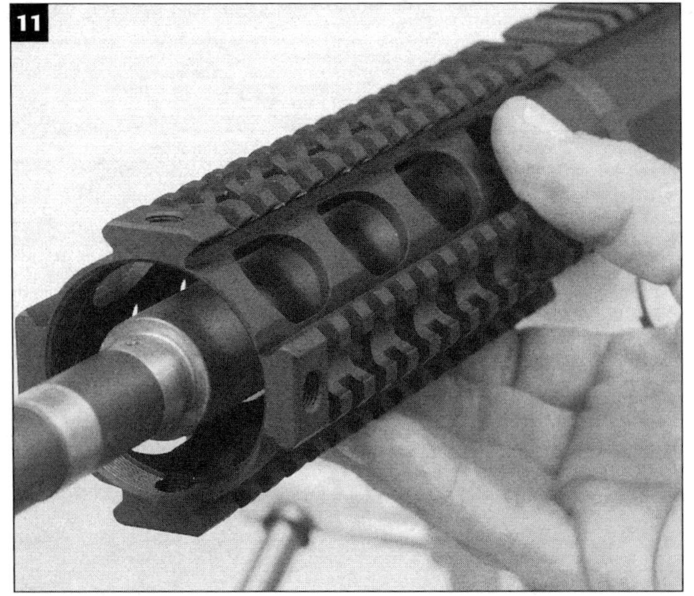

FLOAT TUBE

After confirming gas tube alignment **8.** thread the (degreased) forend tube retaining nut all the way onto the barrel nut, and **9./10.** degrease the barrel nut threads and apply "blue" glue. **11.** Thread on the float tube so it engages both sets of threads and the front of the tube is approximately at the edge of the gas block shoulder on the barrel.

Even though there's a retaining nut, I still use glue to make sure the tube stays put. Heat and handling can loosen the retaining nut, especially if there's a vertical grip installed on the tube (along with a lot of accessories), and you do not want a float tube coming loose. The glue is medium-duty "blue" Permatex threadlocker.

CLEARANCE & ALIGNMENT

Since there is a good deal of room for the tube to shift back and forth (threading it farther on or backing it off) it's important to check its position.

Remove the upper receiver from the block and mount it onto a lower receiver. Pivot the upper down as you would to maintain the gun. **12.** Make sure the forend tube isn't touching the lower. If it is, thread the tube forward. This doesn't affect function but will mar the finishes. If you install rail covers that increase its height, some contact won't hurt because it will be cushioned.

Then slip on the gas block or manifold and run it fully back to its shoulder on the barrel. **13.** There has to be clearance between the tube and the block if any portion of the block might be able to contact the tube. It can be very close, just not touching.

The tube now has to be aligned. There are different ways to do this but the rail alignment tool from Brownell's is the best I've seen. Otherwise, it can be eyeballed or set with a bubble-style level. I use the Brownell's tool because I have one, and it's one I really wouldn't want to be without. The Brownell's tool is height-adjustable to allow mounting on the upper receiver rail and then extend it downward to connect with the rail on a float tube. I use it also to set rail-topped gas blocks dead straight. It's sturdy and doubles, for me, as a clamp to secure the part in alignment while the part is then tightened down.

With the tool in place, **14./15.** tighten the retaining nut against the back of the tube using the YHM spanner wrench. If you don't have a spanner, tap it in place using a flat blade screwdriver (and then touch up paint...).

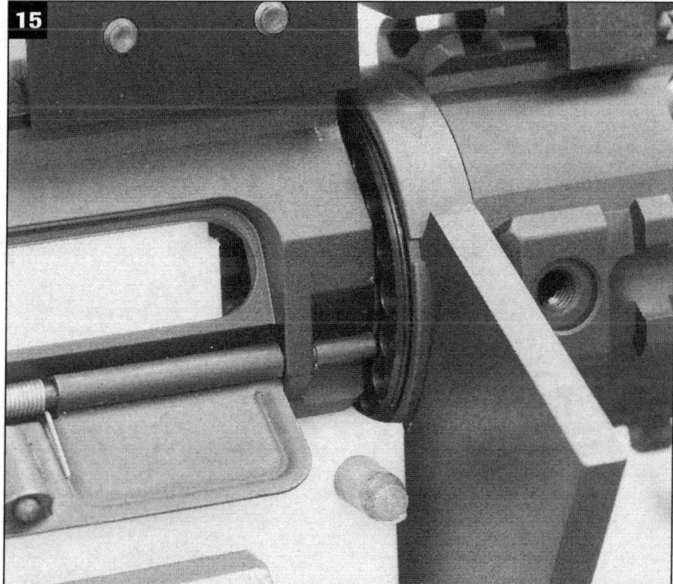

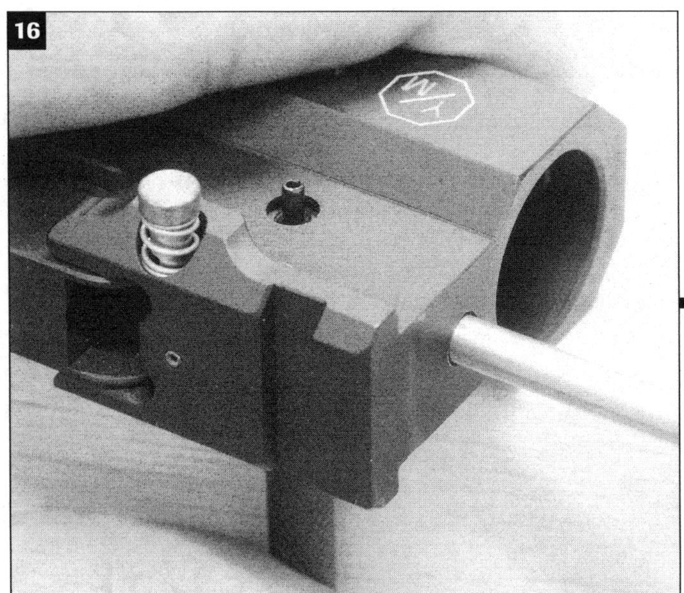

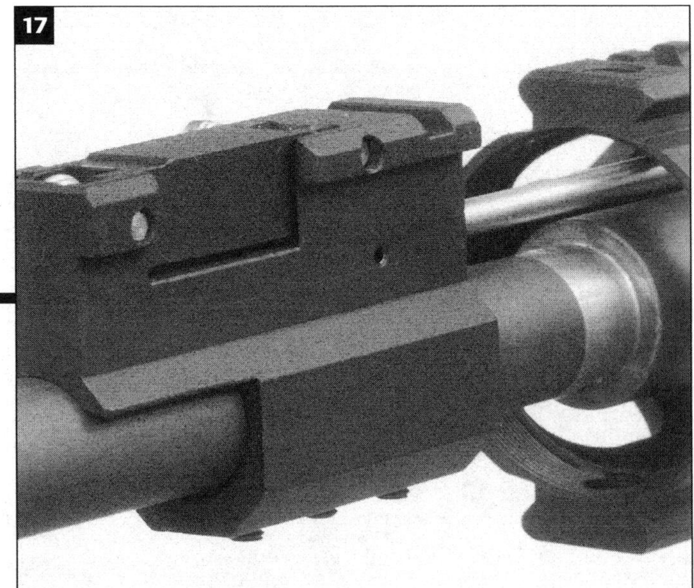

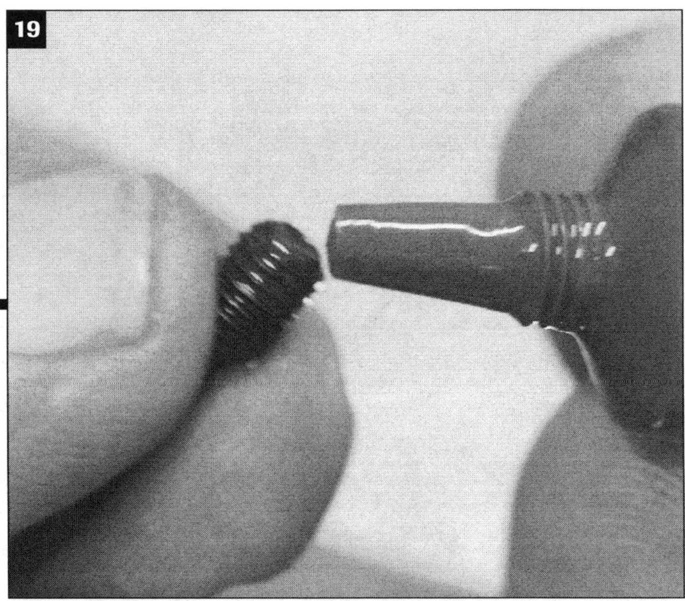

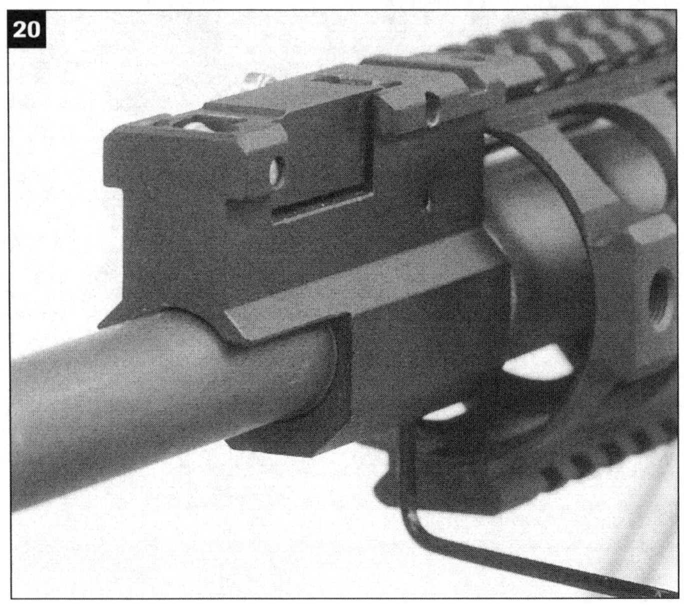

GAS MANIFOLD

16. Since this project got a clamp-style manifold, it's easiest I think to go ahead and pin the gas tube in place before mounting the block. This one uses a standard gas tube roll pin. Seat it to equal depths on either side.

17. Take the gas manifold and run it back through the gas tube outlet portals and to a secure stop against the retaining shoulder on the barrel. **18.** I move the alignment fixture to the front of the forend tube and adjust it to retain the small rail mount on the Young Mfg. manifold. Otherwise, indexing marks on the block and barrel shoulder are helps in getting all oriented correctly.

19./20. Glue the screws (not the block itself) and tighten them securely.

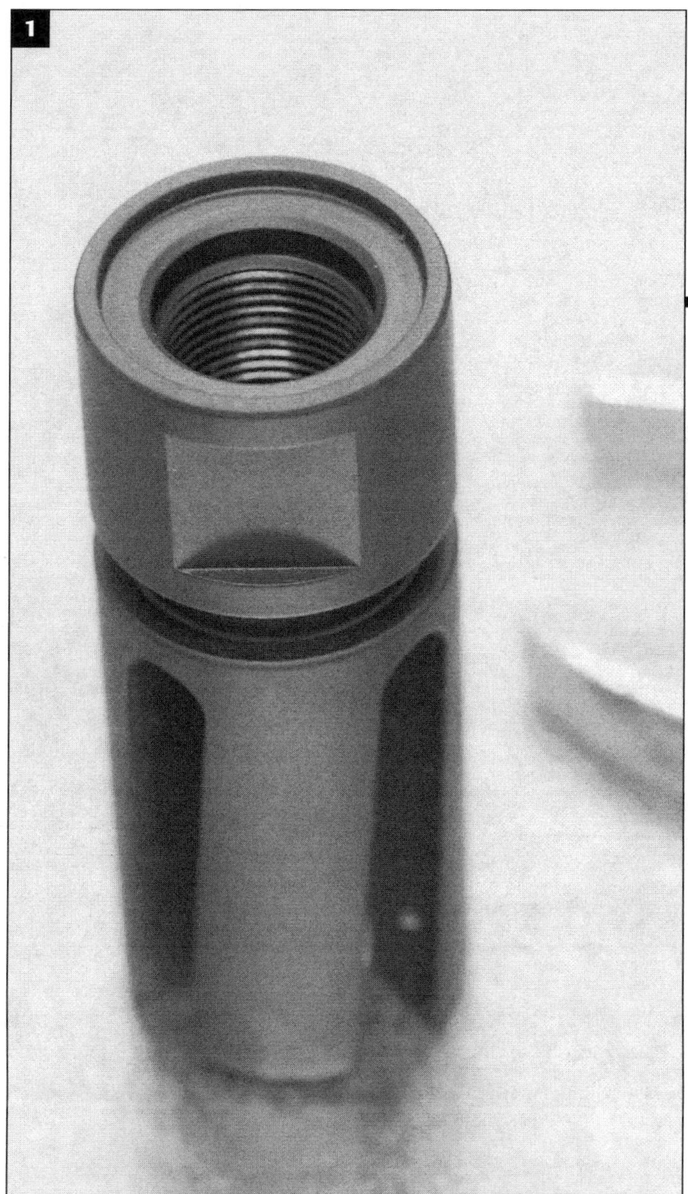

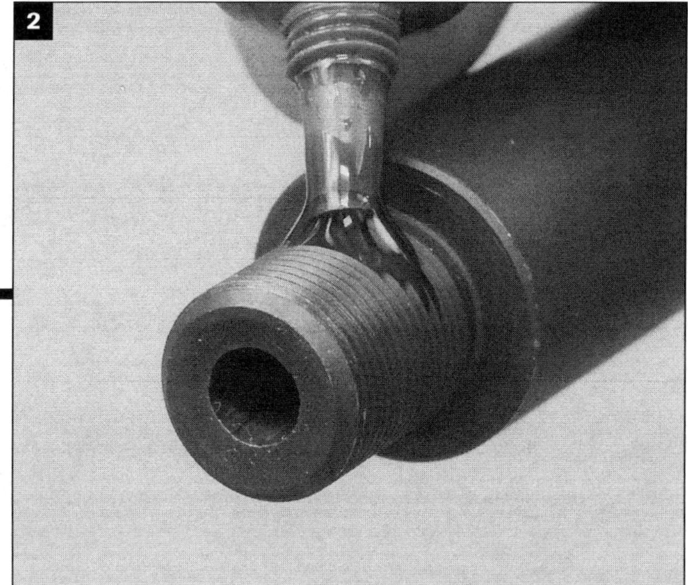

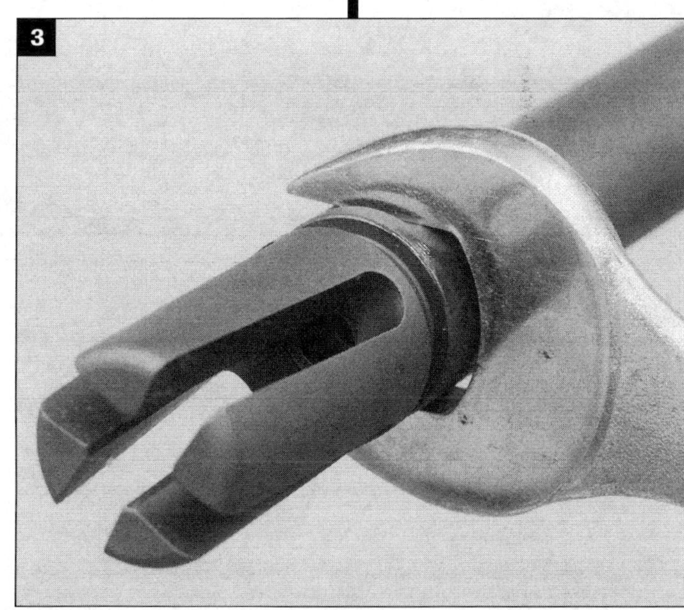

FLASH HIDER (VORTEX)

Not all muzzle-mounted contraptions are as easy as this one, so refer to other project segments to see other means using washers for alignment. The Smith Ent. Vortex is engineered to simply thread on and get snugged down. The installation instructions that come with this device are very specific about that. I still degrease the muzzle and brake threads and glue the part in place ("red") but there are no washers involved.

1./2. Degrease the threads on the Vortex and apply Permatex "red" to the degreased muzzle threads.

3. Use a 1/2-inch wrench and snug it down. It has only to be snug, just a little "oomph" after it stops threading will do it. Glue means it stays put. Alignment will be there.

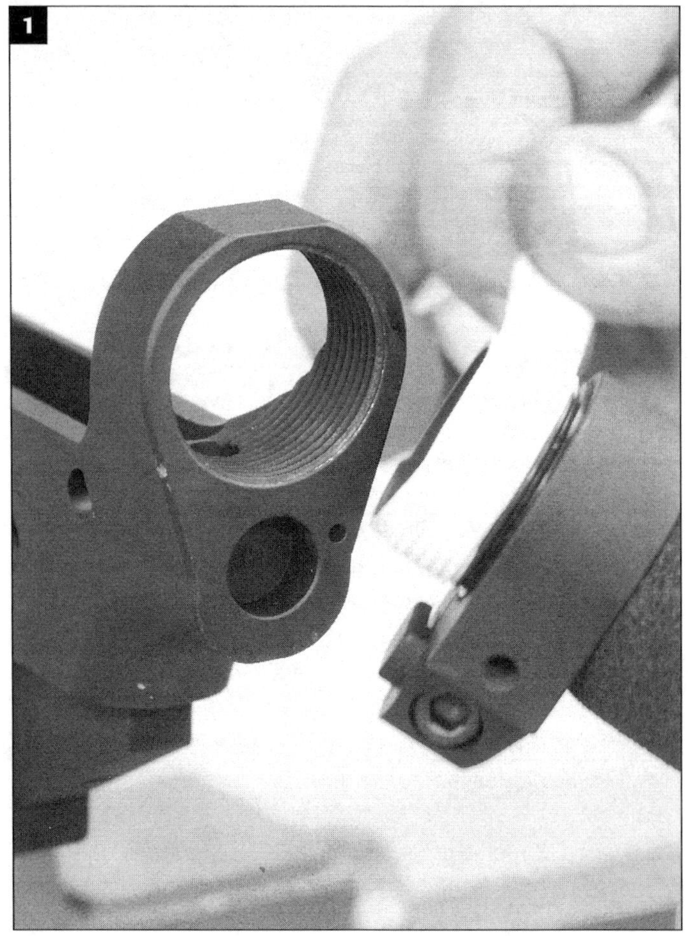

STOCK (SOCOM BOOM-TUBE)

Moving to the lower receiver and back end of the gun, now comes the buttstock. It installs just in the same as a CAR-style collapsible stock, but is actually easier.

Secure the lower using a mag-well insert and a vise.

I've found that the SOCOM receiver extensions seem to be a tad undersized compared to OEM parts, so I use Teflon tape in place of glue for these. **1.** I ran three wraps.

2. Thread on the tube until it nears the edge of the buffer retainer detent hole in the lower receiver.

This stock gives a lot of working room to install the takedown (rear) pin parts. **3./4.** Insert the takedown pin into its hole, making sure the flat area is facing back. Insert the takedown pin detent and spring.

5./6. Place the buffer retainer spring and detent in position, and then push down the detent, turn the tube in a little more until it captures the detent.

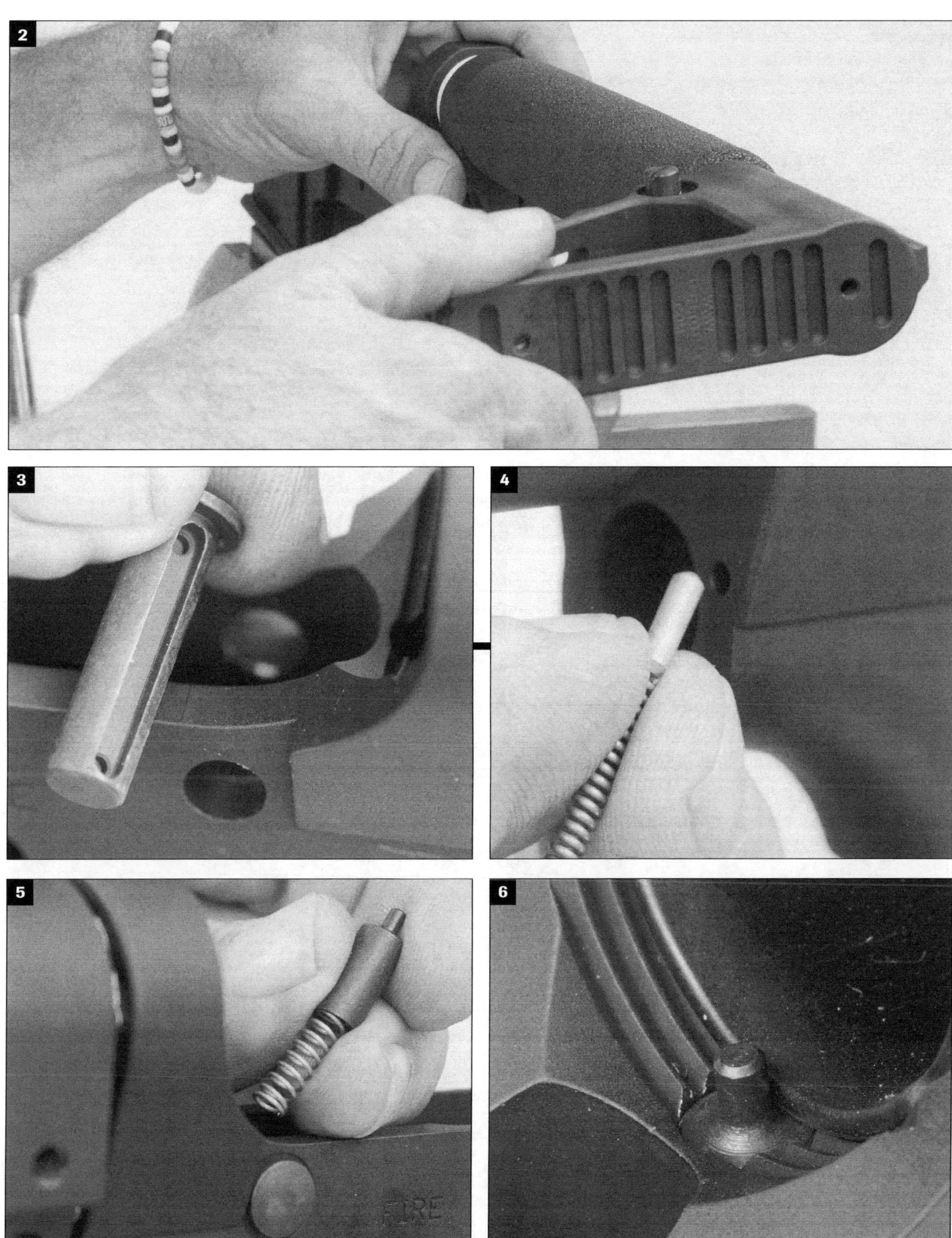

[CONTINUED NEXT PAGE]

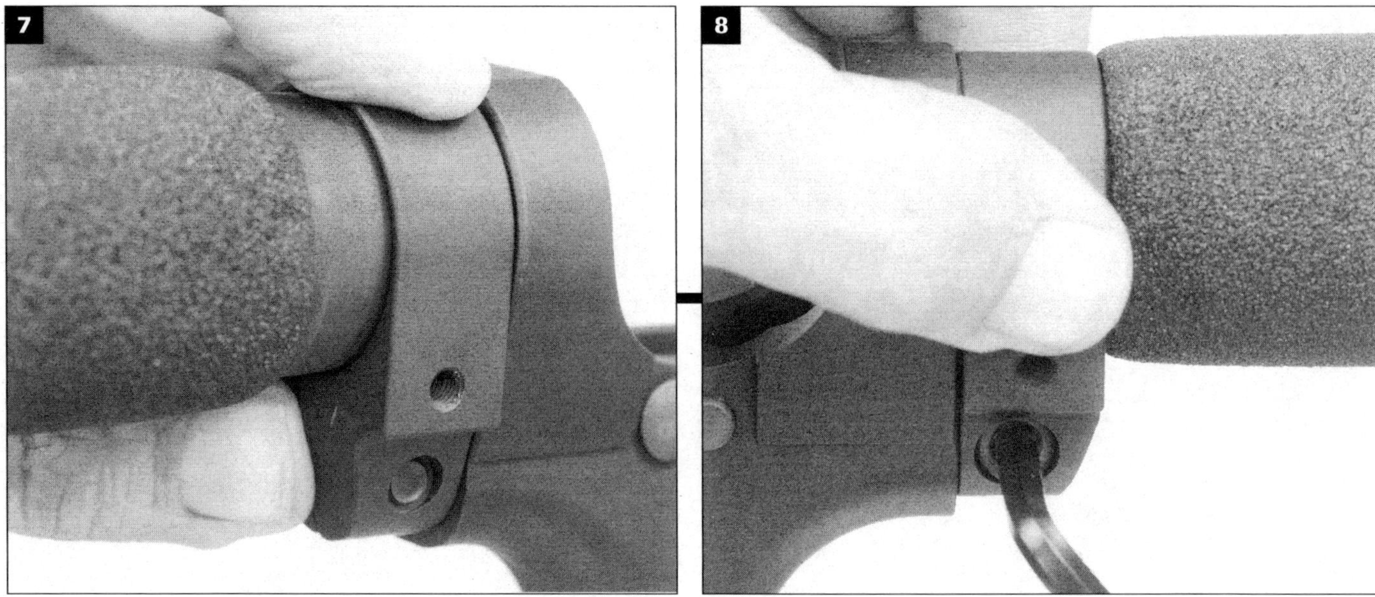

Finish the installation by **7.** sliding the retaining collar on the stock forward, ensuring first the takedown pin detent has found its home in the takedown pin and that the spring doesn't get kinked in the process. **8.** Tighten the screw on the collar.

9. Install the buttpad using the screws and a screwdriver. That's that. (Oh, and put a grip on it! I went with plain old A2.)

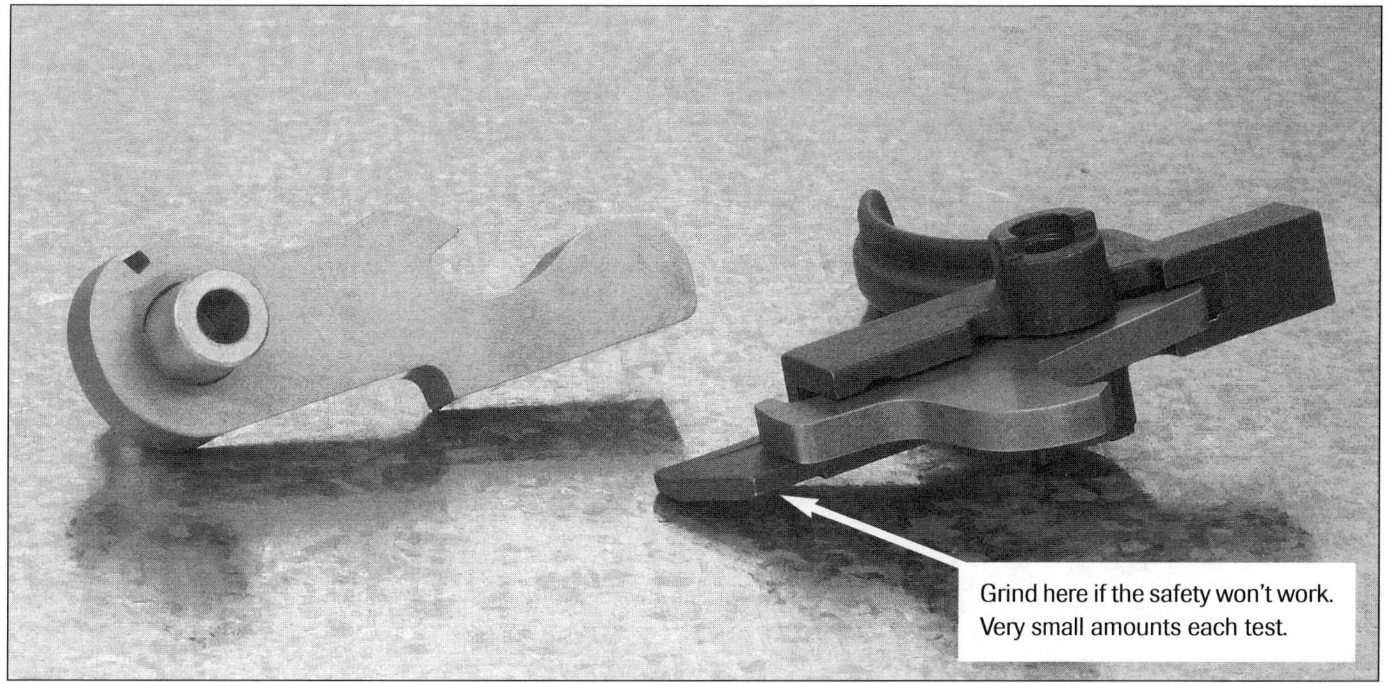

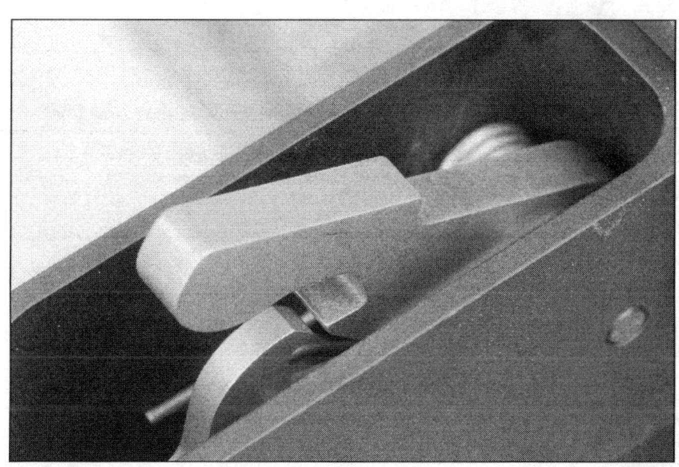

TRIGGER (ACCURACY SPEAKS)

Since I used an Accuracy Speaks single-stage trigger, it installs just the same as a GI. There are no screws, links, and etcetera. Tolerances are radically tighter on this trigger, though, and it's common enough to encounter a need to tune to attain proper function. It dropped into this lower with zero adjustments needed.

First check that the safety is operational. If the safety selector won't move freely, remove some metal from the very top rear of the trigger bar, the portion behind the disconnector. That's the portion of the trigger the safety comes in contact with.

Then do the disconnector checks as described previously. Cushion the contact area so the hammer doesn't hammer the receiver. Release the hammer forward by pulling the trigger fully to the rear and, keeping the trigger held fully to the rear, cock the hammer again, then slowly release the trigger. The hammer should be handed off from the disconnector to the trigger sear. Now, while watching the trigger, push the hammer farther down into its cocked position. The trigger should not jump forward. If there is any movement remove metal from the underside of the disconnector at the point shown. Removing metal from this area lets the disconnector hook move more fully forward to engage the hammer. If any disconnector malfunctions are encountered, remove very (very) small amounts of metal at a time. The trigger has to pass this test. Don't even think about firing the rifle until it does.

Finishing.

Finish it off by adding these parts and pin the receivers together. I used a heavy-weight carrier, otherwise nothing special about it.

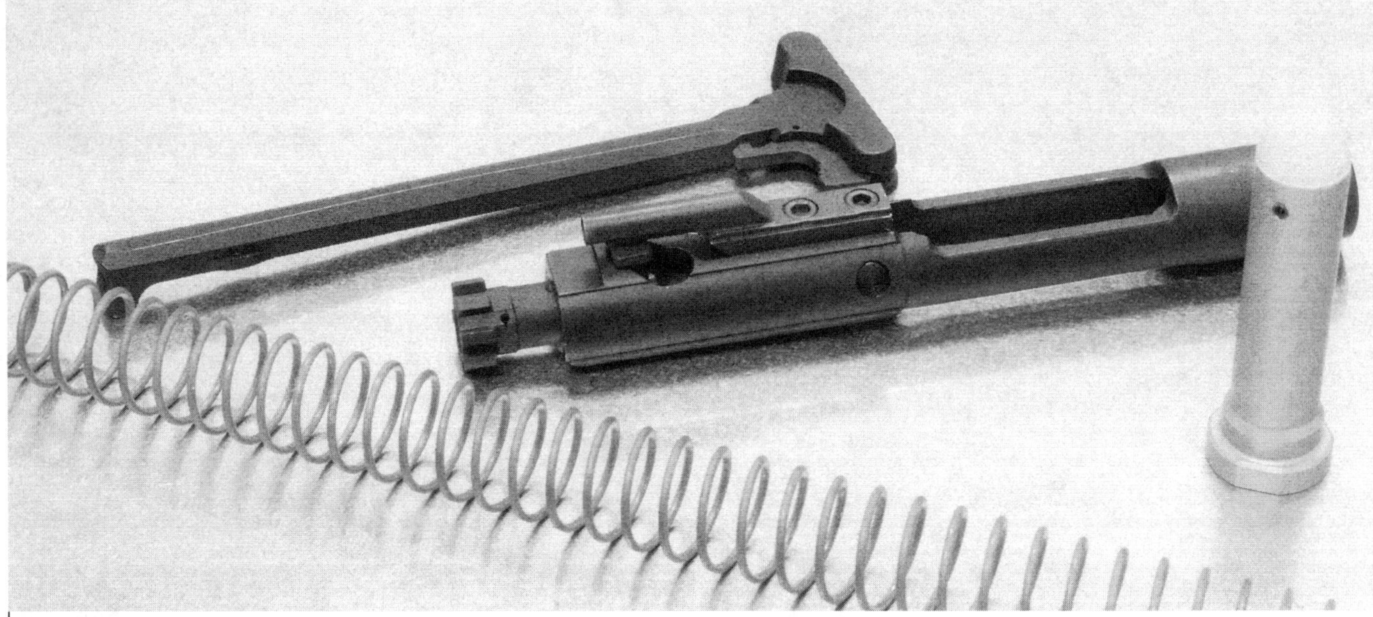

To take away the "porcupine" feel of a Picatinny forend, get these. The addition of Brownell's rail covers, the quality of which are a match to any tube, makes for a comfortable handguard to handle. My hands are on the larger side so its girth might become too much for some, but it's entirely sano. They can be cut to fit and will not come off. I put a snap-on style cover on the top rail.

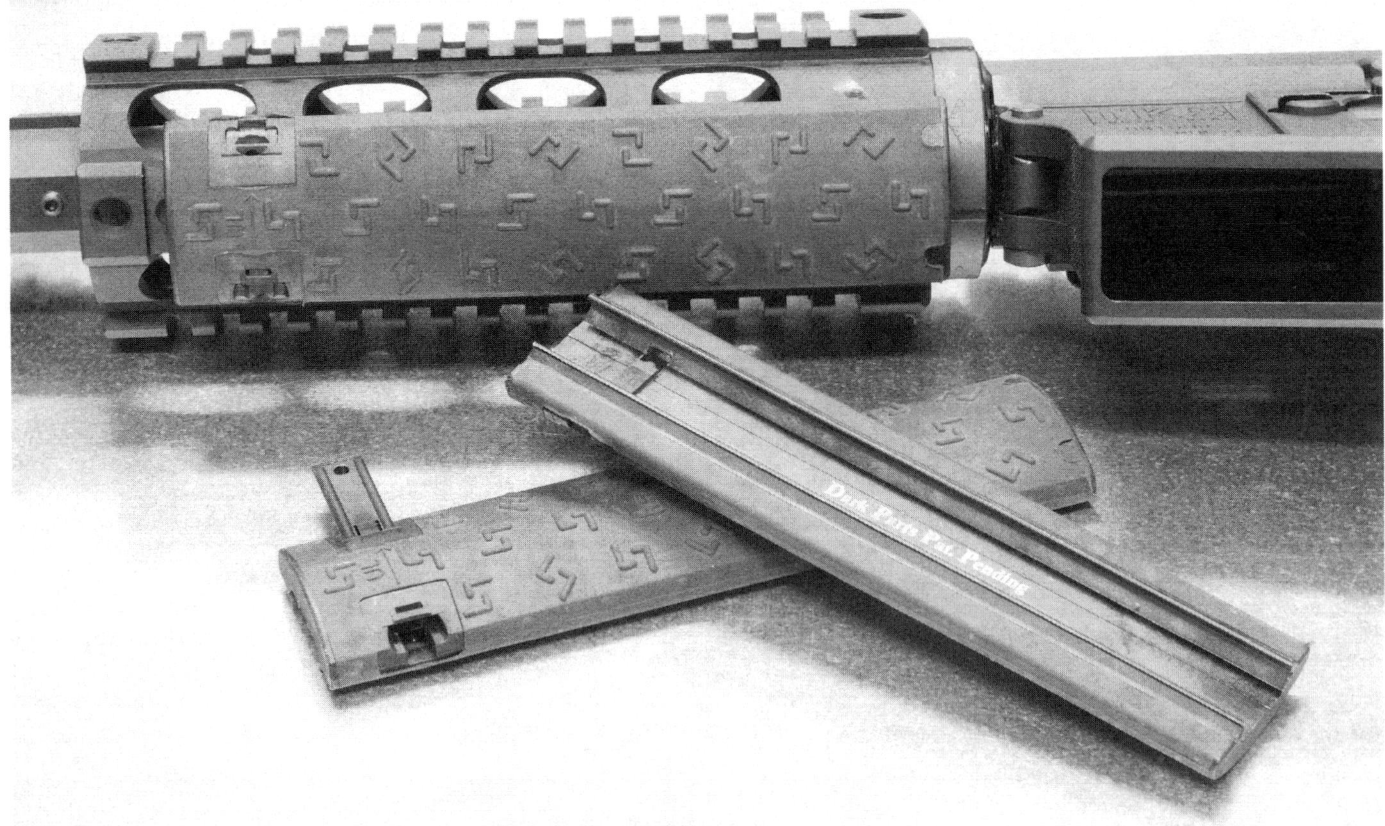

Gun Dun

I really like this rifle. I've built a lot of carbines and this is a favorite. Weight is just at 8 pounds. Function is perfect and shooting feel is very solid. The SOCOM stock has a lot to do with that.

One plug I have to make here for the TKS receivers is that trigger installation was just that easy. I put it in once and that was where it stayed. I've always had to do some tuning in the past using this trigger with different projects. I recollect Charlie Milazzo, inventor of the AR15 two-stage trigger, telling me that was one of his ways of judging receiver quality. Like Derrick Martin's trigger, Charlie's are built with enough precision that they should install with no tuning needed. I used KNS standard pins, and that's a help toward maintaining precision as well.

Take a rifle like this and then, of course, add to it what you will. As said, make sure that the accessory quality is up to the quality of your carbine. That goes double for optical sights. Speaking of, look closely and you'll see a short rail piece on the right side of the upper receiver that sits at a 45-degree angle. It's from YHM. This allows mounting an optical sight and maintaining full view of the irons. I like irons; they're faster for me. To use the optic, just rotate the rifle (cant it). You'll be amazed how well you can shoot with that set up.

I installed a "silent" swivel from SOCOM on this rifle and it worked, but was a bear to do. Very small parts and tricky alignment. I really like the Grunt Loop shown earlier from Superior Shooting Systems Inc. It required less than one minute and virtually no thought to install. Right-handers usually like the sling swivel on the left side, but this was installed on the right for a left-hander.

Nice looking little gun. *Nice working little gun too. I clamped a Versa-Pod combination forward grip and bipod on it. That's neat. Versa-Pod also makes a very nice more conventional forward grip. If you get one of these, get a good one. Many are flimsy, or become so after much use.*

18.0 NRA MATCH RIFLE

LONG GUN

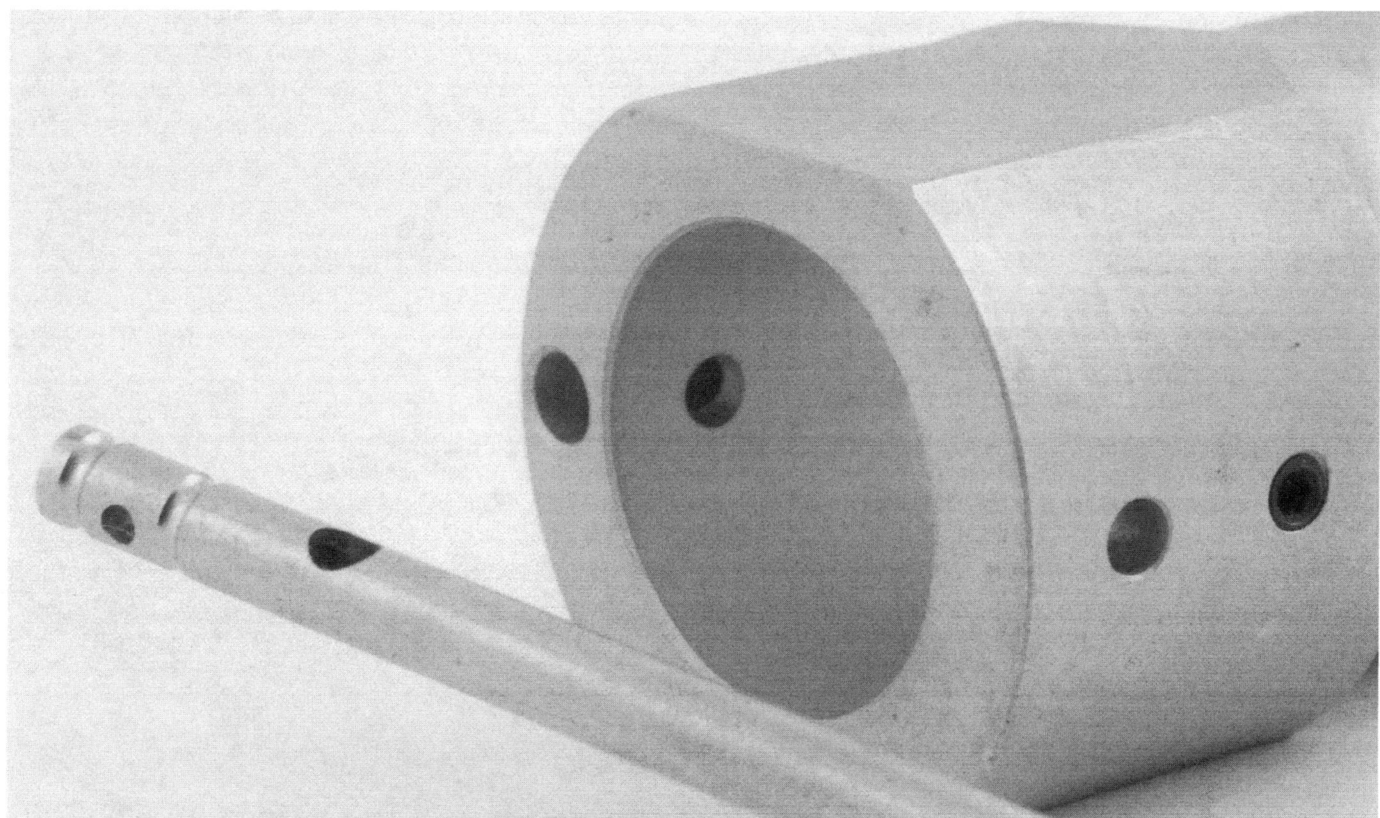

[An NRA Match Rifle isn't much more than a varmint rifle with iron sights, in one way of looking at it. In another way, which is my way, it's an opportunity to produce a target rifle that leaves its owner with no machinery excuses for anything but a clean score. Put some thought into it and stoke up your checkbook, because nothing we're going to purchase is cheap. This, despite its appearance, was actually the easiest build in this book. There's not much to work around.]

Outrageous as they look, a match rifle is comparatively easy to build. Component selection is the center of a Match Rifle project. Not that it's not central to any AR15 rifle build, but component selection ultimately determines how much you can get from this rifle platform.

SEGMENT CONTENT

209   **Barrel Selection**

211   **Float Tube Selection**

212   **Receiver Options**

213   **Trigger Selection**

214   **Stock Selection**

216   **Parts & Tools**

217   **Barrel & Float Tube Installation** (plus gas manifold)

222   **Sight Extension Tube Installation**

224   **Stock Installation** (Medesha Firearms adjustable)

BARREL

As always, and always, the barrel is the heart of any firearm. Options on barrels abound, and there's a ready supply of varmint rifle style barrels that find their way onto many of these rifles. A varmint barrel, however, is not what we're looking for.

This essential common-form mail-order varmint-rifle barrel is 24 inches long and has a muzzle diameter ranging upward to 0.922 which is preceded by a parallel section of that diameter beyond the gas manifold. From barrel extension to gas manifold, diameter is usually "full" or right on 1.000 inches diameter.

Chambers are usually short; many are SAAMI minimum. Common twist rate is 1-9, although 1-8 is becoming more available, and that is a very good thing. You'll find this sort of barrel on many factory-built "target" AR15s, but, at 1-9 twist, that barrel is too short, too heavy, too slow in twist, and not carrying the chamber we want.

It's the barrel that is a central reason most NRA Match Rifle shooters go custom. A good builder starting with a barrel blank can give you what you want back, if you know what you want.

Another help we get with a custom-done barrel is a chance to improve operating manners. Just about any experienced builder will move the gas port hole location another 1-2 inches forward of where it is on a standard A2. That requires a longer gas tube, another custom part.

We're doing this to reduce gas port pressure. The longer barrel means there's more gas under more pressure for a longer time. Moving the port forward gives some extra time for the pressure to lower before the burning gases funnel through the gas port hole. The result is that the bolt stays locked a little longer, spent case condition improves.

What we need for better is a 26-inch barrel contoured in steps. The barrel I chose is a little heavier than I would have requested from a custom job, but, again, is as close to a standard contour as what most custom builders provide without customer input. It's about 0.990 to the gas block and 0.900 from there out (I would prefer it to have been about 0.820). Another step, literally, that we need to make

life and shop work easy (related) is a muzzle-area diameter that fits commonly-sized front-end accessories, like barrel-band style front sights. That's going to be 0.750 at best, and 0.813 second. One of the proven best chambers for High Power use is the "Wylde," and reasons for that were detailed in the NRA Service Rifle segment. Folks, it's hard to find that barrel.

Match Rifles can get heavy, really heavy, without some thought put into weight-savings. Right. Unlike the NRA Service Rifle which myself and a slew of others think should be bulked up, by reason of barrel length and, to more or less extent depending on choices made, stock hardware, building up a gun with the common-form "heavy" barrel described earlier can result in a rifle 16 pounds or more, and that is very heavy.

I think it's better to add weights, if you need more weight. Most shooters will want a rifle to be heaviest prone and lightest for standing. The extra barrel length makes a Match Rifle decidedly front heavy. Extra weight from adding an adjustable buttstock does shift the balance in the right direction to offset an extra-diameter barrel, but then we're also running up "dead" weight. That is very noticeable, and does few much good, at least not in the standing position.

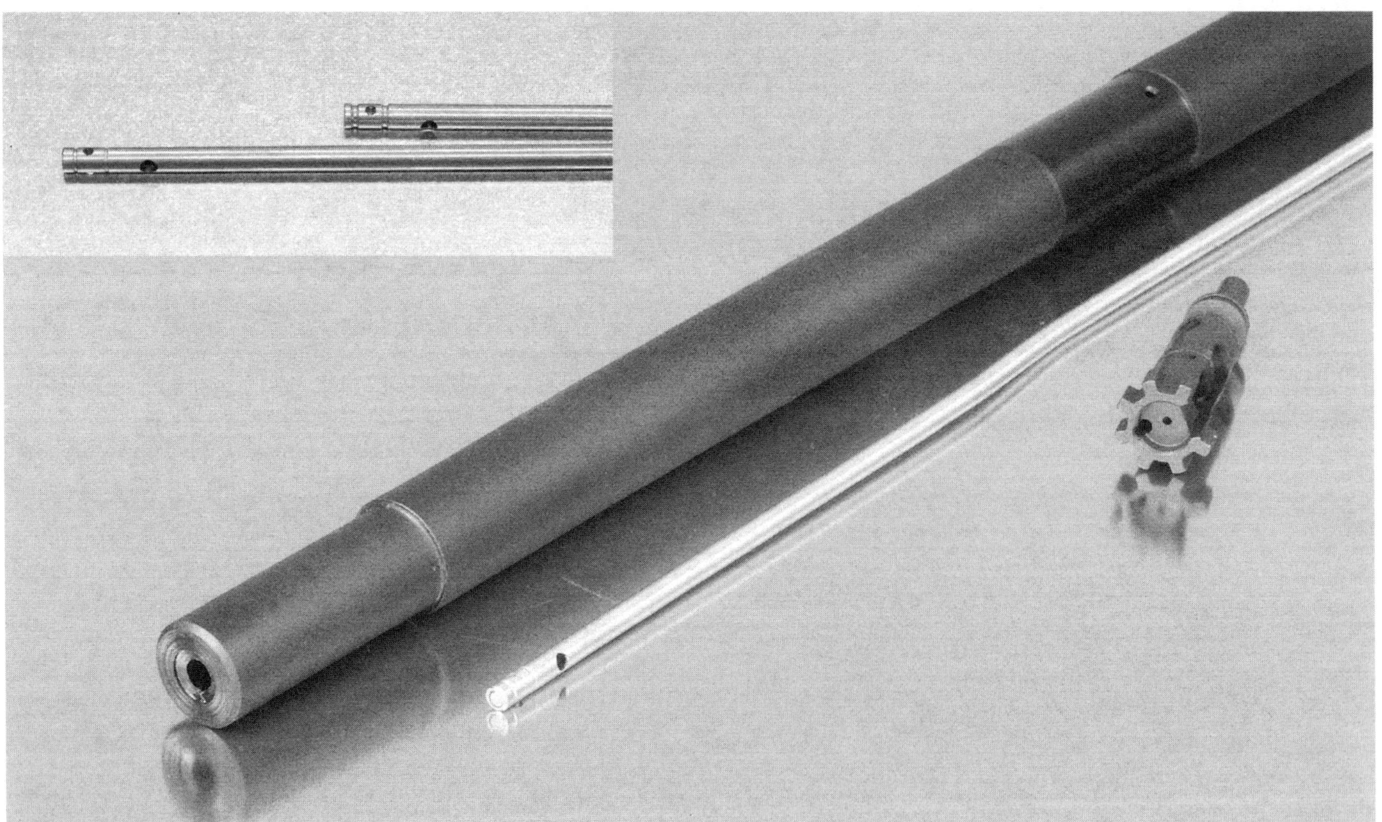

It's about time...

One of the biggest obstacles in putting together a Match Rifle from boxed parts is finding a truly suitable barrel, and that's coincidently one of the main reasons to go to a custom builder. There are good Service Rifle barrels available, and there are good "varmint" style barrels out there, but finding a Match Rifle drop-in with accepted dimensions, twist rate, chambering, and gas port location mods is tough. At the time of this writing, Northern Competition offered this one. It's a 26-inch with a "Wylde" chamber, 1-6.5 twist, and a gas port moved two inches ahead. It came with a stripped bolt so no headspace worries and an extra-length gas tube. It even unpacked with a 0.750-inch parallel section at the muzzle for sight mount ease.

Float tube

I put Medesha Firearms furniture on this one and got it all as a set. Colors match and design and, therefore, installation is stupidly simple. Just tighten down the barrel nut then assemble the float tube after that fact. Due to the intent as well as the execution, no alignment is necessary beyond getting the gas tube centered in the barrel-nut components. This uses a standard barrel nut and a collar that screw-locks to it, and then the tube itself fits over and screw-locks onto it. Medesha tubes have locking external rotation adjustment via the slots and set screws so forend position can be fine-tuned at any time. The Match Rifle tube is on the left in this photo, and, compared to the tactical-style tube next to it, is larger in diameter. There's a trend among competitive shooters to go with smaller and smaller diameter tubes because they catch less wind. The large diameter of most Match Rifle tubes means they will slip right over the gas manifold. Match Rifle tubes are normally 12 inches in length. I like longer tubes that will cover the gas block, and that's a 15. I used a 12 in this build only because, when I needed it, the 15 wasn't available. I don't know how important it really is to cover up the tube. It's not exactly flimsy.

This book shouldn't focus on shooting techniques or tactics, but here it really has to. When we're building up a Match Rifle that's really the entire point: make a rifle fit, and suit, and score as high as you can with it. Nothing is automatic. What really matters is discussed throughout.

FLOAT TUBE

Most builders who do these for their customer rifles will sell you one of their Match Rifle float tubes. They are not hard to get. Although all are similar, they are also decidedly not all the same. There are differences in diameters and "feels" and without your having first-hand experience I can only tell you what I like, and, of course, why.

Many are 2.50-inch diameter, and all will have a rail slot cut into their bottom-side to accommodate a handstop. Most will also have a handstop. *Slings & Things* talked this one to death, and options may not exactly abound but they do exist. I like forend tubes that are smaller rather than larger, within the range I have encountered, and handstops just in the same. Recommended tubes are from Medesha Firearms, Northern Competition, and Gary Eliseo.

What really matters, to us here, is ease of installation, and the only ones to consider are the multi-part systems. Those have a separate barrel nut with external threads, onto which the forend tube his-

Options in "real" ready-to-go Match Rifle barrels can originate with a call to Krieger or Pac-Nor. Others may also be willing to participate. These outfits will produce and then deliver a barrel to your specs, ready for installation. It can cost big, but you will get what you want — if you can tell them what you want. Get all the details in order and fax a sketch. Send a stripped bolt too. If the chambering work is done using your bolt, and respecting that piece, then that barrel will be ready to install with no worries.

Gas blocks

*A Match Rifle gas block or manifold has to serve only as a receptacle for the gas tube. They clamp on. The only trick in this build was finding one that would fit the barrel. It took a 0.922, and that's a fairly large hole. I ended up going with one from Krieger (on right) simply because I liked the looks of it, and I did have to get my local tool and die shop to open it up just a tad. I can tell you that 0.001 inches too small may as well be a foot. The other manifold shown was gotten from Brownell's, which is the only place I saw a 0.922 listed as a catalog item. I ended up putting it on my varmint rifle build because of the rail on its top. No other reason. **Shown floating around** on this page is actually my favorite gas block. It's from EGW and literally clamps on (as opposed to having a set screw). It's steel and there is merit to this material and this design with respect to heat-induced change and also stress applied to the barrel. Unfortunately, the EGW block was too small to incorporate into this project. There's not enough wall thickness there to have opened it up to necessary diameter.*

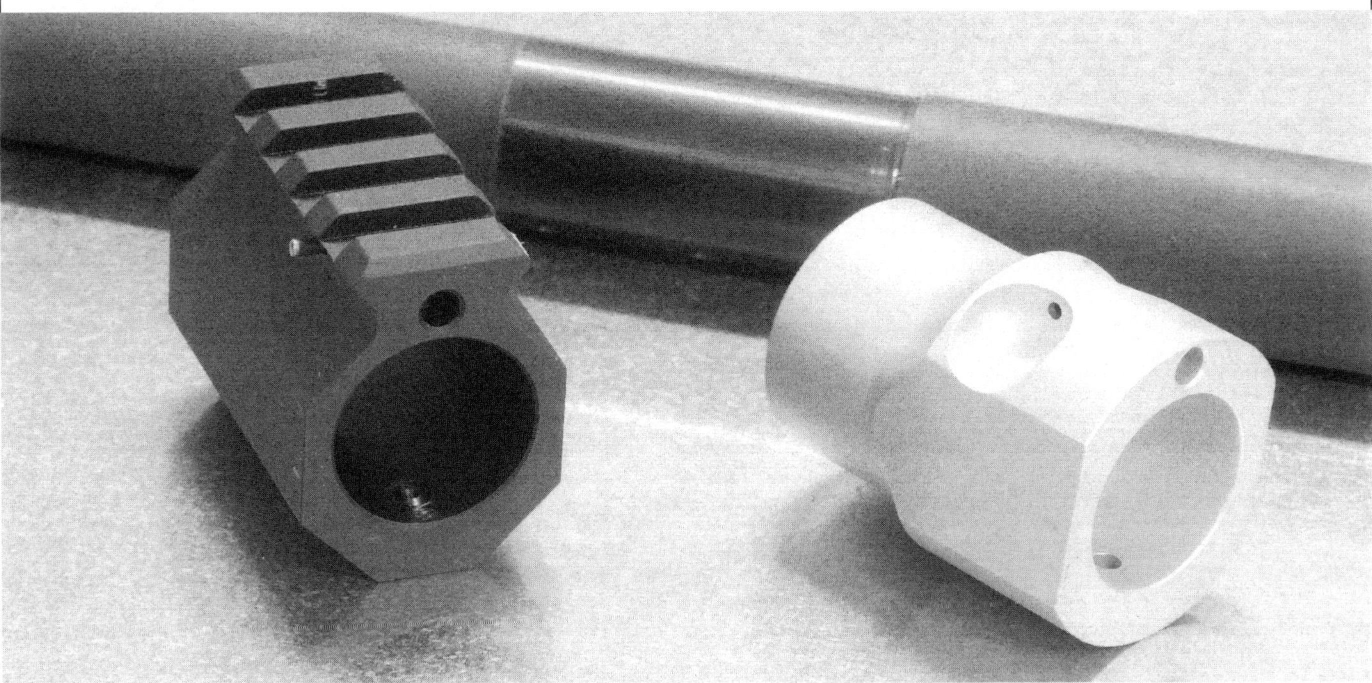

self attaches after the nut is installed to spec. Some tubes then allow for ready angle adjustment through elongated slots and set screws arrangements.

Install the handstop before threading on and finalizing the float tube position. It's easier that way.

Tube lengths are usually 12 or 15 inches. I like the longer ones. Extra-length covers the gas block or manifold. I don't know that that is really all important and there are some who believe that shorter tubes produce less vibration and potentially interfere less with accuracy, but tube length has not really been shown to be a consistent factor in performance by my experience, or those who offer the option. Longer tubes look better to me. Now there's a reason to use one!

RECEIVER SET

Okay, this one was a little non-standard, but you can get the same thing. I wanted a fully-adjustable buttstock on this bad boy, and cheekpiece clearance

factors. Overwhelmingly so. More in a bit.

I picked a favored forged lower from DPMS. It could have been from another outfit, like Bushmaster, and it's nothing tricky, just good.

The upper is another favorite for a Match Rifle build and that is a DPMS "pro" model. This one was a Lo-Pro, which means it has a standard-height rail.

I tried a trick that I don't know helped anything, but there was some effect from it, at least visually. Brownell's at one time cataloged a receiver squaring fixture for AR15 uppers. Throughout these segments I talk about the idea of running the barrel nut on tight, then loosen, then tighten, and repeat another one or two times. The idea is to seat the parts and attain a better fit. This fixture is supposed to accomplish essentially the same thing, but maybe better. I ran it on this upper and did see a small area on the receiver face that had been polished. More than that, though, was that the inside of the upper showed a pretty substantial number of "shiny spots" where the lapping compound smoothed it over. I sincerely doubt this is going to make any

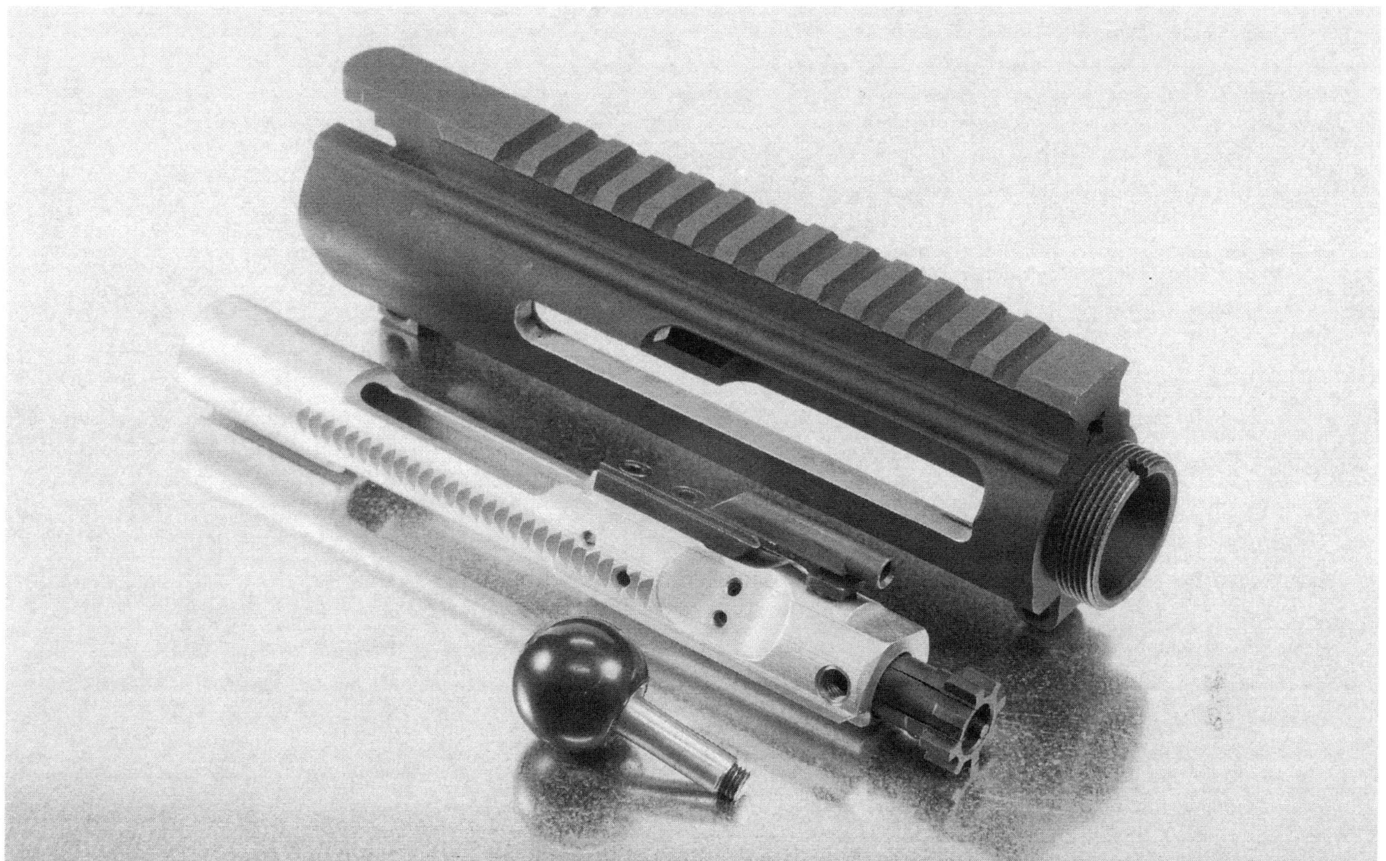

Semi-custom

The upper was purchased from Medesha Firearms with the bolt handle modification to the bolt carrier and corresponding slot cut into the upper receiver to provide clearance for the handle. This is not absolutely necessary because even Medesha's adjustable buttstock can get its cheekpiece run down far enough to clear a standard charging handle, but then it has to be run back up. If there's a problem that necessitates wracking the action during an event, then it has to be run down again. After operating a rifle with a bolt knob, you'll not want to be without it. Package parts included the upper, carrier, and handle. This particular set used a chromed DPMS full-diameter carrier by reason only of its good quality and extra weight. Rail height on this one is the same as encountered on anyone's routine flattop or A3 upper. I also like the Hi-Rider version that adds 0.500 inches to the rail height, but the sight mounting base can address any need for more height in itself.

difference, and I myself don't know of any custom builder who worries about performing any such truing.

HANGERS ON (OPTIONS)

Now here's a chance to install gizmos and doo-dads, and subtle details in definitions aren't necessary. Worthwhile additions include an extended bolt release lever. This lets you trip the bolt release with the trigger finger rather than reaching over the top of the rifle. The only caution here is that this part should be light. If it's too heavy the stop itself (to which this is attached) gets sluggish in operation. Should that happen, boost the spring by stretching it.

Some custom builders will sell you one of theirs ready to install.

I didn't use any other gizmos on this gun. There was no crying need to further abate the influence of excessive gas pressure in this package, due primarily, nearly solely, to the extended gas port location. That works wonders. I wouldn't exactly caution against the installation of a carrier weight or trick buffer on anyone's Match

Rifle, but since this one went together with a heavy carrier to start, the extra weight of the CWS made a freely-seen difference in the forward pitch movement in the rifle when the carrier went home. I noticed it mostly in prone rapid-fire. Try the insert on a similar package if you used a standard-weight carrier, and decidedly, unquestionably run one for a test behind a standard gas port location.

TRIGGER SYSTEM

I chose a Geissele for this one. It's probably the best trigger we have. Lock-time is notably and noticeably faster compared to most others, and adjustment means and routine are respectively extensive and simple.

There's enough said about triggers elsewhere to point out reasons why I chose what I did and also to address other choices. Jewell is good too. To say you get what you pay for is at least half correct because lower-cost triggers can be made to perform well, but at higher cost if that makes any sense. It can take some tinkering, and that

Receiver lapping

Here's the receiver fixture talked about in the text. I ran it at slow speed with 600 grit aluminum oxide lapping compound. Can't say it's all that, but it did do something... There was a noticeable amount of polished high spots, especially on the inside of the upper receiver.

Making things more simple, more better

If you want to skip the rear takedown pin alto-gether, which means there's not even a detent installation necessary, just get one of these. It's from KNS (the trigger pin folk) and provides a slip-fit, push-button installation and removal. There's one for the front end too. I like them.

These can be used on any AR15 intended for any purpose. A sling swivel attachment can go right on the end of it too for tactical users. Of course, there's a chance this pin could get lost, but prob-ably not with a sling hanging onto it...

can take some time. Most will find that a Geissele will install only once and then tune to suit right in the chassis, Jewell too.

STOCK

I chose a Medesha Firearms fully-adjustable "skeleton" style butt-stock. I have to say that now because Scott has had one with the same adjustment apparatus but in a conventional stock shell (shown on page 244). It's really only a looks issue. I like the skeleton. This stock is adjustable for cant or rotation, length, vertical buttpad position, offset or cast, and cheekpiece height. Change the offset by shifting

the mount on the extension tube. Its receiver extension tube screws into the upper just in the same as any other receiver extension tube and performs the same function with respect to securing the rear takedown pin spring and plunger and buffer retainer. I often use Teflon pipe wrap tape on aftermarket tubes. I know it's not necessary but I have seen applications where it made a noticeable difference in how solidly the part assembled. It don't hurt.

There is zero tricky about this installation. After it's threaded into place the adjustment apparatus simply clamps onto it. The Medesha stock takes a carbine-length buffer and spring. Keep that in mind.

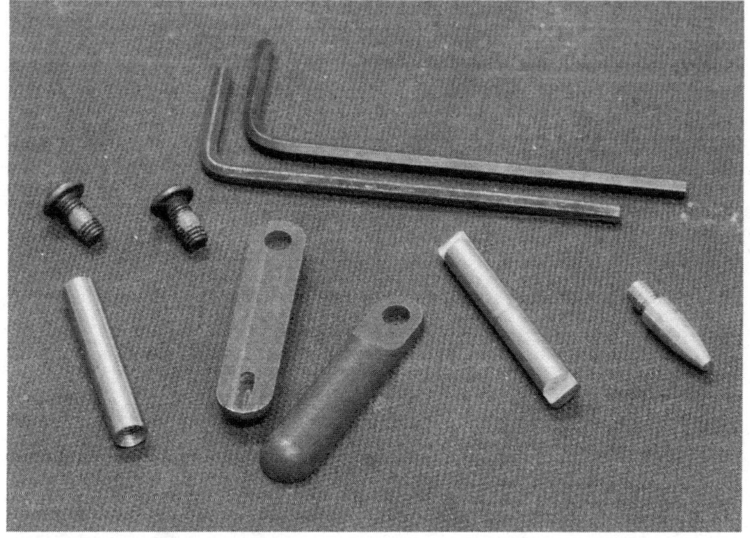

Geissele.

I went with a Geissele two-stage Match Rifle trigger for this one. Reason is easy. It's the best. Locktime is fast, installation is easy, and adjustment is even easier. Geissele offers a lot of trigger options, and the package is incredibly complete. Installation instructions are likewise outstanding, and you'll see my own step-by-step as well. Geissele triggers are "pre-tuned" prior to shipment, so if your lower receiver is what it should be, installation will be painless.

Trick pins

I put a set of KNS locking trigger pins on this one just to show how it's done. Hammer pin has tabs on each end, trigger pin has threads. Geissele comes with its own pin set that is equivalent in quality but the locking mechanism is a nice touch. KNS offers more than one style of locking pins, and also has an oversized set, complete with reamer. Only problem with that approach is that it won't work with all triggers unless the pins are dressed down (only where the trigger and hammer engage them) to get smooth operation. The only time I can think where oversized pins, by the amount oversized these are, might be needed is in a receiver that has enlarged holes.

Magazine

Since it's an NRA Match Rifle and "palm rest" restrictions apply, you'll need to find a short magazine to shoot the standing event. It can't extend more than 3.25 inches below bore centerline. A Bob Sled single-load device is the easiest thing to do, and you'll like it. Otherwise, you'll get along fine with 20-round magazines for rapid-fire and whatever you chose for standing for use also in prone slow-fire. Just make sure the bolt stop is functioning.

Competition Shooting Sports (CSS) *adjustable stock is an outstanding choice and very easy to install. The hardware simply fits over a standard rifle-length receiver extension tube, which means it takes a rifle-length buffer and spring as well. Cheekpiece height adjustment is via click stops as the cheekpiece assembly is rotated on the tube. That means it's easy to run the piece down to allow chambering a round using a standard charging handle, then just click it up into place for the shooting position. This installation is most easily done using one of the "plunger" style takedown pins shown nearby. There are an increasing number of adjustable stocks on the market, but not all are nearly as good as this or the Medesha. Many have cheekpieces located too far back (to allow charging handle use) or otherwise design flaws that preclude best utility on the competitive firing line.*

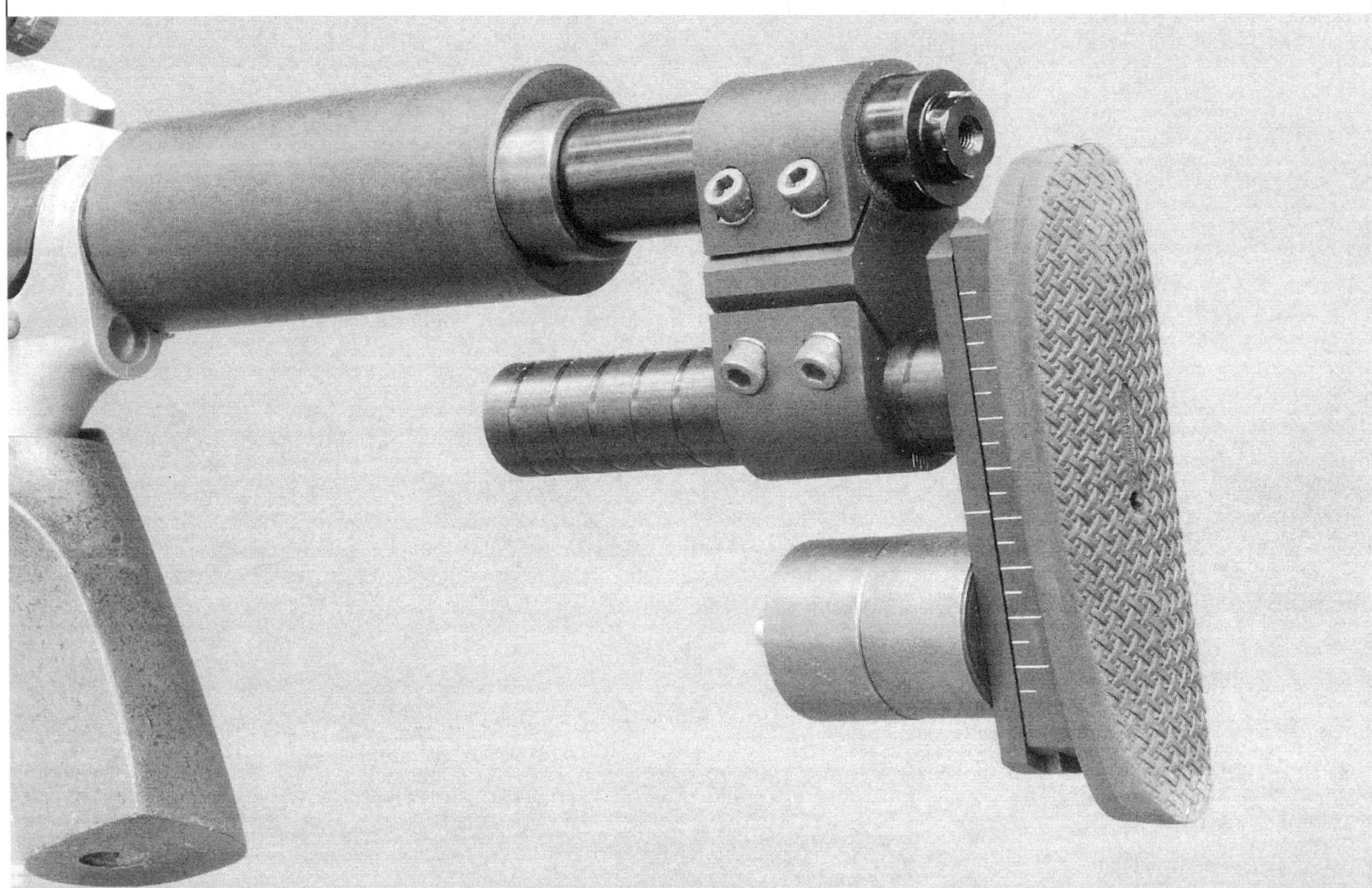

PARTS & TOOLS

I'm going to skip the routine list of parts and tools to assemble the lower receiver (the upper needs nothing), and, depending on components you choose, there could be differences certainly in what you'll need to finish the rifle.

PREPARATION

Vise, very securely mounted

Upper receiver clamp

Tap hammer

Roll pin punch and roll pin starter punch (for gas tube roll pin)

Torque wrench, 1/2-inch drive

Breaker bar, 1/2-inch drive

Barrel nut wrench attachment for the above

Gas tube alignment tool

Lower receiver block (for vise mounting)

Options (good ideas)

Permatex "red" adhesive

Permatex "blue" adhesive

Contact cleaner

Anti-seize compound (I like Loctite C5A)

Gas tube wrench

PRECAUTIONS

Test fit everything! Make sure all parts work together.

19.0 THE BUILD

THIS IS EASY

The Medesha free-float tube consists of a standard barrel nut that slips into a collar. The collar ends up behind the scallops on the barrel nut and then is rotated to get alignment with the gas tube cutout in the collar after the barrel nut has been tightened.

Afterward, three headless set screws anchor the collar to the barrel nut. The float tube itself slips over the outside of the collar at any point after the fact and is likewise retained by three conventional screws.

Note the different style of upper receiver clamp used in this build. No option. The DPMS Lo-Pro upper won't fit into the clamshell-style because its lines are different than a conventional A2-style. **So vise it up and assemble the nut and collar.**

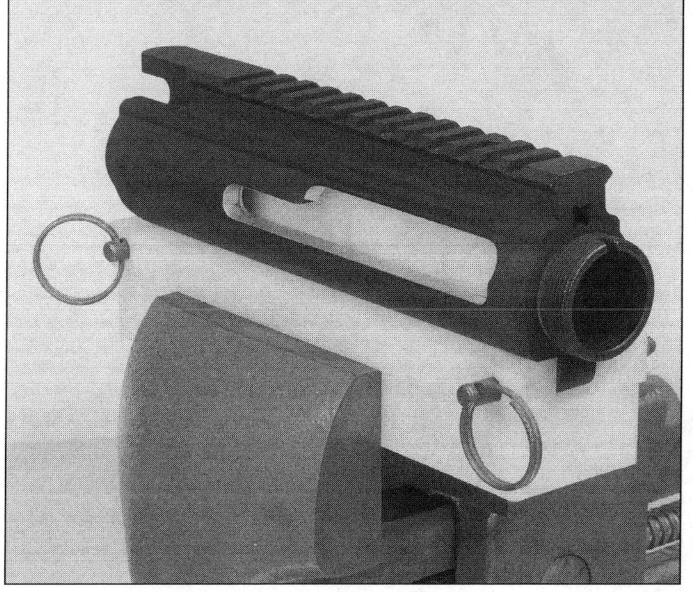

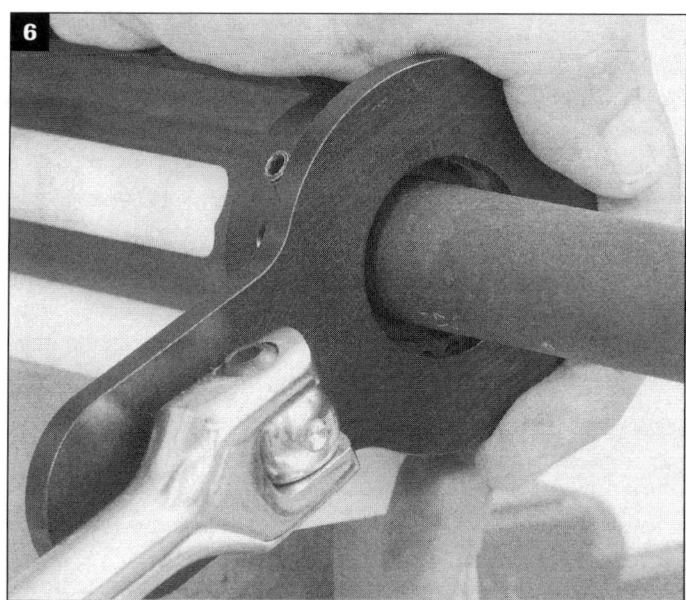

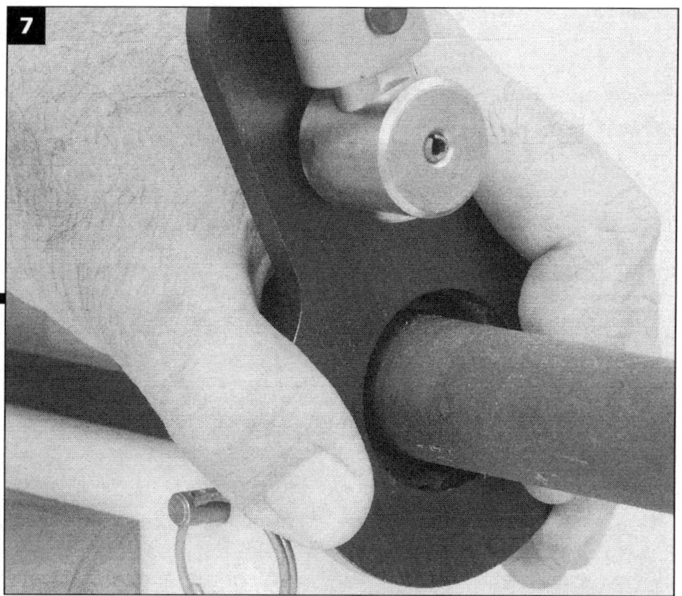

BARREL & FLOAT TUBE

1./2./3.: Degrease the extension and inside the upper, apply anti-seize to the upper receiver threads, apply glue to the barrel extension, position the barrel into the receiver. **4./5.** Thread on the barrel nut with the loosened float tube collar in place. Attach the barrel nut wrench to a breaker bar. **6.** Follow the tighten-loosen procedure described in the basic barrel installation to seat the parts. **7.** Take it to torque (35 ft.lbs.) and hope you got lucky…

8. Check for gas tube alignment using an alignment tool, and then (always) using the carrier. The tube will rattle when it's right. **Get it right! 9.** When gas tube alignment is confirmed, tighten the set screws to fix the collar to the barrel nut. Thread-locker ("blue" duty) is a little extra assurance on these screws.

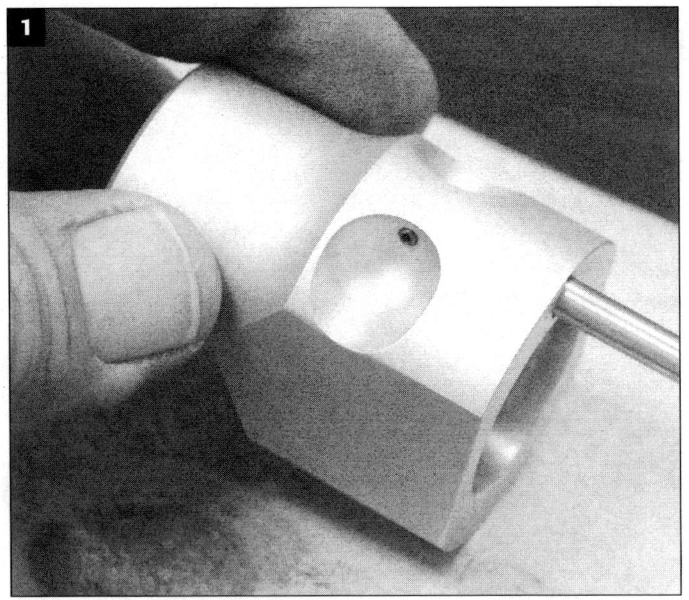

GAS MANIFOLD

1. Since there are no clearance issues due to the large diameter of this float tube, it's easiest to install the gas tube into the gas manifold before installing the manifold onto the barrel.

2./3. Before doing that, it helps get everything lined up if you'll make a couple of index marks, one on the barrel shoulder inline with the gas port hole and one at the gas tube hole on the manifold. There's room for error comparing hole sizes, but why err?

4. Just run the manifold onto the barrel, guide the gas tube into its place, and make sure the manifold is tight against the shoulder. A drop of red glue and a solidly secure tightening of the set screws and that's about that.

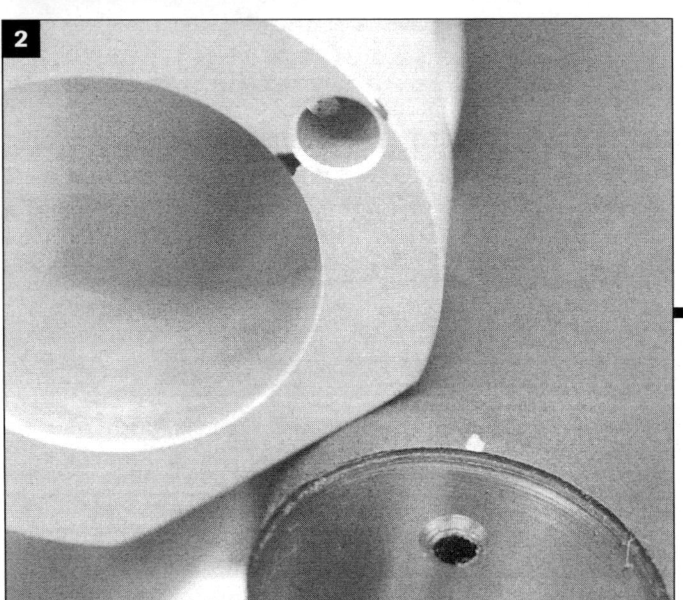

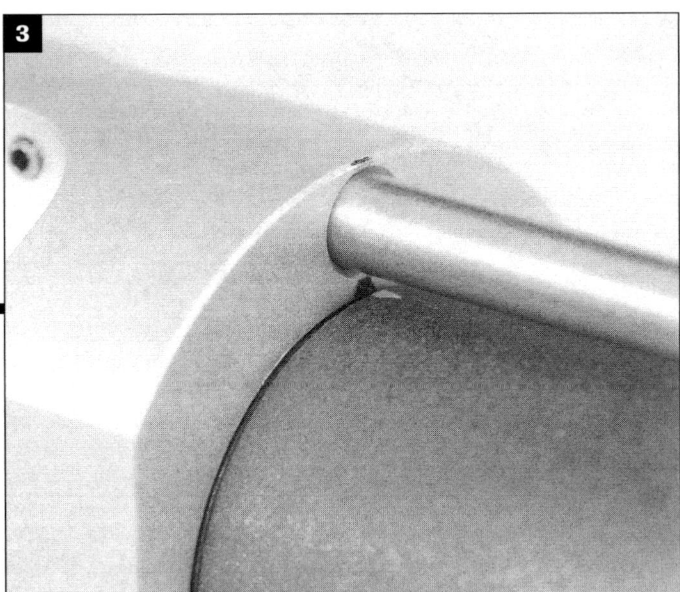

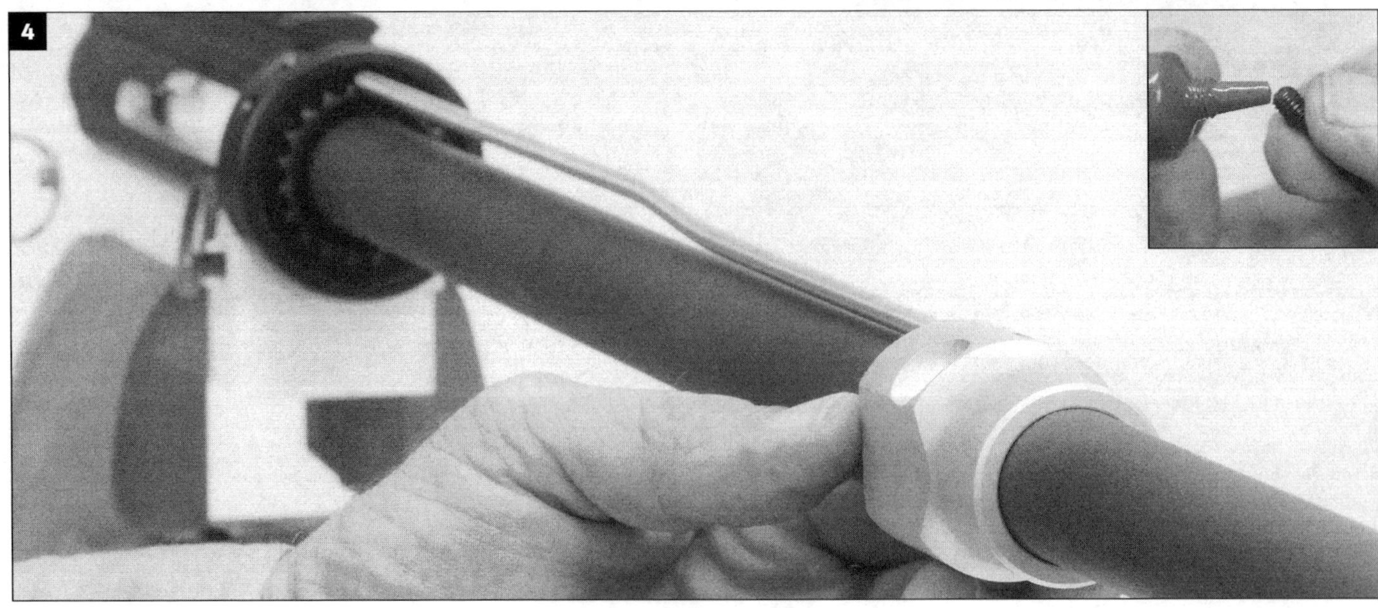

FLOAT TUBE FINISH

The tube itself just slips over the barrel and onto its collar. Line up the screw holes and tighten all equally. This forend is rotation-adjustable and has a pretty good amount built in. You may find it easier, by the way, to install a handstop before slipping on the tube.

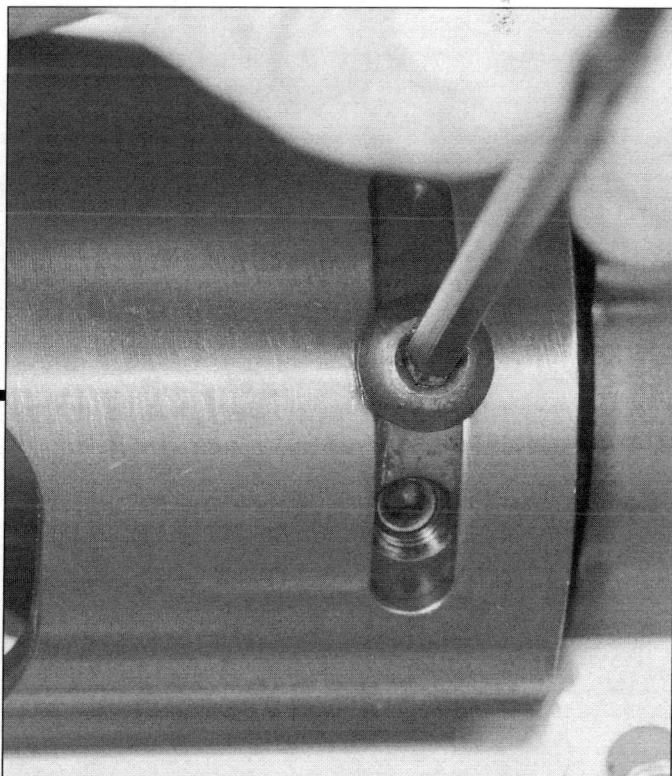

SIGHT EXTENSION TUBE

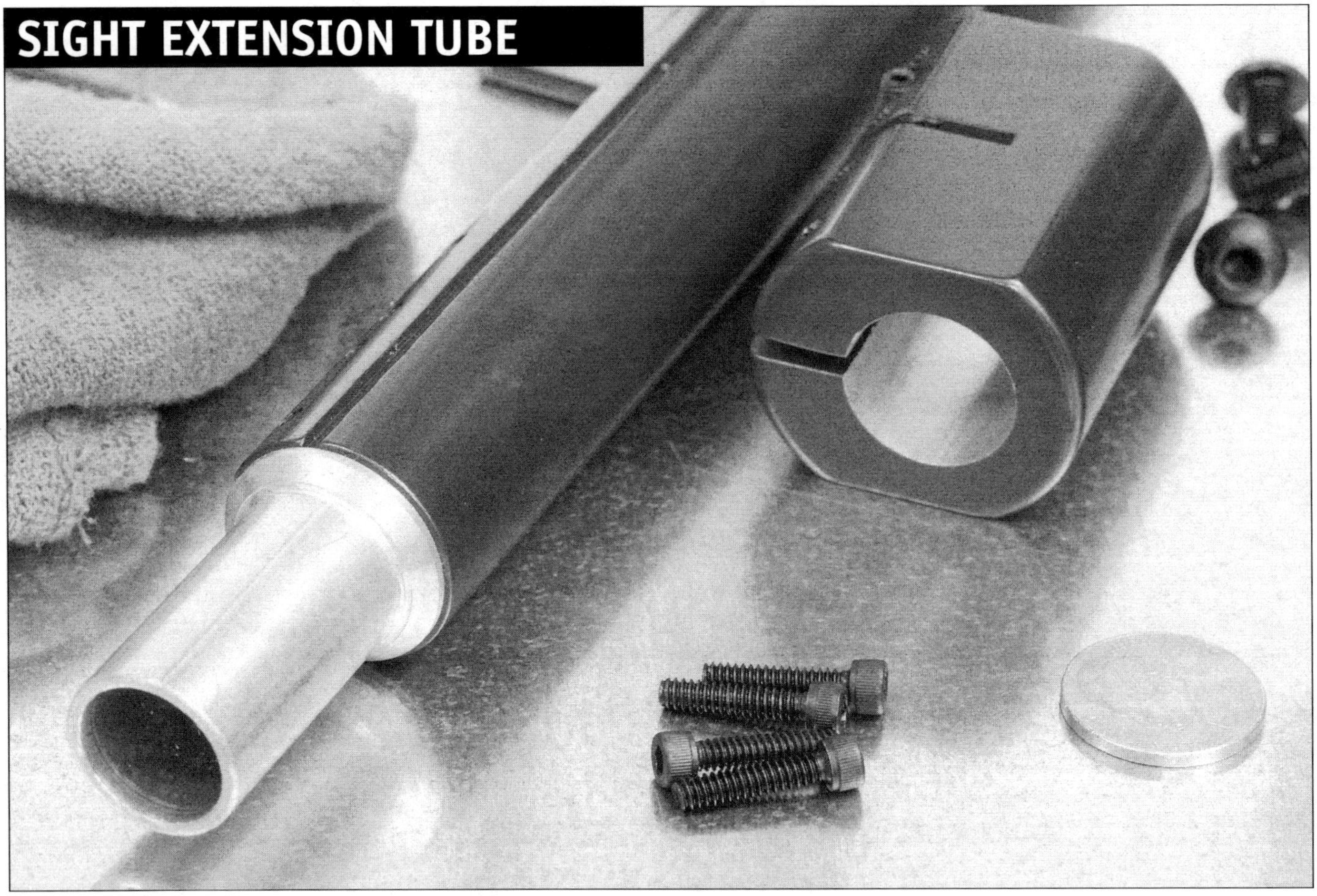

This is strictly optional but thought I'd show it because I had one handy for this rifle. Plus it's a cool trick that might come in handy for other applications for you sometime.

A tight fit is important for this part. If the fit is too tight, which it really should be, then here's a way to open it up just enough to get an easy install.

Remove one screw from the back of the mounting block (the end that goes onto the barrel first). Install the screw in its reverse direction (from the other side).

Take a nickel and place it in the slot so the screw tip will contact it. Thread it in against the nickel and that will provide enough extra gap to get the block started onto the barrel. After it's on, just switch it back like it was, line up the pin on the extension tube with the slot in the block, and snug it all down.

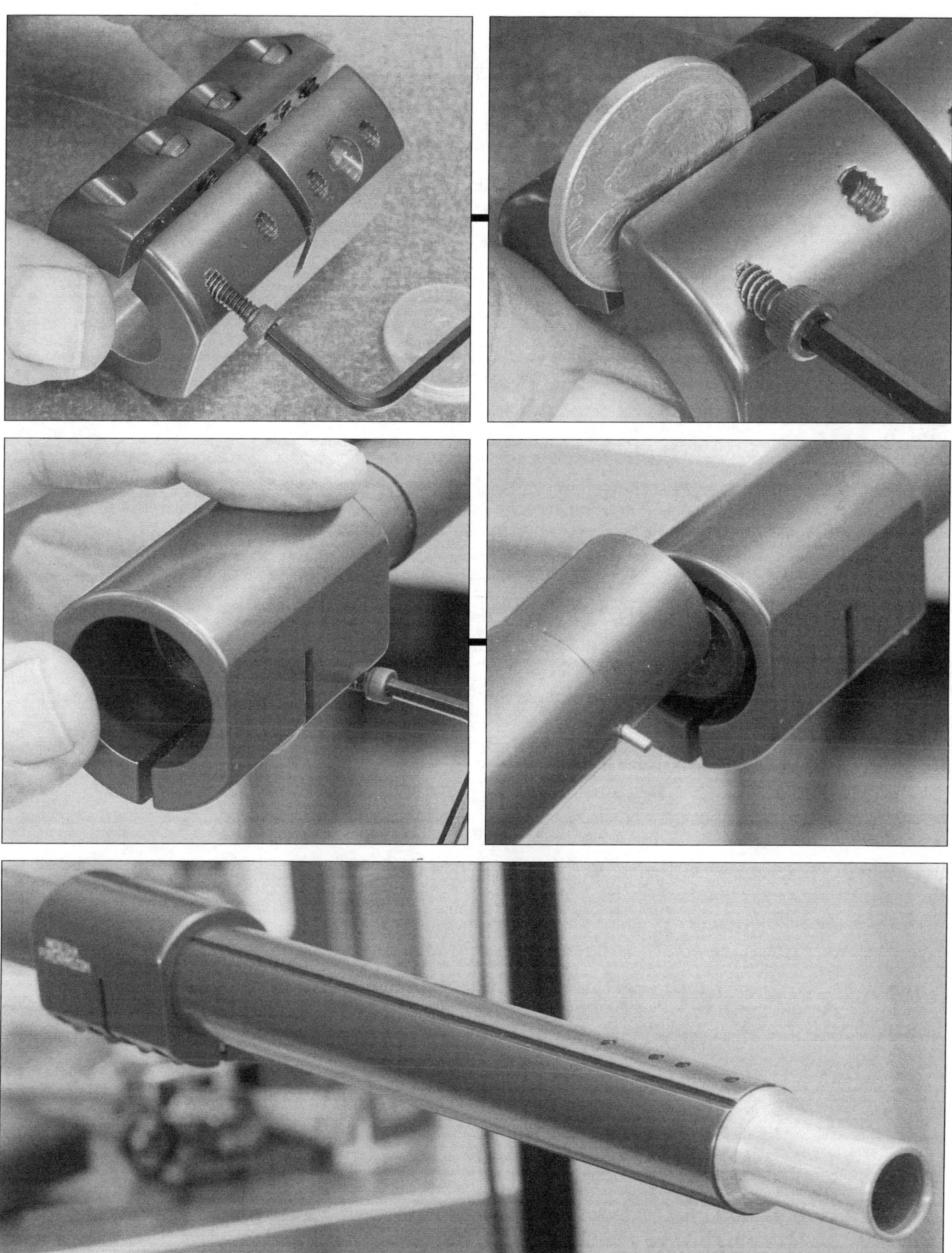

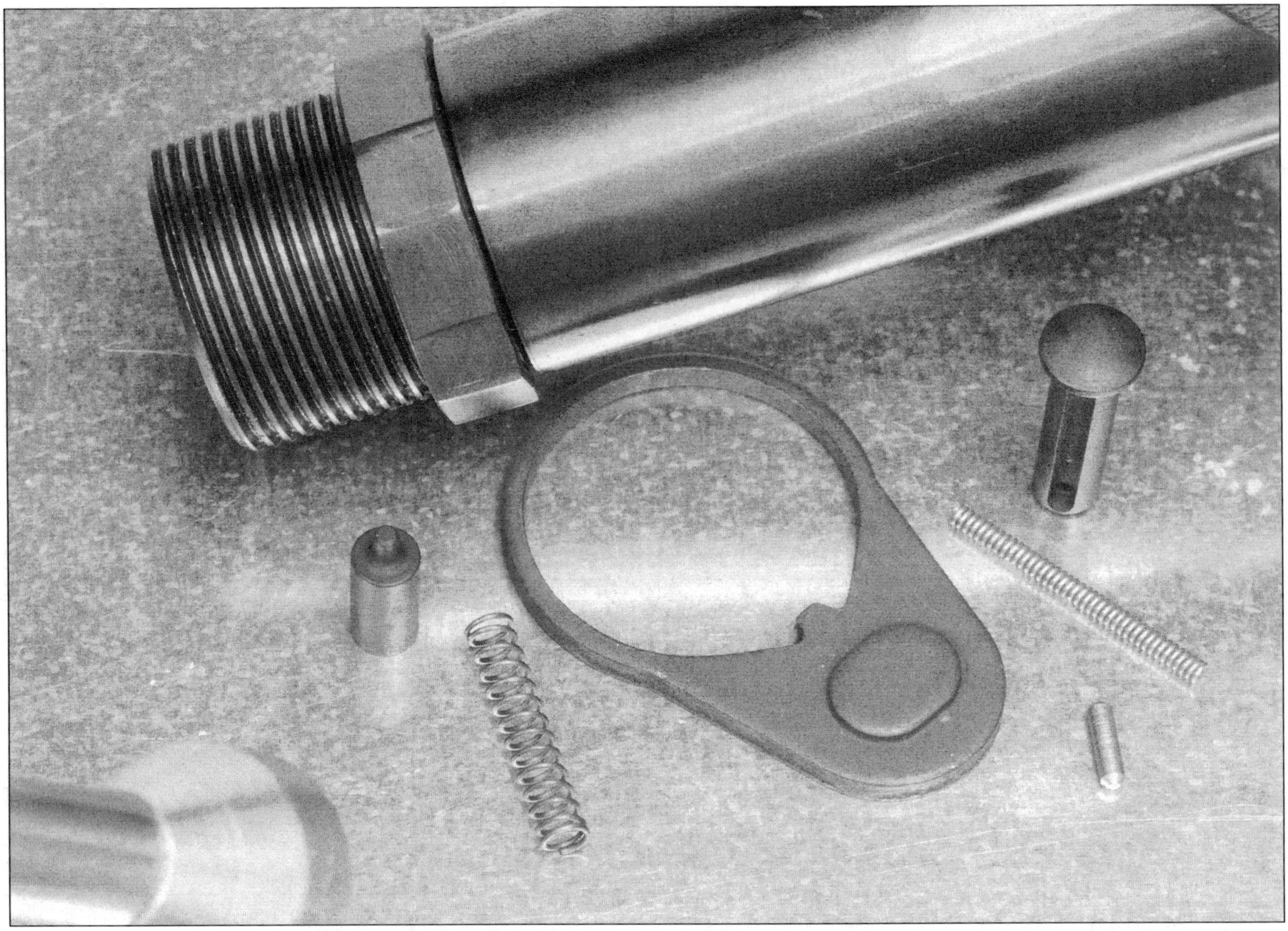

STOCK (ADJUSTABLE)

First remove all the adjustment hardware from the stock. Get it down to only the extension tube.

This stock installs in exactly the same way as a CAR-style collapsible stock, and uses the same back plate to retain the takedown pin detent and spring.

Not trying to shortchange anyone, but it's easiest just to refer to those instructions on page 55 for help.

The only difference between this and the CAR stock is that I use Teflon tape in place of glue on the threads. I ran three wraps.

Make sure the locking nut and back plate are installed and that the plate is oriented like it should be. The locking nut has wrench flats so a honking adjustable wrench is a good tool to use.

After the tube is installed, the adjustment hardware slips on and gets snugged down.

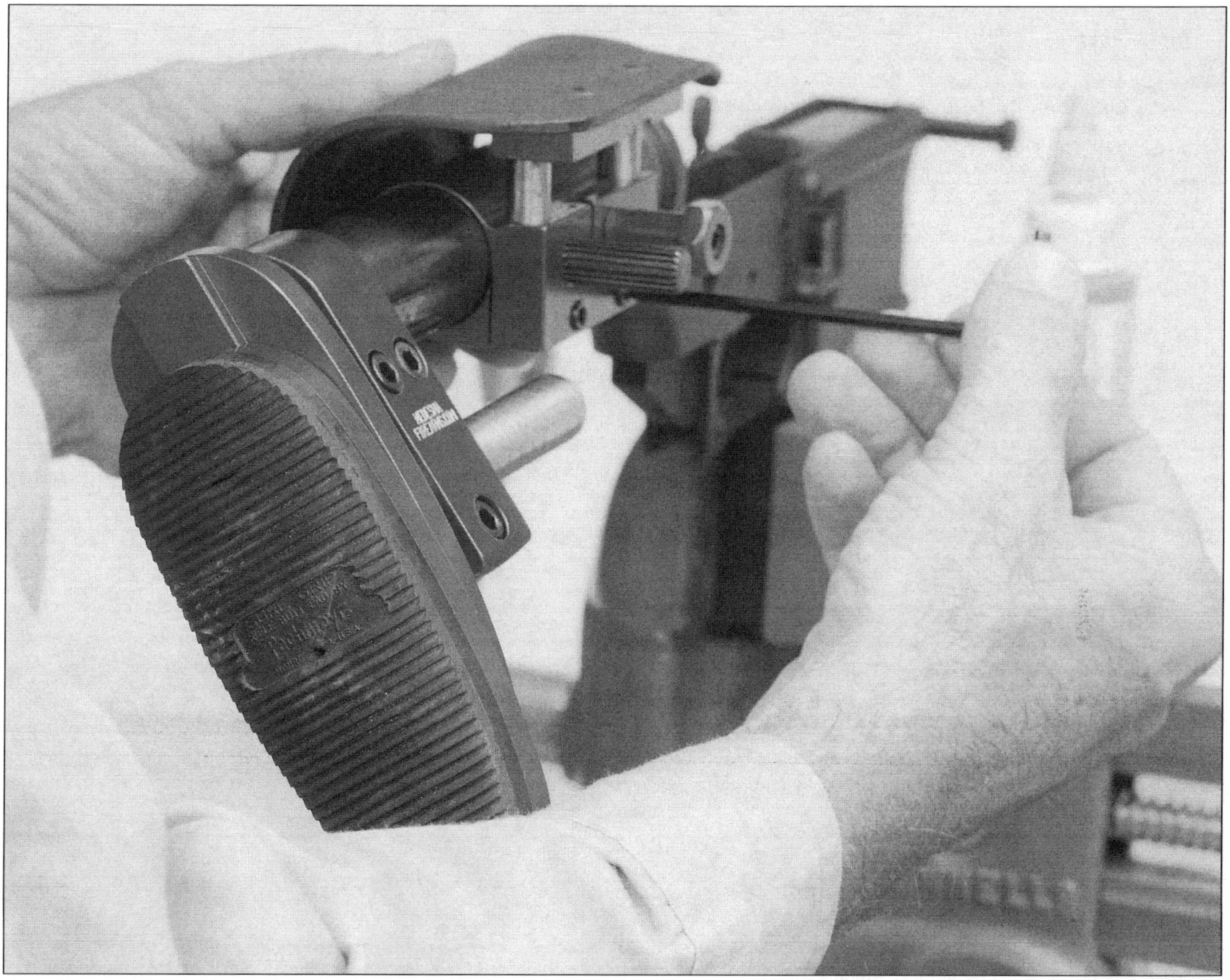

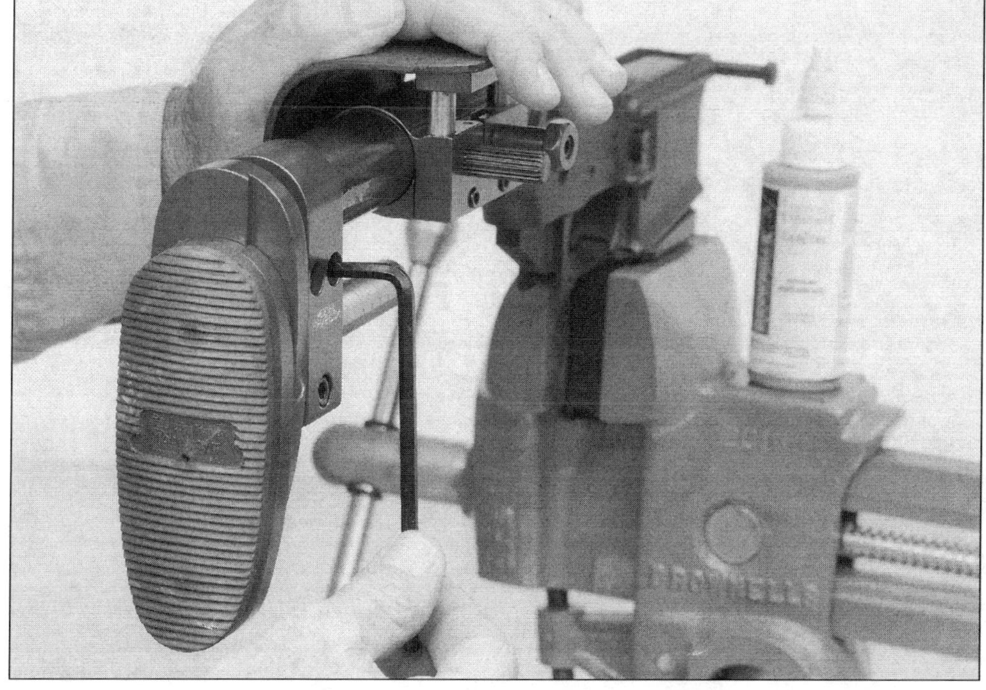

The next step in this project is (and this is a big one) to install a good trigger, accompanied by choice of handgrip.

Then of course incorporate the pieces parts that it needs to finish itself into a functioning firearm: bolt carrier assembly, charging handle. This stock requires a CAR buffer and spring.

Last, and by no means least, is to start searching for sights, slings, and other accessories to get this bad boy to the firing line and in use.

This rifle shoots really, really well by the way! Good barrel.

Buttoning it up.

Finish it off by adding these parts and pin the receivers together.

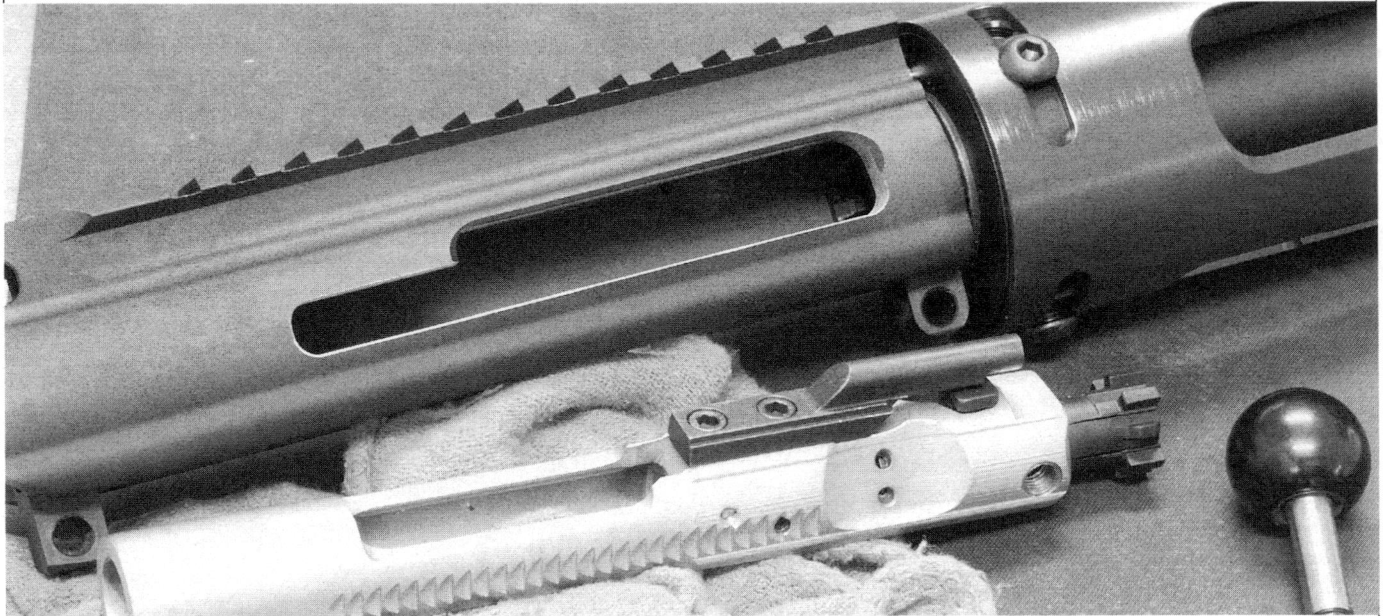

This rifle hammers. The accuracy level of this gun is equivalent to anything other across-the-course rifle I own. If, like the NRA Service Rifle project, we set a goal to beat anything else out there that we could unbox ready to go from a manufacturer, this (as well as that smoke pole) surpassed that goal by leaps and bounds. As said, the only caveat is it's harder for us to really tune rifle specs in a home-shop.

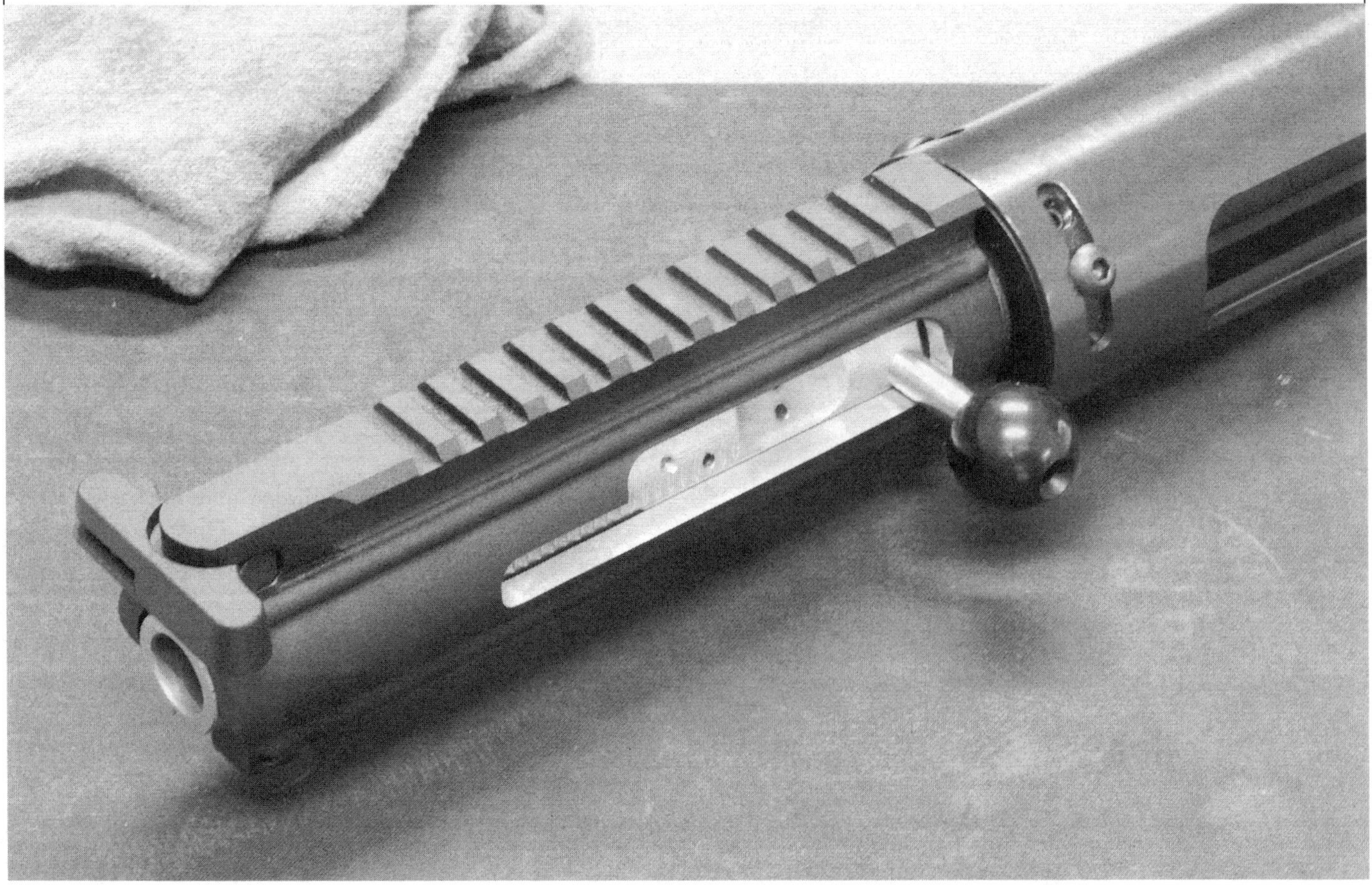

20.0 GEISSELE TRIGGER

FINISHING THIS PROJECT

[The instructions that come with a Geissele Automatics trigger are good, detailed, and a phone call to the maker will resolve any problems you're having. Geissele offers several options in its trigger systems now, including tactical-application specs. Geissele triggers come pre-tuned out of the package. If your receiver is what it should be, very little to no adjustment may be necessary to have a perfectly functioning trigger. Of course, personal tastes mean tuning, and that's another area where this trigger system really shines.]

This rifle got a Geissele Automatics two-stage. *I truly believe this is the best competition trigger on the market, at least when this went to print. It installs just like a GI trigger, and then adjustment is done with the trigger in place.*

SEGMENT CONTENT

227 **Overview**

228 **Components**

229 **Installation**

232 **Adjustment & Tuning**

234 **KNS Locking Pins Installation**

235 **Match Rifle Project Finish**

PURE PRECISION

Geissele (say "gih-sell-ee") combines sound engineering with top quality to produce an outstanding trigger. I think they had the advantage of seeing so many other two-stage triggers before coming up with their take on it that they were able to overcome many of the shortcomings in other designs, and enhance its action in the process.

These are not cheap. They're pushing three hundred dollars. Yes, it is possible to go with a Rock River, ArmaLite, or other, and have a better trigger than stock. However, you will be able to tell the difference in a Geissele and any other with one pull. I've spoken kindly about Jewell triggers, and take nothing back. The main difference, from an operational standpoint, is the Geissele has a noticeably faster lock-time. I will not tell you that a Geissele "feels" any better than a Jewell, but will say it decidedly is a higher-performance product.

GEISSELE PACKAGE

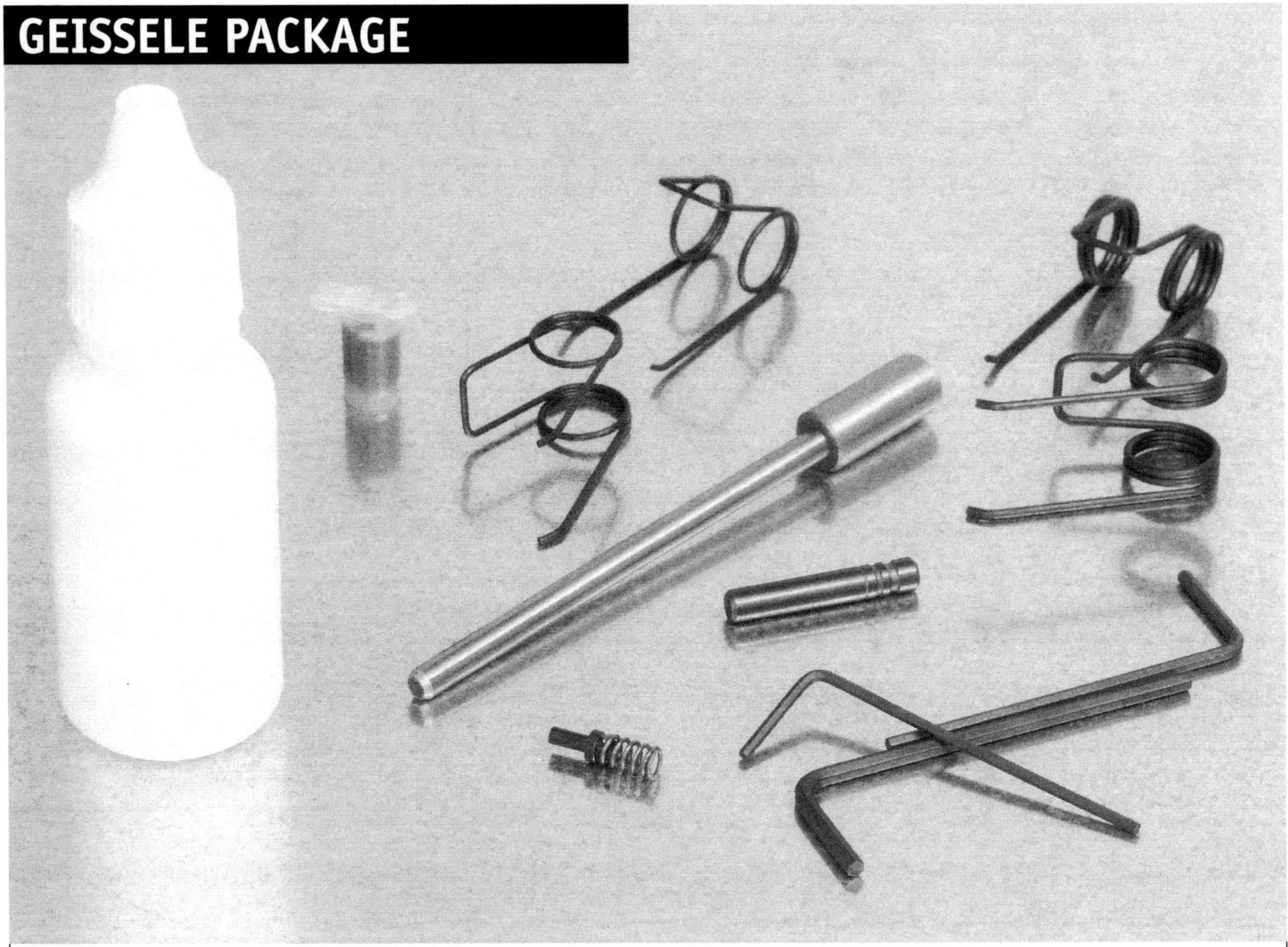

Geissele offers the most complete trigger package on the planet. _Unlike so many others, it's not up to you to hunt down different springs or pins to make the trigger perform as desired. Geissele can do all that for you. This has changed a little over time, and the proper parts are packaged with the trigger you order. I want to show you some of the differences you can encounter in a Geissele trigger. It's the extent Geissele has taken the package components that should impress, and it is impressive. Take a look at the pair of trigger return springs shown below. One is for a Service Rifle application and the other for a Match Rifle. Big difference in the gauge of the wire. That all has to do with first-stage pull weight, which is better higher in a Service Rifle to keep the overall weight what it should be, but produce a lighter feeling second-stage break._

The large photo shows components that either come as part of the package or are options. When I ordered my first trigger (which has been a good while ago) I told Bill I didn't know if it was going to be for a Match Rifle or Service Rifle... The disconnector spring is even different for each.

Also in the package are the proper oil and grease, a trigger-pin-diameter rod to help installation, and an over-sized hammer pin in case it's needed. On that, I substituted a KNS locking pin set because I like those a lot, but the Geissele-supplied pins are equivalent in quality.

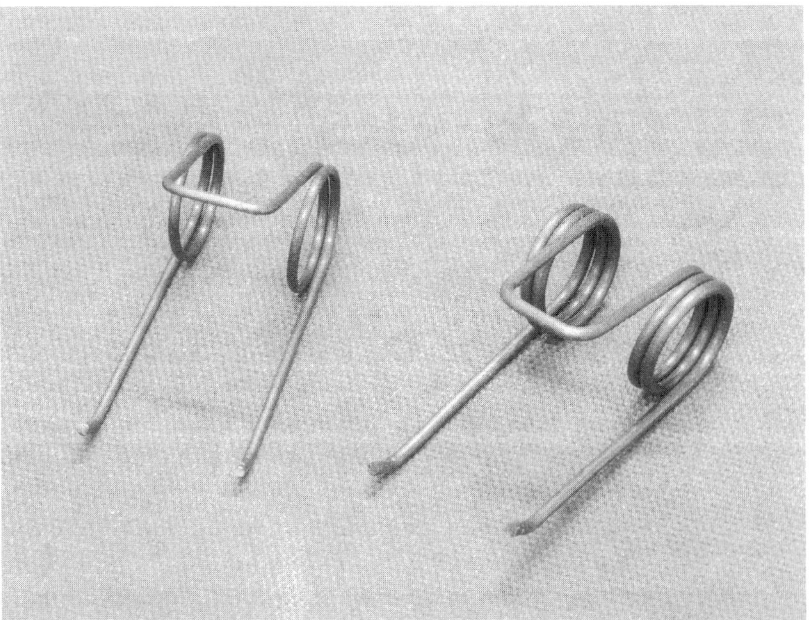

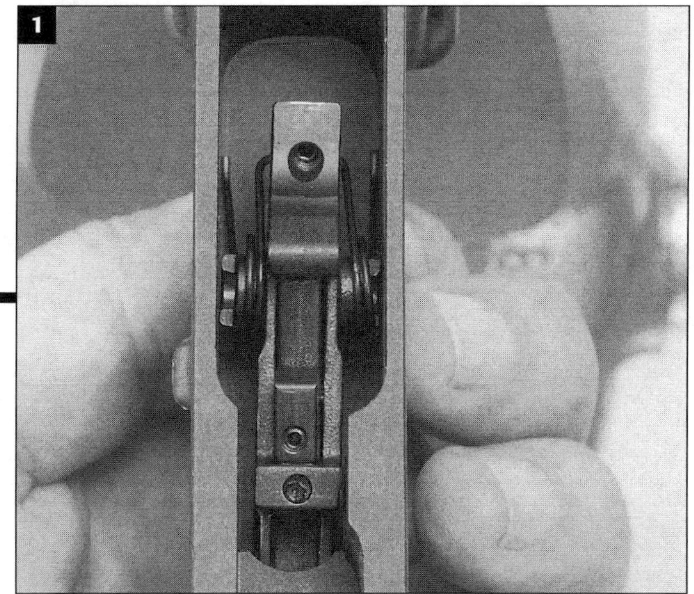

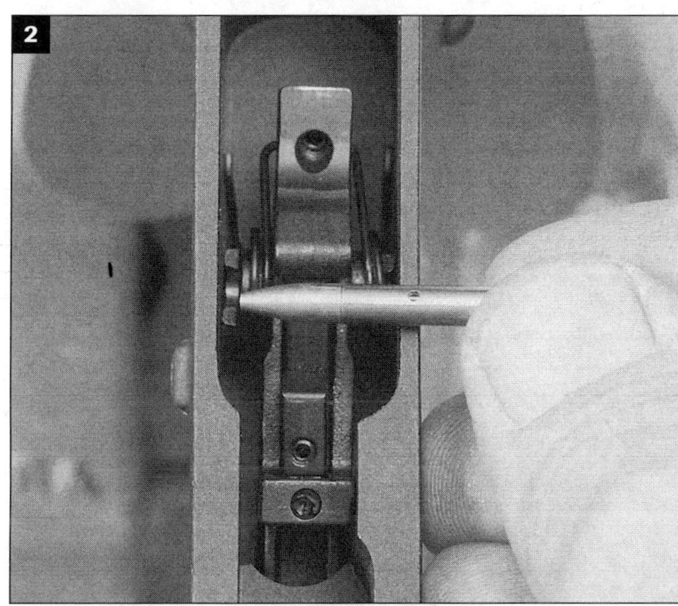

INSTALLATION

You must remove the safety, which means pulling the pistol grip and, along with, the safety detent and spring. Clean out the lower receiver thoroughly while you have all the parts out. **Also,** for reasons of best clarity, I didn't grease all the parts prior to installation. Of course (of course) you should grease them.

1. With the safety removed, place the trigger assembly into the lower. Note that the disconnector assembly is already attached. The KNS trigger pin uses this little screw-in bullet-point to smooth installation and protect the threads. So **2.** thread it on, line up the holes (push down the front of the disconnector), and **3./4.** insert the pin. Otherwise, insert a Geissele pin, **grooved end first** from right to left (that's different). These KNS pins are not grooved because they lock...

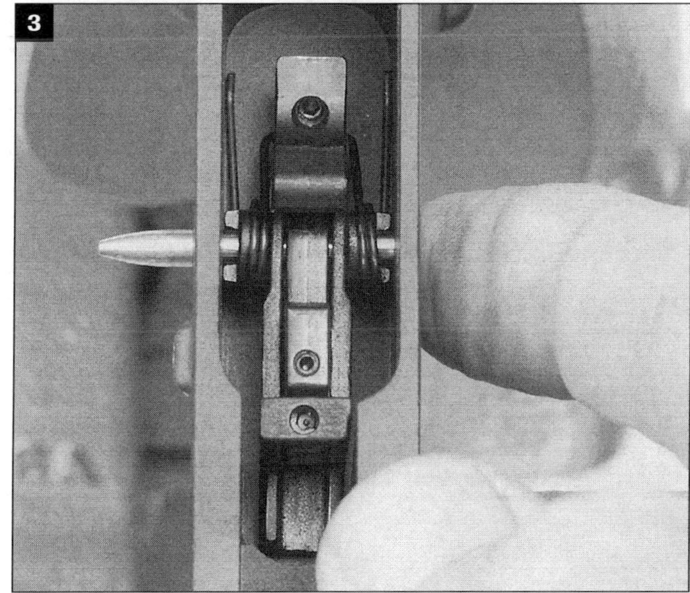

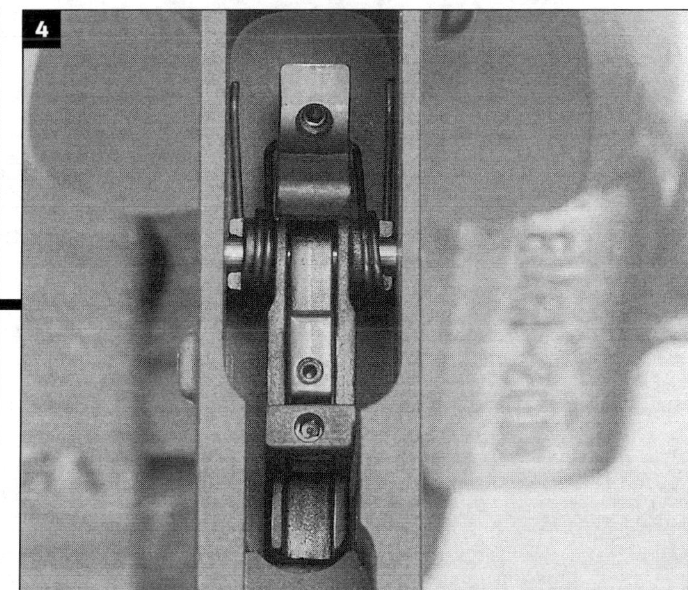

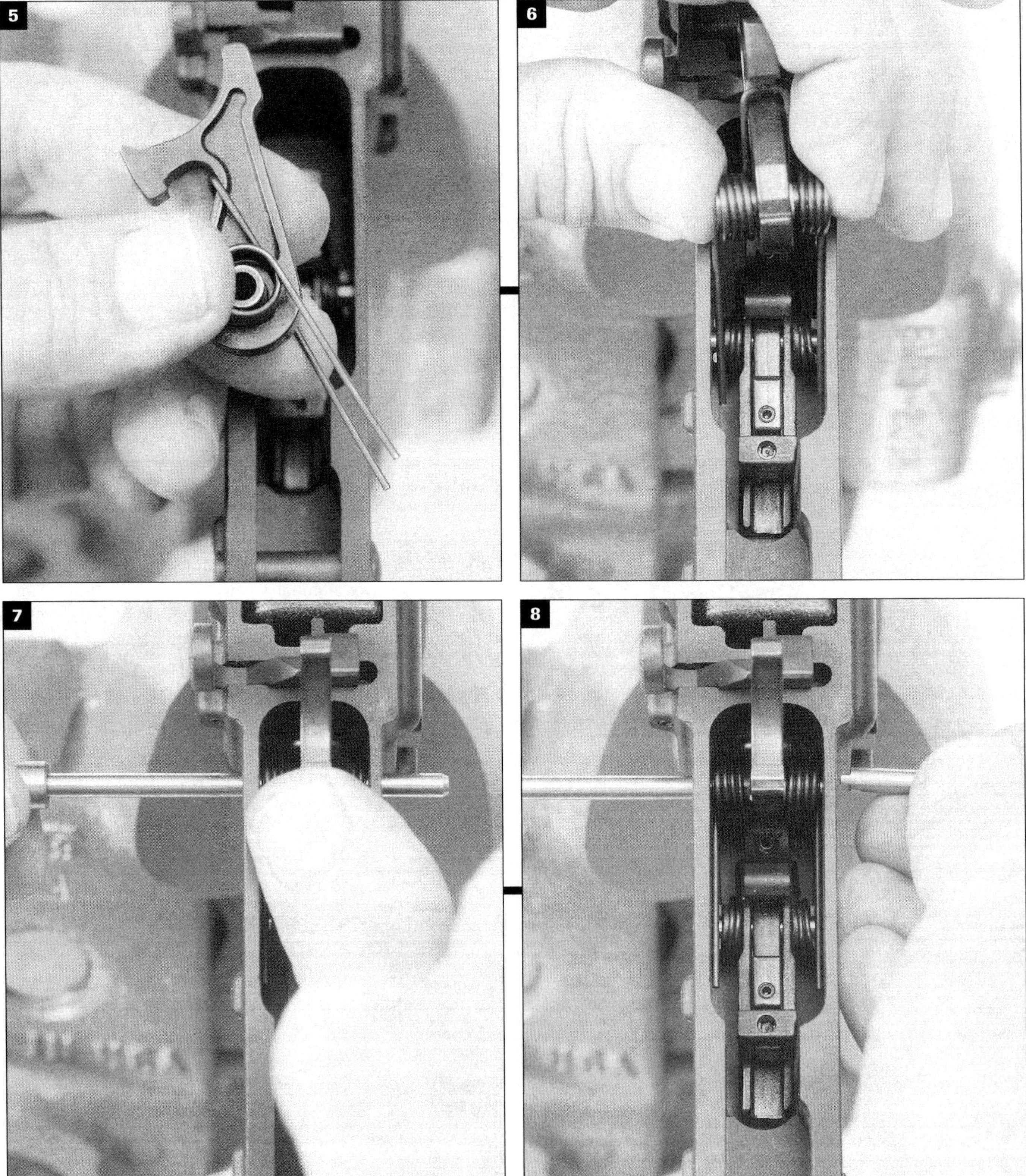

Now install the hammer. **This is not easy.** The spring is stout.

5./6. Position the hammer so that the legs on the spring are on top of the trigger pin. Get the capture punch handy.

7. Push down and forward on the hammer until you get the holes aligned and push the capture punch all the way through.

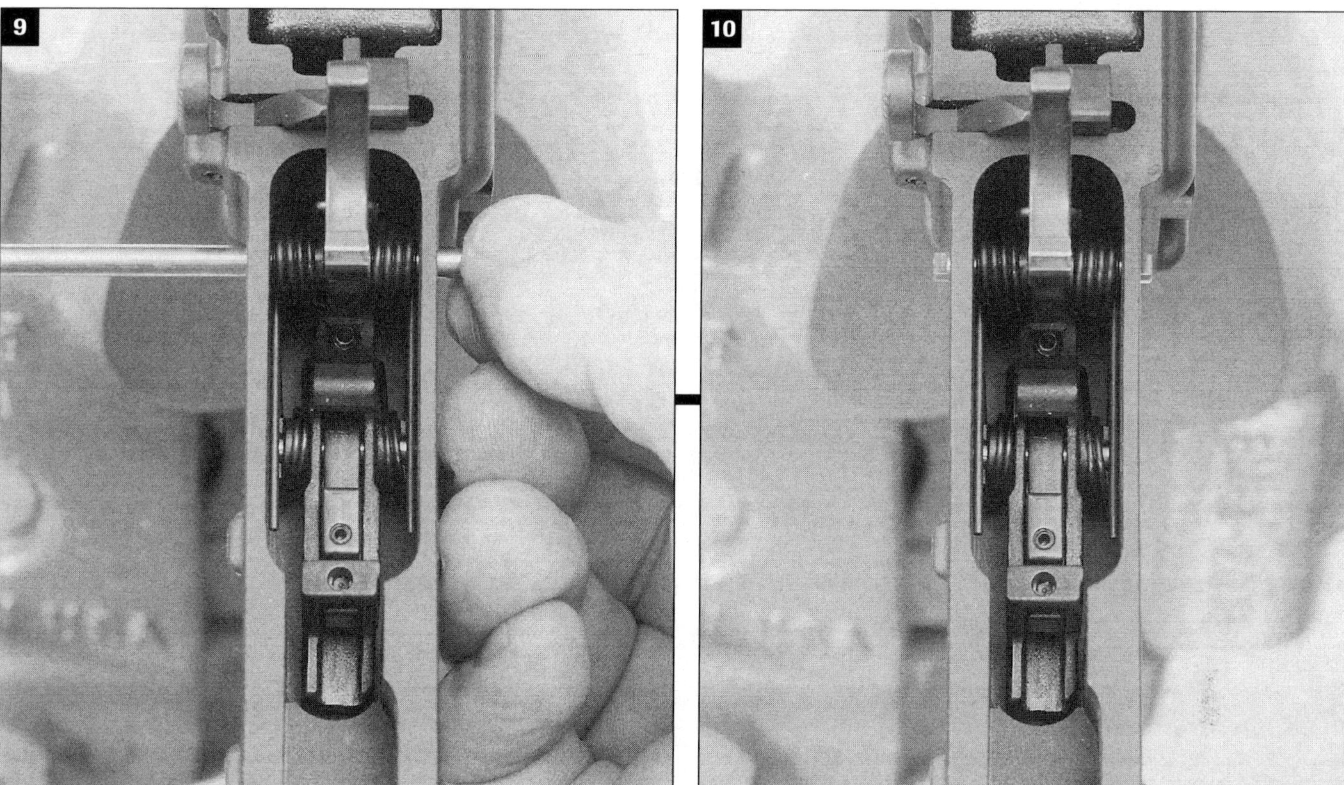

8./9./10. Get a hammer pin and displace the punch with it. Be careful right at the end of its journey into the left-side receiver hole. A little pressure down on the hammer helps. This is entirely a hand-done process. The Geissele pins are a tight fit, but just push harder. I don't think it is ever a good idea to tap either a hammer or trigger pin with a hammer!

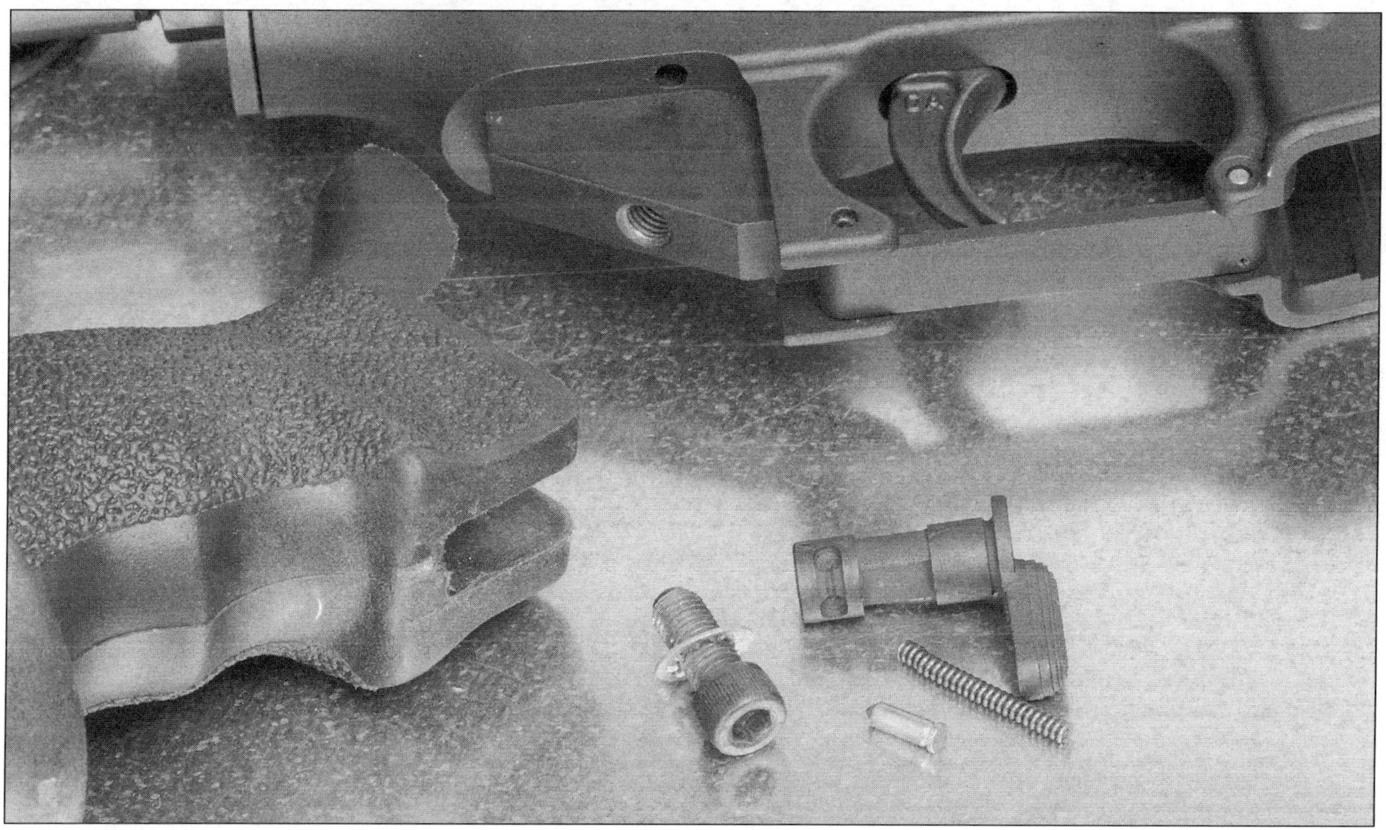

Locate the safety, safety detent and spring, pistol grip and grip screw and turn the page...

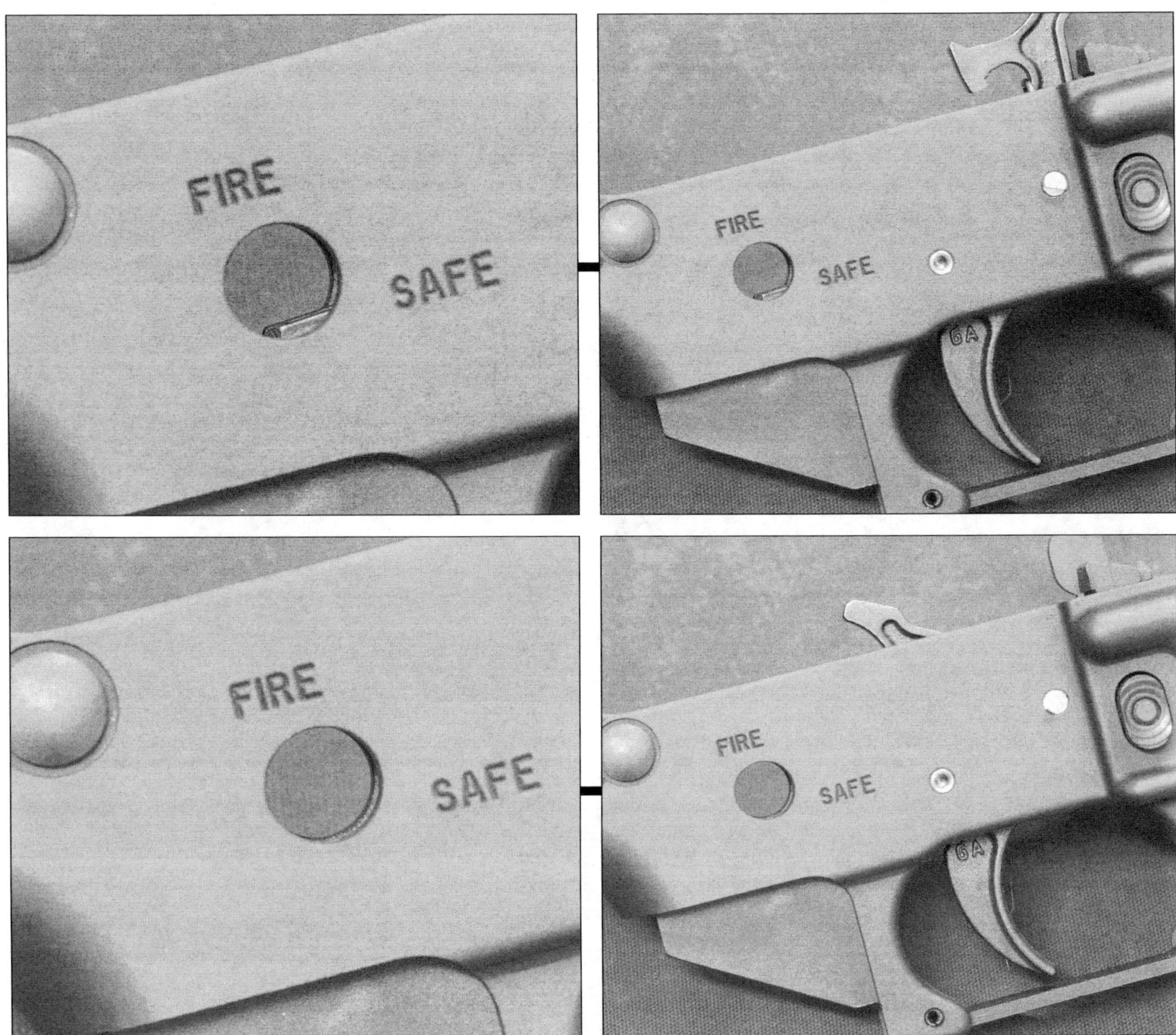

ADJUSTMENT & TUNING

When you go to install the safety, you'll notice it won't go into its hole in the receiver because the trigger is blocking it. That's easy. Just cock the hammer... Get the safety parts and grip installed. Now we can do the safety checks and tune the trigger.

Prior to assembly, it's very important to have test fit all the components. Use the trigger pin tool (Geissele calls it a "fitting pin"). It indeed fits well enough to see what's going on with the trigger and hammer function. Make sure the trigger and hammer move freely.

I skipped following the adjustment routine as outlined by Geissele using the fitting pin because I wanted to go ahead and show you how to get the trigger installed into the lower receiver. That first step, by the way, is to measure and set first stage weight using only the trigger and fitting pin. First-stage weight should be checked and set prior to hammer installation. If it's too low, bend the legs on each spring downward an equal amount. For a Service Rifle, it should be about 4 pounds and can at this point be measured with a trigger pull gage. If it's a Match Rifle it should be at least 1-1/2 pounds because any less may not allow the trigger return spring to, well, return the trigger and do its functional job. Geissele springs are pre-set at approximately 4 pounds for Service Rifle and a little under 2 for Match Rifle. That last is light. **Now** we'll look at making the other adjustments to truly tune the feel of this trigger.

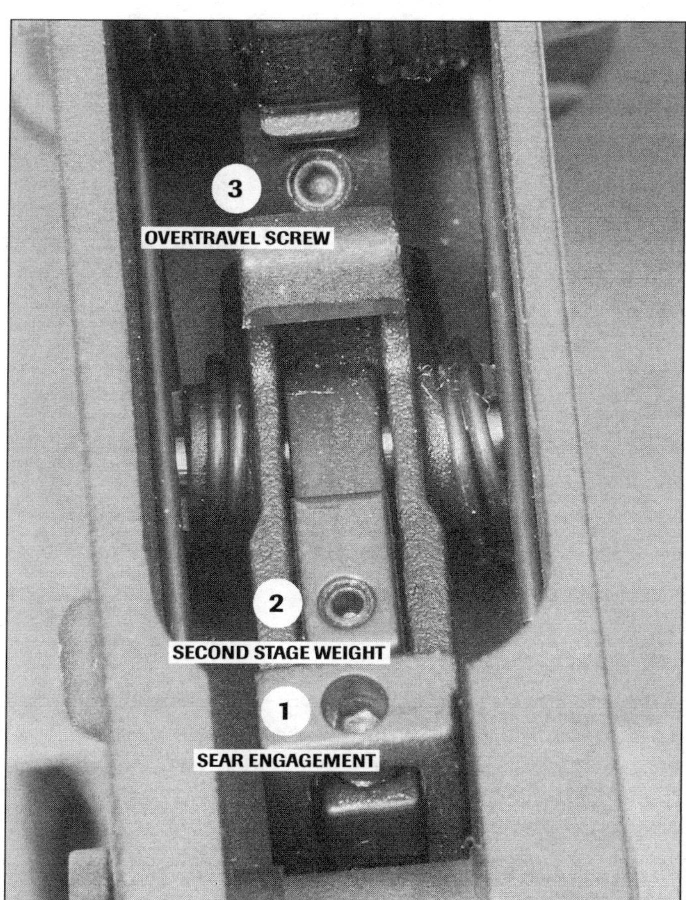

Make sure the hammer fall is cushioned!

Cock the hammer and pull the trigger. Check for the presence of a second stage. **1.** Loosen the #1 screw until there is no second stage. It's a small amount usually. Take small bites at a time. Then take the screw in the "tighten" direction until the second stage is regained. Then take the screw another 1/4 turn. Don't shortcut this one little bit. Check the feel of the trigger. It should pull through the first stage very smoothly. If there's any notchiness or a stutter just before it hits the end of the first stage, there's not enough sear engagement, so head to screw #1 and adjust sear engagement. Clockwise increases engagement, so take the screw in teeny increments in that direction until the trigger pulls smoothly all the way through to the second stage. If there is too much sear engagement, then you'll detect creep as the hammer is released. Turn the screw just a teeny bit backward from the 1/4 turn suggested earlier.

2. Now the second stage weight adjustment screw #2. Clockwise makes it heavier. Minimum weight is when the top of the screw is about 1/32 inch above the disconnector top surface.

3. Overtravel adjustment is screw #3. It's subjective, but minimum according to Geissele is 0.010 inches clearance between hammer sear edge and the trigger sear edge. Rotate the hammer (cocking direction, then releasing) to see the clearance.

Don't lock the pins until you are positive this trigger is just like you want. Big pain to remove and reinstall otherwise. Make sure you've done the disconnector function check with this trigger as has been outlined with all other trigger installations, and safety check too.

Installation is simple with this set. The ends of the hammer pins fit into the notch on the locking bar and the screws thread into the trigger pins. It's a little tricky to get the "winding" action right threading the screws into the pins so the pin doesn't try to come back out; just use two wrenches at the same time and it's easier. Once this is set, it's set. The pins won't move.

FINISHING (IT NEVER ENDS)

Since there are few restrictions on an NRA Match Rifle, there are likewise a large number of options.

You'll need a handstop and sling, of course, and sights. There's a lot (lot) more about all these options in some of my other books, like *Slings & Things* and *The Competitive AR15: the ultimate technical guide.* The essential advice I will give here is you usually get what you pay for. A good rear sight in particular is worth what it costs.

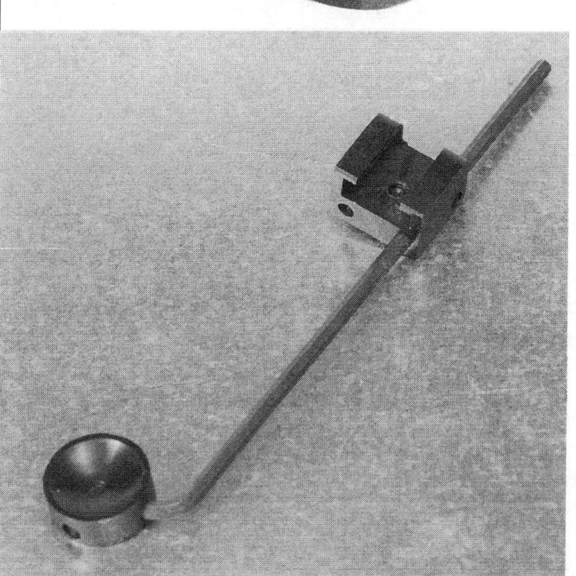

Handstops come in a variety of shapes and sizes. Make sure the attachment mechanism is compatible with the rail slot in your forend.

Handgrips are all about feel. My choice, for myself, is the Superior Shooting Systems grip (shown). Others like Sierra Precision, and the one installed on this rifle was what the user preferred. It's a little smaller diameter than most others.

An **extended bolt stop release** is a right handy gadget. This lets you trip the bolt stop using the trigger finger instead of having to reach over the top of the rifle. Only trick is finding one that fits. Satern Custom (shown) and Bob Hahin (Bob Sled man) make one that's easy to install but neither would work with the upper I chose. They'll fit and function fine on a mil-spec contour flattop. Gary Eliseo's works well with all I've seen.

Sights

I like the **Centra** rear sight. It's a fair price and they are hard to find fault with. There is a model that will mount right on a Picatinny rail, and that's easy.

The **best front sight** I've found thus far is from Superior Shooting Systems. It fits a 0.750 diameter tube, has a level built in, and is elevation adjustable. It's also expensive. See what I mean...

There is probably always something that can be changed, tweaked, tuned, or added onto a Match Rifle. The sight package, for best example, can be outfitted with a number of different rear iris styles, including those with lenses and filters. Front sights also can house different irises and inserts. Do, by the way, make sure you order those parts when you get your sights. Champion's Choice is a good outlet, as is Champion Shooters' Supply.

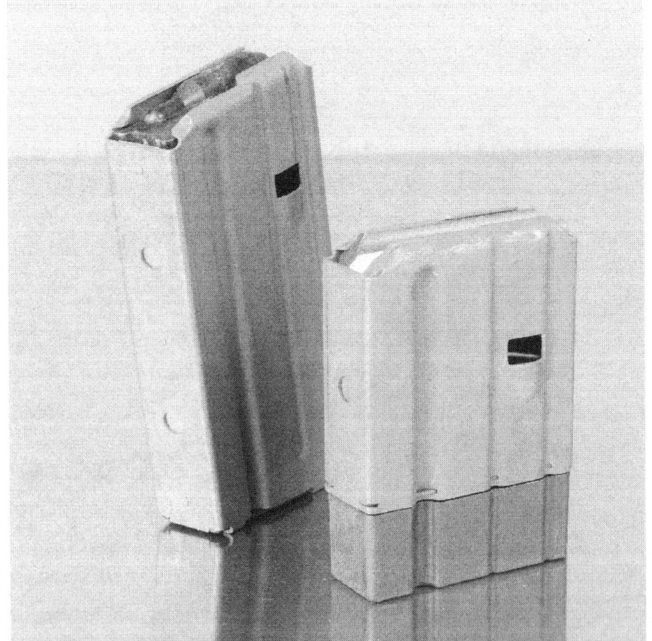

An easy way to go for Match Rifle rules compliance is a Bob Sled. This is a Delrin block for the slow-fire events that has a spring-loaded plunger to activate the bolt stop. Feels good in the hand too.

Magazine

This is a Colt-brand 5-round magazine (on right, top photo). It's fair for use in an NRA Match Rifle for the standing event (200-yard Slow Fire). It's just shy of the 3.25-inch maximum drop from bore centerline allowable, but it's good to go. Know what you're buying because some 5-round magazines are 20-round bodies with a block that restricts capacity. Shop around because they can be hard to find, and when you find them plan on a follower replacement and function overhaul.

21.0 FIELD GUN

IT CAN BE A LOT OF THINGS

[In the NRA Match Rifle segment I said that gun wasn't much more than a varmint rifle with iron sights, and so now I'll say that a varmint rifle isn't much more than an NRA Match Rifle with an optical sight. There are a lot of options with a field-use rifle as well. There is really no difference in assembling one of these rifles than the NRA Match Rifle project. It's really more about component choices and understanding that purpose-built doesn't have mean it's good for only one thing. Very much unlike that NRA Match Rifle.]

SEGMENT CONTENT

237 **Field Gun Defined** (construction/purpose options)

239 **Barrel & Float Tube Selection**

241 **Receiver Selection**

242 **Trigger Selection**

245 **Component Compatibility**

246 **Parts & Tools**

248 **Barrel & Float Tube Installation**

250 **Gas Manifold Installation**

Any more builds, from this point forward, would start to get redundant. Once you're able to install a free-float tube over a good barrel, and see how triggers, stocks, and all the rest go together, then it's all about parts choices, and knowing what you want. This segment here will talk mostly about learning to know about what you want, and then being able to get it...

FIELD GUN DEFINED

That's what I've started calling these. It's an AR15 that is intended to be used on targets and distances that aren't precisely defined. The differences among varmint, practical or tactical, and what I call "fun" guns, pretty much revolves around barrel specifications and diameter, which is really weight. A longer, very heavy barrel and we can call it a varmint rifle; a sort of "standard" heavy barrel and then it's a practical or tactical rifle; and something at least a little shorter, if not also lighter and smaller in diameter, can be a practical competition rifle or a fun gun. Optical sight choices likewise either suit or defeat its intended purpose and since we're all about suiting, that's easy enough to set up as well.

The rest of the build sheet essentials are about all the same.

There's no difference in the components we'll replace stock parts with: triggers, furniture options, float tubes, and so on. What, decidedly, is different was also what was just said: different applications demand different expressions of those parts. Keep in mind, though, that it is mostly the barrel specifications that makes any custom rifle truly satisfy its owner's expectations.

For what it's worth, I am not a big fan of excessively-dimensioned barrels. Some varmint hunters really like a heavy rifle because they do, I'll agree here, sit very still from bench- and bipod-supported platforms. They don't shoot any better though, and I think limit the use of the rifle to rested positions. Again, that may be just exactly what some want, but just don't make the mistake of thinking that any AR15, or any rifle for that matter, needs a "full diameter" barrel to shoot its best.

One barrel I chose for the installation example here was a Krieger ready-to-go dimensioned to specs I provided. It's essentially the same as what I'd put on an across-the-course rifle. It's as big as a barrel ever needs to be. Technically, this particular barrel had specifications that aren't suited to most varminting or field applications. My idea with this project was to put together a bench-top rifle for ammunition testing. That's not really important, and doesn't matter a whit to satisfying the illustration purposes for this book segment.

It's not hard to find a suitable (specification-wise) barrel for a field gun. It's probably best to go ahead and order it with muzzle threads. Many like to install brakes for shooting close to the ground. If the vents are located as they can be, a brake can actually help reduce the amount of dust kicked up by muzzle blast. Otherwise, a brake honestly has virtually no noticeable effect on a heavy-barreled AR15.

Unlike what I've said about other applications, there's really nothing wrong with going with a SAAMI minimum dimension chamber for a varmint rifle. If the application is varminting proper, most will be firing short, lightweight bullets from a magazine, and that favors the shorter jump to the rifling found in such a barrel. Twist rate? You're plenty okay with 1-9 for any bullet up to and including a 69-grain Sierra MatchKing. Factually, if not ideally, something with an even slower twist would work best with bullets up to 55 grains. You are not likely to find anything, however, that's slower than 1-9 ready to install.

If you want to go to a custom-done barrel for later installation using your own tools, I would strongly suggest thinking about going with a 26-inch barrel for a varmint rifle with a smaller than usual

Clearance

Free-float tubes are functionally identical on any rifle with a free-float tube. Said many times that the tactical-style float tubes are smaller diameter, and there's a time when that can be bothersome. When I mounted a rail section to the back-most hole set on the tube (nearest the upper receiver) to use the Brownell's alignment fixture, the screw through the rail section made contact with the gas tube with only this much protrusion. That's tight. To install the section it would be necessary to shorten this screw, which could be done after the fact using a flex-shaft hand grinder.

Float tube

This free-float front is from Yankee Hill and is far and away my favorite for a field gun. This outfit makes good things, well. I've yet to encounter a problem in either fit, form, or function with a YHM component. This model is their "customizable" tactical tube, and that refers to the threaded and well-located holes that receive varying length and format rail-top blocks or stud mounts. The diameter of the tube is smaller than that of a n NRA Match Rifle design, which myself and others think works better for bipod and hand-held firing. I like being able to position a rail segment where I want without being forced into having rails where I don't want.

This tube is easy to install (installs exactly like their carbine tube shown in that project segment) and has a locking collar to fix the forend at the desired rotational orientation. Check clearance with the lower receiver to make sure the rifle can pivot open without banging into the tube. It's possible to thread it on too far.

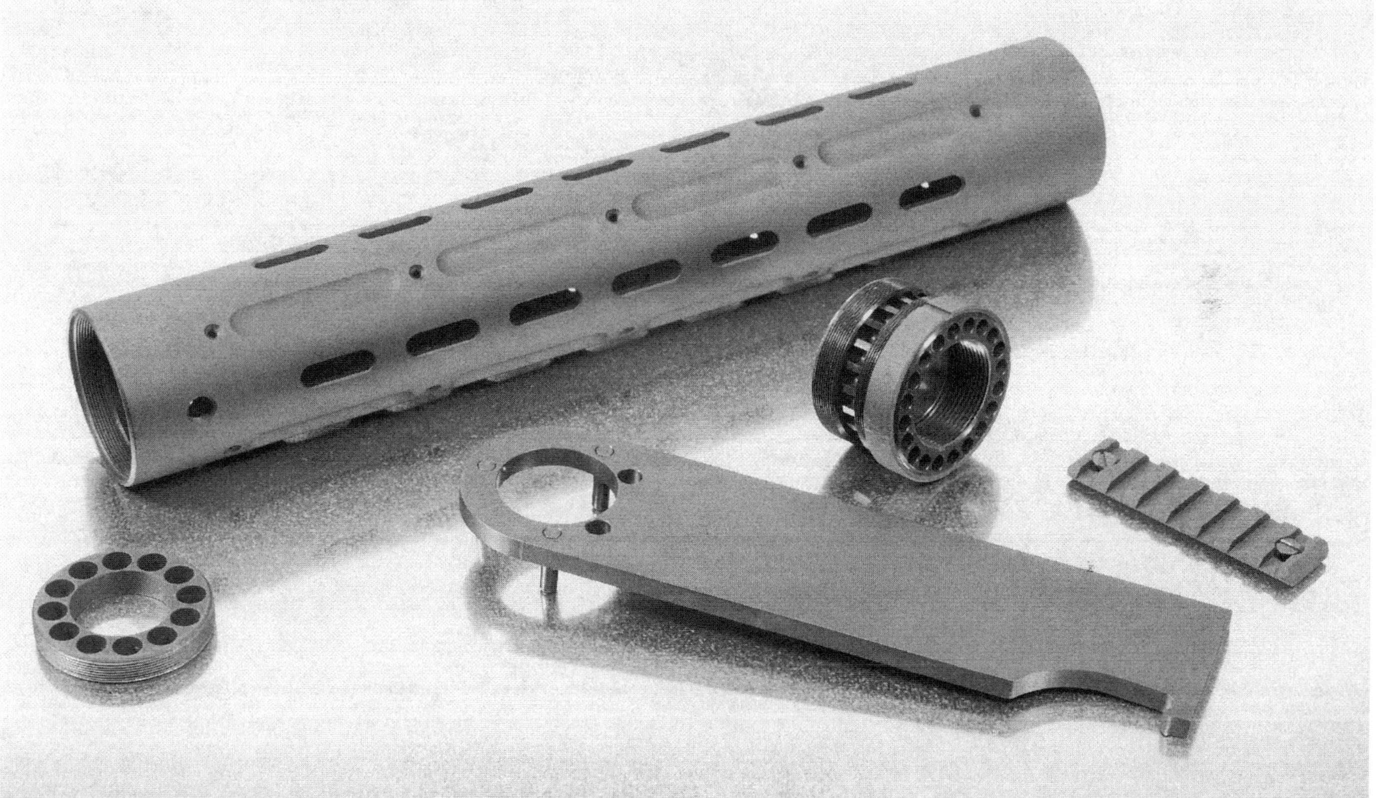

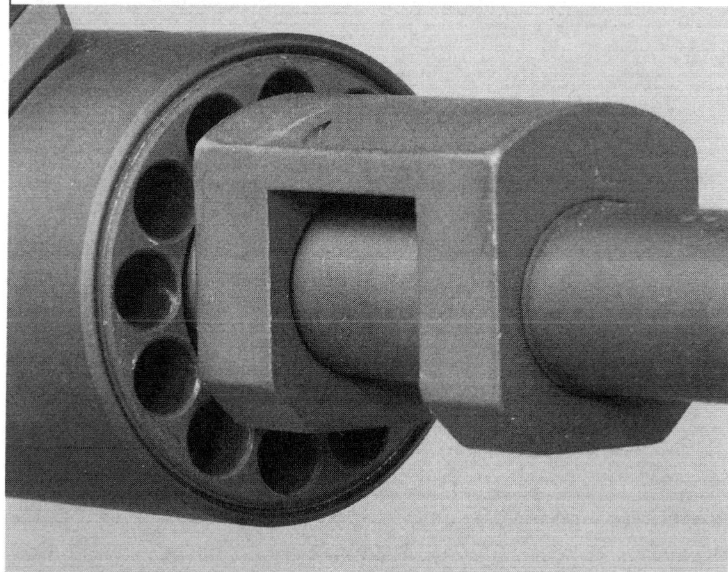

The little end cap might be bogus, but it's cool looking. No, it honestly has no function. It must be glued into place, and positioned precisely for gas tube passage. Against maker instructions, which called for "red" glue, I "wicked it" with Permatex Green. I had to "sand" the threads to get it started. That's common with fine-pitch threads in aluminum parts, and that's frequently the easy solution. Use 600- or 800-grit emery, very, very lightly.

A gas block I have always had good luck with is the YHM "low profile" because it will fit most any application (as long as its mounting diameter is compatible with the barrel's). It's a good steel part that can install under most any float tube, or certainly ahead of it. It's a good choice for an NRA Match Rifle also.

Like this one, many tactical-use tubes are smallish diameter. The split rings allow for glue-insurance under the front band, which I apply. Red.

Best on bags

The Badger Ordnance forend is unique among float tubes in that it's flat on its bottom. This sits atop sand bags way on better than a round tube ever will. It's got a full-length rail on its top and is drilled and tapped for thread-in rail pieces or sling swivel mounts on its sides and bottom.

I chose this tube for the installation example because it is so different. All the others I know of install just in the same as all the others in this book. This one is different because you can't use a conventional barrel nut wrench on it.

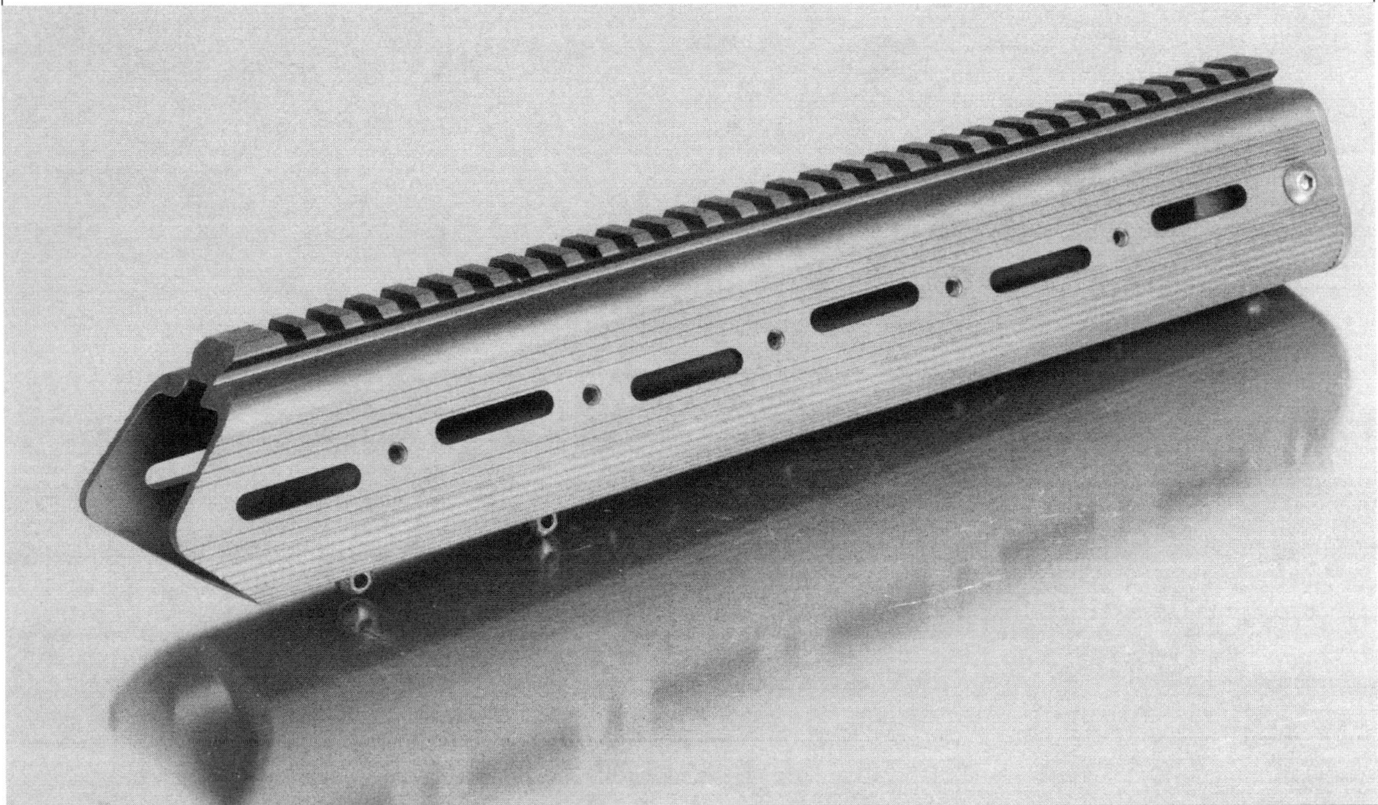

diameter ahead of the gas block shoulder. Make it 0.820 or so and I think you'll like it better. That will give you a little higher bullet speed and no weight penalty, and also, I promise, no accuracy penalty at all.

Bolt carrier assembly likewise nothing special, although chromed. It's from DPMS and AR15 configuration.

I put in a single-stage trigger because, again, that's what many tend to go with. I think a competition-style two-stage might really be better for those looking to get the effectively lightest break weight, as might someone who's looking for a full-blown prairie dog gun. The distance the trigger must be moved forward after firing to reset is considerably shorter with a single-stage.

I think a serious field shooter ought to look into incorporating some adjustment or at least fit improvement mechanisms for the buttstock. An "ultimate" varminting rifle might include a fully

adjustable stock, which means that it has a height-adjustable cheek-piece. Problem there, as was discussed in the NRA Match Rifle build segment, is finding one that has the cheekpiece located forward as it should be and still be able to use the standard charging handle. That's tough, but better is better and it probably matters more just to elevate the shooter's head position enough for comfortable viewing through the scope. Hate to be so low-budget, but even a length of 3/4-inch i.d. foam pipe insulation taped to the buttstock will provide additional height.

Since these rifles are essentially always fired from the prone shooting position, replicated by benchrest or bipod of course, some length extension at the buttplate will suit most better. Given the chance to experiment with a fully adjustable stock, anyone and everyone will find that it's for prone where they'll realize the longest stock setting.

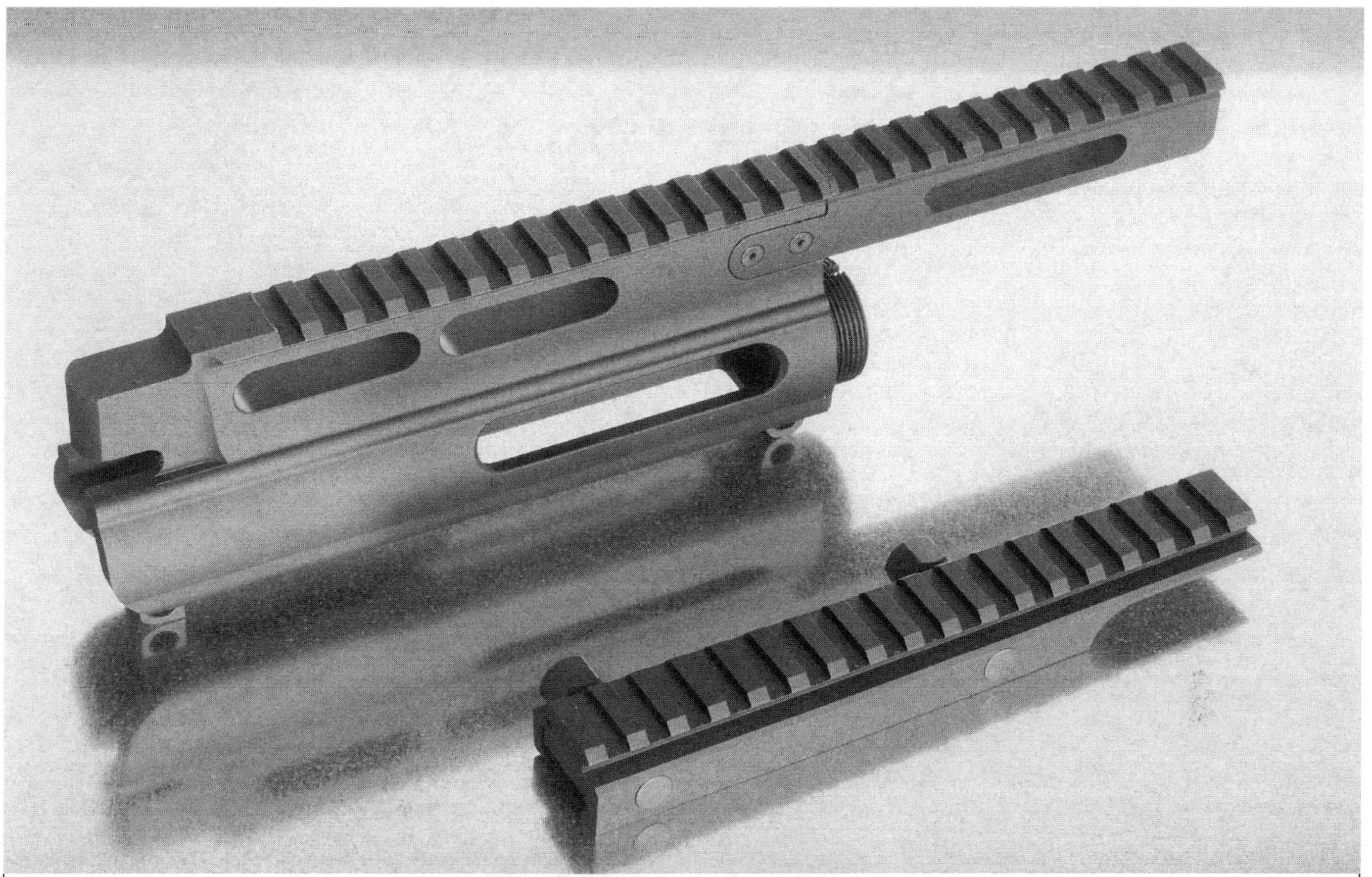

Extended rails

Good ideas. I strongly suggest going with an extended-length mounting rail for any rifle that will have a standard-size optical sight installed on its top.

Shooting from a benchrest all the time, meaning it's a seated position, there **may** be enough distance forward atop the receiver to move the scope ahead enough to have correct eye relief after attaining a natural head position on the stock. When the shooter stretches out and goes prone, though, the scope needs to go forward to prevent the shooter from unnaturally holding his head back just to see through the scope. Literally given the room to work, many will be amazed at just how far forward they'll shift the scope to be fully relaxed and comfortable.

If you want to build a rifle with what must be the honking biggest upper on the planet, here it is. This is a DPMS Hi-Rider with an extended rail. Not everyone likes these, and primarily due to the added 1/2-inch rail height. Suits me but I think requires an elevated cheekpiece to make it work best. The extended rail is decidedly worthwhile and I think should be on any such rifle built for optical sight use. The extension piece in the DPMS just unscrews and slips off for barrel installation.

Other extended rails are available, and the one from EGW also shown in the top photo is a good choice.

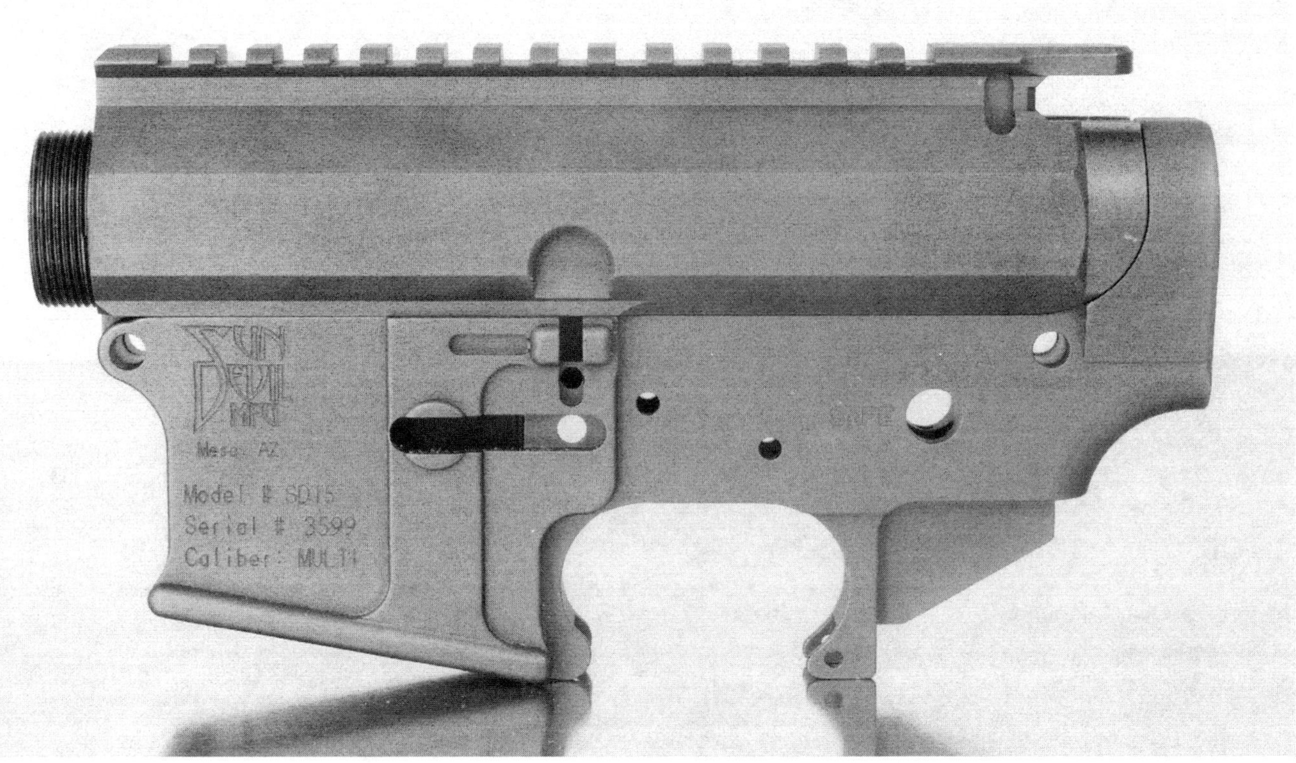

Premium parts

Here's a great choice for someone looking for a little more wahoo factor. This receiver set from Sun Devil is billet-made and essentially eliminates worries over quality correctness. It has all its holes in the right place, and that honestly is an influence on the quality of trigger installation.

Trigger, easy way

*I put a **Timney** trigger in, mostly, or at least originally, to see how it worked. It worked, so here it has stayed. It's a very good trigger, and the best of the "drop-ins" I've yet seen. I know of others I haven't tried, but this one leaves little honestly to want for. It's good. The little screw on the housing bottom is to fit the unit snugly to the lower receiver.*

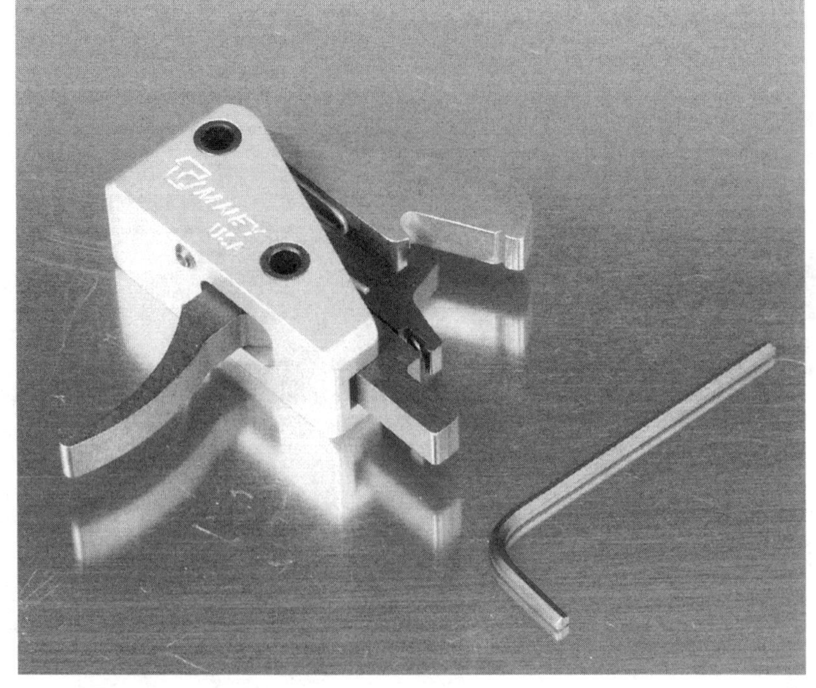

Drop-in triggers are the easiest, by far, to install. Of course they are. The trick is finding one that honestly suits. It's a "better is better" thing, and I'm famous, around my house infamous, for saying that. Anything is better than stock. I think you won't be disappointed with the Timney as long as, as is true with all the drop-ins I know of, you can like and live with what they give you. For competition shooting, I want more because I like to rig them myself, and I'll say the same there, and have: there are two I'll use and those are the only two I'll recommend. They are the only two that have the quality and capacity to be fully and finely tuned to satisfy most anyone after installation and without machine shop ops. Yes, there are people who are never happy.

One huge help with any drop-in trigger, or any trigger that doesn't come with its own precision-made set, is the installation of aftermarket trigger and hammer pins. A locking set from KNS ensures this trigger will be its best.

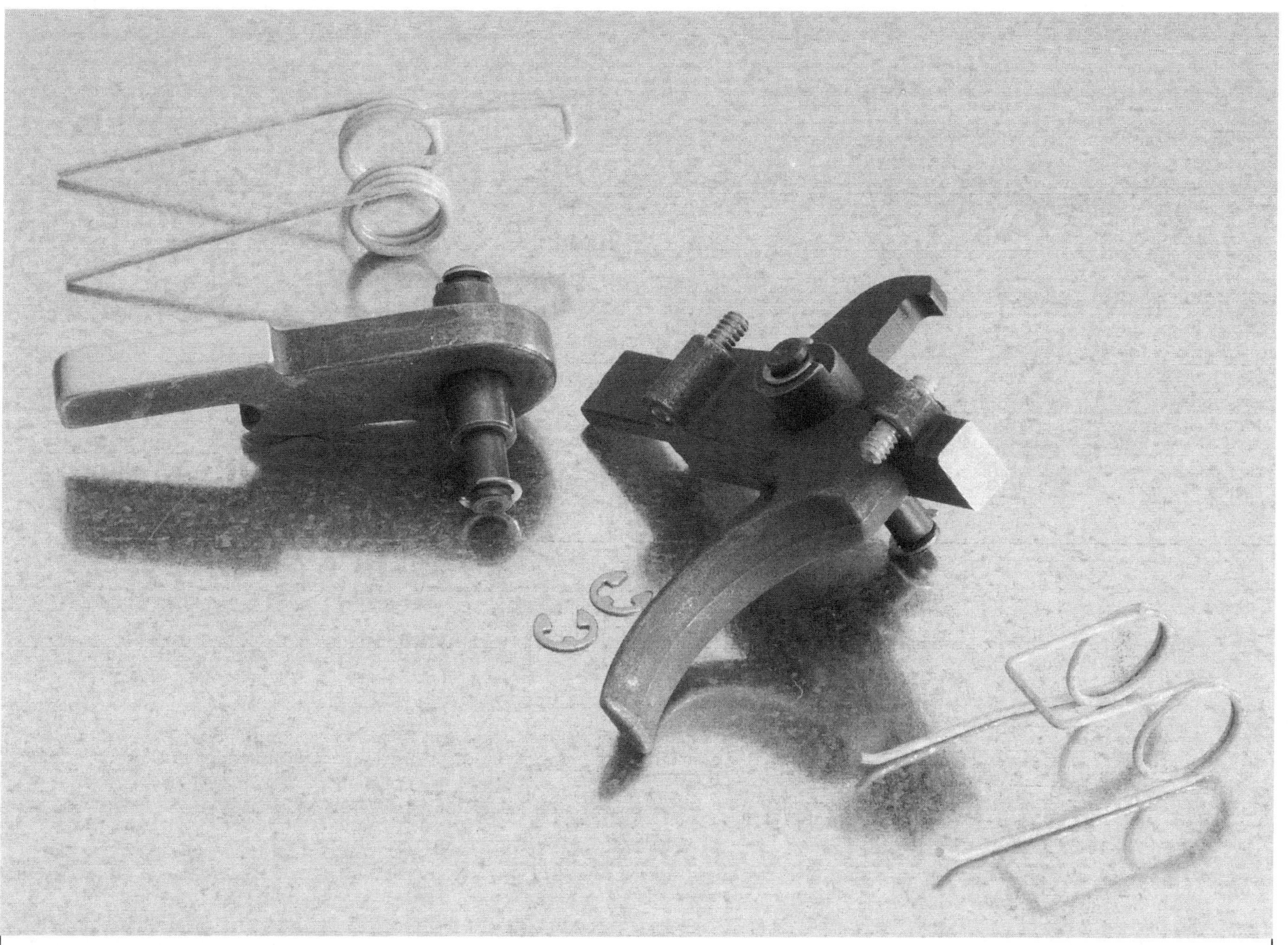

Trigger, better way

Another, and perhaps best, choice in a single-stage is JP Ent.'s. This is a fully-adjustable trigger set with the capacity to deliver a low break weight. It's also fast to ignition thanks, mostly, to its lightweight hammer.

I don't suggest messing with spring replacements on this trigger. Extra-power hammer springs and the like aren't going to ultimately accomplish anything. It's a balanced set, if that makes any sense, and it's well balanced. And make sure the trigger return spring has the little bend on its right-hand leg. That's necessary for correct clearance.

JP supplies a DVD instruction disc that honestly precludes reason for me to elaborate. I can say that I followed instructions and am pleased with the results. All I can add is to keep that trigger heavily lubricated. After installation mine break at 12 ounces. Despite the overwhelming preference I have for two-stage triggers on the firing line, I like the single-stage from the bench. I can't really say why, but I've tried it both ways and return to the single every time on a rifle used in that venue.

Notice the little circlips in the photo. They notch into corresponding grooves on the trigger and hammer pins after installation. Make sure they notch! Believe me, you don't want to try to find one on a shop floor.

*JP has a **modular trigger** shown floating in space here on this page. I've not used one but have confidence it will work just fine. As with any, every, and all triggers, especially modulars, a good set of correctly fitting trigger and hammer pins is a real key to getting the most from it, and locking-style pins are especially beneficial to drop-ins. All the pieces are in place in these type trigger systems, we just have to keep them there.*

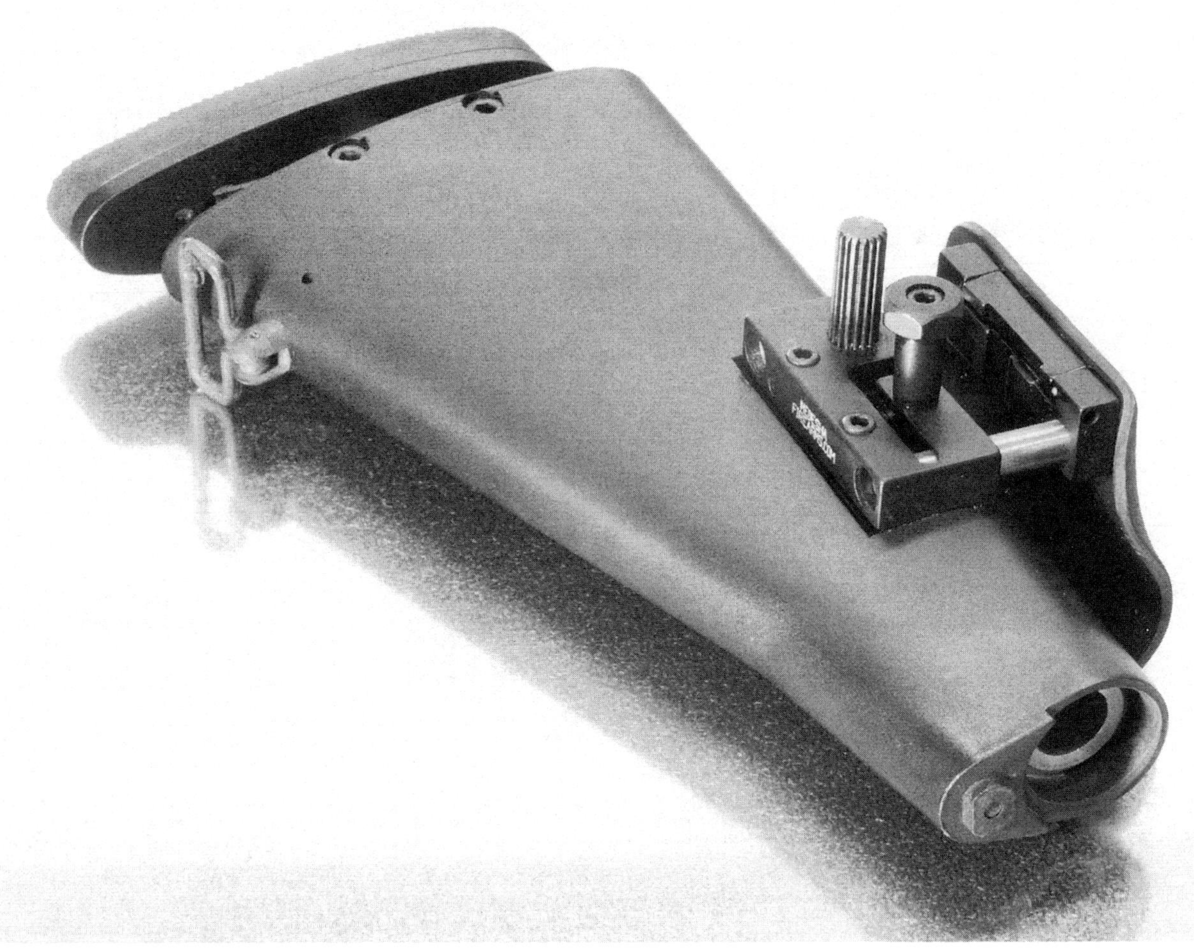

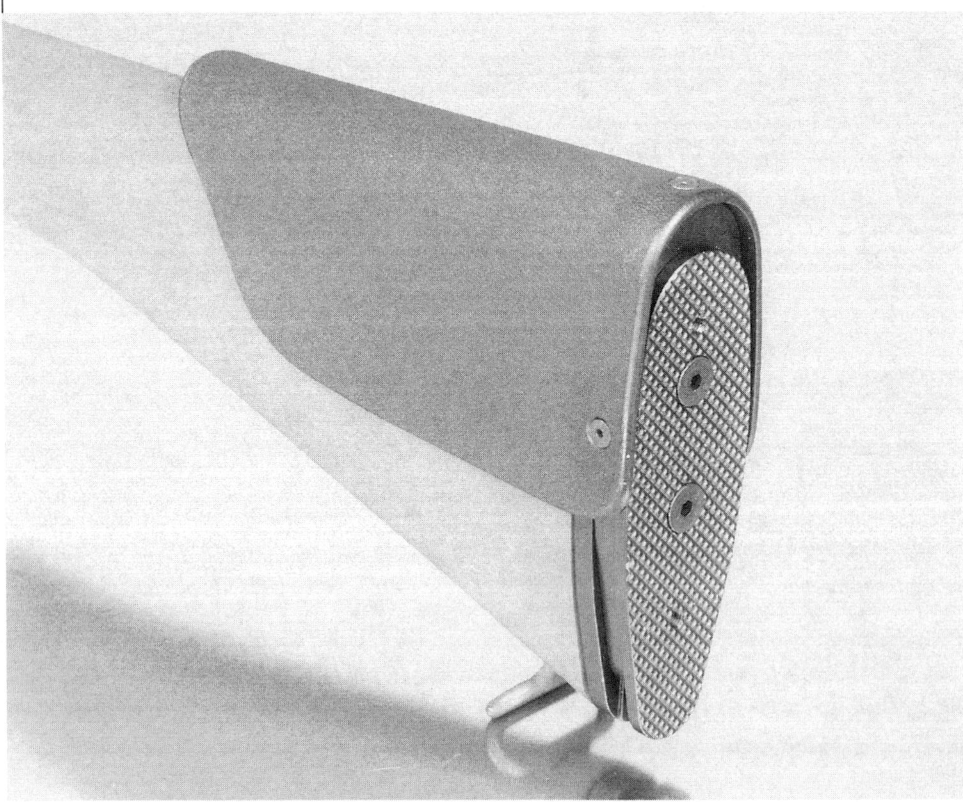

Field rifle furniture.

Even though fully adjustable stocks are most often associated with position shooting, it's better than good to get any rifle to fit you. Medesha has had an adjustable stock that carries the same hardware as the one shown in the NRA Match Rifle build segment, but with a conventional appearance and sling swivel location.

Otherwise, I would suggest at least adding a cheekpiece like this shown from Bell & Carlson to improve your head position when using a scope.

Unless you go with a fully-adjustable stock, though, it's going to be harder to get all you want, if that includes buttplate flexibility and a higher cheekpiece because, no, most single parts will not work together.

COMPONENT COMPATIBILITY

This segment is as good a place as any to address a very important point, and issue, with selecting parts for your build. Said elsewhere, but hammered on here, not all parts from all suppliers will work together.

It is way wise to select as many parts from the same manufacturer as you can, and that's especially true with the forend and barrel parts. It's time consuming and costly (even if it's just shipping being paid) to find out that the gas block you wanted to use won't fit under the free-float tube you wanted to use. This is a particular problem with tactical-application parts. I've experienced particular frustrations, as you might have guessed, mating gas manifolds to forend tubes. This is a bigger problem, to be sure, when you want an extra-long float tube to cover the gas block. Sometimes the problems are as simple, but equally frustrating, as finding out that set screw access is blocked after a tube is installed, and then that the tube has to be installed before the part with the set screw can slip into place.

The easiest way to make sure your parts are going to be compatible is, again, get them with the same brand name. Even that is not an assurance because some manufacturers offer radically different takes on components they produce. Ask before you buy.

I've also more frequently run into parts that won't work with the increasingly available "slab-sided" upper receivers. I will not and cannot blame accessory manufacturers for designing their parts around common-form receivers. That only makes sense. There are a few accessory parts you might encounter that won't work with upper or lower receivers that have a different structure or contour than mil-spec. I've run into that with extended bolt stop releases and even a free-float tube.

Here's a funny one, well it wasn't "ha-ha" funny, but glaring oversights befall all. I got that DPMS Hi-Rider Extended Rail upper all ready for the Badger Ordnance float tube, got it installed, and then realized that the forend wouldn't mate up with the upper because the extended portion of the extended rail on the upper wouldn't clear the first few rows of rail cuts in the Badger float tube. Gack. So I replaced that upper with a Sun Devil and it's all nice like, except now I have to add an extended rail. For reasons, I'm not 100-percent secure with attaching the front scope mount on the forend and the back on the upper receiver rail. Anyhow, that's the upper sitting in the vise on assembly because I just didn't want to shoot the same photos over again.

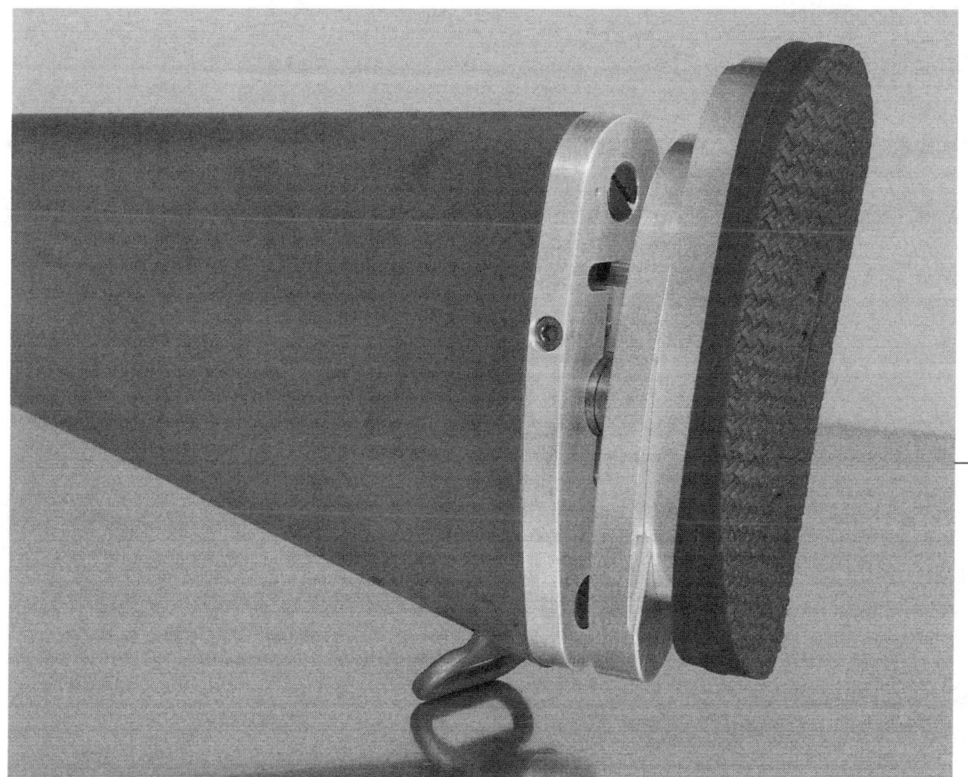

Buttplate

If you want an adjustable buttplate that's truly adjustable, this would be it. David Tubb adapted his 4-way buttplate to an AR15 stock shell cap, and it's the best of its kind on the market. Rotation, height, length, and cast.

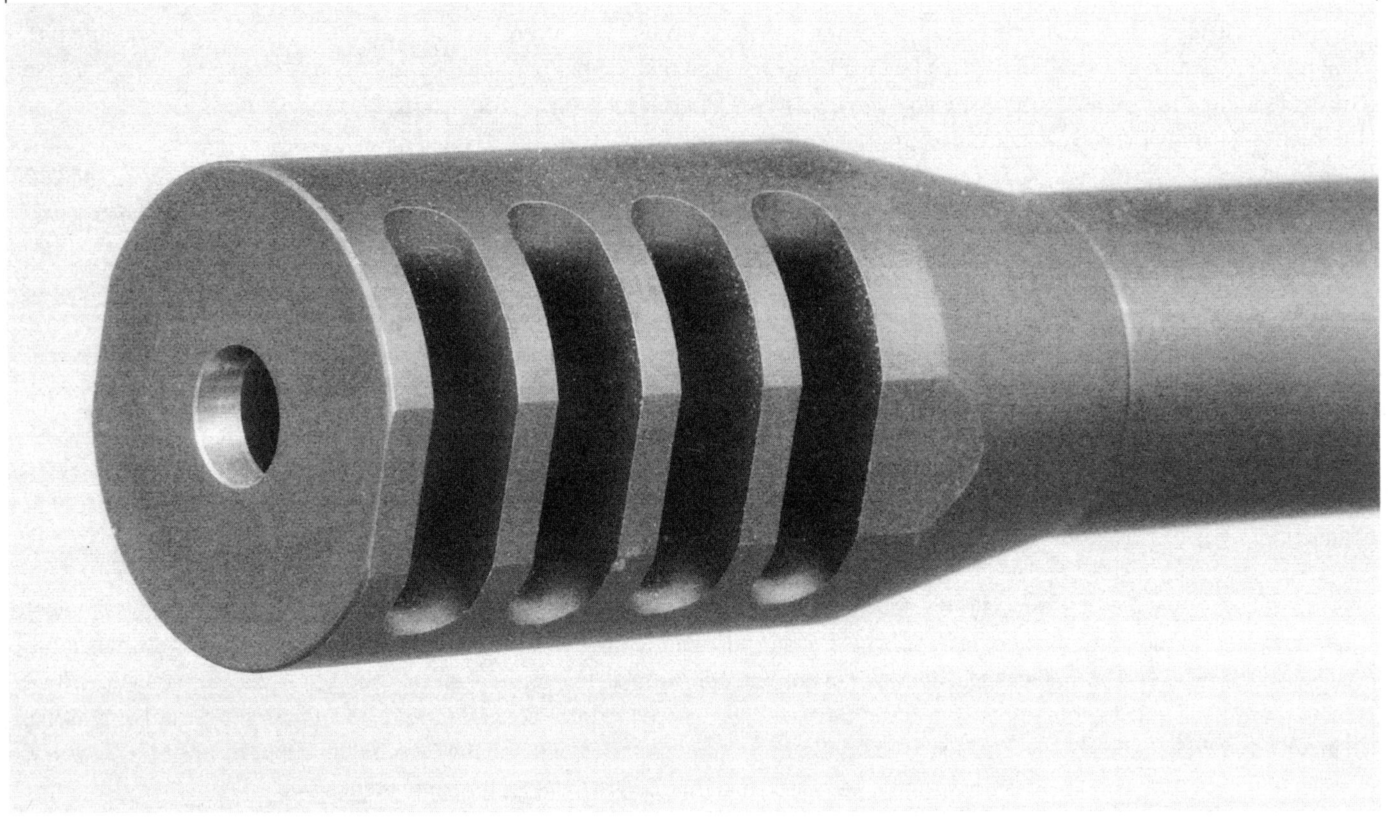

PARTS & TOOLS

I'm going to skip the routine list of parts and tools to assemble the lower receiver (this upper needed nothing), and, depending on components you choose, there could be differences certainly in what you'll need to finish the rifle. For the most part, all the tools necessary to install a free-float tube on any of the other projects shown will apply here. This one is a little different. I had to find an adjustable wrench with a 2-1/4 inch jaw opening. Auto parts store had it. Wasn't cheap.

PREPARATION

Vise, very securely mounted

Upper receiver clamp

Tap hammer

Roll pin punch and roll pin starter punch (for gas tube roll pin)

Torque wrench, 1/2-inch drive

Breaker bar, 1/2-inch drive

Barrel nut wrench attachment for the above

Gas tube alignment tool

Lower receiver block (for vise mounting)

Options (good ideas)

Permatex "red" adhesive

Permatex "blue" adhesive

Contact cleaner

Anti-seize compound (I like Loctite C5A)

Gas tube wrench

PRECAUTIONS

Test fit everything! Make sure all your parts work together.

Get up with a high-quality set of screwdriver blades and hex-head wrenches (for installing accessories)

22.0 THE BUILD

A LITTLE DIFFERENT, BUT NOT REALLY

Installation of this float-tube is different. When I first looked at it I thought I was going to be in for a tedious session, but it actually was amazingly easy to get gas tube alignment. The only trick is that you have to know you're installing it with an adequate amount of torque. If you've done installations of more conventional float tubes or barrel nuts before you will have a pretty good idea, but if not it's easy to imagine many people not applying enough cinch against the barrel. I can tell you to get it a little tighter than you think it should be and there should be no worries. I found a huge adjustable wrench at a local auto parts store and also respected the amount of leverage it could assist me with. Again, pull pretty hard on the wrench handle and it will be on tight enough. Keep in mind also what's been said elsewhere about barrel nut torque. The 35-foot-pounds is a minimum, and that what any installation requires is that plus attaining gas tube and gas tube opening alignment. That always means going in the "tighten" direction to finish. And that means that even better than double the 35 ft.lbs. can be necessary. Do not worry about installing a barrel nut too tightly. The mistake is made going in the other direction.

ROUTINE (WITH A CATCH)

Vise it up. As with other builds in this book, the DPMS upper will not fit into the clamshell-style receiver clamp so it's necessary to use the "plug-in" style alternative. As also said a few times and places, after using this type block, you may find you like it better. It's faster to get the upper in and out because it stays in the vise. Something I didn't mention but will now is that the easiest way to work this holder is to fix the front pivot pin and then lower the back of the receiver down to fix the rear takedown pin in place. It's harder to line up both holes if you pin it after it's flush down on the block.

Same old, same old. Degrease the extension and inside the upper, apply anti-seize to the upper receiver threads, apply glue to the barrel extension, position the barrel into the receiver.

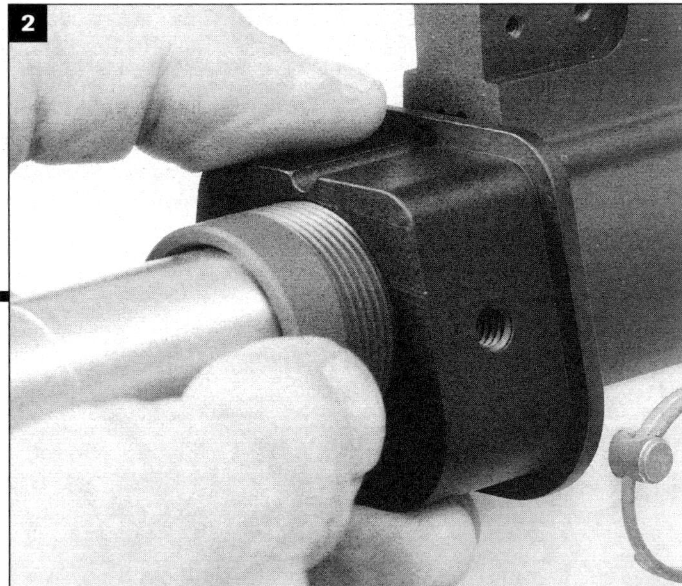

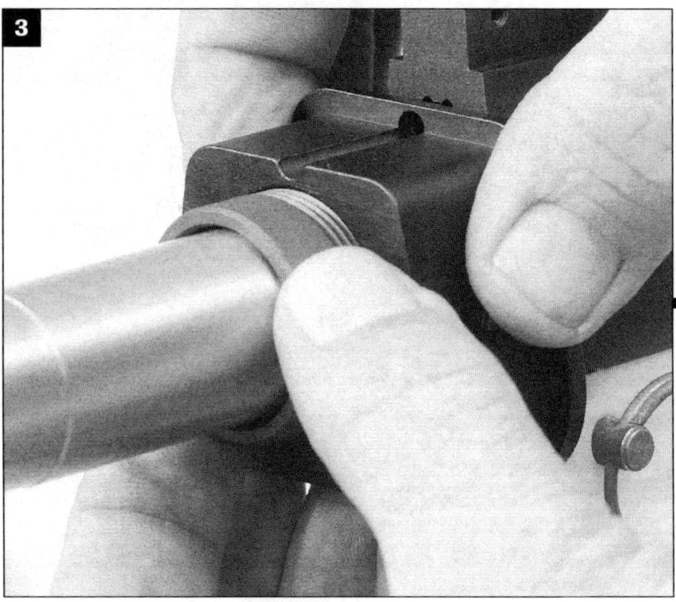

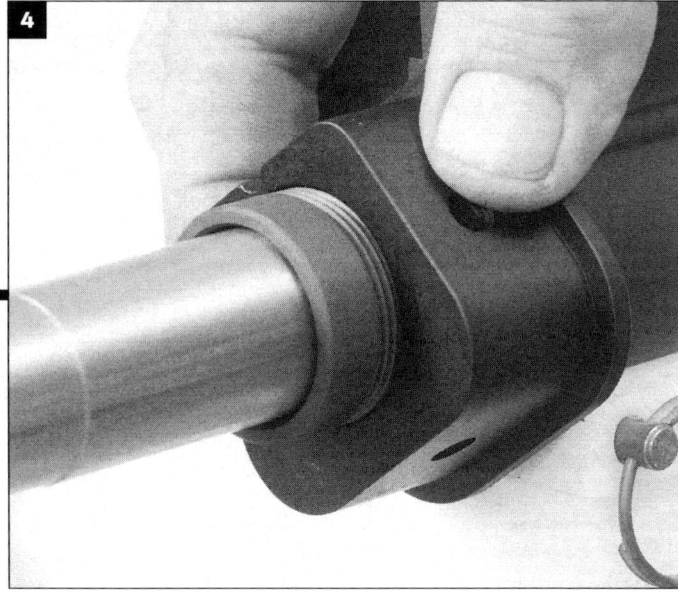

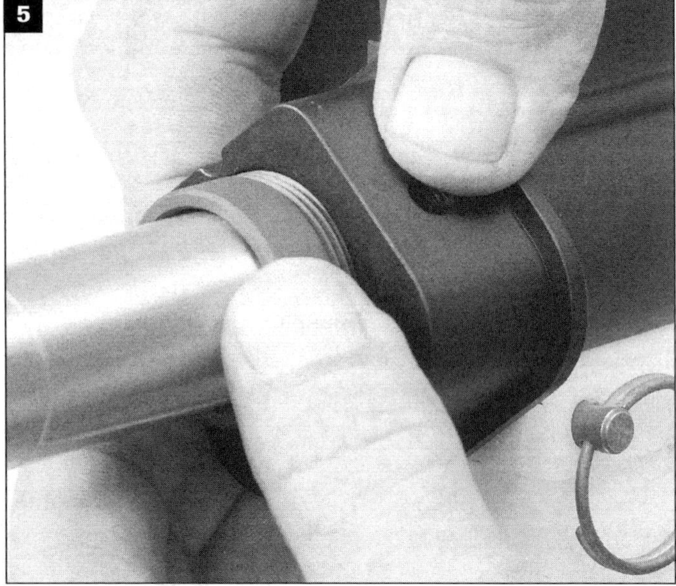

1. Thread the mounting block onto the receiver all the way. Then back it off until the gas hole aligns with the cutout in the top of the block. Make the gap between upper receiver and rear of the block as small as possible.

2./3. Next thread in the barrel nut piece and turn it in until it contacts the barrel extension shoulder.

4./5. Now, turn the mounting block in the "loosen" direction about as shown, and then, while holding it in place, thread the barrel nut in a little more to ensure it is contacting the barrel extension shoulder.

6. No, I said get the **big** wrench! And then tighten that bad boy down until you, one, know you've attained at least 35-foot-pounds, and, of course, two, the gas tube holes are aligned.

BARREL & FLOAT TUBE

This may or may not happen easily for you. Just keep at it until it does. It's actually easy in a way because you can control how the pieces wrench down and avoid excessive the torque necessary as it can be with other installations.

7./8. As always, make sure gas tube alignment is perfect. Check with the gas tube alignment tool, then with the bolt carrier, and then with the gas tube itself. Oh, and by the way, this is another installation where it's fine and wise to go ahead and install the gas tube into the gas manifold.

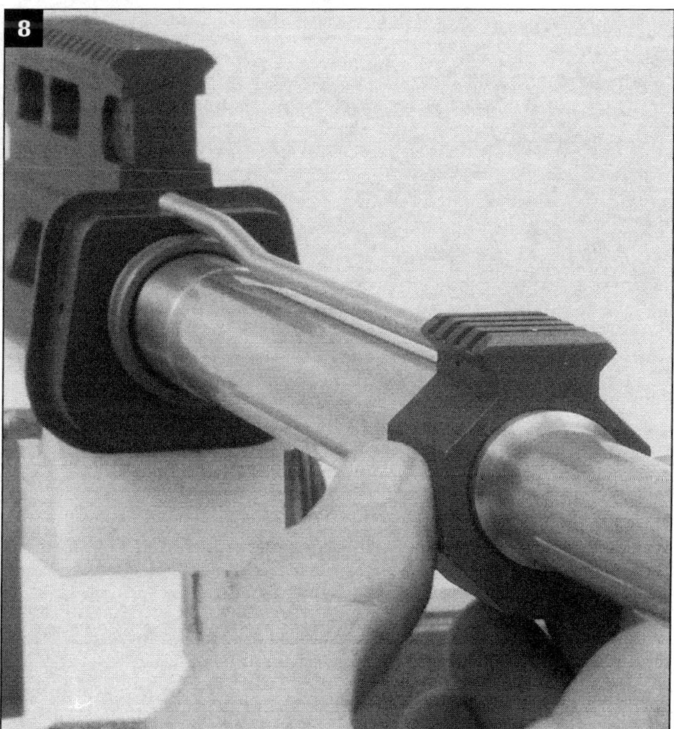

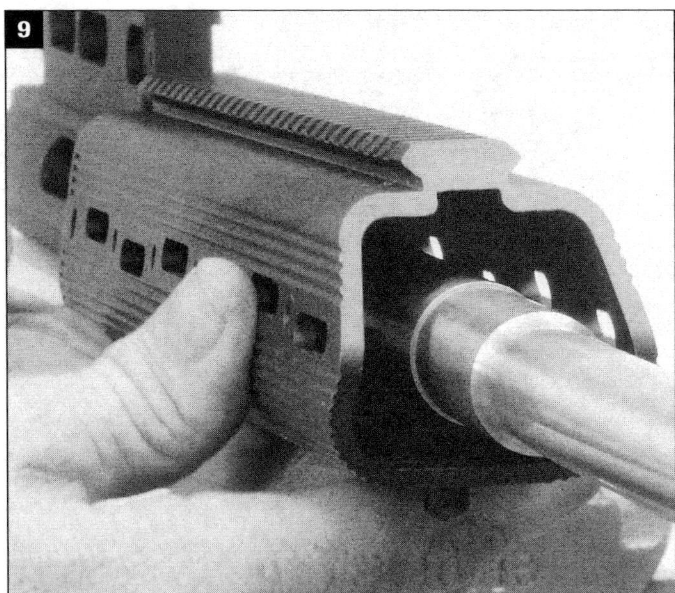 

9. Slip the forend over the barrel and into place on the mounting block. Tighten the screws that hold the forend to the mounting block **securely**. Use glue ("blue" is enough). This is a relatively heavy forend tube.

10. Do the same for the gas manifold.

11. I used a Brownell's rail alignment fixture to set the manifold into place. Make sure the manifold is all the way back against the barrel shoulder and install its ("red" glued) set screws

FOREND & GAS MANIFOLD

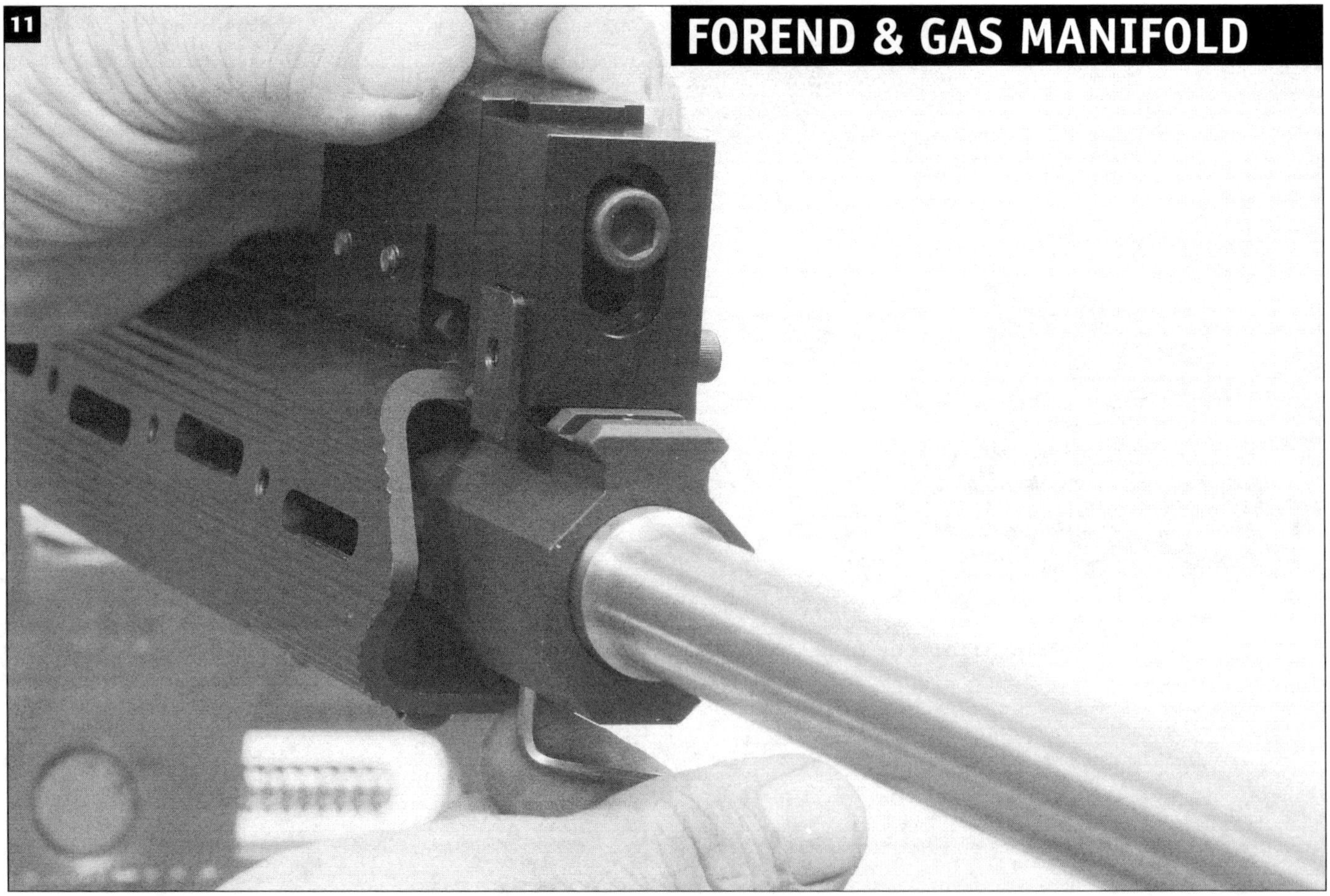

23.0 MAGAZINES
REBUILDS

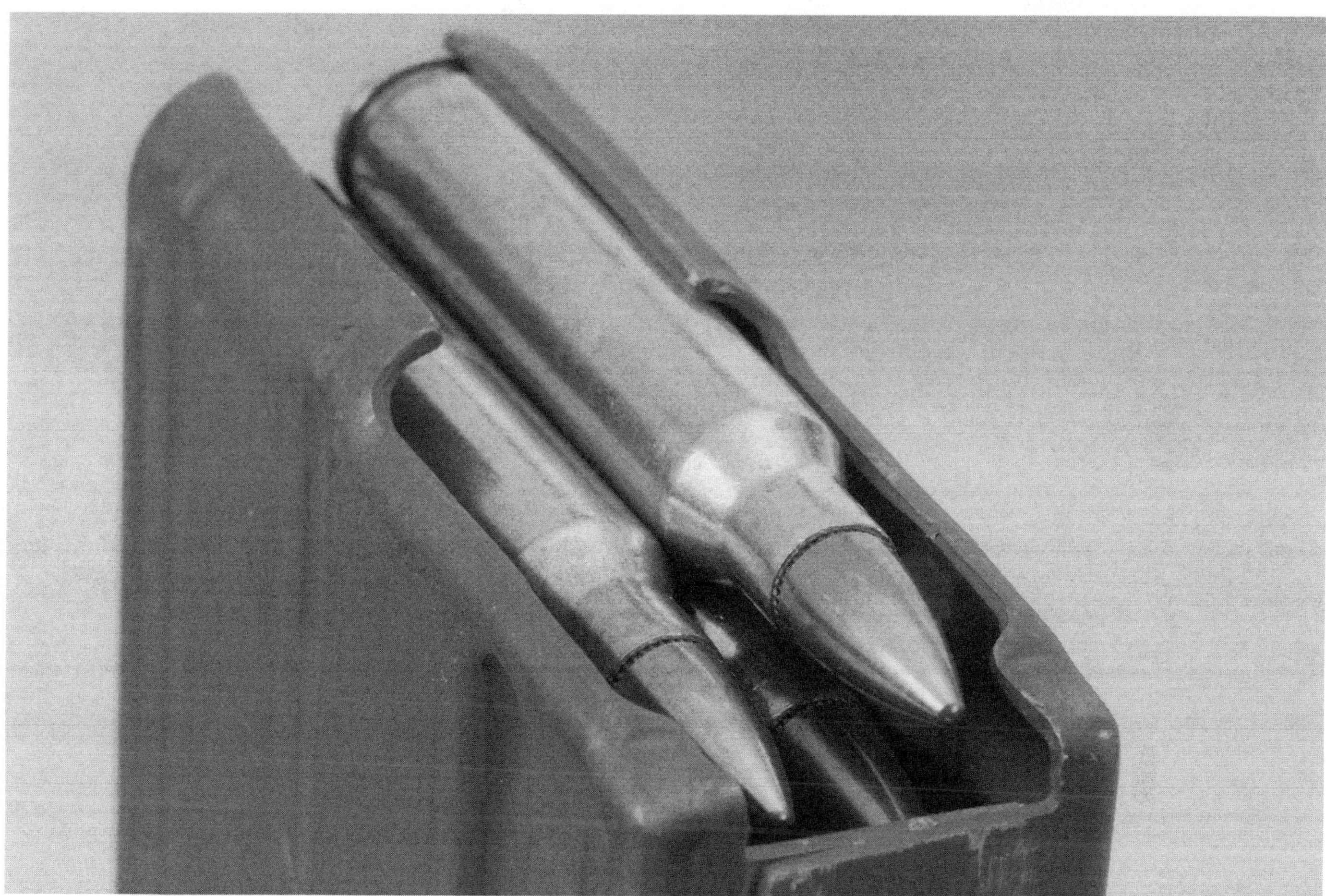

[In years (long years) past my answer to a magazine that didn't work was to throw it away. Now they can be fixed, and I'm holding a belief now that they should be fixed even if they are not faulty. The performance enhancements from improved followers and springs are worth having.]

If it's not broken don't fix it, but if a magazine is not working then your AR15 can't feed itself. It may be possible to fix a magazine until it is broken, but not likely.

SEGMENT CONTENT

251   **Essentials** (choices)

252   **Followers**

253   **Springs**

254   **Base Plates**

255   **Top Coil Orientation**

257   **Base Plate Removal**

258   **Rebuild Process** (installing a new follower and spring)

ESSENTIALS

Any magazine that works all the time is a good magazine. That, of course, is the trick. The world is full of both kinds, and getting good ones is relatively assured dealing with name-brand suppliers. Brownell's, for good instance, hasn't let me down. You can get a magazine from them with a good follower, correct body, and chrome silicon spring all ready to go for a fair price. That's all we need. The "super duty" magazines out there are indeed good. I don't advise against them but also don't use them. As said, if a magazine works all the time it's a good magazine.

If you have no compelling need for the extra rounds, I'm a big believer in 20-round configuration magazines. That box size is necessarily used in Service Rifle competition (30s are allowable but

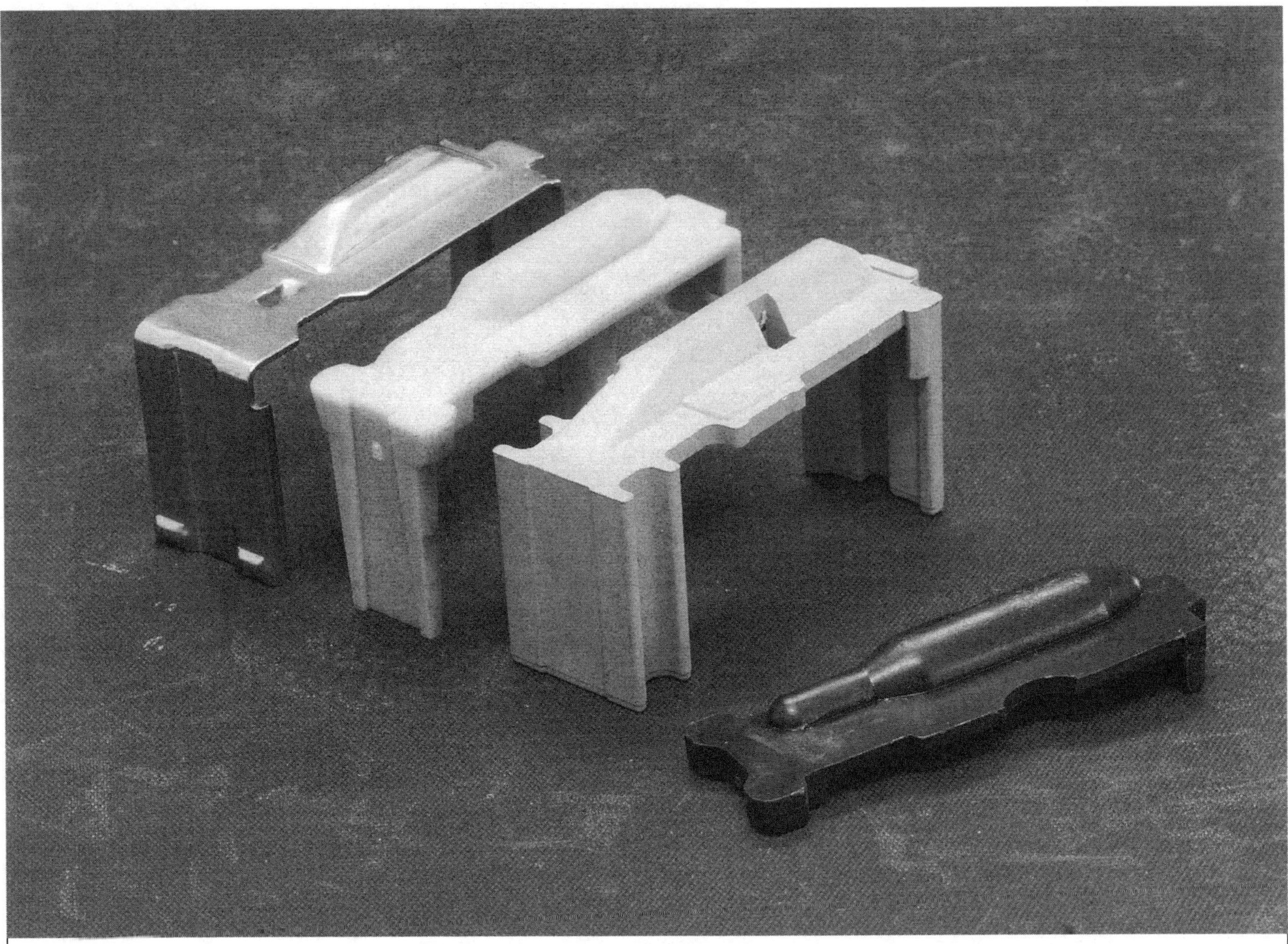

These help

Better followers come from our own gumment-mandated redesign as well as aftermarket engineers. (Left to right: stainless follower from Brownell's; "green" follower also from Brownell's; Magpul; old-school design in plastic, avoid this one...) The Magpul followers in particular make it nearly impossible to stick a follower into anything but its correct flat orientation. The "green" followers, which are available for 20s and 30s, add some support to the front and back to help allay rocking, as does the stainless steel part. Put any of those three under a good spring and you should have a magazine that works.

awkward) but is also a great combination of length to aid removal and insertion and round capacity, in my opinion, and, in my experience, are less problematic than higher capacity magazines. Shorter magazines have their place, and one of those places is to comply with NRA High Power Rifle rules regarding standing event "palm rest" dimensions. I've encountered shorter boxes that don't seem to work as well as the 20s. I am no engineer but believe it has to do with the spring pressure level and consistency along with the straight-walled body on the 20s. It was, after all, the original form.

FOLLOWERS

What normally causes a magazine-fault malfunction is follower orientation, which is ostensibly round orientation, and namely it's the front of the follower (or the cartridge bullet) sticking "down." This is easily solved, and especially so if more than one solution is thrown at

it. And why not? An improved, compared to original form, follower that's sitting atop a correct or corrected spring will do it.

The deal behind the "green" follower the armed forces went to was to solve a common problem of failures to feed the last two rounds from some bum 30-round magazines. The change was to add a skirt to the front and extra length to the skirt at the rear to help stop the follower from tilting. These ride against the inside front and rear box walls. This follower is better.

There are other followers that followed the "green" idea and added additional surface area, which is what it really amounts to, to front and back. They are all better than the original form.

Most of the improved followers have to be "snapped" into place, and installation is not hardly always a breeze. It can be difficult, sometimes not possible, on older magazines, and we'll talk about that shortly.

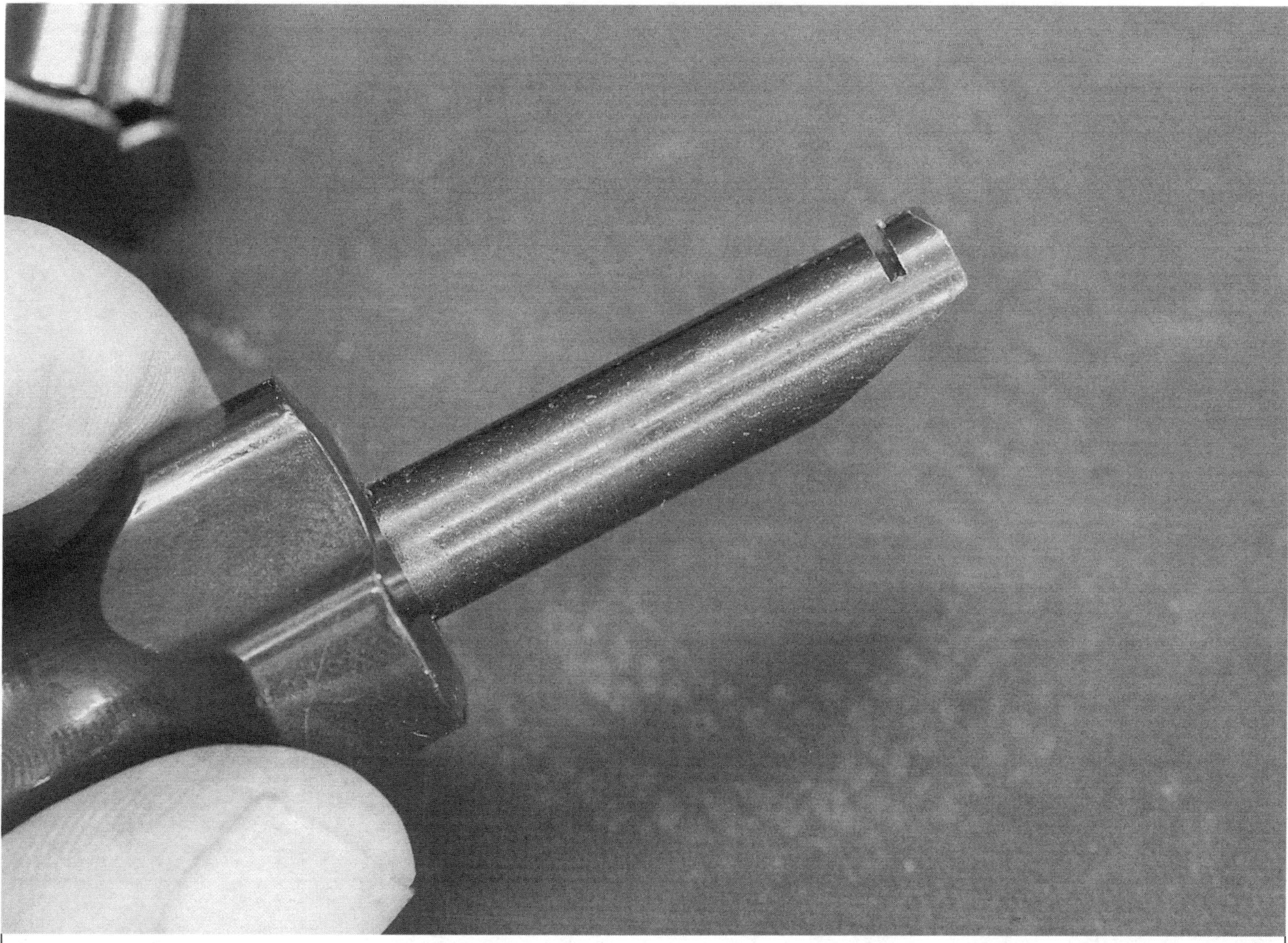

Don't use pliers...

More serious problems can sometimes be solved using a feed-lip tool from Brownell's. It's the tool for the job, and the only tool for the job. Pliers aren't the right tool for the job. This has a simple function to help make the lips what they should be in case they're bent or unequal. The feed lips mostly control how and how much the round pivots upward as it's starting into the chamber. Call it the "release." Aluminum magazines crack easily, so don't get greedy on any one bend. As mentioned elsewhere, I can use this tool also to "adjust" the tabs on the bottom of the magazine that the base plate slips under. Another application is tweaking on the latch cut-out on the magazine body, should that ever be deemed necessary.

SPRINGS

I say, however, that there are fewer follower problems and more spring problems. A box magazine is spring-driven, and there are differences in springs. Chrome silicon (CS), when exploited to its capacity by the spring maker, is always better than music wire. CS magazine springs put a more consistent, constant load on the follower and, this is big, just don't change. They can be left sitting compressed without getting "soft." This is a huge concern and potentially disastrous circumstance for a tactical pro. If you want to check a spring for adequate load, put two rounds in the magazine and push them down only about one-eighth-inch. They should spring right and fully back up. If they don't the spring is sacked.

The correct orientation for the follower atop the spring (when neither is installed in the box) is either dead level or with the bullet-end sitting a little higher than the back end. Dead level is good. If you install enough different springs in enough magazines, you'll notice that not all springs will do this. You have to help. Pliers are the tool and the idea is to get the follower sitting flat or a little nose-up. That's done by tweaking on the top coil. It's common in many that I've disassembled to see the follower nose down when it's installed onto the spring. When it gets back into the box and buttoned up, that orientation can return. Even if the spring is crammed in there and it seems that there's plenty of pressure against the follower to keep it up and sitting level against its stops, when the rounds are dealt in and the follower goes down, the shift or tilt happens.

While you have the box empty, clean it out.

Any roughness on the inside front and back can be smoothed away using emery. Rub a little under the edges of the feed lips also. Don't put a cutting edge on them, just smooth them out.

All this gets detailed starting on the next page.

Base Plates

The first thing that has to happen to work on a magazine, or work a magazine over, is getting the base plate off.

The pair of magazines shown on the right have latching mechanisms, the Brownell's mag shown under that photo doesn't. Doesn't is easier.

It can be a chore installing a new-age follower into an old-school magazine (bottom-left photo). The hang up is the base-plate latching piece. Most newer magazines I've seen, including all those I had on hand, either skipped the latching mechanism altogether or incorporated a detached piece that attached via the bottom spring coil. When the bottom of the magazine is open, the newer followers install pretty easy. The original-type follower designs have nothing protruding downward on the bullet-end of the piece so it's easy to work it in around the spring-style latch. Hate to give this advice, so let's not call it advice and go with "factual reality," but the little spring clip on, for example, an original Colt-brand magazine, has to be removed. I don't think the base-plate really needs to be latched into place, but that assurance is, well, assuring.

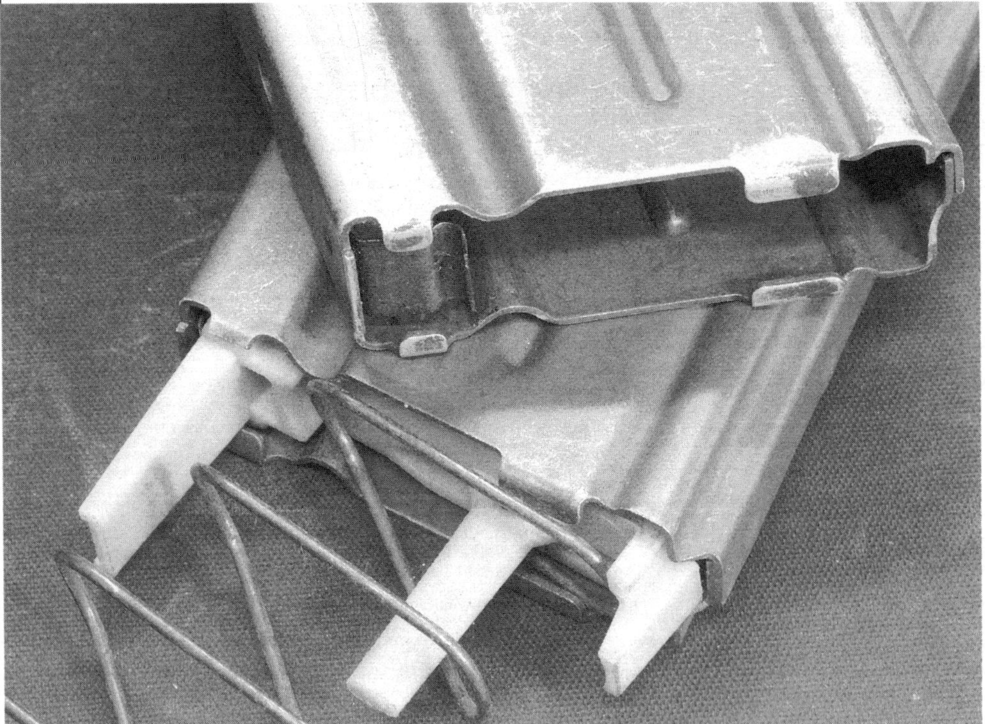

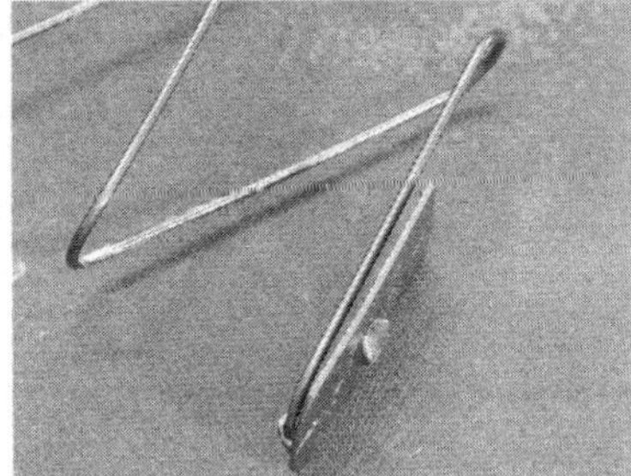

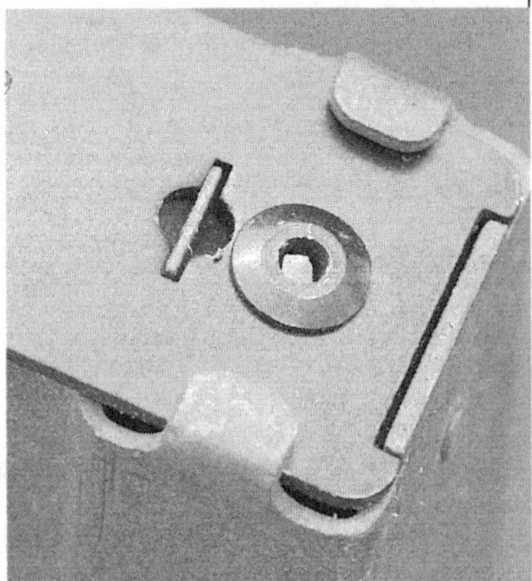

Point is that there's no way I've yet found to install a "green" follower, for example, into a Colt, or Colt-clone, magazine box without getting the plate latching spring out of the picture. I've used a Brownell's AR15 feel lip tool to add a little extra tension to the fit between the plate and the four tabs it slips under. That, plus the magazine spring tension itself against the plate, will keep the plate in place. I've yet to have one come off in use.

The top small photo on the right shows a spring-mounted latch. The one underneath it is a base-plate that's been "permanently" installed by riveting the latch to the plate. Can't advise on what to do if you want to take this apart, but looks like about a quarter-inch diameter.

Top coil tech

*The top coil orientation is critical. That's the one that loops into the fol-
lower. The turned-in end of the top of the coil should be level with the
straight section on the opposite side of the spring. If it's angled down, it
shouldn't be.*

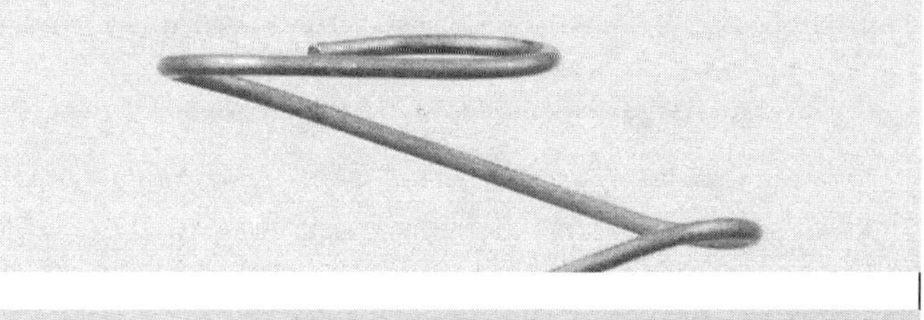

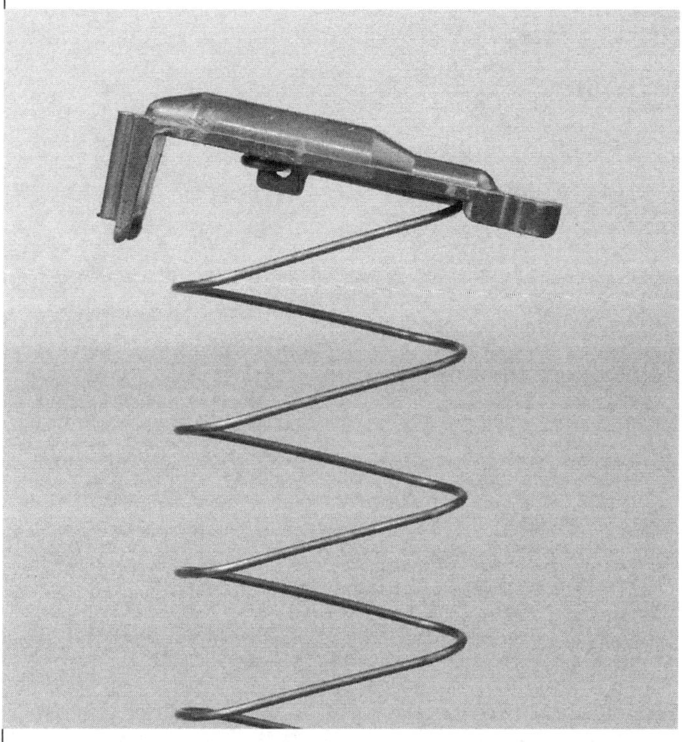

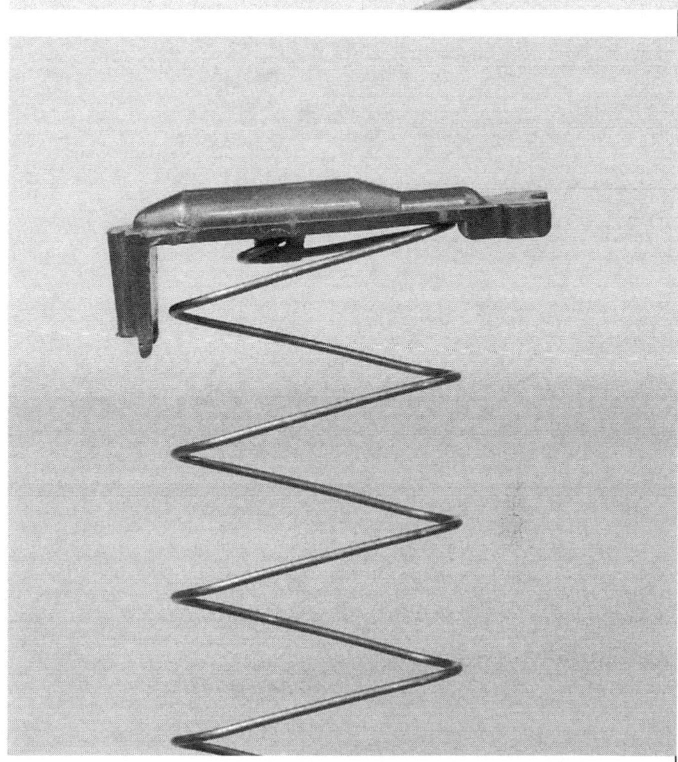

*This one is ugly, but common. It's dead nose down. On right it is fixed, plus a little. Bending the wire to support the follower at its nose goes a long
way toward preventing it from tipping that direction.*

Here's another pair: the one on the left is sitting pretty; the other isn't. Pliers!

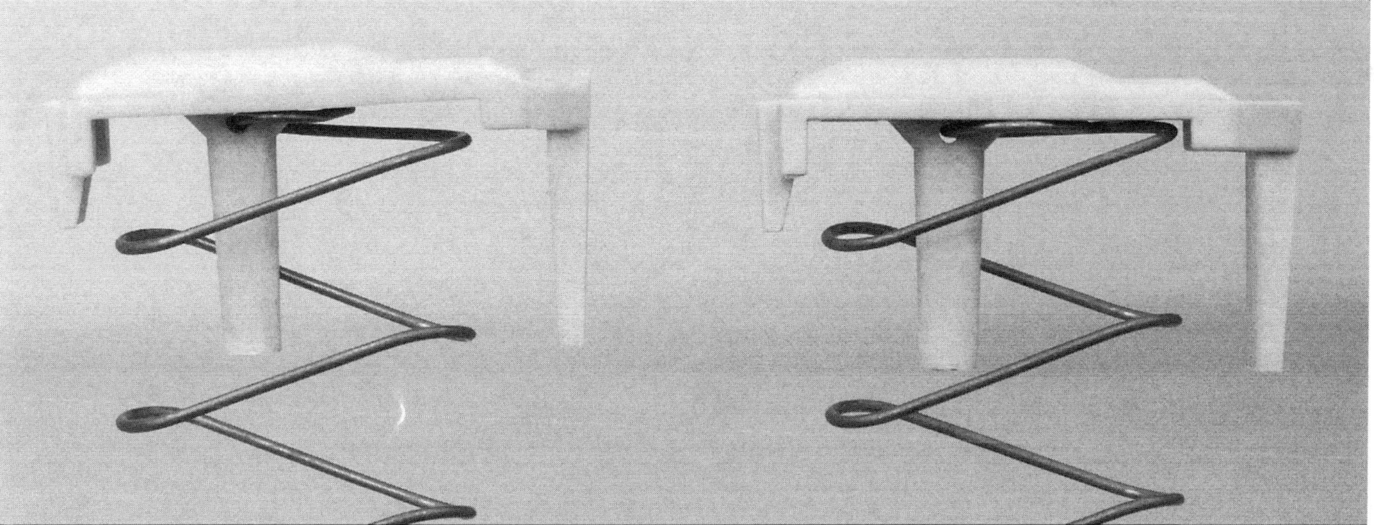

Test it

The test, or my test, for a magazine prior to range firing is pushing the follower down as far as I can and letting it up quickly. It
ought to smoothly return to its original orientation. Next push the front end (bullet side) down. It should release level. If the front
end sticks down, that's a malfunction waiting to happen. Check the follower orientation on the spring and adjust as needed.

Some accessory followers might have the spring loop attachment eyelet located in different positions.

The first loop on the spring (white arrow) should support the front end of the follower (bullet-side). Here's how it looks from an original Colt-brand.

Most followers will use a standard mil-spec location for the spring attachment eyelet, and that will be exactly compatible with most accessory springs. I've seen followers, though, that literally put this eyelet on either extreme end, and the photo on the right is one of those. Its spring was designed to accommodate this location, but also notice the really poor job it's doing of supporting the front end of the follower. It's dead nose down

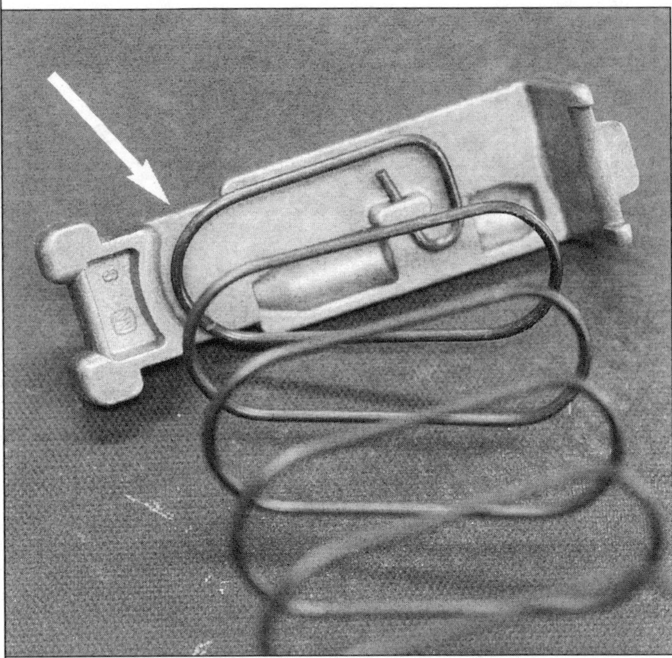
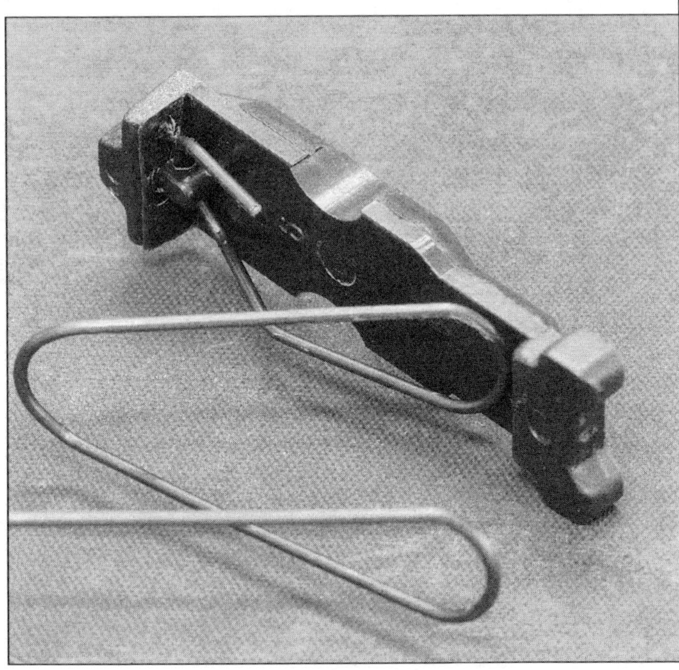

and no way to fix it with a spring tweak. This was the follower and spring set that came out of the "project magazine." Don't be afraid to bend on a spring if you have to, but that's not always necessary as long as the spring can be oriented correctly.

Speaking of springs. I really am a believer in chrome silicon magazine springs. They don't "sack" like conventional springs can, and actually can and do improve magazine function. The load against the follower is more consistent due to the nature of the compression and rebound of this material.

PARTS & TOOLS

I'll assume you're going to replace a magazine spring and follower, or at least one of those pieces.

PARTS
Magazine box
Follower
Spring
Baseplate

PREPARATION
Punch (for base-plate removal)
Pliers

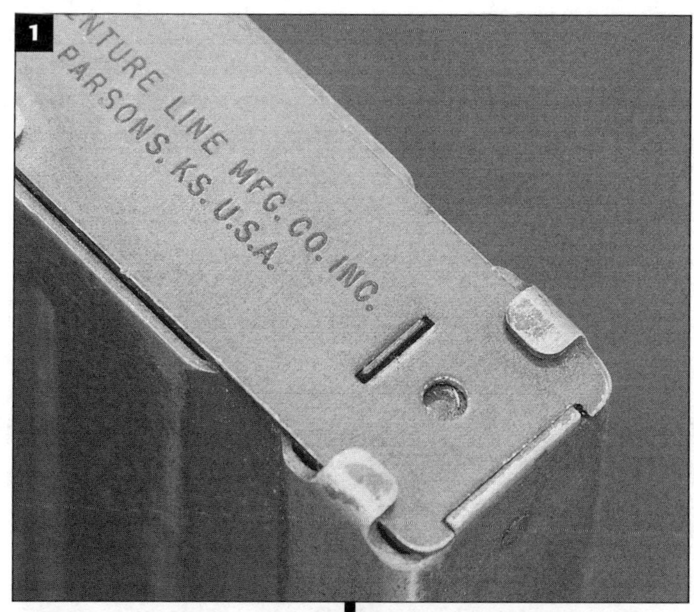

BASEPLATE (CONVENTIONAL)

Here's a very old GI-style aluminum 20-round magazine, and here's how to take it apart.

1./2./3.: Locate the latch on the base plate, depress the latch using a punch or similar until the little latching tab is fully disengaged from its slot, and push the plate forward so the latch stays down.

4. Keep the coil spring compressed while you remove the plate. It's unlikely that a spring will fly out from one of these because it's wider than the tabs the base plate rides under, but use caution.

Remove the spring by rocking it left and right as you pull it clear. The follower can then be extracted.

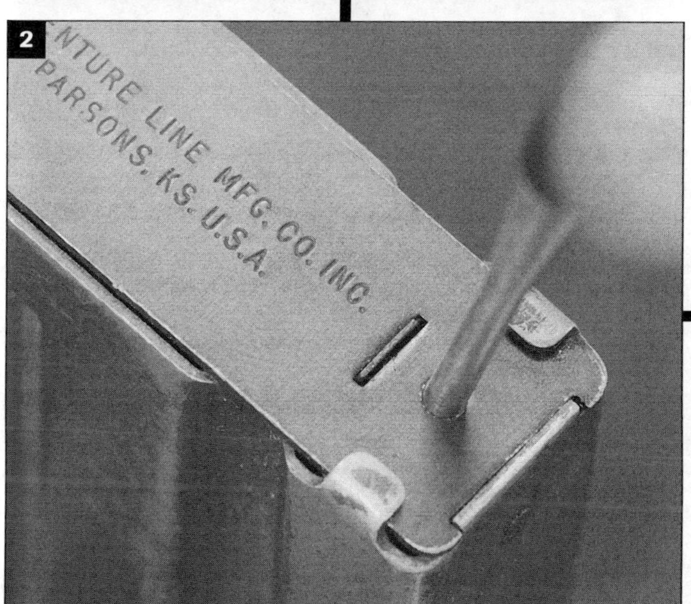

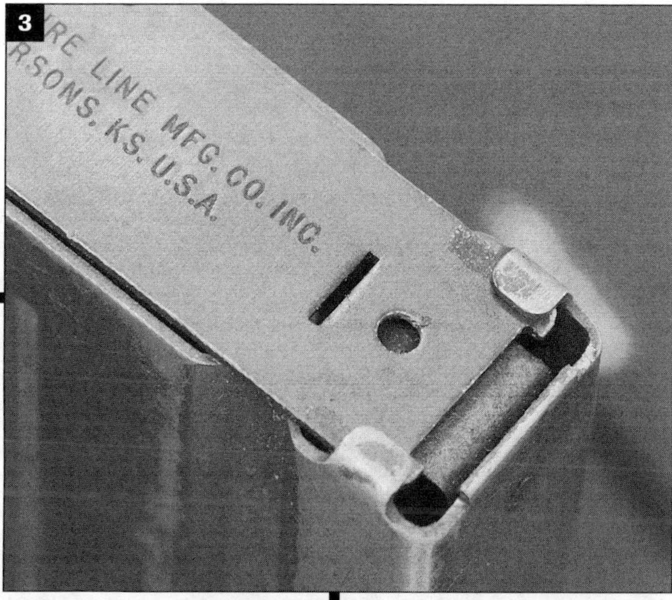

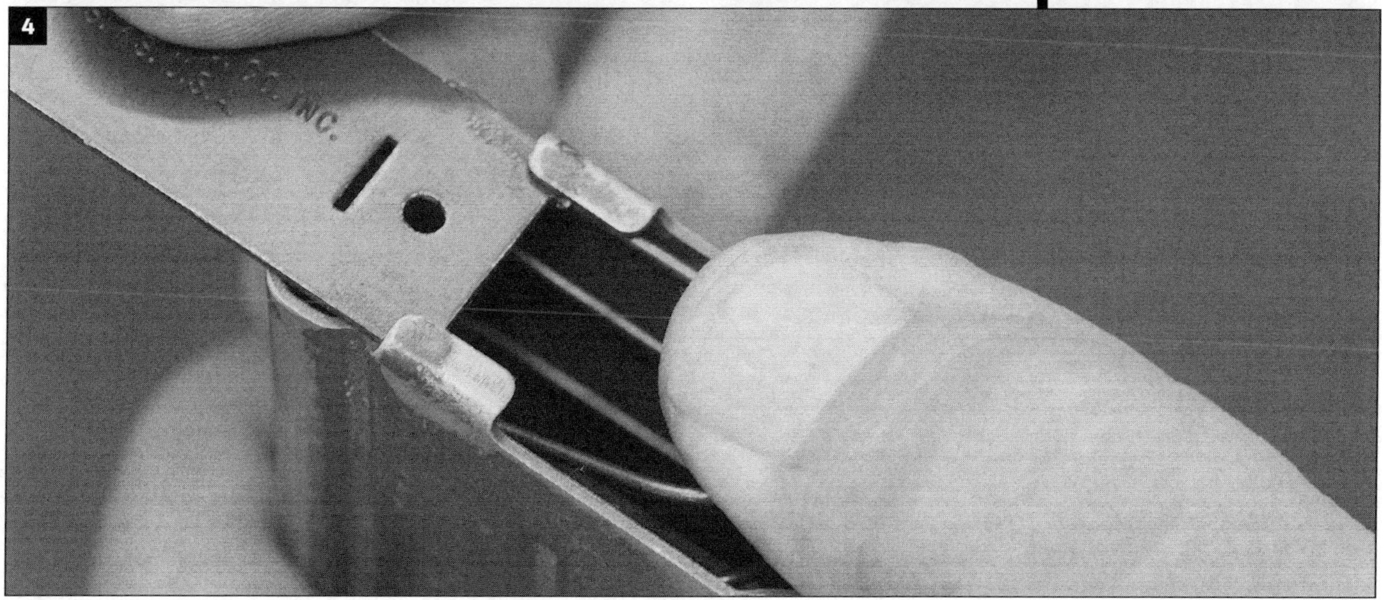

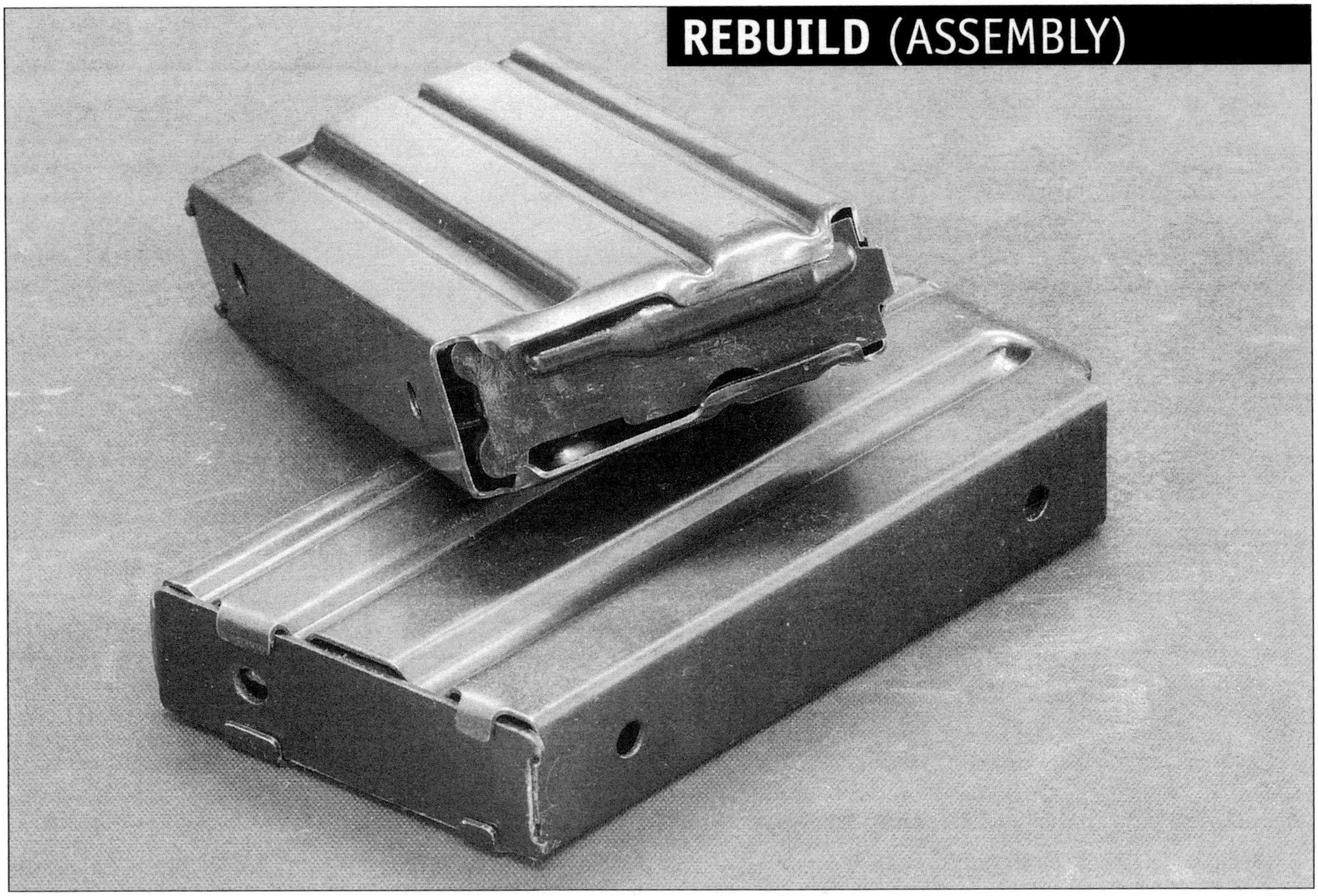

Here's a good example, or at least the best I had to offer, of a magazine in need of help. This is a brand new, just out of its baggie, steel-bodied standard-configuration 20-round with an old-style follower.

This little project is to replace spring and follower and give a better shot at giving service.

Releasing the plate latch was easy: just poke it down with a punch. Removing the base plate on this one was a bear. There was an extreme amount of tension between the tabs and plate, so much so I actually had to tap it off after inserting a punch into the hole on the base plate. I later "tuned" the friction-fit between the tabs and plate by bending the tabs just a little upward. I don't like hammering on magazines.

I chose a chrome silicon spring from Superior Shooting Systems Inc. and a stainless steel follower from Brownell's. All good.

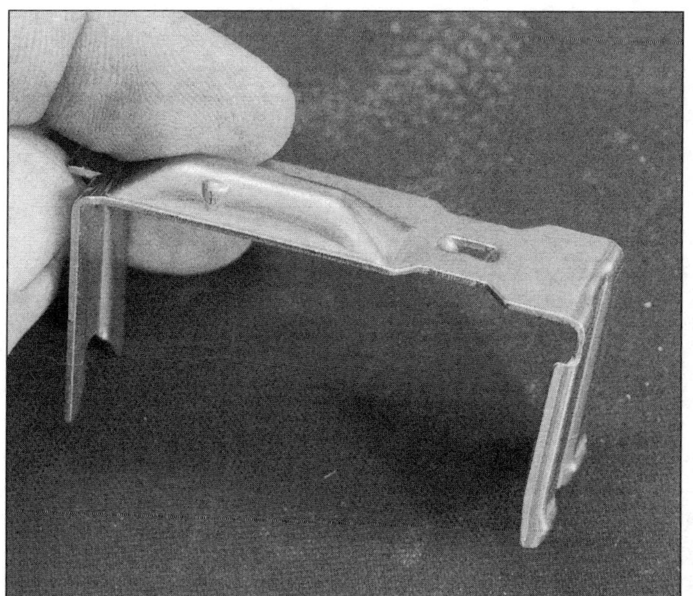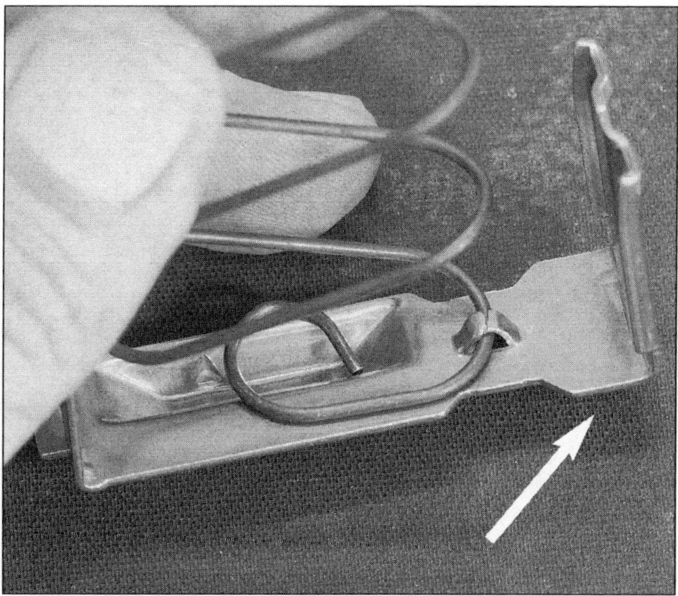

The eyelet location on this follower is not standard. Notice the top loop orientation between spring and follower. I had to wind it around this way to get the correct orientation for the top loop (toward the bullet-tip end of the follower, white arrow) and also along with that the correct orientation for the bottom of the spring (bottom photo).

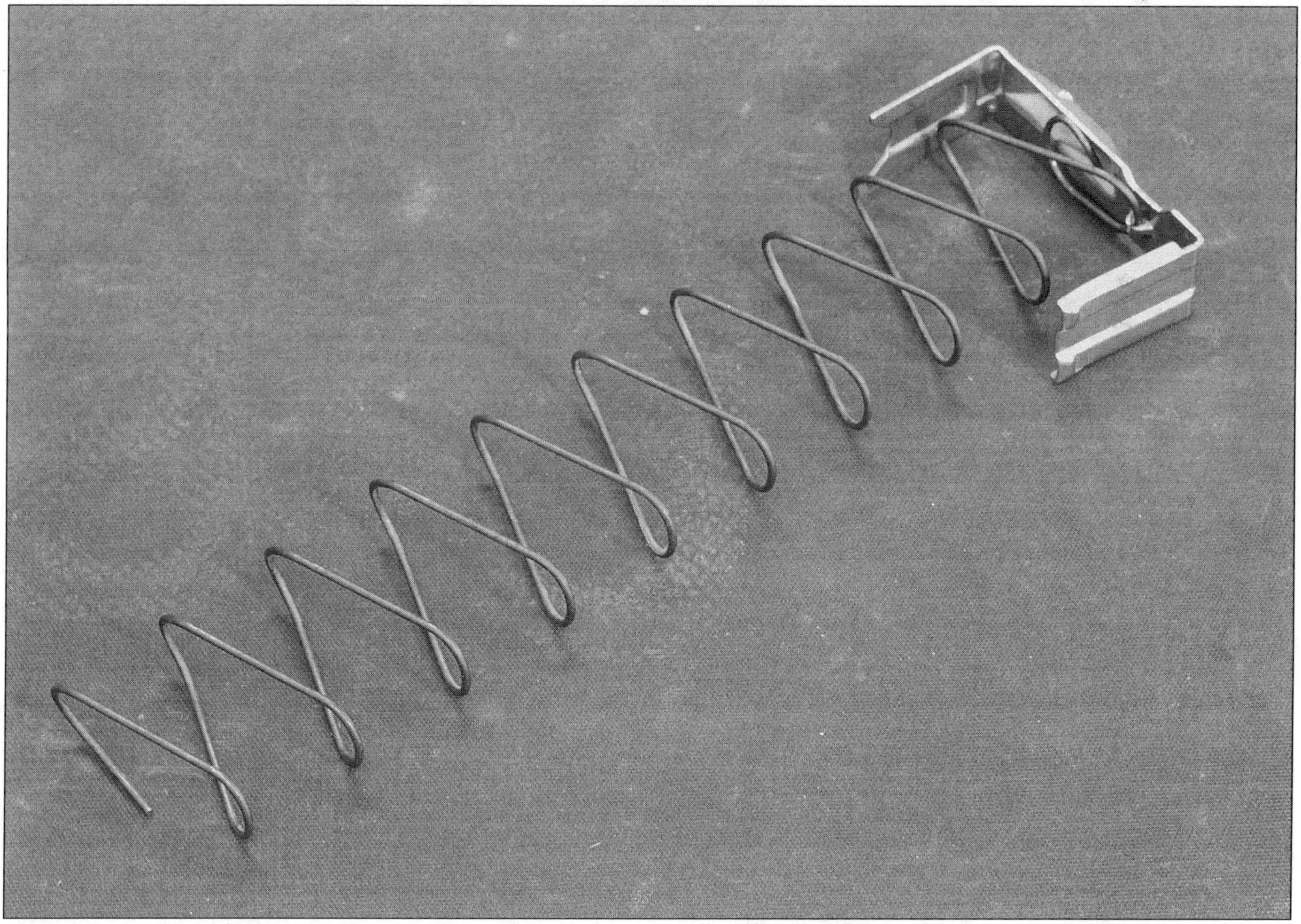

[**On a plastic follower**, a little flex and twist with a pair of pliers is the best way to connect the spring to the eyelet. Sometimes, though, it may be necessary to bend the spring loop a bit to get it installed. Pinch it back afterward.]

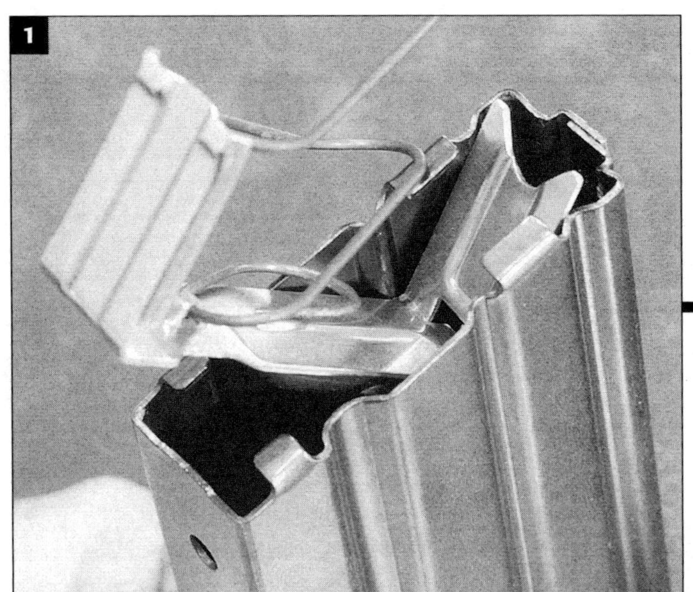

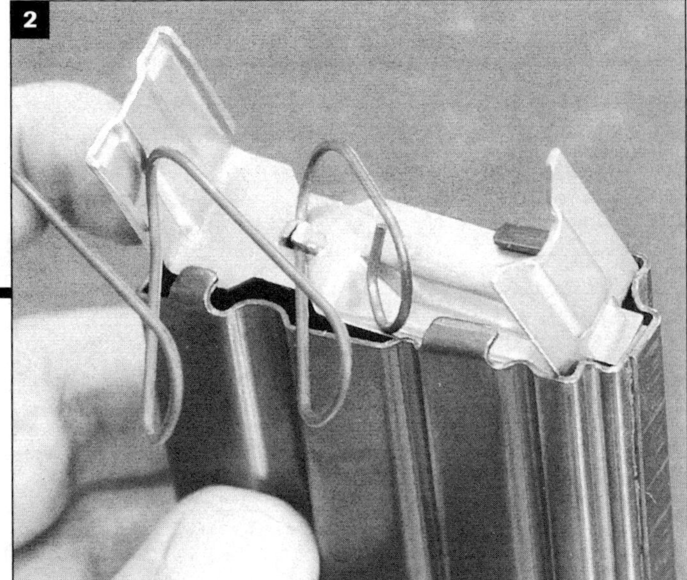

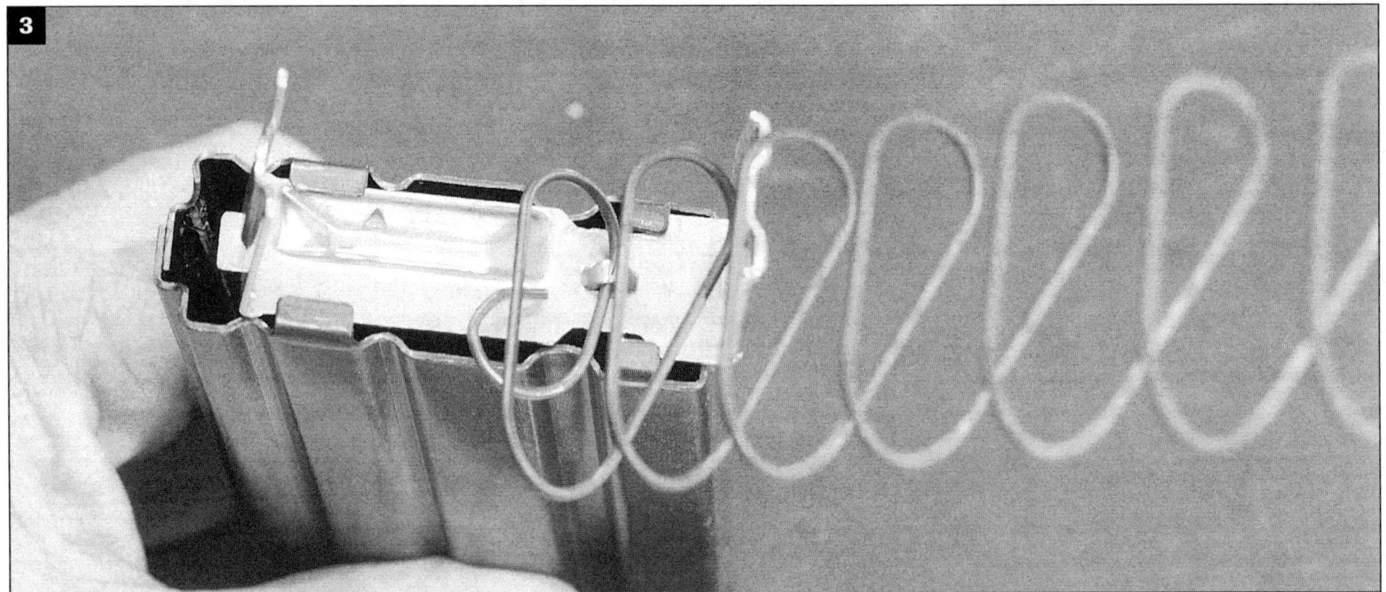

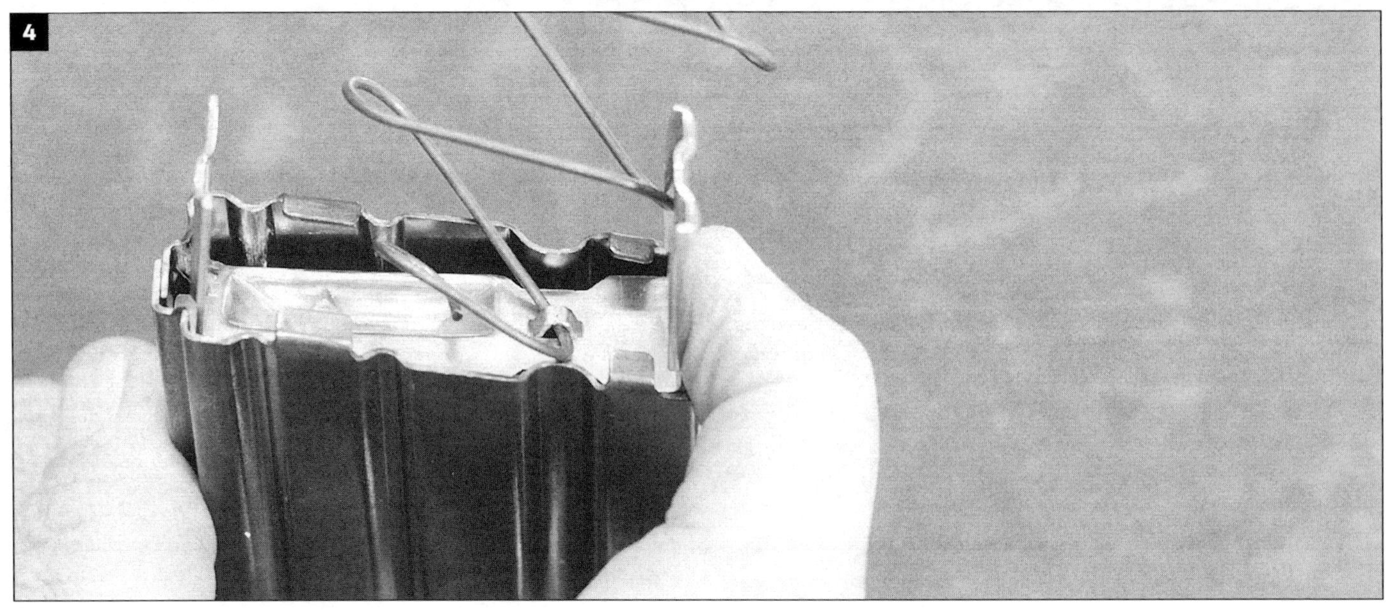

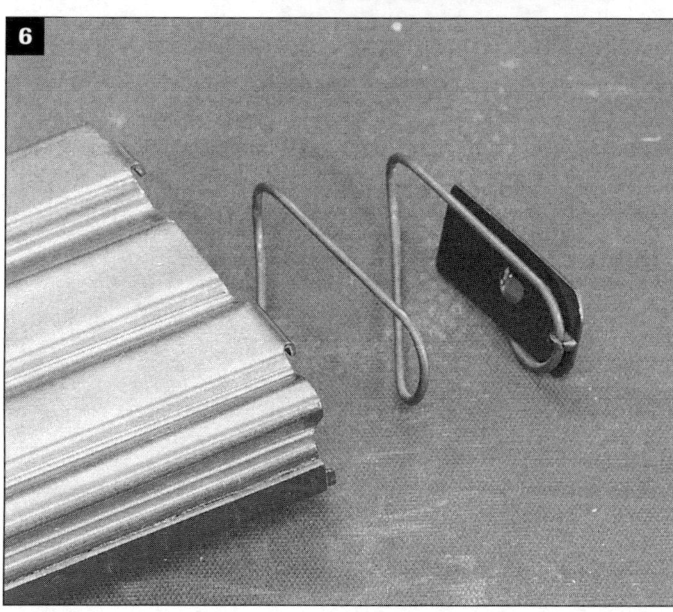

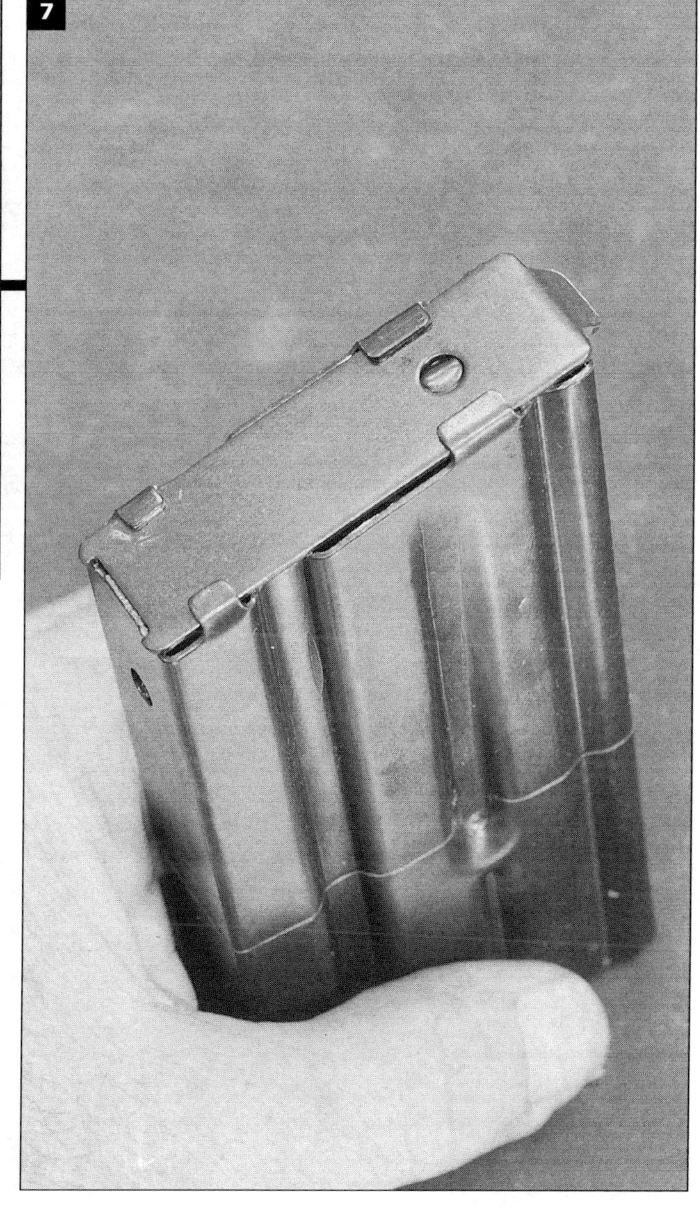

Install followers, usually, front first (bullet end). There are rounded guides on either side that look like bumps and those need to fit into their corresponding recesses in the magazine body. The newer-style followers sort of snap into place, so a little twist is usually necessary to seat the front end into its recesses.

This follower installed best back-end first...

1./2./3.: Drop the back end into the mag bottom opening and then jiggle and fitz to get the front end of the follower sitting underneath the base plate tabs.

4. A push forward snap-seated the front of the follower in place.

5. Rock in the spring, **6.** reattach the base plate retainer (was really optional on this one) to the magazine spring, and **7.** slip the plate back where it came from. Done.

Woo hoo. *Now it works.*

I don't recommend lubricating magazines *because they collect enough fuzz as it is. Once it's working correctly, the spring is a big key to keeping it going, and that's the reason for spending up a little extra for chrome silicon.*

Hope this helped...

Folks there are a few more pages, but we're done with the builds and projects. There's always something more to talk about and new things to show, but the basics won't change much, if any, as far as getting the part or parts onto and into an AR15-style firearm. I don't know if you'll ever call yourself a gunsmith, and I don't think that matters. Much in the same as auto mechanics are now "technicians," building AR15s is more about installing parts rather than digging into the essence of a system and improving it. As said to start, don't do anything you're not confident in and don't be afraid to call for help. It's not a sign of weakness and the only stupid question is the one that isn't asked. Get answers from manufacturers in particular should you encounter difficulties with any of their parts. Any one of them worth a flip will want you to be happy and make a problem go away forthwith. **Never, ever break a "rule" on safety checks.** Always ensure trigger and safety function and do headspace checks.

24.0 TROUBLESHOOTING

CHECKS AND FIXES

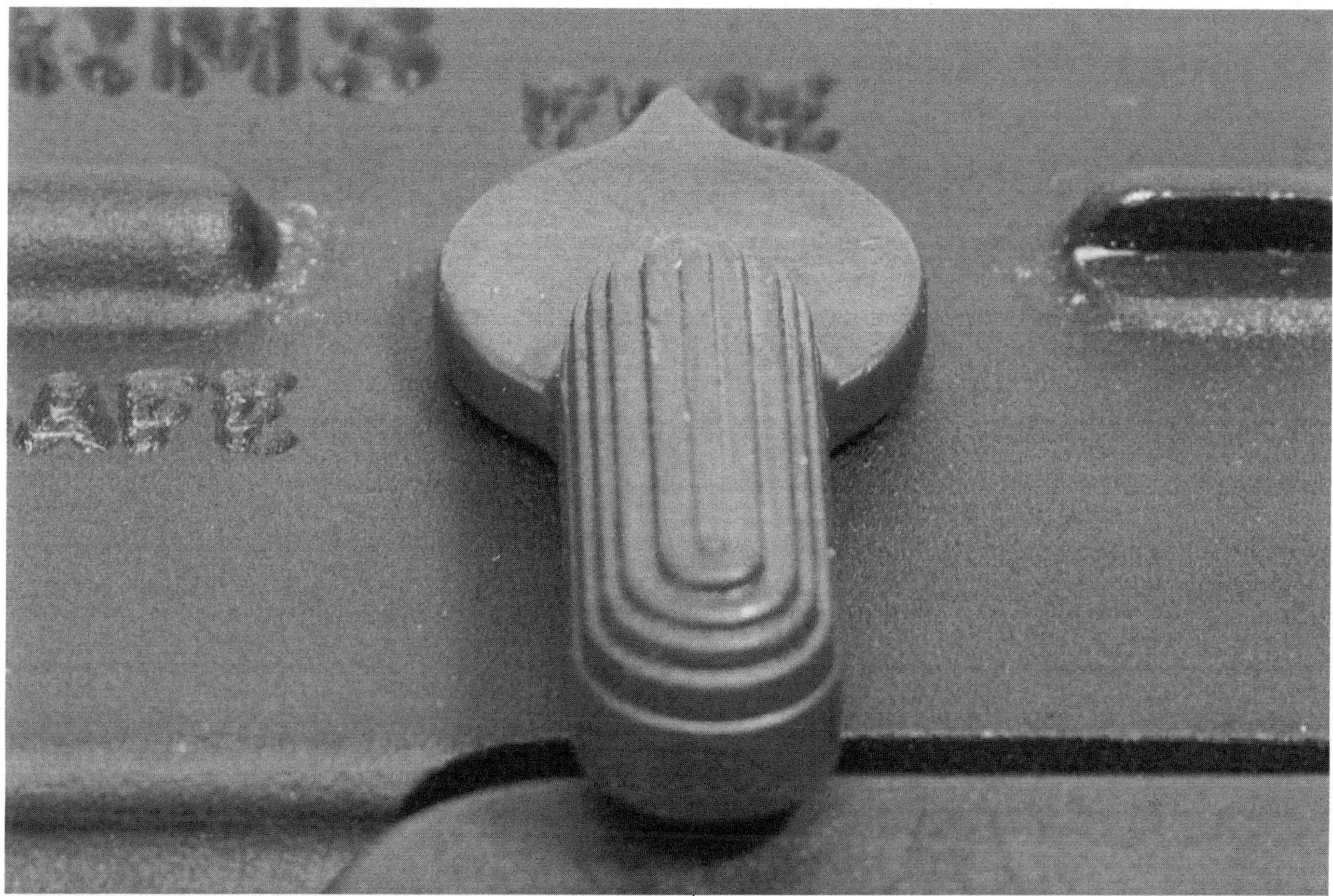

[All through a build process components should be checked for proper function. That's about test-fitting first and then testing again after assembly. Of course, when the rifle is completed is when function is really tested. You have the opportunity to be critical of parts fit and, without a doubt, should know if there is a problem anywhere within a parts system. Like the trigger for good example, and that should have already been fixed.]

If an AR15 won't work, well, it has to work... Here are a few places to look and a few things to try if yours goes on the fritz. This segment is concerned with problems immediate in nature, like first-time-out malfunctions, sporadic performance, and the like.

SEGMENT CONTENT

263 **Rifle Preparation**

265 **Ammunition Concerns**

266 **Gas Manifold**

266 **Testing**

267 **Inspecting**

267 **Barrel Break In**

268 **Common Problems** (plus cures)

RIFLE PREPARATION

We're not going into routine cleaning and maintenance, but do make double sure that your newly-assembled rifle is clean and has been lubricated with a suitable spooge. Make sure the barrel is clean, and has been cleaned, along with its chamber.

Always take more than one magazine along, and these magazines should be known-good. You're also going to confirm that you have a reliable set for your new rifle.

Take along ammunition that you plan to use most often. Nothing else makes any sense. If you are a handloader, I strongly suggest taking along rounds that have been sized to minimum headspace.

Even if the barrel was supplied with a bolt, chamber headspace

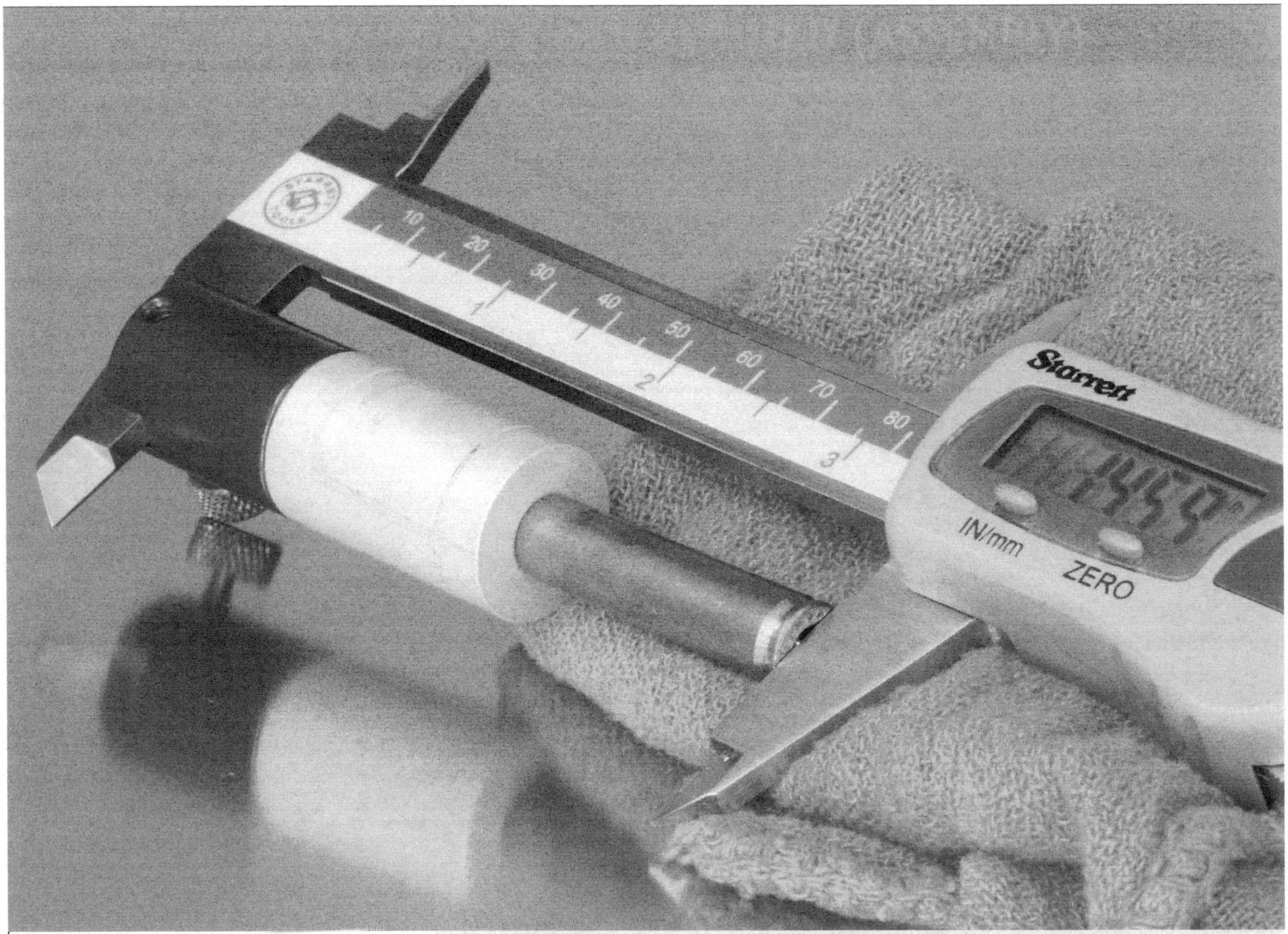

Cartridge case headspace.

Gage any and all ammo prior to use in your rifle. It's easy. Cartridge headspace should at the least be known on new ammunition and the best way to measure it is with a **Hornady LNL** gage. This gives the same number as what is actually read in the rifle chamber.

Size cases to minimum headspace before use. 1.460 inches, or a little less, is a good place to take one. Then after one firing, measure again and you'll know how much sizing (case shoulder set-back) to give the case for reuse in that chamber. Take fired dimension and reduce it by 0.003 inches.

A "drop-in" headspace gage won't give an accurate reading (no numbers) but will show minimums and maximums. Size to the minimum (case head level with the bottom step) for first use. This gage makes it easy to check new (factory) ammo on the spot.

Overrun

This is hardly ever an issue with stock parts, but there are combinations of accessory parts that can create inadequate overrun, which is the gap between the bolt stop and the bolt when the carrier is fully retracted. If it's too little the stop may not extend and therefore engage quickly enough.

should have been checked and certainly confirmed, but even correct chamber headspace (within tolerance) can, without a doubt, influence function. It's the relationship between the cartridge case headspace and the chamber headspace.

If you measure enough examples of new cases from enough different makers, and even different lots from the same source, you'll see differences in out-of-the-box dimensions. If a cartridge case is on the longer side and fitted to a chamber that's on the shorter side, guess what, it may not fit.

A case that's too tall for the chamber, including one that has the same dimension, creates a circumstance where the bolt may not close easily. It should close easily by the way. It can also create a potential for an out-of-battery-firing, which means the bolt wasn't fully locked down.

Get a gage and use it. After one firing, a spent case gives a fair representation of chamber dimensions. There's always a little "spring back" in brass alloy so it won't be exact, but you'll at the least know how "tall" a case shoulder can be and still chamber correctly. This book isn't about reloading, but here's a topic in that subject that decidedly matters to a build project.

About the only thing a longer headspaced chamber does, as suggested, and we're talking about one still within spec, is force extra expansion in a cartridge case. Certainly, a chamber that's too long and outside of spec may cause a rupture in the case, usually at the head, because it allowed too much stretch and expansion. But we checked for that.

Critical checks include the bolt catch, magazine latch, and, most important, disconnector function. That was covered pretty thoroughly in the trigger installation segments, as was safety function. Certainly the bolt, bolt carrier, charging handle should be ginning. This group can be a little stiff for a while, and sometimes for a good while, but should move freely — no snags or grabs. Bolt parts, if you assembled the bolt yourself, should be known-operational. Checks are easy in the shop using dummy rounds. The range gives the verdict. Gas tube checks should already have been done when that apparatus was installed. About the only other thing, and it's unusual to be

Always, always, always check spent case condition, especially on the first few rounds. Be alert for dings, creases, or worse.

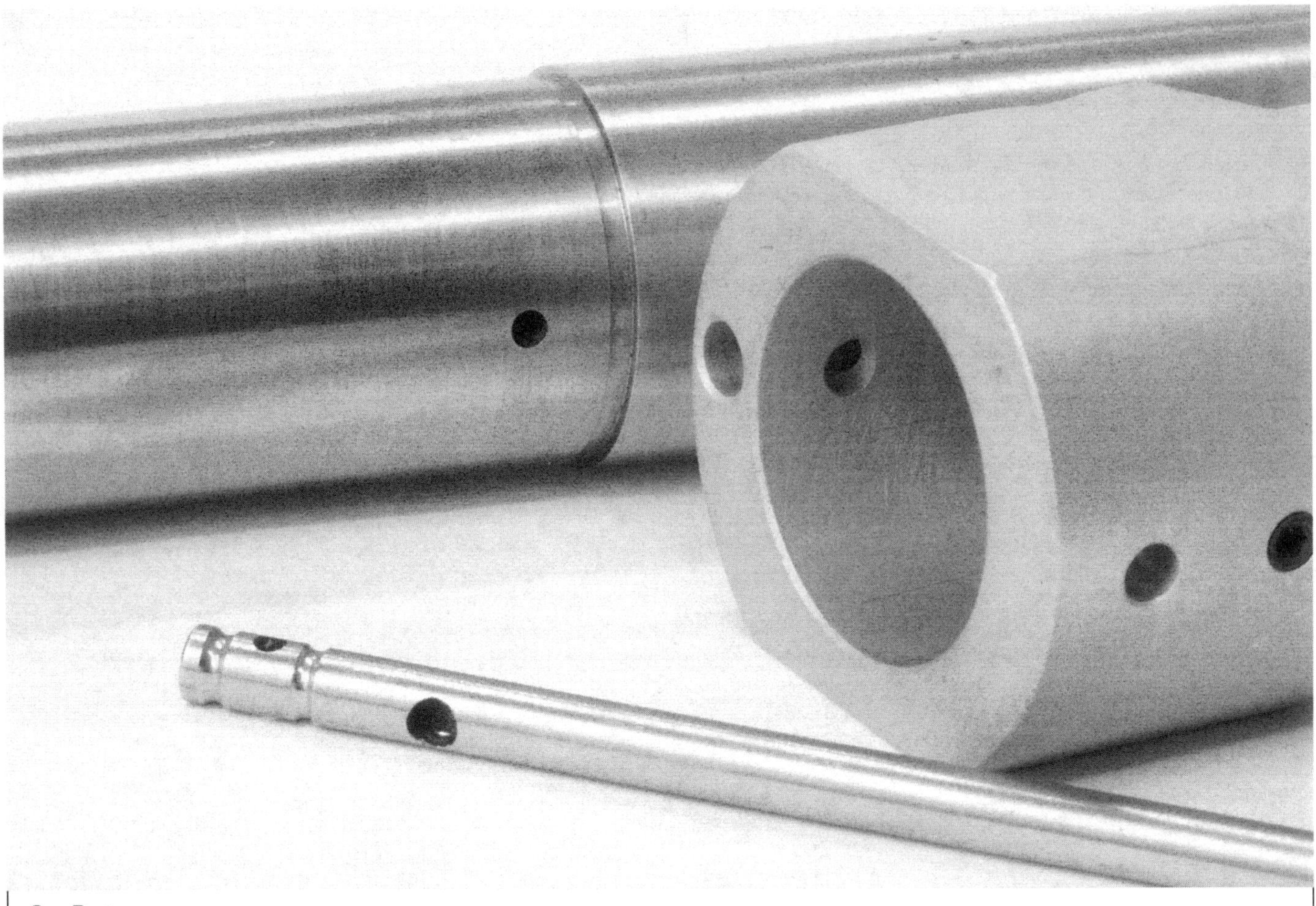

Gas Ports

Folks, if the holes aren't lined up right enough there just about will be a problem. Aside from the obvious change in gas flow (which actually may or may not be noticed in function) the hole that accepts the gas tube matters here too. Ideally a gas tube ought to sit "straight" and stress-free over its entire length. Misalignment at the manifold translates to the tube.

Now. The good news, and this is the best news, I've yet to see a gas manifold block that didn't have an oversized acceptance hole compared to the port hole in the barrel. There is leeway, but, keep in mind the other point which may be most important, that is a block installation that aligns the gas tube acceptance hole dead ahead for a "straight" gas tube installation. If it's off much, and I can't say how much matters, the gas tube may be technically stressed.

It doesn't have to be perfect, for function. All the manifolds I had on hand, and there are a good number that aren't shown here, allowed for approximately 0.050 inches leeway, which is actually 0.025 on either "side," to fully cover the gas port hole cut into the barrel. Still, though, make a strong effort to get it dead centered. Again, it might matter more to rifle performance than to rifle function.

a problem, is to ensure that the buffer moves fully to the rear, and returns. A warped extension tube can create binding and limit travel or create draggy return, and I've seen a couple of receivers that had threads cut in poor alignment with center such that the buffer was bound at its fully rearward position.

Make sure a loaded magazine latches fully.

Certainly, if any aftermarket parts were installed, give them a close eye. Especially buffering system related pieces need to be checklisted and then checked off. Things on that list include adequate bolt stop overrun. I have seen this be a problem with aftermarket stocks and accessory buffers, and sometimes following an installa-tion of a CWS (accessory insert that adds weight to the carrier).

If there's not enough overrun, trim the cushioned end of the buffer. About a nickel's worth of gap is usually adequate, but function tells the tale.

TESTING

When you get to the range, don't be bashful. Start with a single-loaded round or two, but then put a full magazine in there and see how she runs. If attention was paid to deburring pieces and breaking edges and lube was applied all around, malfunctions are going to indicate parts problems.

Check that the bolt locks back and can be released. Do more trigger checks, especially after running a box or two of ammo through the barrel. Fire, keep the trigger back, and release it very slowly (sometimes a quick release masks a borderline-functional disconnector). Repeat several times. Any problem, and that means one, means stop immediately and fix it.

Of course you're going to get a zero and group that bad boy and check your handiwork. Problems there are big indeed, but the first check on sleuthing a poor group from an otherwise anticipated-good performer should be the gas tube.

I fire the gun a few times "from the hip," which really only means not shouldered, to look for gas leaks at the manifold and watch the cycling and ejection. Speaking of, if ejection performance wasn't what you want, consider and then refer to the piece in this book on ejector tuning. If you already took a stab at it as suggested, all should have been fine. It may not really be necessary for all, but if you're seeing dings and dents on the cases it's about mandatory if you want to reuse them. Sometimes ammunition factors. I've had loads show differences in how they leave the gun. I'm hoping you went with ammunition that you'll actually use later on in the rifle. If you didn't then take that as advice, and also as reason.

INSPECTING

Get it home, break it down, and clean it up. Here's where more checks on parts come in. There won't be much wear, of course, but if you see areas where there anything has the appearance of being nicked, dinged, or unusually scuffed, find out if there's a reason. There may or may not be, but avoiding unnecessary wear is logically prudent.

Clean the barrel when back home in whichever fashion you deem proper.

Relube before next use and just keep going.

If you don't take the advice of lubing the fool out of your AR15, then failures related to insufficient lubrication, which is then excessive friction, well, you figure it out.

Be safe

Especially if it's an aftermarket trigger, continually check correct disconnector function after putting several rounds through the rifle. It has to work... Back home, double-check break weight too, especially if it's an NRA Service Rifle. There's no substitute for live firing to break in parts!

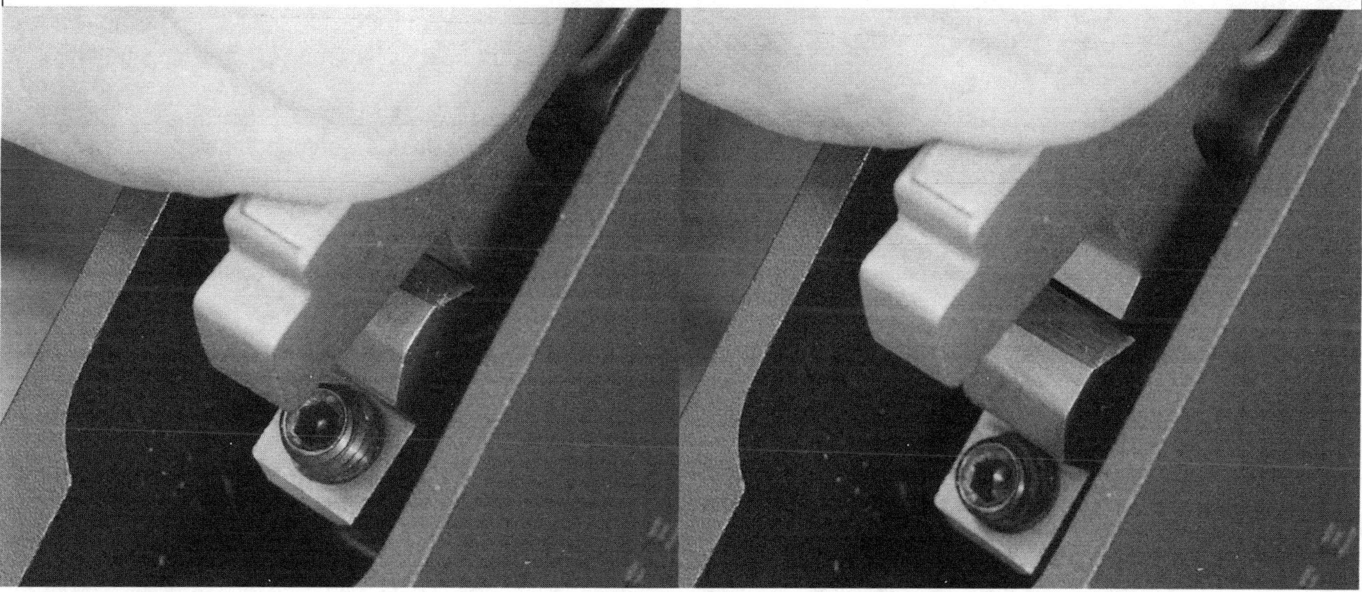

Barrel Break In

It's never too late, or too early, to break in a barrel. However, I think function checks come first. If you want to break in the barrel first trip to the range, haul your cleaning gear along with you. The tried and true way to break in a barrel is to shoot one round, only one, and then clean the barrel. Rinse and repeat, or actually repeat and then rinse. Clean after each round. Use a wet patch only for the cleaning. No need to brush. Do this for 10-15 rounds and the barrel will be broken in about as much as that process can allow. You know the barrel is as good as it's going to get when you see no bullet jacket fouling on the patch (it's bluish). I use copper solvent for cleaner. If you have fired even as many as 20 rounds and still see fouling (it's blue-green) then consider one of Superior Shooting Systems FinalFinish products. There's a break-in package that works well for any barrel. The "patch test" is actually a good way to know if it will help your barrel. As suggested, it will if you're seeing blue after 20 rounds of shoot-and-clean.

COMMON PROBLEMS + CURES

Tips and tidbits are discussed throughout the book on solving operational issues, and, inadvertently, possibly creating them. One thing, though, is that building up your own rifle gives you a big head start in learning to solve problems because you will develop a better understanding of how it all works together. And that ultimately is the trick. Process of elimination, and it starts with the firing cycle. Some of the most common conditions and their potential cures are given. These are far from the only reasons your AR15 won't do something like it's supposed to, but they are, as suggested, common maladies and frequent remedies.

FAILURES TO CYCLE

Action won't operate, or lock back —

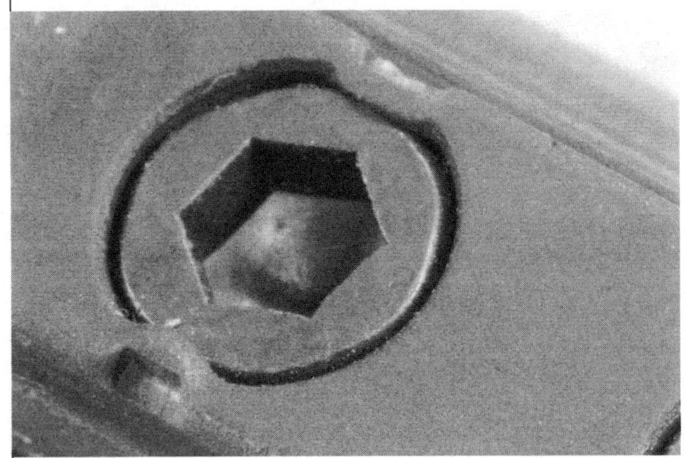

Loose carrier key. These need more torque than most ever think, and they should be **staked** in place, not glued. If this is the problem the reason is because gas bleeds off.

Loose gas manifold. Sometimes there are fit problems, sometimes it's just a loose or loosened screw. Fit problems aren't easy to fix, but the screw is. The symptom is an unusual amount of visible carbon deposit.

Bolt stop. If it's broken that's your problem right there, otherwise look for a rounded edge and check for free function. Always check with more than one magazine. Bolt stop function can be a little draggy if there's been an extended release added; "boost" the bolt catch spring by stretching it.

Now, there is one problem that can look like other problems. It's pervasive in carbine-length-barreled rifles, as has been discussed more than once. **Excessive carrier velocity.** Right. If the carrier gets too much too soon and heads rearward more forcibly than we want, it can "bounce" after bottoming out against the buffer. This can happen fast enough for the carrier to ride right over the bolt stop and even right over a round in the magazine. It looks like it's not getting enough "power" but it's actually getting too much. Weighting the carrier, aftermarket buffer, heavier buffer spring, reducing gas flow, and other tricks, can delay bolt unlocking enough (they all ultimately have about that same effect) to slow the carrier's rearward travel speed enough to get all the "timing" working like it should. These things happen so quickly that only a little bit makes a big difference, or can, and that's because only a little bit causes the problems.

EXTRACTION & EJECTION

These problems were discussed throughout, and excessive gas pressure ultimately leads most causes, however it manifests itself. Another is a filthy, corroded, or otherwise rough chamber, and also, related, dirty cartridge cases or improperly sized cases.

The last reason for extraction problems is likely to be insufficient extractor tension. If the action is cycling as it should, failures to extract are unusual indeed. One sign is bent or even ripped case rims. That means the bolt carrier is moving while the case is still "stuck" in the chamber, again, it's stuck either because it's too swelled up from too much pressure or because it's unwilling to move due to chamber/case condition prior to firing. The same excessive-pressure results can be related to failures to eject. Too much carrier velocity upsets the "timing" of this simple operation.

After dealing with malfunction symptoms as much as I have, and having then seen them so significantly reduced after the fact, and the fact that reduced them was getting literally more time involved in the firing cycle. Things like carrier weight and buffer springs tame cycling to a point that, as explained, allows everything to work in sequence as designed. There are a many, many who seem dead set on disagreeing with me on this, and that's fine. I believe in cures rather than fixes.

X.X 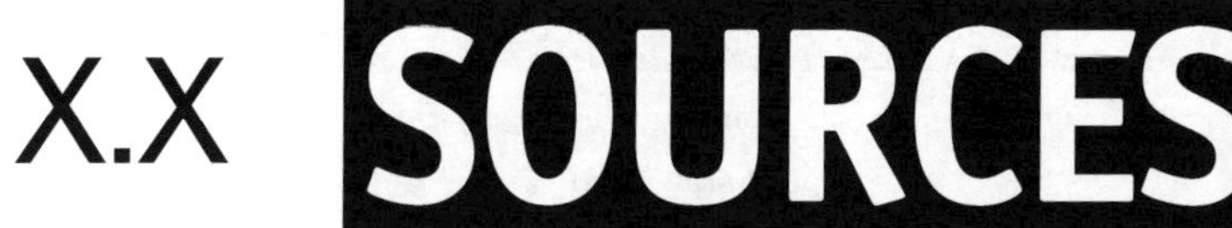 SOURCES
AND ACKNOWLEDGEMENTS

THANKS...

These are the people who help me get my hands on the tools, parts, and insights that make these books possible.

Frank Brownell and Larry Weeks, of Brownell's of course, kept me up on all things old and new and contributed mightily to making sure all this was as easy as it was. Brownell's has been and will be my first suggestion to all for parts and tools. A lot of good help comes from that outfit too. Check out their AR15 material on-line.

Scott Medesha has contributed mightily to all these books and that's a been over a lot of years now. He's never failed to answer all my questions. He has a knack for seeing through the haze.

Gary Eliseo is without peer as a rifle builder, and his insights and tips have been priceless.

Randy Luth at DPMS helped hugely with my project rifles. DPMS has become my first "go to" source for major components.

Jack Krieger makes what I think are the best barrels on this planet, and I sincerely thank him for helping me to get what I wanted to work with. These guns wouldn't shoot so well otherwise.

Derrick Martin at Accuracy Speaks helped as always. He's a pioneer always looking for a new path.

Thomas Spithaler at Olympic Arms helped with boxes full of project rifle parts.

Much thanks too to Geissele Automatics, Midway USA, David Beaty at Sun Devil, Northern Competition, Bushmaster, Satern Custom, JP Ent., Yankee Hill, and Young Mfg.

Thanks much to Pat, my UPS man, and John C. Easley for toting and transferring every box received for this book.

Now to the addresses...

TOOLS & PARTS [WHERE I GET MOST OF MY STUFF]

Brownell's Inc.
200 South Front St.
Montezuma IA 50171
800-741-0015
www.brownells.com

Sinclair International
2330 Wayne Haven St.
Fort Wayne IN 46803
800-717-8211
www.sinclairintl.com

MidwayUSA
5875 West Van Horn Tavern Rd.
Columbia MO 65203-9274
800-243-3220
www.MidwayUSA.com

CUSTOM BUILDERS & ACCESSORY PARTS

Medesha Firearms
Scott Medesha
10326 E. Adobe Road
Mesa AZ 85207
480-986-5876
www.MedeshaFirearms.com

Northern Competition
Lee Penzkowski
8332 196th Ave.
Bristol WI 53104
(262) 857-8762
www.NorthernCompetition.com

Gary Eliseo
Competition Machine
3121 E. La Palma Ave. Unit Q
Anaheim CA 92806
714-630-5734
www.compctitionshootingstuff.com

EGW Inc.
George Smith
48 Belmont Ave.
Quakertown PA 18951
(215) 538-1012
www.egw-guns.com

Accuracy Speaks, Inc.
Derrick Martin
3960 N. Usery Pass Rd.,
Mesa AZ 85207
(480) 373-9499
www.accuracyspeaks.com

Fulton Armory
8725 Bollman Place #1
Savage MD 20763
(301) 490-9485
www.Fulton-Armory.com

SIGHTS AND TRIGGERS

Leupold & Stevens, Inc.
14400 NW Greenbrier Parkway
Beaverton OR 97006-5790
800-LEUPOLD
www.leupold.com

Warner Tool Company
18 Lucinda Terrace
Keene NH 03431
(603) 352-9521
www.warner-tool.com

Zelenak Firearms Co.
P.O. Box 182
Painesville OH 44077

Bob Jones
5115 E. Edgemont Ave.
Phoenix AZ 85008
(602) 840-2176
www.BJonesSights.com

Geissele Automatics
603 Caroline Drive
Norristown PA 19401
(610) 272-2060
www.geissele.com

Jewell Triggers
3620 Highway 123
San Marcos TX 78666
(512) 353-2999

Timney Triggers
3940 W. Clarendon Ave.
Phoenix AZ 85019
866-484-6639
www.timneytriggers.com

JP Enterprises, Inc.
P.O Box 378
Hugo MN 55038
(651) 426-9196
www.JPrifles.com

*Call **Sinclair** for Centra sights and accessories, Champion Shooters Supply and Champion's Choice have much of interest too.*

MAJOR COMPONENTS [RECEIVERS, BOLT CARRIERS, PARTS KITS, ETC.]

DPMS Firearms, LLC
3312 12th Street SE
St. Cloud MN 56304
800-578-3767
www.DPMSinc.com

Bushmaster Firearms International
999 Roosevelt Trail
Windham ME 04062
800-998-7928
www.bushmasterfirearms.com

Rock River Arms, Inc.
1042 Cleveland Road
Colona IL 61241
(309) 792-5780 www.rockriverarms.com

Olympic Arms Inc.
624 Old Pacific Hwy. SE
Olympia WA 98513
(360) 459-7940
www.olyarms.com

Titan Ordnance
2110 NE Cornell Road Suite E
Hillsboro, OR 97124
503-701-2670

Model 1 Sales
P.O. Box 569
Whitewright TX 75491
(903) 546-2087
www.Model1Sales.com

BARRELS & PARTS

Krieger Barrels, Inc.
2024 Mayfield Rd.
Richfield WI 53076
(262) 628-8558
www.kriegerbarrels.com

Pac-Nor Barrels
P.O. Box 6188
Brookings OR 97415
(541) 469-7330
www.pac-nor.com

Superior Shooting Systems Inc.
800 N. Second St.
Canadian TX 79014
(806) 323-9488
www.SuperiorShootingSystems.com

Satern Custom
320 West 5th Avenue North
Estherville IA 51334
(712) 362-4991
www.saternmachining.com

Sun Devil Manufacturing
663 West 2nd Ave. Suite 16
Mesa AZ 85210
(480) 833-9876
www.sundevilmfg.com

Yankee Hill Machine Co., Inc.
20 Ladd Avenue, Suite 1
Florence MA 01062
(413) 584-1400
www.yankeehillmachine.com

Numrich Arms
226 Williams Lane
W. Hurley NY 12491
(845) 679-2417, 866-Numrich
www.gunpartscorp.com

Young Manufacturing
3613 N. 35th Ave.
Phoenix AZ 85017-4409
(623) 915-3889

Sabreco, Inc.
2013 Voit Drive
Skippack PA 19474
(610) 584-8228
(custom barrel work)

MOACKS [carrier key staking tool]
Michiguns
Box 42 Three Rivers MI 49093
www.m-guns.com

You need these books!

THE COMPETITIVE AR15: the ultimate technical guide comes 10 years after the first book, and the original set a new standard. This one raises the bar even higher! It's not a rewrite or update but takes an all-new approach, with all-new material. There is a lot to talk about! Each and every AR15 rifle system and component is discussed and dissected, with each parts group given its own separate segment. Barrels, triggers, stocks, sights, receivers, gas system, and more. All totalled, there are 30 separate section and segment headings. It's easy to read and clearly organized. There are 475 large and clear photographs to accompany the text. No squinting! Learn which tricks work and which are a waste of time and money. Also see what enough money thrown at an AR15 can produce!

PARTS — barrels, receivers, bolts, carriers, magazines, gas system, triggers, and more...

PROJECT RIFLES — an even dozen AR-type rifles are dissected and evaluated, four factory-made target rifles and eight custom-built. The custom rifles featured a "best-of-the-best" approach with the goal of building The Competitive AR15 ultimates!

AMMUNITION — chambering, bullets, loads, cartridge alternatives, tools, tricks, troubles

MAINTENANCE — cleaning and caring for your AR15
and **MORE** and **MORE** and **MORE**.

$34.95 softcover, include $7.00 shipping (order total $41.95).
470 pages, 475 photos.

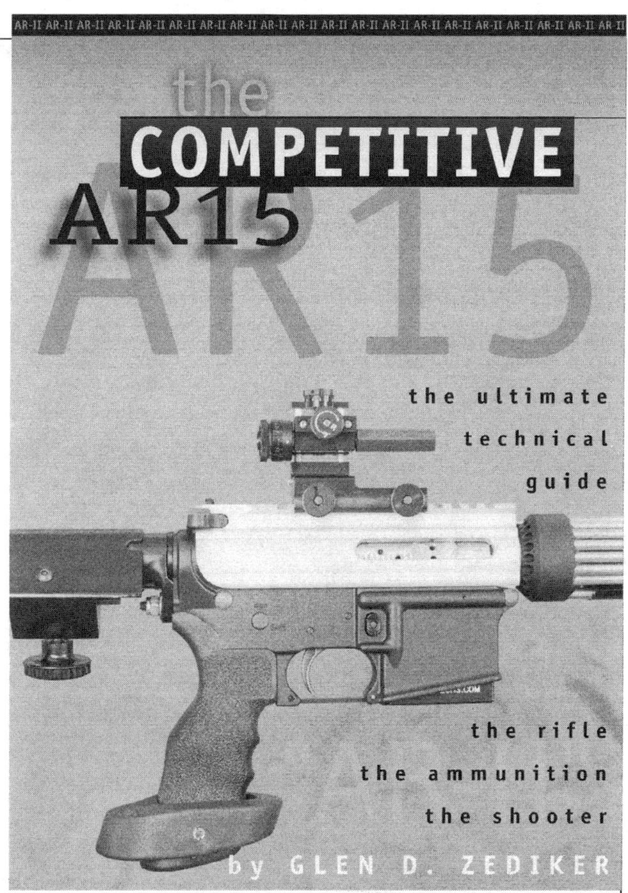

THE COMPETITIVE AR15: the mouse that roared

This book is the original source for information and insight for getting the maximum from this rifle. There is no other book that took it to this level, and discusses the totality of the process.

The Rifle, the Gear, the Ammunition, the Shooter.

All angles and elements are covered with special sections and chapters by some of the world's top shooters, like Lew Tippie and David Tubb, and gunsmiths like Derrick Martin and Bill Wylde. It's in a class of its own, just like the folks who produced it!

$29.95 softcover, include $7.00 shipping (order total $35.95).
288 pages, 300 photos

Zediker Publishing
P.O. Box 1497
Oxford MS 38655
(662) 473-6107
[or use your credit card and order on-line: www.**zediker**.com]

your new source for shooting information
www.zediker.com

X.X SPRINGS + PINS

+ PUNCHES

JUST IN CASE YOU SPILL ALL THESE...

The outlines may help you identify which goes where, and the appropriate pin punch size to choose for the installation.

ROLL PINS

Gas tube		#1
Rear sight base		#3
Forward assist		#3
Ejector		#1
Extractor (not a roll pin)		#1
Rear sight wind. knob		#1
Bolt catch		#2
Trigger guard		#4

SPRINGS

- Rear sight base
- Rear sight detent
- Magazine catch
- Pivot/takedown pins
- Ejector
- Extractor
- Bolt catch
- Forward assist
- Disconnector
- Safety detent
- Buffer retainer

PUNCH SIZES

#1 1/16	#2 5/64
#3 3/32	#4 1/8
#5 5/32	#6 3/16

Roll pin starter punch numbers correspond with an "S" suffix, #1S for example

Hex wrench for elevation wheel   1/16